ANIMAL SCIENCE, ISSUES AND RESEARCH

CRITICAL RESEARCH TECHNIQUES IN ANIMAL AND HABITAT ECOLOGY

INTERNATIONAL EXAMPLES

Animal Science, Issues and Research

ANIMAL SCIENCE, ISSUES AND RESEARCH

CRITICAL RESEARCH TECHNIQUES IN ANIMAL AND HABITAT ECOLOGY

INTERNATIONAL EXAMPLES

MICHAEL O'NEAL CAMPBELL
AND
KAUSHALENDRA KUMAR JHA
EDITORS

nova
science publishers
New York

NOTICE TO THE READER

Additional color graphics may be available in the e-book version of this book.

Library of Congress Cataloging-in-Publication Data

Names: Campbell, Michael O'Neal, 1965- editor | Kumar Jha, Kaushalendra, editor.
Title: Critical research techniques in animal and habitat ecology : international examples / Michael O'Neal Campbell, PhD., (editor), Professor, Finance, Universitat Politècnica de València, Valencia, Spain, Kaushalendra Kumar Jha, PhD. (editor), Professor, Economics, Peter the Great St. Petersburg Polytechnic University, St Petersburg, Russia.
Description: New York : Nova Science Publishers, [2021] | Series: Animal science, issues and research | Includes bibliographical references and index. |
Identifiers: LCCN 2021031342 (print) | LCCN 2021031343 (ebook) | ISBN 9781536198461 (hardcover) | ISBN 9781536198591 (adobe pdf)
Subjects: LCSH: Animal ecology--Research--Methodology. | Habitat (Ecology)--Research--Methodology.
Classification: LCC QH541.2 .C75 2021 (print) | LCC QH541.2 (ebook) | DDC 577.072--dc23
LC record available at https://lccn.loc.gov/2021031342
LC ebook record available at https://lccn.loc.gov/2021031343

Published by Nova Science Publishers, Inc. † New York

CONTENTS

INTRODUCTION

This is edited by two environmental scientists with interests in GIS and remote sensing applications, forest and habitat change and large animal ecology. It examines the cutting-edge issues related to animal and habitat ecology research and management, with cases studies across Asia, the Americas, Africa and Europe. The topics are based on research and reviews of specific and general topics covering the habitats as well as the species of importance in selected case studies, and the overall general scenarios. The chapters of the book are written by leading academic and field experts, who discuss their skills and research findings. The field covered are vast, so selectivity enters, based on concurrent and relevant subjects, such as field research techniques, nature society relations and chemistry in conservation biology and policy. The chapters focus on cases as varied as vultures, storks, waterbirds, pumas or cougars, and elephants, and research techniques such as genetics and GIS.

Technological developments, such as GIS and remote sensing, and some genetic methods have altered the nature of ecological research. These include the utility of GIS, and the related techniques of remote sensing, which allow more precise and accurate measurements and consequently more informed and reliable results. Species distribution modelling enables evaluations habitat suitability and the impacts of habitat alteration and the requirements for the improvement of animal conservation. Integrated research, including the interfacial studies of social and natural sciences are increasing important in ecological research, as disciplinary boundaries beak down and hybrid disciplines emerge. Simultaneously, chemical and genetic studies are increasing in importance, with applications in the interfaces of the ecological, social and medical sciences.

The topics covered in this book may contribute the scientific understanding of different, relevant topics on research methods on ecology and conservation biology. This is especially the case, considering the wide selection of research topics in widely varying contexts. Strands may emerge from these selected topics that may inform further research and development in varied areas. These research findings may be replicable in the different contexts to contribute to the objectives of ecological sustainability. The results and conclusions presented, and the strategies recommended in different chapters will help the policymakers and decision implementers, scientists, resource managers, research scholars and other stakeholders to attain effective and sustainable animal conservation and habitat management.

Michael O'Neal Campbell

PREFACE

This book covers selected topics on research methods in modern ecology, focussing on animal ecology, landcover assessment and habitat change. This is plainly a vast area, due to the multiplicity of research foci and the great range of research techniques available. Two main clusters of research methods have emerged: laboratory studies and field techniques. Both are now totally reliant on computer hardware and software and associated technology. Recent developments in ecological research methods largely concern the adoption of enhanced computer software techniques and the syntheses of these into pre-existing research methods, the better to record, analyze and explain the behavioral patterns of the studied animals and/or plants, and their ecologies. Chemical analyses, including studies of animal dietary and foraging patterns are also important, given reductions in food resources, environmental degradation and animal intrusions into human life spaces. This volume focusses on specific examples that typify some of the recent developments in research methods. In this regard, the research methods are spread over several fields of research and the interests of several researchers. Animal and integrated ecology in these specialisms is also considered broadly, to include habitat analysis, animal behavior, landcover, habitat and plant ecological change and even human/animal relations, and international case studies along these lines. Chapters on plant ecology and genetics are included, to indicate that animal ecology is inherently linked to vegetation assessment and plant dynamics, as part of the animal species interactive environment and animal bodies and related genetic studies run through any studies of animal ecology.

Technological and methodological advancements in the applied sciences may appear to be the main drivers in the development of animal ecology, but the latter discipline, based on people's understanding of animals remains the main focus and guide for research method development. Remote sensing and geographical information systems are emerging as relevant, cutting-edge research methods, at small, medium and large-scale levels. Animal behavioral studies at the more detailed, individual subject level are also complementing this trend. To better explore animal behaviors and the relations between these and the environmental context, the key developments concern closer, less intrusive and observable recording of animal life activities. This includes more critical human observation of animal behavior, and more detailed study of animal environments, including plant ecology. From a technological perspective, developments include more accurate positioning systems, more sensitive tracking systems, the removal of obstacles to clearer observation and species identification, such as darkness and poor lighting, dense vegetation and coarse image resolution and more comparative studies across different local contexts and global ecosystems. More detailed ecological studies and shared analyses of the research results may serve as guides for the development of more sophisticated research tools. The syntheses of these new developments with older research methods such as field foot surveys, coarse resolution satellite imaging and aerial photography, georeferencing with hand-held devices and eyeball observation of animals in the field may ensure the continuance of constructive research engagements.

The current book examines these issues through with some carefully selected and relevant case studies, each occupying one chapter. The first chapter, titled *Urban Vulture Ecology, Tracking Technologies, and Human Quality of Life* by Michael Campbell, looks at drone-based research methods and their application to studies of the urban ecology of Black vultures (*Coragyps atratus*, Bechstein, 1793) and Turkey vultures (*Cathartes aura*, Linnaeus, 1758), and the related issues of urban human quality of life in North American cities. This topic is new in that cities are expanding, animals are adapting to these developments and new, previously unresearched topics are becoming increasingly relevant. Research methods, based on drones (unmanned or uncrewed aerial vehicles, UAV or remotely piloted aircraft, RPA, which are defined as remotely controlled aircraft without onboard humans, but computer software controls) may be used with global positioning systems (GPS) and cameras for detailed, dynamic ecological and related social research. The capabilities of recent advanced drone designs are described with examples from drone manufacturers. Such technological developments exemplify the core issues of current ecological research, the requirement of more detailed applications for more comprehensive research findings.

The application of drone-based survey methods to vulture urban ecology is very current, as vulture foraging and interactions with human life spaces may be recorded with detailed location positioning, high resolution image capture for species identification and subject animal behavioral assessment for intra-species analyses. Highly mobile vultures may be studied with similarly mobile remote survey drones, which can track vultures due to their high visibility, mobility, soaring altitudes and flight speeds, remote nesting locations, vast foraging areas and frequent urban impacts, including urban pollution, building damage, livestock attacks and aircraft collisions. The chapter looks at the factors for the paucity of drone-based studies of vultures, which results from legal, logistical, financial and social issues and restrictions. Drone-based urban ecological studies are also assessed through examination of applications to non-vulture species ecologies. The chapter also looks at the parameters examinable in drone-based studies, including flight speeds, soaring duration, foraging flight patterns, courtship, nesting habits and locations, physical attacks on infrastructure, roosting, defecation and aggressive incidents, and human impact incidents. It is concluded that there are positive impacts from drone-based studies, including greater accuracy in the ecological recording, but there are some negative issues such as social privacy laws restricting drones, and vulture avoidance of drones. Hence, smaller, quieter drones and more flexible laws may determine the future of drone-based research.

The second chapter, *Facts Around Vultures: An Attempt to Join the Puzzle Pieces* by Kaushalendra Kumar Jha a member of most distinguished technical Indian Forest Service and vulture ecology researcher Radhika Jha, examines vulture populations, which are diminishing, with consequent impacts on their invaluable environmental services. This chapter notes that there is much research on vulture strongholds in South-west European, South-east African, and South Asian countries, and on subjects related to vulture feeding behaviour, diseases, diclofenac story, global strategy, name changes, captive breeding, population assessment, extinction prediction and artificial feeding. The chapter's objective is to integrate these varied components of scientific knowledge, to enable the discovery of gaps for further improvement in knowledge. It also summarizes the incomplete, conflicting, and complementary information on such selected topics, the better to improve the knowledge of vulture ecology.

The third chapter, by Kaushalendra Kumar Jha and Radhika Jha, titled *Brown World of Vultures: A Case of Vulture Restaurant*, looks at the concept of mapping Vulture Restaurants (VR) as a technique for mapping the variables in vulture status and ecology, and its role in the conservation biology and protection of vultures, which may narrow the gap between policy actions and vulture ecological status. The justification of this study is the current endangered status of vultures, especially in India. The brown world refers to their feeding system, which is based on the diversity of mammal carcasses, especially ungulates. It is argued that while the vultures' natural prey, dead mammals are declining, the artificial vulture restaurants are also beset by problems. Such issues include problems with contaminants or competitors. The positive possibilities for such supplementary feeding stations include the augmentation of the vulture populations, but negative impacts may include reduced fecundity, changed social behaviour and even life threats. These may impact negatively on conservation attempts.

The fourth chapter, *Environmental Chemistry and the Conservation Biology of Raptors and Waterbirds* by Michael Campbell, examines current issues surrounding the chemistry of pollution and conservation of raptors and waterbirds. It argues that the pollution chemistry of the environment, especially chemical accumulation at the upper levels of the food chain is fundamental to the conservation biology of raptors (birds of prey) and waterbirds, a situation that is developing with human-induced pollution of ecosystems and food chains, leading to local species extinction and affecting species recovery efforts. Chemical pollutants are defined as elements or compounds that affect the physiology and health of animals. Industrial, agricultural and urban technology, and related human activities, are factors for these events, and poisonous chemical elements and compounds include mercury, lead, copper, arsenic, chromium, copper, rodenticides and insecticides, nonsteroidal anti-inflammatory drugs (NSAIDs) and polychlorinated biphenyls -PCBs. Raptors and waterbirds are strongly affected, as they eat animals that have ingested such poisons from infected materials, with accumulations at the upper levels of the food chain. Raptors and waterbirds are studied in this context by variable techniques, and these methodologies are also examined. The chapter argues that the contributory factors for toxic chemical impacts of birds, agricultural and industrial development, are in some cases managed by technological, ecological and biological knowledges. Integrated research methods are increasingly, fields of knowledge have advanced into the interface between environmental chemistry and conservation biology. This study contributes insights into the ornithological, chemical, ecological, social and policy sciences.

The fifth chapter, *Vulture Foraging and the Chemistry of Putrefaction* by Michael O'Neal Campbell, looks at the unique sense of smell of the *Cathartes* vultures, the Turkey, Greater Yellow-headed (*Cathartes melambrotus*, Wetmore, 1964) and Lesser yellow-headed (*Cathartes burrovianus*, Cassin, 1845) vultures, and the implications for the foraging ranges, inter-species competition and survival of these species. These species are compared with the Black vulture, which lacks a sense of smell but has an advantageous social ecology. The chapter looks further at current research on vulture foraging, comparing the *Cathartes* vultures with the non-*Cathartes* species, the chemistry of putrefaction, the physiology of vultures' digestion of carrion and non-food items and briefly examines the recent diseases that have affected vultures. It is noted that ingestion of dangerous articles poses a grave threat to vulture health and remains among the more serious ecological threats to *Cathartes* and Black vultures in the current age.

Current evidence indicates that *Cathartes* vultures have only a minimal advantage over other species that lack a sense of smell, due largely to the shadowing abilities of other species and the conflicts that may emerge with larger or more social species at carcasses. The main advantage conferred by olfactory ability is a wider possible range into dense canopied forest, which may be moot given recent intensive deforestation under urbanization and agricultural expansion, resulting in more open, mixed landcover. The chapter concludes that new research methods, such as hidden stationary or drone-based cameras, high resolution strong global positioning systems (GPS) and critical field techniques may be required for further research into these issues.

The sixth chapter, titled *South and Northeast Brazil: Does A Shared Species Arise Shared Human Dimensions Of Conflicts With Wildlife*? by leading researchers Francine Schulz and Cláudia Sofia Guerreiro Martins, examines the relationship between large carnivores and human action in changing, increasingly modified habitats, within the paradigm of human-wildlife-conflicts (HWC). The case study is of pumas (*Puma concolor*, Linnaeus, 1771) in two different Brazilian biomes: the hinterland of Northeast and the Southern Atlantic Forest. The first area, the hinterland of the Brazilian Northeast includes the Caatinga, which is largest area of the Seasonally Dry Tropical Forest and Woodlands (SDTFW) biome in the Americas. The climate comprises an extended dry period with low rainfall. The second area, which includes the southern limit of Atlantic Forest in Brazil, includes patches of *Araucaria angustifolia* forest within south Brazilian natural grassland. The subtropical climate has a high annual precipitation, with no strong dry season. Both areas include protected areas and experience conflicts between resident carnivores and human actors. However, few studies document these local dynamics, and the authors argue that more such studies are needed to solve the problems of such ecosystems in South and Northeast Brazil. The objective of the study was to examine the human dimensions of human-puma-conflicts, using South and Northeast Brazil examples. In both areas, local residents and farmers appeared suspicious and threatened by puma conservation and consequent proximity and did not seek help from wildlife management institutions. The authors conclude that these results contribute to knowledge of conservation and wildlife management strategies, and the design and implementation of solutions to human-large carnivore conflicts.

The seventh chapter, titled *Conflicts with Pumas In Southern Brazil: The Farmers' View* by Francine Schulz, Rodrigo Cambará Printes and Larissa Rosa de Oliveira, examines the severe conflicts between large carnivores and people, due to livestock predation, in a mosaic of Protected Areas in southern Brazil. It is argued that habitat loss, partly due to the increasing use of traditional pasture systems, has made livestock vulnerable to puma attacks in the case study. The hypothesis was that farmers' perception, knowledge and attitudes towards the conflicts with pumas was affected by socioeconomic variables, such as age, education, monthly income and farming experience. The results of a structured questionnaire indicated the majority of the local population considered puma conflicts a serious problem and required the involvement of government authorities. Farmer opinions were more related to social variables such as age and educational level, than to financial losses. The study suggests a long-term educational program with local farmers about possible mitigations tools, and the use of both academic and traditional knowledge to connect the community and Environmental Agencies.

The eighth chapter, by world class researcher Ida Dutt, titled *Reformation of African Elephant Ecology and Human Quality of Life Through GIS and Remote Sensing technologies: A Case of African Elephant Ecology in Botswana* looks at the conservation of the African elephant using multiple research methods. This chapter examines the possibilities of African elephant ecology reformation through GIS and RS technologies within Botswana. The author notes that the first Great Elephant Census, a pan-African survey conducted in 2016, showed a decline in the African elephant population by at least 144,000 between 2007 and 2014. Botswana has approximately one third of Africa's remaining population of elephants, with records indicating that from 2014 and 2018, the number of elephant carcasses within the country increased by 593%. Of these, 385 elephants were confirmed poaching cases between 2017 and 2018. There were also some 350 mysterious elephant deaths between May and June of 2020. Factors for population losses include habitat loss, increasing human-elephant conflicts, and the lifting of a hunting ban by Botswana's government. The elephant range in Botswana has expanded by 53% over the past 20 years, and this has increased concerns on biodiversity and associated species and ecosystems, and the status of people and their social spaces. The author argues that in-depth information about the movements of elephants, their social dynamics, habitat use, and impacts on people around these periphery regions are currently limited within the country. It is also argued and concluded that the improvement of understandings of the African elephant's ecology is vital not only for ecological reasons, but also for improved human health, quality of life (QOL), and productivity in agricultural systems.

The ninth chapter, titled *Bio-environmental Monitoring and Assessment using eDNA: An International Approach to Research and Future Directions,* by leading research scientist Phillipa Dutt, looks at the general direction of research on Bio-environmental Monitoring and Assessment using eDNA. The author argues that Genome-based monitoring has revolutionized the scope of environmental monitoring and contrasts genomic-based monitoring which provides more detailed and accurate information by obtaining DNA samples via non-invasive techniques, with the traditional monitoring methods which emphasize more error-prone, visual and quantitative assessments. The more advanced, genome-based monitoring methods of environmental DNA (eDNA) applications enable flexibility of information access for terrestrial and marine species. The eDNA is obtained through samples from terrestrial and aquatic sediments, air, faeces, fresh water, sea water, urine, ice cores, soil, snow, and epidermal cells. It is noted that sample extraction may be dependent on the rate of material degradation, as sample materials may last in the environment for a few weeks to hundreds of thousands of years. Additionally, eDNA extractions in ecological monitoring can provide more detailed temporal and spatial information on the studied species. This contrasts with traditional monitoring methods, especially as eDNA samples may provide information on higher trophic levels and invasive species from water samples, and extract information from both terrestrial and marine environments. Further noted advantages and problematics of the use of eDNA include economic and efficiency monitoring benefits, issues around false positives, the degradation of samples before analyses, and data misinterpretation on studied species biomass. The study concludes that the integration of traditional methods with eDNA, provides a more robust approach to species studies.

The tenth chapter, titled *Wilding or Worrying? Place Meaning in a Rewilding Landscape* by Rachel Orchard and Carijn Beumer, leading researchers in Global Health at Maastricht University, Netherlands, examines public acceptances of 'place change' as an issue of increasing significance in the UK. Arguably, this is results as post-Brexit policy changes encourage more landowners to provide ecosystem services. The chapter examines the conservation project of Knepp Wildland, with the aim of contributing to the understanding of the effects of place change on 'place meaning' for the local community. Place meaning is defined as looking beyond specific attitudes, to examine the emotive subjective experience of places for people. The research methodology used a qualitative questionnaire distributed through purposive sampling online community groups and finding other contacts through snowballing. Theoretical thematic analysis was also conducted, using Jacobs and Buijs' (2011) place meaning categories: beauty, functionality, attachment, biodiversity and risk, with two additional categories: emotions-in-place and temporality. The findings indicated that in the case of transitioning from an agricultural to a rewilding landscape, place change affects place meaning by triggering renegotiation of place meaning for people themselves and their relationship to the landscape. This can be a factor for change in cultured naturalness, which through time reflects a community's local identity and way of life. The authors argue that this renegotiation of place meaning includes an increasing sense of ecological loss, a changing view of visible stewardship, and an increasing appreciation of wildlife. The influences on how place change affected place meaning included perceptions on access, wilderness, and belonging. These findings also indicated over time the place meaning derived from Knepp Estate's rewilding landscape was predominantly positive, evidenced from the emotions-in-place expressed.

The eleventh chapter, by leading ecological researchers Azimah Abd Rahman and Aimi Shahirah Mohd Nazri titled *Environment Effects on Milky Stork Habitat Selection in Matang Forest Reserve, Perak* examines the impacts of human activities and landuse change on the global Milky stork (*Mycteria cinereal*, Raffles, 1822) populations in the Matang Forest Reserve in Malaysia. Arguing that the Milky stork population has seriously declined over the past decades, the authors give evidence of a rapid, global population decline to a current estimated population of 5,500 individuals. There has also been a recent, local decline of over 90% in Malaysia. GIS applications are used for the analysis of habitat selection and presence frequency of Milky storks in a case study of the Matang Mangrove Forest Resort, the parameters being land use, vegetation and food availability. The findings were that the highest density of milky storks was at Pulau Gula and the lowest density was at Pulau Kelumpang. Storks were recorded mostly in disturbed areas, especially in locations with aquaculture activity, as food sources were more readily available in such areas than in wild habitats. Evidence indicated that captive bred storks from the Department of Wildlife and National Parks (DWNP) breeding program may have lost their natural foraging ability. Captive bred stork presence in human presence indicated these birds preferred human proximity for food sources. Environmental factors, including vegetation, food and climate had less than expected impacts on the density of Milky storks, contrasting with published research findings that indicated that released captive birds frequented less modified and more natural habitat. These findings hence make an important contribution to stork conservation and avian conservation planning.

The twelfth chapter, *The Landscape Ecology of Urban Avian Scavengers: An International Perspective* by Michael O'Neal Campbell, is based on a field study conducted in three countries, Canada, the UK and Ghana, taking the case of eight urban avian scavengers. The premise of this comparative methodology is the paucity of multi-species, multi-national studies of the role of integrated research methods in urban avian studies. The locations and species are as follows: (1) Glasgow UK, Carrion crows (*Corvus corone cornix,* Linnaeus, 1758), rooks (*Corvus frugilegus*, Linnaeus, 1758), jackdaws (*Corvus monedula*, Linnaeus, 1758), magpies (*Pica pica*, Linnaeus, 1758), Lesser Black-backed gulls (*Larus fuscus graellsii*, Linnaeus 1758) and Black-headed gulls (*Larus ridibundas*, Linnaeus 1776); (2) Kumasi and Accra, Ghana, Pied crow (*Corvus albus* Muller 1776), Hooded vulture (*Necrosyrtes monarchus* Temminck 1823), and Black kites (*Milvus migrans* Boddaer 1783); and (3) Ottawa, Canada, the American crow (*Corvus brachyrhynchos* Brehm, 1822), the

Turkey vulture (*Cathartes aura* Linnaeus 1758), Blue jay (*Cyanocitta cristata*, Linnaeus 1758), and the Ring-billed gull (*Larus delawarensis*, G.Ord 1815).

The chapter also assesses the methodology for the studies, which used point counts, landcover parameter measurement and regression analysis of dependent and independent variables as employed in the three case studies. The findings of the field research revealed generalizable results for the target species, in terms of measurable avian reactions to landscape configuration parameter change (green space area and proximity, water body proximity, urban features). There were also inter- and intra-species differences in avian reactions to human behavior and presence, and localized avian adaptation to variable landscape configurations. The chapter concludes that studies of urban avian scavengers must target multi-species and multi-city comparisons, to shed light on generic and unique urban avian ecologies. In this study the case studies were widely separated, and some of the target species were strongly dissimilar, but further studies may employ more case studies to facilitate trends in species size and ecology.

In: Critical Research Techniques in Animal …
Editors: Michael O'Neal Campbell et al.
ISBN: 978-1-53619-846-1

Chapter 1

URBAN VULTURE ECOLOGY, TRACKING TECHNOLOGIES AND HUMAN QUALITY OF LIFE

Michael O'Neal Campbell*
Simon Fraser University, British Columbia, Canada

ABSTRACT

Research methods, including drones (unmanned aerial vehicles or uncrewed aerial vehicles-UAV or remotely piloted aircraft-RPA, defined as remotely, computer controlled, non-human occupied aircraft), global positioning systems (GPS) and attached cameras increasingly benefit ecological and socio-environmental research. Drones, like satellites and airplane aerial photography, originated as military hardware, later transposing into recreational, observational and even law-enforcement supports. Sophisticated drones contribute accurate location recording, high resolution image capture and subject animal behavioral assessment to ecological studies. This article looks at the use of cutting-edge drone-based research methods for studies of the urban ecology of Black vultures (*Coragyps atratus*, Bechstein, 1793) and Turkey vultures (*Cathartes aura*, Linnaeus, 1758), and current and potential applications to related urban human quality of life issues in North America. Vultures are key subjects for high mobility, remote survey machines, due to their visibility, high mobility, remote nesting, wide foraging and migratory ranges, and human social impacts (urban pollution, building damage, livestock attacks, collisions with aircraft). Vulture intrusion into urban areas is currently socially relevant, but few studies have been undertaken using drone-based methods, possibly due to legal, logistical, financial and social issues. Hence, the utility of these tools may be surmised through their application the study of other species, and the implications of human landscapes. Measurable parameters include flight speeds, soaring duration, foraging flight patterns, courtship, nesting habits and locations, physical attacks on infrastructure, roosting, defecation and aggressive incidents, and human impact incidents. Recent findings indicate positive results (increased accuracy in the documentation of ecological

* Corresponding Author's E-mail: ecol55@hotmail.com.

dynamics) and negative issues (animal avoidance, social privacy issues). Future development of drone-assisted studies requires smaller, quieter more elusive drones, and a more supportive legal framework. This makes an important contribution to urban raptor ecology and urban planning for human quality of life.

INTRODUCTION

Urban ecology, an increasingly vital research cluster given rapid global urbanization, increased human/animal interactions and developing interest in human quality of life and urban ecologies, currently requires more sophisticated and precise research techniques (Campbell, 2016; 2017, 2018). Animal, especially avian, intrusion and adaptation to densely populated urban contexts has greatly increased human/wildlife tensions and spawned conflicted debates on conservation, tolerance and human quality of life (Campbell, 2015; Konijnendijk, 2017; Koh & Wich, 2018). Impacts of such developments are disputed, with acknowledgment of insufficient research findings for supported conclusions (Whitehead & Hugenholtz, 2014; Ivošević et al. 2015; Vas et al. 2015). An important, disputed impact of dense animal presence is social disruption and a consequent reduction in human quality of life (Campbell, 2017). Human quality of life, variously defined generally includes: (a) emotional well-being, (b) interpersonal relationships, (c) material well-being, (d) personal development, (e) physical well-being, (f) self-determination, (g) social inclusion, and (h) human rights, all variably defined within age-related, cultural and social settings (Schalock et al. 2002). Related to these ideas is the acknowledgement that animals and may either degrade or enhance human quality of life (Schaltegger & Beständig, 2012).

Vultures are a dominant urban foraging guild, frequenting urban substrates in some cases over natural landcover (Campbell, 2015). Serious conflicts emerge with vulture intrusion into areas significant for human quality of life, including both dense urban substrate (including downtown areas, sidewalks, large gutters, airports, parking areas, rooftops, plazas or shopping malls, recreational facilities) and green spaces (greenbelts, greenhearts, parks, gardens, treed avenues, grassed beaches) (Campbell, 2014, 2015). Such green spaces are currently recognized as vital for urban health and human quality of life (Park & Lee, 2000; Chace et al. 2003; Hostetler & Knowles-Yanez, 2003; White et al. 2005; Konijnendijk, 2017). They are defined as areas were the landcover includes trees, shrubs and grasses, which may surround dense urban built areas (as greenbelts), or be surrounded by such urban areas (as greenhearts) or be scattered though an urban area (as parks and gardens) (Kuhn, 2003; Campbell, 2006). Green spaces are expanding with urbanization, with the recreational and health interests of urban populations (Savard et al. 2000; Chiari & Seeland, 2004; Cornelis & Hermy, 2004). Urban people perceive vultures as cleaners or aggressive polluters or food thieves (Campbell, 2015).

Recently, drones (unmanned aerial vehicles or uncrewed aerial vehicles, UAV or remotely piloted aircraft, RPA) have emerged as important, even leading tools for such research (Anderson & Gaston, 2013). More sophisticated drones include autopilot and computer systems, Global Positioning System (GPS) receivers and compasses, and sometimes barometric altimeters, video cameras and photographic technology, and complex, optical sensors (infrared and/or ultraviolet capable), physical sensors (measuring temperature, humidity etc.) and sound sensors (Ivošević et al. 2015; Ditmer et al. 2015a). Figures 1, 2, 3 and 4 illustrate some of the recent developments of drone technology, quadcopters and hexacopters (drones with four or six rotors for lift and propulsion) (DJI, 2019).

When drones are used for ecological studies they may be termed "eco-drones" or "conservation drones" (Ivošević et al. 2015, 113). This application may have far reaching consequences for environmental research. Ditmer et al. (2015a, 2278) hypothesize that "unmanned aerial vehicles (UAVs) have the potential to revolutionize the way research is conducted in many scientific fields" as drones "can access remote or difficult terrain, collect large amounts of data for lower cost than traditional aerial methods, and facilitate observations of species that are wary of human presence" (see also Anderson & Gaston, 2013; Marris, 2013; Vermeulen et al. 2013; Lucieer et al. 2014).

Figure 1. Matrice 200 Series V2. Source: Reproduced with permission of DJI (2019). This quadcopter drone, a cutting edge machine with a design that responds to the current needs for a versatile and maneuverable aircraft, is described by DJI (2019) as "the ultimate platform for aerial productivity" as it "combines a rugged design and simple configurability" with improved "intelligent control systems, flight performance, and add flight safety and data security features."

Figure 2a. Mavic 2 Enterprise. Source: Reproduced with permission of DJI (2019). This is also a foldable quadcopter drone. It may be strong enough to be used for police work, fire response, infrastructure inspections and search and rescue. The Mavic 2 Enterprise Dual version adds thermal imagery. By extrapolation, this would also be good for ecological work.

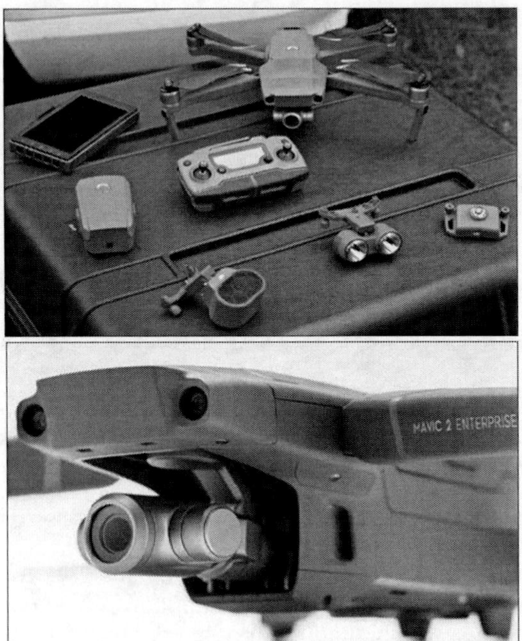

Figure 2b. Mavic 2 Enterprise. Accessories. Source: Reproduced with permission of DJI (2019).

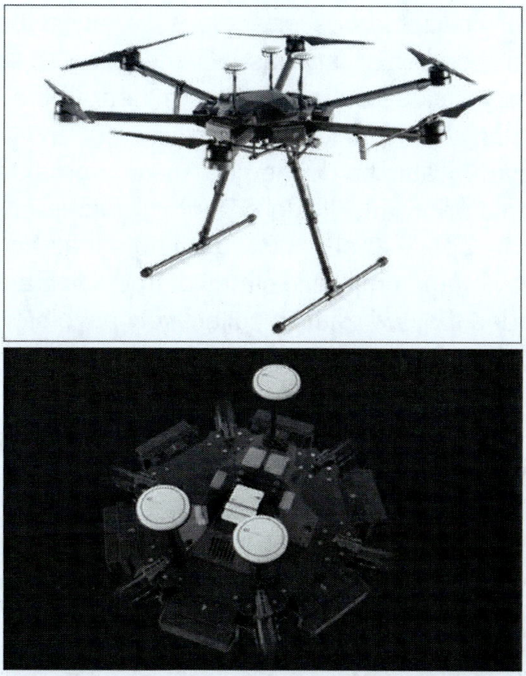

Figure 3. Matrice 600 Pro. Source: Reproduced with permission of DJI (2019). This is a hexacopter or a drone with 6 rotor blades for lift and suspension. This drone can compare sensor data from three Global Navigation Satellite System (GNSS) units and has a dust free dustproof propulsion system.

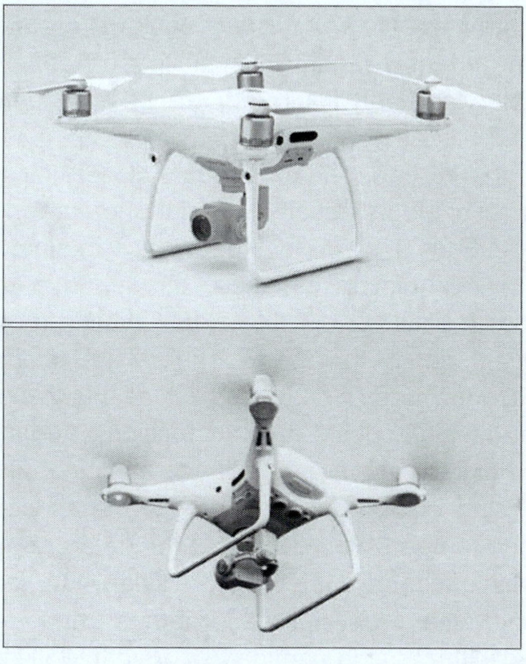

Figure 4. Phantom 4 Pro. Source: Reproduced with permission of DJI (2019). This drone has a forward vision, backward vision downward vision system, and an infrared sensing system.

GPS technology is vital for drone operations because the location of observed subjects is crucial to ecological studies (Campbell, 2018). Sophisticated GPS technology systems may include biotelemetry, defined as the "remote detection and measurement of a human or animal function, activity, or condition" (Merriam-Webster, 2018). GPS technology may also be defined as "the instrumental technique for gaining and transmitting information from a living organism and its environment to a remote observer" (Slater, 1965). Cooke et al. (2004: 334) note that the "remote measurement of the physiology, behavior and energetic status of free-living animals is made possible by a variety of techniques that we refer to collectively as 'biotelemetry'." This includes transmitters that send signals over varying distances and to orbiting satellites and data devices. GPS may also include bio-logging, defined as the "on-animal deployment of miniaturised electronic data recorders' allowing "for the study of location, body position, and physiology of individuals throughout their ontogeny."

GPS systems vary in accuracy and resolution, usually between 100 meters (m) and 4 kilometers (km) (Campbell, 2018; Farve, 2018). Technology includes "VHF (very high frequency) radio-tracking, GPS tracking, satellite tracking using the Argos system and a combination of the advantages of GPS and the Argos satellites into a GPS/satellite telemetry method" (Farve, 2018). Telemetry includes hand-held antenna to receive VHF transmissions from a VHF transmitter, which may be attached to the body of the observed animal. Triangulation of the transmissions from several receivers and connections with signals from some of the GPS satellites orbiting the earth determines the location of the subject animal or object. The satellites are the Argos Satellite system operated jointly by the United States (National Oceanic and Atmospheric Administration, NOAA and National Aeronautics and Space Administration, NASA) and the France (French Space Agency) and data processing centers in Landover, Maryland and Toulouse, France (Tomkiewicz et al. 2010; Farve, 2018).

One rationale is that drones enable much closer image capture and video recordings than other technologies such as aerial photographs, satellite images and hand-held camera recordings These older geomatics techniques lack the detail and frequency (e.g., patterns changing by the second) required for many ecological studies (Campbell, 2018). Nowak et al. (2019: 56) argue that "remote sensing technology, which enables the observation of the Earth's surface and is currently very common in environmental research, has many limitations such as, insufficient spatial, spectral and temporal resolution, and a high cost of data acquisition." Anderson and Gaston (2013: 138) also make this point: "to date, heavy reliance has been placed on obtaining such data from remote-sensing instruments mounted on spacecraft or manned aircraft, although the spatial and temporal resolutions of the data are often not suited to local-scale ecological investigations." Objectively, "flying low and slow, UAVs offer ecologists new opportunities for scale-appropriate measurements of ecological phenomena. Equipped with capable sensors, UAVs can deliver fine spatial resolution data at temporal resolutions defined by the end user" (Anderson & Gaston, 2013: 138). More sophisticated drones have high resolution cameras for detailed photography. Recent technological developments have allowed "improvements in navigation and the miniaturization of measurement technologies, allowing the study of individual organisms and their spatiotemporal dynamics at close range" (Anderson & Gaston, 2013: 138).

Drones also enable detailed georeferencing of selected locations, when equipped with GPS tracking software (Ivošević et al. 2015; Hammill et al. 2017; Johnston et al. 2017; Lucas, 2018). Therefore, another rationale is that drones are more accurate than other tracking methods for ecological assessments (Ditmer et al. 2015a; Hodgson et al. 2018). Another rationale is that drones vary in size, from a few centimeters across to the size of airplanes, this allowing purchasers and operators to choose a variant that suits their needs, which could be photographing or filming butterflies or a whole lake. Drones also vary in flying ranges from a few kilometers to over 500 kilometers. There are also different designs, with some having fixed wings and flight patterns like airplanes and others with one or more rotor blades with flight similar to helicopters. These flexibilities have led to an increase in the number of drones used for variable tasks (Ditmer et al. 2015b; Franke, 2015; Kennedy & Rogers, 2015; Federal Aviation Administration FAA, 2018).

For these reasons, Hodgson et al. (2018, 1160) argue that "ecologists are increasingly using technologies, such as remotely piloted aircraft... unmanned aerial systems or unmanned aerial vehicles), for wildlife monitoring applications." Johnston et al. (2018, 178) agree: "unoccupied aircraft systems (UAS) are now frequently used in wildlife research, including studies of marine species. Ivošević et al. (2015, 113) justify the increased interest in drones for environmental jobs: "a drone can provide a low-cost and low-impact solution to environmental managers working in a variety of ecosystems." Therefore, drones are "reliable tools for conducting population assessments... and establishing life history stages of animals. These new tools provide flexibility in operations and may reduce costs and human risk in some cases" (Johnston et al. 2018, 178). Applications of drone technology include the observation and tracking of highly mobile wildlife such as large birds, their frequented foraging areas, prey species and nesting (Campbell et al. 2015). Drones may also participate in data acquisition for animal population estimation and poaching prevention (Koh & Wich, 2018; Barasona et al. 2014; Whitehead & Hugenholtz, 2014; Mulero-Pázmány et al. 2014; Ratcliffe et al. 2015).

However, drone technology is both relevant and problematic for the field study of large raptors. Such birds may be easier than others to spot, due to their large size, nests and foraging ranges, fairly predictable feeding habits (large trackable prey and food sources such as carcasses), and conflicts in human life spaces, including attacks on livestock and companion animals, and aircraft collisions. Large eagles and some vultures may attack or aggressively approach drones (Fessenden, 2015; Rockwell, 2016). Such issues justify in-depth studies on the potential and discovered advantages and pitfalls of using drone-based technology for studying large raptors. This article examines the current and potential roles of drones in the study of Turkey and Black vultures in urban North America, justified by the crucial role of such ecologies to examples of human/vulture conflicts and human quality of life. Key issues examined are the adaptation of the vultures to changing urban dynamics, collisions with aircraft, pollution, attacks on humans and companion animals and interspecies differences in these activities. As the introduction of drones into such studies is recent and often rudimentary, this review may be introductory to later studies of more detailed applications. As such the potential of drones as tools of investigation is also examined and the requirements for more effective drone applications, including legal and administrative issues conclude the article.

URBAN TURKEY AND BLACK VULTURE ECOLOGY

The co-evolution of the human–avian relationship within the urban areas is a central, developing research problem. Implicit in this paradigm is the acknowledgment that urban development relates to vultures positively through food provision (organic products, road kills, vulnerable domestic animals, human discards), convenient roosting and perching structures, and negatively through habitat loss, noise, poisoning, shooting and in some places of high sanitation, a relative absence of food. Vultures in urban areas must adapt to intensive human presence and also to total landscape change, and in some cases new competition from corvids and *Larus* gulls (Campbell, 2014, 2015). The Turkey vulture (*Cathartes aura*, Linnaeus, 1758) and the Black vulture (*Coragyps atratus*, Bechstein, 1793) occupy the niche of medium-sized scavengers in North America (Figures 5, 6). However, according to Holland et al. (2017) "knowledge of Black vulture... and Turkey vulture... spatial ecology is surprisingly limited despite their vital ecological roles. Fine-scale assessments of space use patterns and resource selection are particularly lacking, although development of tracking technologies has allowed data collection at finer temporal and spatial resolution." In urban areas, they may be the dominant avian scavengers, competing in presence and foraging aggression with the gulls *Larus* (*Linnaeus,* 1758). No other vulture species are common in the urban areas of North America (Campbell, 2015).

As shown in Table 1 below and Figures 5 and 6 below, the Turkey and Black vultures are very similar in appearance, size and general habits, but there are also some pronounced differences that impact on their foraging habits and general behavior towards people. Their adaptability can be seen in their variable behavior in relation to human activities and landcover modifications, and human reactions based on public perceptions of their real or perceived behavior. In this it is necessary to compare the vultures with other species such as *Larus* gulls, eagles and even feral dogs. These are species that may either compete with vultures in their scavenging role or be perceived as similar in a predatory role (Ferguson-Lees & Christie 2001; Campbell, 2015).

Figure 5. Turkey Vulture. (Permission Chris F from Pexels).

Figure 6. Black Vulture. (Permission John Barnard from Pexels).

The Turkey vulture differs from the more social Black vulture in having a sense of smell, allowing scent-based foraging through the detection of ethyl mercaptan, the gas emanating from such rotting organic matter. The olfactory lobe, which processes smells is larger than that of most bird species (Snyder & Snyder, 2006). Lowery and Dalquest (1951) found that Turkey vultures could locate decaying meat, even when it was located under logs or tree base hollows, hence leading the other vultures to such sources. This sense of smell allows the Turkey vulture to detect carrion not in sight, below the forest canopy or under roofs, in gutters or under coverings. Black vultures may follow Turkey vultures to such food sources, which they could not have detected, and displace the latter through their greater tendency to create flocks, in some cases up to 50 individuals (Wallace & Temple, 1987; Muller-Schwarze, 2006; Snyder & Synder, 2006; Campbell, 2015).

Table 1. Characteristics of the Turkey vulture and Black vulture

Attribute	Turkey Vulture	Black Vulture
Length (cm)	62–81	56–74
Wingspan (cm)	1.60–1.83	1.33–1.67
Weight (kg)	0.8 - 2.3	1.18 - 3.5
Sexual dimorphism	Monomorphic (similar)	Monomorphic (similar)
Social behavior	Solitary, small groups	Flocks
Roosting	Solitary, small groups	Flocks
Sense of smell	Strong	None or weak
Foraging	Urban, open land, forest	Urban, open land
Soaring level	Solitary, few, lower	Flocks, high
Habitat	Urban, open land, forest	Urban, open land, farms
Urbanization	Attracted	Attracted
Human tolerance	Strong, medium	Strong
Population	Not threatened	Not threatened
Human conflicts	Fecal pollution	Fecal pollution, livestock

Sources: Ferguson-Lees and Christie (2001); American Ornithologists' Union (2009); Campbell (2015). Black vultures are generally heavier than Turkey vultures (Wallace & Temple 1987: 293). Avery et al. (2011) note that "Black vultures flew at higher altitudes than did Turkey vultures in all seasons except summer when altitudinal distributions (above ground level) did not differ."

The adaptation of the Black vulture to urban substrates and human presence is well documented (Buckley, 1999; Lowney, 1999; Avery 2004a, 2004b; Avery & Cummings, 2004; Blackwell & Wright, 2006; Novaes & Cintra, 2013; Campbell, 2014). Hill & Neto (1991: 173) argued that Black vultures have benefited greatly from human post-Columbian colonization of the New World, due to the substantial urban development, availability of human waste and opening of closed forests. The opportunities created include possibilities of feeding on the organic materials in garbage dumps, fish killed in polluted waters, livestock carcasses in pastures and road kills. As a result of this development, the Black vulture's population "numbers have increased so dramatically that it is now considered the most abundant bird of prey in the Western Hemisphere" (Hill & Neto, 1991 173) In the United States, foraging and roosting in urban areas is common (United States Department of Agriculture USDA 2003). For example, Carrete et al. (2010: 390) describe the Black vulture as an example of a winning species, which is responding positively to human transformations, such as land development change. Despite the lack of detailed research on the continental and national level range development of this species, there is evidence that the Black vulture is moving beyond its previous tropical landcover and has advanced into most parts of North and South America (Carrete et al. (2010: 390).

Nickens (2008) notes that Black vultures are among more adaptive among the members of the vulture family, this despite their slightly smaller size than Turkey vultures. Crucially, Black vultures can better tolerate human presence, which allows their foraging in spaces of close human proximity. Additionally, although most Turkey vultures migrate, Black vultures are largely residents and hence participate in all year foraging. Nickens (2008) further notes that Black vulture numbers and range expansions are positively correlated with human population and urban area expansions. Similarly, the Turkey vulture is the most wide-ranging avian obligate scavenger in North and South America, ranging from the coniferous forests of Canada and the Northern United States to the Andes mountains and the Pampas plains of southern South America, adapting easily to increasing trends in deforestation, urbanization and mammal population extinctions (Campbell, 2015).

Urban ecology for the Turkey vulture differs from that of the Black vulture in that the former bird is less communal, flying, foraging, perching and roosting in smaller numbers than the Black vulture (Campbell, 2015). Fewer Turkey vultures than Black vultures, nest in urban areas, hence it is suggested that urbanization may reduce Turkey vulture populations in California (Kiff, 2000, see also Garrett & Dunn, 1981; Unitt, 1984). The findings of several studies however indicate pronounced adaptation to urban settings in North America. These include nesting in abandoned houses in southern and south-central Saskatchewan in 1982, in the Aspen Parkland and Boreal Transition ecoregions of Saskatchewan with similar success as nesting elsewhere (Houston et al. 2007).There was also nesting in old trees in William Land Park in urban Sacramento, California, in which the vultures ignored "low-intensity human activities beneath and adjacent to roosting trees, including use of a walking and running trail and golf course, and the feeding of ducks in the pond" (Airola, 2011: 5).

Turkey vultures may be attracted to urban roosting sites by the presence of tall trees and warmer air rising from flat open ground that facilitates soaring (Airola, 2011). Nevertheless, in citing an example of an urban park in Sacramento heavily used by roosting Turkey vultures, Airola (2011: 7) questions why the resident Turkey vultures travel to urban roosting sites, which are quite distant from suitable rural foraging areas to roost in an urban area. More logical roosting areas would probably be nearby foraging habitats. Airola (2011: 7) also argues that "the Land Park site and adjacent residential areas with extensive turf and ornamental plantings does not seem to trap heat that would reduce thermoregulatory energy load during cool time of year." A further argument is that "perhaps given the relatively low energetic cost to vultures of long-distance flights, seemingly minor advantages of lower predation risk at an urban park site make it worth travelling to as a nightly roost" (Airola, 2011: 7).

Black and Turkey vultures' adaptation to urban food sources, such as rubbish dumps, human discards in parks and streets, small livestock and road kills are the key issues (del Hoyo et al. 1994; Carrete et al. 2010; BirdLife International, 2019a, b). Communal roosts, which are frequented because of predation risk and social interactions are associated with these invasions (Devault et al. 2004; Campbell, 2015). Major problems that may evolve from these invasions include aircraft collisions (DeVault et al. 2005; Blackwell and Wright 2006), attacks on livestock (Lowney, 1999; Avery & Cummings, 2004) and damage to property (Lowney, 1999). Black vultures are generally blamed for more such incidents than Turkey vultures (Campbell, 2015).

Attacks on wildlife include "plucking the eyes and eating the tongues of newborn, down, or sick livestock, disemboweling young livestock, killing and feeding on domestic fowl, and general flesh wounds from bites" (United States Department of Agriculture, USDA, 2003).

Roosting problems for both species include fecal droppings, which pollute water, and affect electricity supplies. Both species are protected by the Migratory Bird Treaty Act (USA), the Migratory Birds Convention Act (Canada) and the Mexico-United States Convention for the Protection of Migratory Birds and Game Mammals. These prohibit the killing of vultures, leaving non-lethal methods as the only means of controlling their urban impacts (Avery et al. 2002; Seamans, 2004). Few non-lethal methods, such as harassment or removal have been proved to be successful, due to the high mobility and adaptability of vultures (Blackwell et al. 2006).

Urban contacts between these vultures and people are likely to increase (Campbell, 2015). Similar to the Black vulture, the Turkey vulture has also significantly expanded its range northwards, especially after the 1920s" (Kiff 2000: 175 - 176), including its wide range migratory behavior (Jackson, 1983). Turkey vultures also have the "most catholic habitat preferences of any member of the *Cathartidae*, perhaps owing in part to their ability to use olfactory clues in locating carcasses" (Kiff, 2000: 175 – 176). Devault et al. (2004: 710) found that "Turkey vultures are able to adjust their home ranges, movement patterns, and flight behavior to local conditions, especially those related to habitat structure and foraging opportunities," this being defined as "plasticity in ranging behavior" allowing significant range expansion. In addition to urban areas (including gardens, roadsides and rooftops), Turkey vultures breed, roost and forage in farmed areas, open and closed natural vegetation (Hilty, 1977; Kaufman, 1996; USDA, 2003; Carrete et al. 2010). Communal (sometimes several hundred birds) or single roosting may be on both natural and artificial features, including water or microwave towers, backyard trees or on suburban rooftops (Lowney, 1999; Lovell & Dolbeer, 1999; Kiff, 2000; Houston et al. 2007; Campbell, 2014). Turkey vultures may nest in buildings and even abandoned cars (Igl & Peterson, 2010). In southern and south-central Saskatchewan, they nested in abandoned houses, and populations increased with expanding urbanization, revealing no decline in nesting success (Houston et al. 2007).

The USDA lists the general nuisance behavior and social damage incurred from vultures: "unwanted congregations of these birds around areas of human activity (homes, schools, churches, shopping areas) resulting in accumulations of feces on trees and lawns, residential and commercial buildings, electrical and radio transmission towers, and other structures." The USDA also lists the physical problems vultures bring to urban areas, as the justification for ameliorative action: property damage, including roosting, tearing of "window caulking, roof shingles, vent seals, rubber roof liners, and pool covers"; damage of vehicles by "scratching paint, removing rubber seals and wipers, and ripping vinyl seat covers from boats and tractors"; impacts on human quality of life through accumulation of "the birds' feces and vomit... especially on roofs of houses, office buildings, communication towers, and electrical transmission structures," in the last case sometimes resulting in expensive arcing and power outages (USDA, 2016). Additionally, "droppings near homes and drinking water sources may pose health concerns for individuals and municipalities. Many people find offensive the white-wash effect of vulture droppings and associated ammonia odors" (USDA, 2016).

Vulture/human contacts and conflicts have spawned several methods to reduce vulture presence, even to the extent of influencing competing private companies that advertise their abilities in this regard (USDA, 2016). Of course, these companies must adhere to the legislation in the United States and Canada that protects vultures though the Migratory Birds Act (Campbell, 2015). The serious problems justify the intervention of professional agencies: vulture management is defined as complicated and site-specific, with detailed ecological problems that may be solvable with the assistance of a wildlife professional. Solutions may be dependent on the advice on vulture dispersal, evaluation of damage and use of ameliorative tools such as sound- and light-devices, including propane cannons or pyrotechnics and lasers (USDA, 2016). One such professional company is named Bird-Be-Gone (Bird Be Gone, 2018). This company lists vulture problems:

> Turkey vultures can damage homes and commercial buildings by tearing up window caulking, roof shingles, vent seals, rubber roof liners, and pool covers. They…damage vehicles by scratching paint, removing rubber seals and wipers, and ripping vinyl seat covers from boats and tractors. Their droppings and vomit can damage communication towers and electrical transmission structures, resulting in arcing and power outages—a huge expense for utility companies. Turkey vultures can also seriously injure livestock, as they often target the eyes and soft tissues of these animals, so animals must often be euthanized. Vulture droppings near homes, pools and drinking water sources may pose health problems for individuals and municipalities. Finally, vultures loitering around airports can be hazardous to aircraft due to their soaring behavior.

Conflicts between Turkey vultures and Black vultures may also increase as the two species occupy new urban foraging spaces. Turkey and Black vultures come out even in most conflicts, although flocks (sometimes numbering over 200 birds) of the more social Black vulture may exclude solitary Turkey vultures from food sources (Stewart, 1978; Buckley, 1996). Wallace and Temple (1987: 293) found that in one-on-one situations, Turkey vultures won more than half of their aggressive encounters with Black vultures, which showed that they had no real advantage over the Black vultures (56% wins in 863 encounters). Other species may also effectively compete with vultures. For example, Turkey vultures may be dominated at food sources by feral dogs (*Canis familiaris,* Linnaeus, 1758), domestic cats (*Felis catus,* Linnaeus 1758), raccoons (*Procyon lotor*), red foxes (*Vulpes vulpes,* Linnaeus 1758) and coyotes (*Canis latrans,* Say 1823) and Red-tailed hawks (*Buteo jamaicensis*, Gmelin 1788) (Prior & Weatherhead, 1991).

Behaviors leading to conflicts with people include intrusion into food stalls, windows, roofs, butcheries and markets and greater tolerance of human proximity (Campbell 2014). Campbell, 2014 found in Central America that most vultures were recorded in downtown areas (41%), town suburbs (22%), fallow farm and grassland (15%), savanna (11%), open forest (8%) and closed forest (4%). They frequently engaged in observational and possibly anticipatory perching, even in the absence of clear food sources. Such possibly learned behavior was only observed in urban areas and was absent from natural ecosystems. Campbell (2014) also found that the higher percentage of urban substrate was a significant positive predictor of Turkey vulture presence, this applying to markets (including rubbish loaded streets, abattoirs and fish unloading zones), residential and financial centers, hotel districts and transportation yards, plazas, rooftops and electric poles.

Airline collisions or Bird-Aircraft Strike Hazard (BASH), are an extremely important issue for urban birds, but few studies have examined the role of vultures compared with other species (Camphuysen et al. 2011; Campbell, 2015). For example, the FAA recorded 57,702 bird/aircraft collisions in the United States, from 1990 and 2004, costing over $600 million (Gard et al. 2007). Airline collisions occur mostly at the cruising stage of aircraft flight, but some occur during descent, mostly below 200 m. Fighter planes are among the more likely aircraft to be involved in collisions, due to their high speed at low latitudes. The United States Department of Agriculture (2003) notes that birds with either large body size or those that move in large flocks are especially likely to be involved in such collisions, criteria that comfortably fit both the Turkey vulture and the more social Black vulture (Campbell, 2015). Concerning collisions between Black vultures and airplanes in the United States, costs have totalled over US$25 million to the US Air Force (USAF 2009). Collisions with Turkey vultures have similar costs (Satheesan & Satheesan, 2000).

Bird migration patterns and bird attraction to prey sources in the grassed areas near the airports are particular issues for most species, but for vultures communal thermal soaring flight, gliding among thermals, slow bird reactions to approaching airplanes, flocking around individuals that detect prey and roosting in nearby trees are also important factors (Gard et al. 2007; Campbell, 2015). As in tropical cities, including those in the southern United States, vultures are among the dominant foraging guilds, these species are of important research concern (Borrow & Demey, 2001; Gbogbo & Awotwe-Pratt, 2008). This is largely because tropical cities have more food sources for scavengers than temperate urban areas, these including open food markets, large garbage disposal areas, poorer sanitation and faster rates of decomposition of organic discards (Mundy et al. 1992). These problems are accentuated by "the expected increase in human populations and urbanization in the tropics and the rich biodiversity that characterizes this region" consequently "more studies are desperately needed to inform public policy so that the negative consequences of human development are mitigated" (Marzluff et al. 2001: 1; also Schochat et al. 2006).

DeVault et al. (2010) examined mid-air collisions between Black and Turkey vultures and both military and civilian aircraft at the Savannah River Site (SRS) in South Carolina, USA. The hypothesis was that weather and time variables predicted vulture flight patterns. The findings were that Black vultures average flight altitude (169 ± 115 m) was higher than that of Turkey vultures (163 ± 92 m), higher than the averages found in other studies, and with minimal impacts from weather and time variables. Factors were food availability, inter- and intra-specific interactions, and physiological demands. The conclusion was that solutions to vulture/aircraft collisions should rely on variable flight behaviors of Black and Turkey vultures and factors contributing to these differences across regions), to a greater extent than on the effects of local conditions on flight patterns. Suggested operational methods to reduce collisions include removal of food sources from within 200 km of military airfields and ranges and 100 km of civil airports (Satheesan & Satheesan, 2000). Commoner methods include shooting, but these are mostly against *Larus* gulls (Dolbeer, 1998).

Avery et al. (2011, 1581) looked at Black and Turkey vulture flight behavior and activities at the Marine Corps Air Station in Beaufort, South Carolina, using solar-powered Global Positioning System (GPS) satellite transmitters. The findings were that Turkey vultures had larger seasonal home ranges than did Black vultures, Turkey vultures spent more than twice as much time in flight than did Black vultures and Black vultures generally flew at higher altitudes than did Turkey vultures, with equality during the summer. Both species generally flew at low altitudes; in fact more than 60 percent of the flights of both species "occurred from 4 hr to 9 hr after sunrise at altitudes below 200 m" (Avery et al. 2011, 1581) These findings may reveal a serious risk to aircraft during the take-off and landing phases, especially for fast military aircraft, as other studies have found (Campbell, 2015). The study suggests that potential collisions may be avoided using a "continuation of aggressive harassment coupled with flexible training schedules to avoid times and altitudes of high vulture activity" (Avery et al. 2011, 1581).

Urban vultures also encounter dangers. For example, one measurable issue concerns the food ingested by vultures in urban areas (Campbell, 2015). The findings of one study revealed that 39.1% of the defecated pellets examined included synthetic products, mostly plastic bag remains (Inigo Elias, 1987). Electrocution by electric power lines is another serious problem. Research findings indicate that since the late twentieth century, hundreds of raptors were electrocuted in North America (APLIC, 2006). Such collisions are commoner when power lines bisect vulture foraging or nesting areas (Hunting 2002). Solutions may include alternative perch availability, vegetation form and carcass presence. Power lines and towers may also be used as replacements for nesting or roosting trees, especially deserts, open plains and intermontane basins (APLIC 2006). However, Lehman et al. (2007) note that despite studies from the 1970s, in the United States, where there has been detailed research, with product testing and design standards development but most efforts have not reduced raptor electrocution (Franson et al. 1995; Melcher & Suazo 1999; Harness & Wilson 2001). Hunting (2002) also found that bird electrocutions cost energy suppliers billions of dollars annually through power interruptions, equipment repairs, revenue losses and statutory compliance issues. Electrocutions occurred when a bird connected two energized components or an energized and an earthed (or 'grounded') component of the electric pole structure, allowing an electric flow through the birds' body. Larger birds connected components more effectively than smaller birds (APLIC, 2006; Ferrer & Janss, 1999).

In the United States, of the 31 diurnal raptor species, 26 are recorded as victims of electrocution (Johnsgard, 1988, 1990; Harness & Wilson, 2001). Current evidence indicates that the design of powerlines is a major factor for raptor electrocution. Low-voltage distribution lines (<69 kV) for private use in the United States contribute to most of the electrocutions. Arguably, this is because older systems used less insulation with less wire separation. Wooden poles are less conductive than metal structures, hence may reduce electrocutions. As noted by AltaLink (2019) "a bird is at risk of electrocution on a structure if it contacts two energized components or an energized component and a grounded component.

Birds are typically at risk on distribution poles where the electric and grounded components are closer together, rather than on larger transmission structures that AltaLink operates where they are further apart." Therefore, electrocution is a greater risk with a likelihood of a link between energized components, e.g., with transformers or other auxiliary equipment (fused cutouts, capacitors, reclosers, jumper wires) (Olendorff et al. 1981; APLIC, 2006; Harness & Wilson, 2001). American powerlines are less likely than European ones to contribute to electrocutions, as European lines generally have steel or steel-reinforced concrete support, which unlike wood is conductive and grounded, allowing current to pass through both the materials and the bird's body. This can occur even when it is perched on a cross ram and it touches the wires (Bayle, 1999; Janss 2000). Birds may also be killed through collisions with wires or even poles (Campbell, 2015).

Vulture perching on poles (and electrocutions) increases when poles are on higher topography, partly for survey opportunities (Campbell, 2015). Other factors for perching may include daily weather and seasonal variation (wet feathers increase the electrocution risk ten-fold (Nelson, 1979, 1980), nesting periods (Harness & Wilson, 2001), and wind direction, as wind flow diagonal or parallel to the crossarms, may contribute to more accidents than wind flow perpendicular to the crossarm, as landings and takeoffs are more difficult (Nelson & Nelson, 1976; Benson, 1981). The fledging period may also see more accidents, due to the increased bird populations (Harness & Wilson 2001). Other factors include bodily contacts between birds during nest defence or mating, which may link conductive components (Dickerman, 2003). Large nests may lie across conductive equipment and cause electric flow through any birds in the nest (Vanderburgh, 1993; Campbell, 2015). Young birds are particularly likely to be electrocuted, due to inexperience (Campbell, 2015).

Vulture failure to detect electrical structures is blamed for the frequent deaths (Campbell, 2015). Martin et al. (2012) found that despite vultures' excellent eyesight (necessary for long-range foraging from great heights, "the visual fields of vultures contain a small binocular region and large blind areas above, below and behind the head." Martin et al. (2012) also found that these enable coverage of the ground, while protecting the vulture's eyes from the sun and enables lateral vision for foraging. Vultures therefore have narrow, frontal binocular vision, which may fail to detect structures such as electric poles and wires, and even wind turbines Martin et al. 2012: 5) argue that by "erecting structures such as wind turbines, which extend into open airspace, humans have provided a perceptual challenge that the vision of foraging vultures cannot overcome." If the structures cannot be changed, the solution would be to attract vultures away from such areas, by creating alternate perching, nesting and foraging areas and structures, a difficult proposition due the extremely wide foraging ranges of vultures (Campbell, 2015).

In the United States and Canada, social opinions and attitudes to Black and Turkey vultures are partly based on the belief that the vultures may attack farm animals. The perceived behavior of vultures towards livestock in suburban and rural areas, affected general perceptions of vultures even in urban areas (Campbell, 2015). Kiff (2000: 180) observed that "throughout the first half of the present century, many ranchers and farmers associated Turkey and Black vultures with the spread of diseases, especially anthrax, and therefore often sought to eliminate them, usually by trapping." Flocks of Black vultures were also blamed for attacks on newborn calves in open fields. Avery and Cummings (2004: 59) argue that the Black vulture has historically been "a problem species for many livestock producers": with an accumulation of historical evidence (Campbell, 2015). For example, Baynard (1909) noted that in Florida "hundreds of young pigs, lambs, etc., are annually devoured by them.... I have had them come into my yard and catch young chickens" (Avery & Cummings, 2004: 59). More recently, the United States Department of Agriculture's Wildlife Services Program recorded increased attacks by vultures on domestic animals after 1997. During the period 1997 to 2002, 84% of reports of attacks on mostly young cattle, were recorded in the states of Virginia, Florida, Texas, South Carolina, and Tennessee, the rest in 13 other states (Campbell, 2015).

In Virginia during the period 1990 to 1996, 115 incidents of Black vulture contact with 1037 livestock animals were recorded (Lowney, 1999). One typical attack involved an action by about 20 to 60 vultures, attacking the eyes, rectum, genitals and nose of calves and lambs, and even calving cows. Black and Turkey vultures were observed on a Florida farm near areas of active calving, in some cases feeding on afterbirth and fresh calf excreta (Avery & Cummings, 2004). Some newborn calves appeared to be killed and eaten by up to 40 vultures. However, it was unclear if the calves were killed by the vultures or were stillborn. In another event, three Black vultures unsuccessfully attacked a calf during the birth process (Avery & Cummings, 2004).

Turkey and Black vultures consequently were persecuted, with the elimination of hundreds of thousands of vultures in Florida and Texas during the 1940s and 1950s. Later the Migratory Bird Treaty Act, signed by the United States, Canada and Mexico protected both Turkey and Black vultures (Kiff, 2000). However, the general perception of roosting vultures as a social problem led to varied attempts to breakup roosting vulture flocks, which slowly declined due to conservation efforts in the USA (Kiff, 2000: 180). Research by the National Wildlife Research Center (NWRC) (Avery, 2004), found that roosting vultures impacted on human health and wellbeing in several States, through nuisance and physical impacts such as electrical power outages, in addition to attacks by Black vultures on livestock.

One action was the dispersal of a large communal Black vulture winter roost in Virginia, which had limited effects as vultures relocated and still had access to cattle and sheep carcasses in vicinity. Collaborative research involving the NWRC, the Virginia Wildlife Services and the Florida Farm Bureau remained, in an effort to ameliorate vulture predation on new livestock.

Drone Research on Urban Vultures

Urban vulture ecology is an area for intensive drone-based research (inclusive of attached cameras, video recorders and GPS equipment), with requirements for intrusion into micro-spaces, tracking of individual birds, close observation of nesting areas, intra- and inter-species conflicts, monitoring of airfields and spaces of human presence and detectable individualism among the observed vultures. Table 2 below shows several measurable parameters recordable by drones, for individual and group assessment, and statistical measurement. These parameters, when numerically defined and tabulated, allow statistical analyses on drone-collected data (Ivošević et al. 2015; Ditmer et al. 2015a, b; Campbell, 2018).

Table 2. Parameters for drone-based studies of urban vultures

Parameter	Important Issues
Flight	Head/wing movement on sighting food, circular flight pattern, flight angle
Foraging	Soaring, gliding, flapping flight, perching, food sources, urban structures
Urban nesting	Location (tree, built structure), nestling feeding, rivalry, fledging, predation
Reactions	Alert and flight distances on human/aircraft approach, nearby structures
Feeding conflicts	Interspecies, intraspecies interactions at food sources, flocking, exclusion
Territory conflicts	Interspecies, intraspecies interactions over urban structures or urban forests
Arrivals	Arrival speeds over sighted food sources, interspecies differences
Roosting	Locations in urban built structures or urban forests, classification of cover
Identification	Gender and age differences in observed subject birds

Source: Campbell (2015).

Table 3. Classification of urban vulture ecologies

Landcover variable	Issues
Dense urban	Small food discards, roof top perching, conflicts with feral dogs
Industrial urban	Large scale rubbish dumps, open drains, butcheries
Residential suburbs	Garden visits, roof perching, attacks on pets, defecation, low flight
Suburban farmland	Cultivated land foraging, livestock attacks, defecation,
Airports	Aircraft collisions, proximate roosting, structure perching
Greenbelts, parks	Defecation, pollution, noise, attacks on pets, discard targeting
Abandoned buildings	Nesting, roosting, pollution, noise, general nuisance

Source: Campbell (2015).

Table 4. Hypotheses concerning Turkey and Black vulture behavior in urban areas

Hypotheses	Actual Facts
Black vulture sociality creates problems	Flocks pollute, collide with airplanes, attack livestock, create dense roosting sites. Lack of olfactory glands affects foraging
Black vulture sociality dominates Turkey vultures	Evidence indicates equality between the species, but large flocks may dominate. Black vulture hampered, with no sense of smell.
Turkey vulture sense of smell creates problems	Flocks may congregate based on smell, attract Black vultures and collide with airplanes. Forages even when no food is sighted.
Turkey vulture olfactory ability beats B. vultures	Turkey vultures first at food, forages near semi-covered sources. Black vultures follow, flocking, may dominate later.
Vultures clean human social spaces	Twofold effect, in clean cities, sanitation replaced scavenger functions, in less clean cities, vultures are still needed.
Urbanization reduces vulture numbers	Evidence that urbanization increases vulture numbers, good sanitation may change this.

Source: Campbell (2015).

These parameters may be recorded as drone data and allow interspecies and intraspecies comparisons, and predictors of vulture presence. Record reliability will depend on the accuracy, mobility and resolution of the mounted cameras, the precision of the GPS location measurements, the power of sound detection equipment and the speed, maneuverability, control and noise of the drone. Vultures may react to drone presence, hence the detectability of the drone (based on its size, design, color, operating noise and speed) and interspecies and individual differences in drone awareness may determine the effectiveness of any drone-based study. Locations for urban-based vultures may be divided into dense urban substrates (including downtown areas and urban industrial areas), suburban residential areas, farmland close to urban locations, airports and food marketing areas. The measurable parameters are shown in Table 3 above. These may support or reject hypotheses, as is illustrated in Table 4 and the following notes.

Drones are a new research method that may contribute knowledge to the events and issues described above. There of course will be serious problems. For example, Junda et al. (2015, 234) argue that "investigating raptor nests with UAVs in urbanized areas may also raise some permit issues, arising from safety and privacy issues with the general public." Nevertheless, in a review of applications to the general field of environmental biology, Nowak et al. (2019: 56) note in most of the studies they surveyed drones were used to record measurable parameters such as "crown height, volume, number of individuals and quantification of the spatio-temporal dynamics of vegetation changes"; additionally, there were animal counts, "detecting, assessing and predicting threats on vegetation, measuring the biophysical parameters of vegetation, quantifying the dynamics of changes in plants and habitats and population and behaviour studies of animals" (Nowak et al. 2019: 56). This level of detail reveals the potential for studies of urban wildlife, as urban structures are generally less complex than vegetation, and avian urban activities may involve foraging in refuse dumps or car parks, or simply vegetated urban green spaces that may be easier to measure with drones (Campbell, 2015).

Drones may also be used to count vultures, although few studies have been conducted on Turkey and Black vultures in this regard. Other species such as Cinereous vultures and *Larus* gulls have been assessed by these methods, in contexts more challenging than urban areas, indicating strong potential applications in North America (Hausheer, 2016; Rush et al. 2018). The former study, describing work by Tuguldur Enkhtsetseg, using drones to search and find Cinereous vulture nests hidden among granite outcroppings, a method that located 33 raptor nests. The latter study on *Larus* gulls developed protocols and derived findings that would be useful to any drone-based study, similar to studies by Fretwell et al. (2012, 2014) and Grenzdörffer (2013).

The survey method was robust, in terms of the remote, relatively inaccessible location of the seabird colonies, minimal disturbance of birds, high-resolution imagery, expert drone control, a GIS-based semiautomated classification process for counting the gulls, and differentiation of the *Larus* gulls, with greater accuracy than human counts (Rush et al. 2018). These studies therefore established parameters for searching, accessing, elusiveness, recording, observing and analyzing data captured by drones, in contexts more remote than urban areas. These parameters are linked to possible urban issues in Table 5 below.

As noted in Figures 1 to 4, the development of drone technology parallels the needs of industry, including tracking, police work, fire assessment and search and rescue. These functions require increasingly "intelligent" designs that can react to the complex parameters that must be measured (DJI, 2019). As vultures also increasingly adapt to environmental change, the ecological relations increase in complexity. Hence, much may be learned by comparing non-vulture or different vulture species, the better to understand the ecologies of the focus species (Campbell, 2015, Table 1).

Drones are also frequently used to monitor and record waste management sites and issues. For example, Guedim (2017) notes that "some U.S. counties are now using unmanned quadcopters to take aerial photos of landfills and provide real-time data for waste management operators." Issues concern the remaining space at landfills and the potential developments. Examples are cited from Idaho and New Mexico. Drones may also be used to 'control' troublesome vultures. Bird-X (2019) describes a new product: "The most effective removal techniques to get rid of Turkey vultures include Avishock Bird Shock Track or the Extra Tall version of our Bird Spikes to keep the buzzards off of building edges, and Bird Netting to physically block their entrance into larger areas. If larger areas need coverage, the BroadBand PRO and Critter Blaster PRO have proven effective against Turkey Vultures. These birds have very few natural predators, but adults and youth alike can fall prey to various eagles and great horned owls, which these devices utilize." Bird -X (2019) further cautions that "All vultures are protected in the United States by the Migratory Bird Treaty Act and may not be harmed by law. Humane control solutions, however, are allowed for their removal from unwanted areas. All products manufactured by Bird-X, Inc. are legal, humane, green, and effective."

Table 5. Studies of other species linked to vulture ecology

Parameter in study	Urban application for Turkey/Black vultures
Remoteness: Seabird colonies difficult for human intrusion (1)	Vulture nesting on rooftops, tall buildings/trees similarly remote
Detection: automatic detection of birds using spectral signatures (2)	Vultures may be more visible against urban substrate than gulls against natural cover.
Searching: Cinereous vulture nests in Mongolian mountains, detected (3)	Vulture nests on ledges and abandoned buildings, urban forest trees
Elusiveness: drones approaching gulls, avoidance of bird reaction (4)	Vulture urban ecology requires elusive drones, avoiding reactive behavior
Data analysis: high precision drone images, analysed for species/age (5)	Vulture breeding in urban areas require detection of juveniles, breeding, species
Observation: bird behavior detected, by high resolution, close images (6)	Intraspecies and interspecies behavior needs closeup images in urban foraging/nesting
Drone control: knowledge of bird behavior linked to drone movements (7)	Vulture urban movements unpredictable, reactive to human actions
Drone image resolution: Detail, e.g., Albris UAV has a resolution of 1 cm (8)	Good for identification purposes, especially differentiating Black from Turkey vultures

Note: The numbered points in column 1 above link to the numbered points below.

1) Rush et al. (2018). The findings were that seabird colonies were generally located outside human access, and other image capture methods were too coarse in resolution to identify species. Drones, that are well managed according to bird ecology enable the study of inter- and intra-species differences.
2) Fretwell et al. (2012, 2014); Grenzdörffer (2013); Rush et al. (2018). The findings were that automatic detection of species using spectral signatures allowed more accurate species detection, and these were as accurate as close human observation. For the current study, this would be applicable to vulture detection in urban situations, and with high resolution distinguish vulture species.
3) Hausheer (2016, on Enkhtsetseg). The findings were that drones were more effective in tracking remote mountain nests than aircraft (that might disturb birds) or human action (the mountain locations were remote). For the current study, these would be a similar to vulture nests in high rise urban structure and tall, dense trees.
4) Rush et al. (2018). The findings were that drones may maneuver to avoid detection, disturbance or subject reactions, enabling close observation and image capture impossible with close human observation. For the current study, this would be crucial for the observation of vulture abilities at food detection, especially distinguishing the Turkey vulture (with a sense of smell) from the Black vulture.
5) Fretwell et al. (2012, 2014); Grenzdörffer (2013); Rush et al. (2018). The findings were that high-resolution images were sufficiently detailed to detect intra-species differences relevant to breeding patterns. For the current study, this would be crucial for the detection of juveniles and adults, the better to determine urban vulture breeding patterns.
6) Fretwell et al. (2012, 2014); Grenzdörffer (2013); Rush et al. (2018). The findings were that drones could detect ecological behavior, that could highlight species feeding and foraging patterns. For the current study, urban vulture feeding, roosting, defecation and damage of urban structures is crucial for ecological studies.
7) Hausheer (2016, on Enkhtsetseg); Rush et al. (2018). The findings were that for optimal information capture, detailed knowledge of the subject species is needed, with the drone being flexible enough to adapt to the requirements of such applications. For the current study, this is particularly important urban vulture studies, where there are extremely flexible ecologies.
8) Martin (2018). The advantages of drone resolution were discussed. For example, the Albris UAV with a resolution of 1 cm, compared with the Airbus Spot 6 satellite with a resolution of 1.5 metres.

LIMITATIONS OF DRONE RESEARCH

It is currently uncertain how Turkey and Black vultures may react to drones. Some studies have shown that eagles may attack drones, as they would bird prey or rival predators (Fessenden, 2015; Cherney, 2017; Wade, 2017). However, as pointed out by Egan (2018) vultures may not react to drones perceived as birds of prey, unless they mistake the drone as a species they would normally avoid. Some studies have demonstrated that vultures avoid or even fear eagles (Campbell, 2015). Possibly, large drones may be perceived by vultures as eagles or larger buteo hawks such as the Red-tailed hawk (*Buteo jamaicensis*, Gmelin, 1788), which may dominate the largest vultures (Stolen, 1996; Kirk & Mossman, 1998). Small drones may be better suited for the study of vultures, but elusive movements may ensure more tolerance from the vultures.

Aside from research methods, drones may be used to control flocking vultures, by herding them away from sensitive areas. However, Egan (2018) argues that this may fail, as vultures may not fear drones and avoid them: "Drones that mimic the appearance and flight dynamics of falcons, operated by mechanically flapping wings, have also been engineered to deter birds and their use has already been adopted by airports. This strategy might deter species that perceive raptors as lethal predators but might not work on larger species such as geese (Anatidae) or vultures (Cathartidae)." More effective tools in this regard would be pistol-fired pyrotechnic shells, that disturb roosting vultures, sometimes termed vulture hazing (Nickens, 2008).

Drones are limited in any use for solving vulture presence at airports in North America. In Canada they may not be used within 5.6 kilometres from any airport listed as Certified ("Cert") in the Canada Flight Supplement and 1.9 kilometres heliports or aerodromes used by helicopters (Government of Canada, 2019a). In the United States, similar limitations arise. Shepardson, (2019) notes that "there has been rising concern about drones near U.S. airports." The FAA informed U.S. airports that only federal agencies with clear authority could install systems for the detection and destruction of drones, while airport authorities would require federal consent to install such systems, especially as "they could pose an aviation safety risk by interfering with aircraft navigation and air navigation services" (Shepardson, 2019). In 2018, the US Congress granted ed the Justice and Homeland Security departments authority to disable or destroy drones (Shepardson, 2019).

Similarly, drones may not be used in Canadian National Parks under Parks Canada supervision. Recreational flying of drones is prohibited, but commercial drone use may be allowed by permit. Permissions, termed Restricted Activity Permits, may be granted for the use of a drone for "natural or cultural resource management and protection, public safety, law enforcement, or park/site management purposes directly relating to park administrative purposes" (Government of Canada, 2019b). In the United States, the National Park Service banned drones from National Parks (Policy Memorandum 14–05 issued by the NPS on June 19, 2014), without a Special Use

Permit, issued only for applications such as search and rescue, research and fire safety (Drone U, 2018).

New regulations applied to drones were implemented in Canada in 2019, which differed from those in the United States of America. These regulations apply to those drones that "weigh 250 grams (g) up to and including 25 kilograms (kg) and are operated within the drone pilot's visual-line-of-sight" (Lee, 2019). Although previously recreational and commercial drone usage was differentiated, the new laws gave equal treatment to recreational and commercial drone usage. In addition to registration and acquisition of relevant certificates, drone operators must identify their usage as basic, where the operator flies the drome in uncontrolled airspace, more than 30 meters horizontally from people and never over people. Less than these three implies that the operator is conducting advanced operations, implying that the operator is flying the drone in controlled airspace, within 30 meters of people and/or over people.

Further rules are that the drone must be flown within sight, below 122 meters elevation, fly away from people, avoid forest fires and social events, stay within 5.6 kilometers from airports and 1.9 kilometres from heliports, avoid other aircraft and respect human privacy. In the United States recreational and commercial uses are still distinguished. Recreational rules include flight within sight, drones weighing less than 25 kg, stay 8 km from airports (without notification or permission) and avoid aircraft and emergency response efforts. Commercial rules include flight below 122 meters, drone weights less than 25 kg, flight during daylight to twilight, less than about 162 km, avoid flight over people and operation of drones from a moving vehicle (unless in areas with low human presence) and yield to manned aircraft (UAV Coach, 2019).

These regulations severely restrict the monitoring of vultures in airports, suburban farms, downtown areas and other places with dense urban substrate and private gardens, and in flight above about 130 meters. Questions may emerge as to the possible flexibility of these rules for studying urban vultures. Such issues may be less problematic for studies of vultures in rural areas, due to the general absence of privacy and conflict concerns. Some of these concerns may explain the low number of drone-based studies of urban vultures in North America, despite the fact that such studies may outweigh studies in rural areas in importance. This is an important problem considering the influx of Black and Turkey vultures into urban areas, and the current paucity of solutions to this problem (Campbell, 2015).

CONCLUSION

The status of drones in the study of Turkey and Black vultures is a complex and problematic issue, with insufficient studies of these species in urban areas, possibly due to legal restrictions and issues. Applications are clearly possible, given the cited studies and findings based on other species in more complex environments. However, the legal, political, social and ecological quagmire that is the urban context is plainly different from natural landcover. A wide range of drone sizes are available, but the legal impediments may exist for all of the different variants. Drones currently stand as the most effective tools for ecological observation, with shaper resolution and superior flexibility to fixed field or motion cameras, satellite images and aerial photographs. Aside from the legal issues, developments based on experiences with other species would be smaller sized, quieter and more elusive machines, more accurate GPS technology and more detailed ecological studies to support the actions of drone controllers. Taking these issues into account may make an incisive contribution to vulture urban ecology, as consequently human quality of life and methodologies for the study of nature/society relations.

References

Airola, D. A. 2011. Dynamics of an urban turkey vulture roost in Sacramento, California. *CVBC Bulletin*, 14: 1 – 8. Retrieved from http://www.cvbirds.org/wp-content/themes/cvbirds/files/ V.14no.1/CVBC_Vol14_No1_pp1-8.pdf.

Altalink 2019. Avian Electrocutions. Retrieved from http://www.altalink.ca/environment/avian-protection-plan/avian-fatalities.cfm.

American Ornithologists' Union. 2009. *Check-list of North American Birds, Tinamiformes to Falconiformes* 7th edition. AOU.

Anderson K. & Gaston K. J. 2013. Lightweight unmanned aerial vehicles will revolutionize spatial ecology. *Frontiers in Ecology and the Environment*, 11: 138-146. https://doi. org/10.1890/120150.

APLIC (Avian Power Line Interaction Committee). 2006. *Suggested Practices for Avian Protection on Power Lines: The State of the Art In 2006*. Washington, D.C.: Edison Electric Institute.

Avery, M. L., Humphrey, J. S., Tillman, E. A., Phares, K. O. & Hatcher, J. E. 2002. Dispersing vulture roosts on communication towers. *Journal of Raptor Research*, 36: 45–50. https://digitalcommons. unl.edu/icwdm_usdanwrc/456.

Avery, M. L. 2004a. *Managing Depredation and Nuisance Problems Caused by Vultures*. Washington: United States Department of Agriculture.

Avery, M. L. 2004b. Trends in North American vulture populations. In Timm, R. M. & Gorenzel, W. P. (Eds.). *Proceedings of the 21st Vertebrate Pest Conference*; 1–4 March 2004, 116–121. Visalia, CA. University of California, Davis, CA.

Avery, M. L. & Cummings, J. L. 2004. Livestock depredations by black vultures and golden eagles. *Sheep and Goat Research Journal*, 19: 58–63. https://www.aphis. usda.gov/wildlife_damage/nwrc/publications/04pubs/avery044.pdf.

Avery, M., Humphrey, J. S., Daughtery, T. S., Fischer, J. W., Milleson, M. P., Tillman, E. A., Bruce, W. E., & Walter, W. D. 2011. Vulture flight behavior and implications for aircraft safety. *Journal of Wildlife Management*, 75: 1581–1587. http://www.doi:10.1002/ jwmg.205.

Baynard, O. E. 1909. Notes from Florida on Catharista urubu. *Oologist*, 26: 191–193. https://www.aphis.usda.gov/wildlife_damage/nwrc/publications/04pubs/avery044. pdf.

Barasona, J. A., Mulero-Pázmány, M., Acevedo, P., Negro, J. J., Torres, M. J., Gortázar, C. & Vicente J. 2014. Unmanned aircraft systems for studying spatial abundance of ungulates: relevance to spatial epidemiology. *PLoS ONE*, 9: e115608. http://www.doi:10.1371/ journal.pone.0115608.

Bayle, P. 1999. Preventing birds of prey problems at transmission lines in Western Europe. *Journal of Raptor Research*, 33: 43–48. https://sora.unm.edu/sites/ default/files/journals/jrr/v033n01/p00043-p00048.pdf.

Benson, P. C. 1981. *Large raptor electrocution and power pole utilization: a study in six western states.* Unpublished PhD Dissertation, Brigham Young University, Provo.

Bird Be Gone 2019. *Vulture Buzzard Control*. Retrieved from https://www.birdbgone. com/bird/ vulture-buzzard-control/.

BirdLife International. 2019a. *Turkey vulture Cathartes aura.* Retrieved from http://datazone. birdlife.org/species/factsheet/22697627.

BirdLife International. 2019b. *American black vulture Coragyps atratus*. Retrieved from http://datazone.birdlife.org/species/factsheet/american-black-vulture-cora-gyps-atratus.

Bird-X 2019. *Got Vulture Problems?* Retrieved from https://bird-x.com/interactive-problem-solver/vulture-control/.

Blackwell, B. F. & Wright, S. E. 2006. Collisions of red-tailed hawks (Buteo jamaicensis), turkey vultures (Cathartes aura), and black vultures (Coragyps atratus) with aircraft: implications for bird strike reduction. *Journal of Raptor Research*, 40(1): 76–80. https://digitalcommons.unl.edu/cgi/viewcontent.cgi? article=1405&context=icwdm_usdanwrc.

Buckley, N. J. 1996. Food finding and the influence of information, local enhancement, and communal roosting on foraging success of North American vultures. *Auk*, 113: 473–488. https://sora.unm.edu/sites/default/files/journals/auk/v113n02/p0473-p0488.pdf.

Buckley, N. J. 1999. Black vulture (Coragyps atratus). In Poole, A. & Gil, F. (Eds.), *The Birds of North America*, No. 411. Philadelphia: The Birds of North America, Inc.

Campbell, M. 2006. Urban parks shared spaces? The utility of alert distances as indicators of avian tolerance of human in Stirling Scotland. *Area*, 38: 301–311. https://doi.org/10.1111/j.1475-4762.2006.00695.x.

Campbell, M. 2014. The impact of urbanization and agricultural development on vultures in El Salvador. *Vulture News*, 66: 16 – 28.

Campbell, M. 2015. *Vultures: Their Evolution, Ecology and Conservation*. Boca Raton: CRC Press.

Campbell, M. 2016. Avian reactions towards human approaches in different urban greenery structures in Nanaimo. *Urban Forestry & Urban Greening*, 19: 47-55. https://doi.org/10.1016/j.ufug.2016. 06.022.

Campbell, 2017. *Biological Conservation in the 21st Century: A Conservation Biology of Large Wildlife.* New York, NY: Nova Science Publishers.

Campbell, 2018. *Geomatics and Conservation Biology.* New York, NY: Nova Science Publishers.

Camphuysen, K., Verheij, J. & Cremer, J. 2011. *Risk Assessment of Bird Strike Hazards: Gulls Laridae.* Oosterend: CSR Consultancy.

Carrete, M., Lambertucci, S. A., Speziale, K., Ceballos, O., Travaini, A., Delibes, M., Hiraldo, F. & Donázar, J. A. 2010. Winners and losers in human-made habitats: interspecific competition outcomes in two Neotropical vultures. *Animal Conservation*, 13: 390–398. https://doi.org/10.1111/j.1469-1795.2010.00352.x.

Chace, J. F., Walsh, J. J., Cruz, A., Prather, J. W. & Swanson, H. M. 2003. Spatial and temporal activity patterns of the brood parasitic brown-headed cowbird at an urban/wildland interface. *Landscape and Urban Planning*, 64: 179–190. https://doi.org/10.1016/S0169-2046(02)00220-7.

Cherney, M. 2017. Bold Eagles: Angry Birds Are Ripping $80,000 Drones Out of the Sky. Retrieved from https://www.wsj.com/articles/bold-eagles-angry-birds-are-ripping-80-000-drones-out-of-the-sky -1506701429.

Chiari, C. G. & Seeland, K. 2004. Are urban green spaces optimally distributed to act as places for social integration? Results of a geographical information system GIS approach for urban forestry research. *Forest Policy and Economics*, 6: 3–13. https://doi.org/10.1016/S1389-9341(02)00067-9.

Cooke, S. J., Hinch, S. G., Wikelski, M., Andrews, R. D., Kuchel, L. J., Wolcott, T. G., et al. 2004. Biotelemetry: a mechanistic approach to ecology. *Trends in Ecology & Evolution*, 19: 334–343. http://www.doi: 10.1016/j.tree.2004.04.003.

Cornelis, J. & Hermy, M. 2004. Biodiversity relationships in urban and suburban parks in Flanders. *Landscape and Urban Planning* 69: 385–401. https://doi.org/10.1016/j.landurbplan.2003.10.038.

Del Hoyo, J., Elliott, A. & Sargatal, J. (eds.). 1994. *Handbook of Birds of the World.* Vol. 2. New World Vultures to Guineafowl. Barcelona: Lynx Edicions.

DeVault, T. L., Brisbin, Jr. I. L. & Rhodes, Jr., O. E. 2004. Factors influencing the acquisition of rodent carrion by vertebrate scavengers and decomposers. *Canadian Journal of Zoology*, 82: 502–509.

DeVault, T. L., Reinhart, B. D., Brisbin, I. L. & Rhodes, O. E. 2010. Flight behavior of black and turkey vultures: implications for reducing bird-aircraft collisions. *The Journal of Wildlife Management*, 69(2): 601–608. https://doi.org/10.2193/0022-541X(2005)069[0601:FBOBAT]2.0.CO;2.

Dickerman, R. W. 2003. Talon-locking in the red-tailed hawk. *Journal of Raptor Research*, 37: 176. Retrieved from https://sora.unm.edu/sites/default/files/journals/jrr/v037n02/p00176-p00176.pdf.

Ditmer, M., Vincent, J. B., Werden, L. K., Tanner, J. C., Laske, T. G., Iaizzo, P. A., Garshelis, D. L. & Fieberg, J. R. 2015a. Bears Show a physiological but limited behavioral response to unmanned aerial vehicles. *Current Biology*, 25(17): 2278-2283. https://doi.org/10. 1016/j.cub.2015.07.024.

Ditmer, M. A., Garshelis, D. L., Noyce, K. V., Laske, T. G., Iaizzo, P. A., Burk, T. E., Forester, J. D. & Fieberg, J. R. 2015b. Behavioral and physiological responses of American black bears to landscape features within an agricultural region. *Ecosphere*, 6(3): 1 – 21. https://doi.org/10.1890/ES14-00199.1.

DJI 2019. DJI. Retrieved from https://www.dji.com/ca/.

Dolbeer, R. A. 1998. Evaluation of shooting and falconry to reduce bird strikes with aircraft at John F. Kennedy International Airport. In *Proceedings of the IBSC* 2, 145–153. Stara Lesna, Slovakia 14–18 September. WP13.

Drone U. 2018. *What Are the Rules for Flying Your Drone in National Parks and National Forests?* Retrieved from https://medium.com/@askdroneu/what-are-the-rules-for-flying-your-drone-in-national-parks-and-national-forests-1d275549 84ad.

Egan, 2018. *Evaluating the Potential Utility of Drones to Deter Birds from Areas of Human-Wildlife Conflict.* Master of Science Thesis, North Dakota State University of Agriculture and Applied Science.

Federal Aviation Administration 2018. *FAA Aerospace Forecast.* Retrieved from https://www.faa.gov/data_research/aviation/aerospace_forecasts/media/FY2019-9_FAA_Aerospace_Forecast.pdf.

Farve, R. 2018. *Demonstration of Satellite/GPS Telemetry for Monitoring Fine-Scale Movements of Lesser Prairie-Chickens.* Retrieved from https://www.fs.fed.us/t-/programs/im/satellite_gps_ telemetry/wildlifetrackingtelementry.htm.

Fessenden, M. 2015. *A Drone Encounters Two Eagles, and the Birds Win.* Retrieved from https://www.smithsonianmag.com/smart-news/drone-encounter-two-eagles-didnt-go-planned-180957336/.

Ferguson-Lees, J. & Christie, D. A. 2001. *Raptors of the World.* London: Christopher Helm.

Ferrer, M. & Janss, G. F. (Eds.). 1999. *Birds and Power Lines: Collision, Electrocution, and Breeding.* Madrid: Quercus.

Franke, U. E. 2015. *Civilian Drones: Fixing an Image Problem?* International Relations and Security Network. Retrieved from https://isnblog.ethz.ch/security/ civilian-drones-fixing-an-image-problem.

Franson, J. C., Siloe L. & Thomas, J. J. 1995. Causes of eagle deaths. In LaRoe, E. T., Farris, G. S., Puckett, C. E. & Doran, P. D. (Eds.). *Our Living Resources: A Report to the Nation on the Distribution, Abundance, and Health of U.S. Plants, Animals, and Ecosystems*, 68. Washington, D.C: National Biological Service.

Fretwell, P. T., LaRue, M. A., Morin, P., Kooyman, G. L., Wienecke, B., Ratcliffe, N., Fox, A. J., Fleming, A. H., Porter, C. & Trathan, P. N. 2012. An emperor penguin population estimate: The first global, synoptic survey of a species from space. *PLoS One*, 7(4): e33751. https://doi.org/10.1371/journal.pone.0033751.

Fretwell, P. T., Trathan, P. N., Wienecke, B. & Kooyman, G. L. 2014. Emperor penguins breeding on ice shelves. *PLoS One*, 9(1): e85285. https://doi.org/10. 1371/journal.pone.0085285.

Gard, K., Groszos, M. S., Brevik, E. C. & Lee, G. W. 2007. Spatial analysis of bird-aircraft strike hazard for Moody Air Force Base aircraft in the State of Georgia. *Georgia Journal of Science*, 65(4): 160–169.

Garrett, K. & Dunn, J. 1981. *Birds of Southern California: Status and Distribution*. Los Angeles: Los Angeles Audubon Society.

Government of Canada 2019a. *Where to fly your drone*. Retrieved from https://www.tc.gc.ca/ en/services/aviation/drone-safety/where-fly-drone.html.

Government of Canada 2019b. *Use of Unmanned Aerial Vehicles (UAV) or drones in Canada's national parks*. Retrieved from https://www.pc.gc.ca/en/pn-np/ab/banff/ info/permis-permit/drone.

Grenzdörffer, G. J. 2013. UAS-based automatic bird count of a common gull colony. *International Archives of the Photogrammetry, Remote Sensing and Spatial Information Sciences*, XL-1/W2, 169– 174. https://doi.org/10.5194/isprsarchives-XL-1-W2-169-2013.

Guedim, Z. 2017. *How Drones are Helping with Waste Management*. Retrieved from https://edgy.app/drones-waste-management-china-dubai.

Hammill, M. O., Dale, J., Stenson, G. B., den Heyer, C., Gosselin, J. F., Leblanc, P. & Johnston, D. W. 2017. Comparison of methods to estimate grey seal pup production at different colonies. *DFO Canadian Science Advisory Secretariat (CSAS) Research Document* 2017/041.

Harness, R. E. & Wilson, K. R. 2001. Utility structures associated with raptor electrocutions in rural areas. *Wildlife Society Bulletin*, 29(2): 612–623. https://www.jstor.org/stable/3784188.

Hausheer, J. E. 2016. *Can Drones Help Monitor Vultures on Mongolia's Eastern Steppe?* Retrieved from https://blog.nature.org/science/ 2016/03/09/can-drones-help-monitor-vultures-on-mongolias-eastern-steppe/.

Hill, J. R. & Neto, P. S. 1991. Black vultures nesting on skyscrapers in southern Brazil. *Journal of Field Ornithology*, 62(2): 173–176. Retrieved from https://sora.unm.edu/ sites/default/files/journals/ jfo/v062n02/p0173-p0176.pdf.

Hilty, S. L. 1977. A Guide to the Birds of Colombia. Princeton, NJ: Princeton University Press.

Hodgson, J. C., Mott, R., Baylis, S. M., Pham, T. T., Wotherspoon, S., Kilpatrick, A. D., Segaran, R. R., Reid, I., Terauds, A. & Koh, L. P. 2018. Drones count wildlife more accurately and precisely than humans. *Methods in Ecology and Evolution*, 9: 1160–1167. https://doi. org/10.1111/2041-210X.12974.

Holland, A. E., Byrne, M. E., Bryan, A. L., DeVault, T. L., Rhodes, O. E. & Beasley, J. C. (2017). Fine-scale assessment of home ranges and activity patterns for resident black vultures (*Coragyps atratus*) and turkey vultures (*Cathartes aura*). *PloS One*, 12(7), e0179819. http://www.doi:10.1371/ journal.pone.0179819.

Hostetler, M. & Knowles-Yanez, K. 2003. Land use, scale, and bird distributions in the Phoenix metropolitan area. *Landscape and Urban Planning*, 62(2): 55-68. https://doi.org/10.1016/S0169-2046(02)00096-8.

Houston, C. S., Terry, B., Blom, M. & Stoffel, M. J. 2007. Turkey vulture nest success in abandoned houses in Saskatchewan. *Wilson Journal of Ornithology*, 199: 742–747.

Hunting, K. 2002. *A Roadmap for PIER Research on Avian Power Line Electrocution in California*. Sacramento: California Energy Commission.

Ivošević, B., Han, Y., Cho, Y. & Kwon, O. 2015. The use of conservation drones in ecology and wildlife research. *Journal of Ecology and Environment*, 38(1): 113-118. http://dx.doi.org/10.5141/ ecoenv.2015.012.

Igl, L. D. & Peterson, S. L. 2010. Repeated use of an abandoned vehicle by nesting turkey vultures (*Cathartes aura*). *Journal of Raptor Research*, 44(1): 73–75. https://doi.org/10.3356/JRR-09-02.1.

Inigo Elias, E. E. 1987. Feeding habits and ingestion of synthetic products in a black vulture population from Chiapas, Mexico. *Acta Zoologica Mexicana*, 22: 1–15. https://doi.org/10.21829/azm.1987.19222040.

Jackson, J. A. 1983. Nesting phenology, nest site selection, and reproductive success of black and turkey vultures. In: Wilbur, A. & Jackson, J. A. (Eds.). *Vulture Biology and Management*, 245–270. Berkeley: University of California Press.

Janss, G. F. E. 2000. Avian mortality from power lines: a morphologic approach of a species-specific mortality. *Biological Conservation*, 95(3): 353–359. Retrieved from http://www.atclave.es/english/publicaciones/descargas/pub_avifauna/47_avian_mortality.pdf.

Johnston, D. W., Dale, J., Murray, K. T., Josephson, E., Newton, E. & Wood, S. 2017. Comparing occupied and unoccupied aircraft surveys of wildlife populations: assessing the gray seal (*Halichoerus grypus*) breeding colony on Muskeget Island, USA. *Journal of Unmanned Vehicle Systems*, 5(4): 178-191, https://doi.org/10.1139/juvs-2017-0012.

Johnsgard, P. A. 1988. *North American Owls: Biology and Natural History*. Washington, DC: Smithsonian Institute Press.

Johnsgard, P. A. 1990. *Hawks, Eagles, and Falcons of North America*. Washington, DC: Smithsonian Institute Press.

Junda J., Greene E. & Bird D. M. 2015. Proper flight technique for using a small rotary-winged drone aircraft to safely, quickly, and accurately survey raptor nests. *Journal of Unmanned Vehicle Systems*, 3: 222-236. https://doi.org/10.1139/juvs-2015-0003.

Kaufman, K. 1996. *Lives of North American Birds*. Boston, MA: Houghton Mifflin Field Guides.

Kennedy, C. & Rogers, J. I. 2015. Virtuous drones? *The International Journal of Human Rights*, 19 (2): 211–227. https://doi.org/10.1080/13642987.2014.991217.

Kiff, L. F. 2000. The current status of North American vultures. In Chancellor, R. D. & Meyburg, B. U. (Eds.). *Raptors at Risk*, 175–189. Berlin and Surrey, Canada: WWGBP/Hancock House.

Kirk, D. A. & Mossman, M. J. 1998. Turkey vulture (*Cathartes aura*), version 2.0. In Poole, A. F. & Gill, F. B. (Eds.) *The Birds of North America*. Ithaca, NY: Cornell Lab of Ornithology. https://doi.org/10.2173/bna.339.

Koh, L. P. & Wich, S. A. 2018. *Conservation Drones: Mapping and Monitoring Biodiversity*. Oxford: Oxford University Press.

Konijnendijk, C. C. 2017. Cities and nature: urban forestry for greater biocultural diversity. In Campbell, M., *Biological Conservation in the 21st Century: A Conservation Biology of Large Wildlife*, 15 – 29. New York, NY: Nova Science Publishers.

Kuhn, M. 2003. Greenbelt and greenheart: separating and integrating landscapes in European city regions. *Landscape and Urban Planning*, 64: 19–27. https://doi.org/10.1016/S0169-2046(02)00198-6.

Lee, I. 2019. *Canada Announces New Drone Laws that Allow Recreational and Commercial Pilots to Operate Under Same Rules | Plans to Implement June 2019.* Retrieved from https://uavcoach.com/canada-new-drone-laws/.

Lehman, R. N., Kennedy, P. L. & Savidge, J. A. 2007. The state of the art in raptor electrocution research: a global review. *Biological Conservation*, 136: 159–174. https://doi.org/10.1016/j.biocon. 2006.09.015.

Lovell, C. D. & Dolbeer, R. A. 1999. Validation of the U.S. Air Force bird avoidance model. *Wildland Society Bulletin*, 27: 161–71. https://www.jstor.org/stable/3783954.

Lowery, G. H. & Dalquest, W. W. 1951. Birds from the State of Veracruz, Mexico. *University of Kansas Publications, Museum of Natural History*, 3: 531–649. Retrieved from https://www.biodiversitylibrary.org/item/21915#page/563/mode/1up.

Lowney, M. S. 1999. Damage by black and turkey vultures in Virginia, 1990–1996. *Wildland Society Bulletin*, 27(3): 715–719. https://www.jstor.org/stable/3784093.

Lucas, T. 2018. *Drones Prove Reliable, Cost-Efficient Tool for Wildlife Surveys.* Retrieved from https://nicholas.duke.edu/about/news/drones-prove-reliable-cost-efficient-tool-wildlife-surveys.

Lucieer, A., Turner, D., King, D. H. & Robinson S. A. 2014. Using an unmanned aerial vehicle (UAV) to capture micro-topography of Antarctic moss beds. *International Journal of Applied Earth Observation and Geoinformation*, 27, Part A: 53-62. http://dx.doi.org/10.1016/j.jag.2013.05.011.

Marris, E. 2013. Drones in science: Fly and bring me data. *Nature*, 498: 156-158. http://www.doi:10.1038/498156a.

Martin, G. R., Portugal, S. & Murn, C. P. 2012. Visual fields, foraging and collision vulnerability in Gyps vultures. *Ibis*, 154(3): 626-631. https://doi.org/10.1111/j.1474-919X.2012. 01227.x.

Martin, G. 2018. *UAVs a useful tool in wildlife conservation*. Retrieved from https://www.defenceweb.co.za/aerospace/unmanned-aerial-vehicles/uavs-a-useful-tool-in-wildlife-conservation/.

Melcher, C. & Suazo, L. 1999. Raptor electrocutions: the un-necessary losses continue. *Journal of the Colorado Field Ornithologists*, 33: 221–224. https://birdconservancy.org/wp-content/uploads/2018/10/CO_birds1999v33_news_from_field.pdf.

Merriam-Webster 2018. *Biotelemetry.* Retrieved from https://www.merriam-webster.com/dictionary/biotelemetry.

Mulero-Pázmány. M., Stolper, R., van Essen, L. D., Negro, J. J. & Sassen T. 2014. Remotely piloted aircraft systems as a rhinoceros anti-poaching tool in Africa. *PLoS ONE*, 9(1): e83873. https://doi.org/10.1371/journal.pone.0083873.

Muller-Schwarze, D. 2006. *Chemical Ecology of Vertebrates.* Cambridge: Cambridge University Press.

Nickens, T. E. 2008. *Vultures Take Over Suburbia.* Retrieved from https://www.audubon.org/magazine/november-december-2008/vultures-take-over-suburbia.

Nelson, M. W. 1979. *Power lines progress report on eagle protection research.* Boise, Idaho Power Company.

Nelson, M. W. 1980. *Update on eagle protection practices.* Boise, Idaho: Idaho Power Company.

Nelson, M. W. & Nelson, P. 1976. Power lines and birds of prey. *Idaho Wildlife Review* 28: 3–7.

Novaes, W. G. & Cintra, R. 2013. *Factors influencing the selection of communal roost sites by the Black Vulture Coragyps atratus (Aves: Cathartidae) in an urban area in Central Amazon. Zoologia.* http://dx.doi.org/10.1590/S1984-467020130 05000014.

Nowak, M., Bogawski, P. & Dzió速b, K. 2019. Unmanned aerial vehicles (UAVs) in environmental biology: A review. *European Journal of Ecology*, 4. 56-74. https://doi.org/10.2478/eje-2018-0012.

Olendorff, R. R., Miller, A. D. & Lehman, R. N. 1981. Suggested practices for raptor protection on power lines—the state-of-the-art in 1981. *Raptor Research Report* No. 4. St. Paul, Minnesota: Raptor Research Foundation, Inc.

Park, C. R. & Lee, W. S. 2000. Relationship between species com-position and area in breeding birds of urban woods in Seoul, Korea. *Landscape and Urban Planning*, 51: 29–36. https://doi.org/10.1016/S0169-2046(00)00094-3.

Prior, K. A. & Weatherhead, P. J. 1991. Competition at the carcass: opportunities for social foraging by turkey vultures in southern Ontario. *Canadian Journal of Zoology*, 1991, 69(6): 1550-1556. https://doi.org/10.1139/z91-218.

Ratcliffe N., Guihen, D., Robst, J., Crofts S., Stanworth A. & Enderlein, P. 2015. A protocol for the aerial survey of penguin colonies using UAVs. *Journal of Unmanned Vehicle Systems*, 3: 95–101. https://doi.org/10.1139/juvs-2015-0006.

Rockwell, A. 2016. *Flying the Phantom 3 Pro with turkey vultures - Nearly Attacked.* Retrieved from https://www.youtube.com/ watch?v=nb3Gfo_0iys.

Rush, G. P., Clarke, L. E., Stone, M. & Wood, M. J. 2018. Can drones count gulls? Minimal disturbance and semiautomated image processing with an unmanned aerial vehicle for colony-nesting seabirds. *Ecology and Evolution*, 8(24): 12322-12334. https://doi.org/10.1002/ece3.4495.

Satheesan, S. M. & Satheesan, M. 2000. *Serious Vulture-Hits to Aircraft Over the World. International Bird Strike Committee IBSC25/WP-SA3.* IBSC, Amsterdam.

Savard, J. P. L., Clergeau, P. & Mennechez, G. 2000. Biodiversity concepts and urban ecosystems. *Landscape and Urban Planning*, 48: 131–142. https://doi.org/10.1016/S0169-2046(00)00037-2.

Schalock, R. L., Alonso, M. A. V. & Braddock, D. L. 2002. *Handbook on quality of life for human service practitioners*. Washington DC: American Association on Mental Retardation.

Schaltegger, S. & Beständig, U. 2012. *Corporate Biodiversity Management Handbook: A Guide for Practical Implementation.* Berlin: Federal Ministry of the Environment.

Seamans, T. W. 2004. Response of roosting turkey vultures to a vulture effigy. *Ohio Journal of Science*, 104: 136–138. Retrieved from https://digitalcommons.unl.edu/icwdm_usdanwrc/385/.

Shepardson, D. 2019. *FAA to U.S. airports: Do not install drone countermeasures on your own.* Retrieved from https://www.reuters.com/article/us-usa-drones-airports/faa-to-u-s-airports-do-not-install-drone-countermeasures-on-your-own-idUSKCN1SE2QZ.

Slater, L. E. 1965. Introduction (special issue on telemetry). *BioScience*, 15(2): 81-82.

Snyder, N. F. R. & Snyder, H. 2006. *Raptors of North America: Natural History and Conservation*. Beverly, MA: Voyageur Press.

Stewart, P. A. 1978. Behavioral interactions and niche separation in black and turkey vultures. *Living Bird*, 17: 79–84.

Stolen, E. D. 1996. Black and turkey vulture interactions with bald eagles in Florida. *Florida Field Naturalist*, 24(2): 43–45. https://sora.unm.edu/sites/default/files/FFN_24-2p43-45Stolen%5B1% 5D.pdf.

Tomkiewicz, S. M., Fuller, M. R., Kie, J. G. & Bates, K. K. 2010. Global positioning system and associated technologies in animal behaviour and ecological research. *Philosophical Transactions of the Royal Society of London B*, 365: 2163–2176. http://www.doi:10. 1098/rstb.2010.0090.

Unitt, P. 1984. The birds of San Diego County. *San Diego Society of Natural History Memoir 13*.

USDA. 2003. *Fact Sheet: Managing Vulture Damage.* United States Department of Agriculture Animal and Plant Health Inspection Service Wildlife Services, Pittstown.

USDA 2016. *Managing Vulture Damage*. Retrieved from https://www.aphis.usda.gov/publications/wildlife_damage/content/printable_version/fs_vulture_damage_man.pdf.

Vas, E., Lescroël, A., Duriez, O., Boguszewski, G. & Grémillet, D. 2015. Approaching birds with drones: first experiments and ethical guidelines. *Biology Letters*, 11: 20140754. http://www.doi: 10.1098/rsbl.2014.0754.

Vermeulen, C., Lejeune, P., Lisein, J., Sawadogo, P. & Bouché, P. 2013. Unmanned aerial survey of elephants. *PLoS ONE*, 8: e54700. https://doi.org/10.1371/journal.pone.0054700.

Vanderburgh, D. C. 1993. Manitoba Hydro accommodates osprey activity. *Blue Jay*, 51: 173–177.

UAV Coach 2019. *Drone Laws in the United States of America*. Retrieved from https://uavcoach.com/drone-laws-in-united-states-of-america/.

Wade, M. 2017. When Birds Attack! — Experienced Drone Operators Discuss This (Surprisingly Common) Mission Threat. Retrieved from https://waypoint.sensefly.com/bird-drone-attacks-avoid-threat/.

Wallace, M. P. & Temple, S. A. 1987. Competitive interactions within and between species in a guild of avian scavengers. *Auk*, 104: 290–295. Retrieved from https://sora.unm.edu/sites/default/files/ journals/auk/v104n02/p0290-p0295.pdf.

White, J. G., Antos, M. J., Fitzsimons, J. A. & Palmer, G. C. 2005. Non-uniform bird assemblages in urban environments: the influence of street vegetation. *Landscape and Urban Planning*, 71: 123–135. https://doi.org/10.1016/j.landurbplan.2004.02.006.

Whitehead, K. & Hugenholtz, C. H. 2014. Remote sensing of the environment with small unmanned aircraft systems (UASs), part 1: a review of progress and challenges. *Journal of Unmanned Vehicle Systems*, 2: 69-85. https://doi.org/10.1139/juvs-2014-0007.

In: Critical Research Techniques in Animal ... ISBN: 978-1-53619-846-1
Editors: Michael O'Neal Campbell et al.

Chapter 2

FACTS AROUND VULTURES: AN ATTEMPT TO JOIN THE PUZZLE PIECES

Kaushalendra Kumar Jha[1,*] and Radhika Jha[2]

[1]Forests UP and Indian Institute of Forest Management, Bhopal, Nehru Nagar, India

[2]Zoology Department, Lucknow University, UP, India

ABSTRACT

Vultures, a group of obligate scavengers of the Old World (Eurasia and Africa) and the New World (North and South America), provide invaluable ecosystem services at no cost. Their populations are diminishing, this event posing risks to the benefits they provide us. However, their unique functioning, status and charisma have prompted a lot of research in vulture strongholds like South-west European, South-east African, and South Asian countries. This chapter examines research findings on varied subjects related to vulture species, including feeding behaviour, diseases, diclofenac story, global strategy, name changes, captive breeding, population assessment, extinction prediction and artificial feeding. Few publications have integrated these broad topics and explored gaps that must be filled for increased knowledge. Incomplete, conflicting, and complementary information on such selected topics has been summarised in a chart (Figure 1). It is assumed that this chapter is going to improve the scholarship on vultures.

* Corresponding Author's E-mail: jhakk1959@gmail.com.

Figure 1. Chart showing knowledge base on vultures of this text.

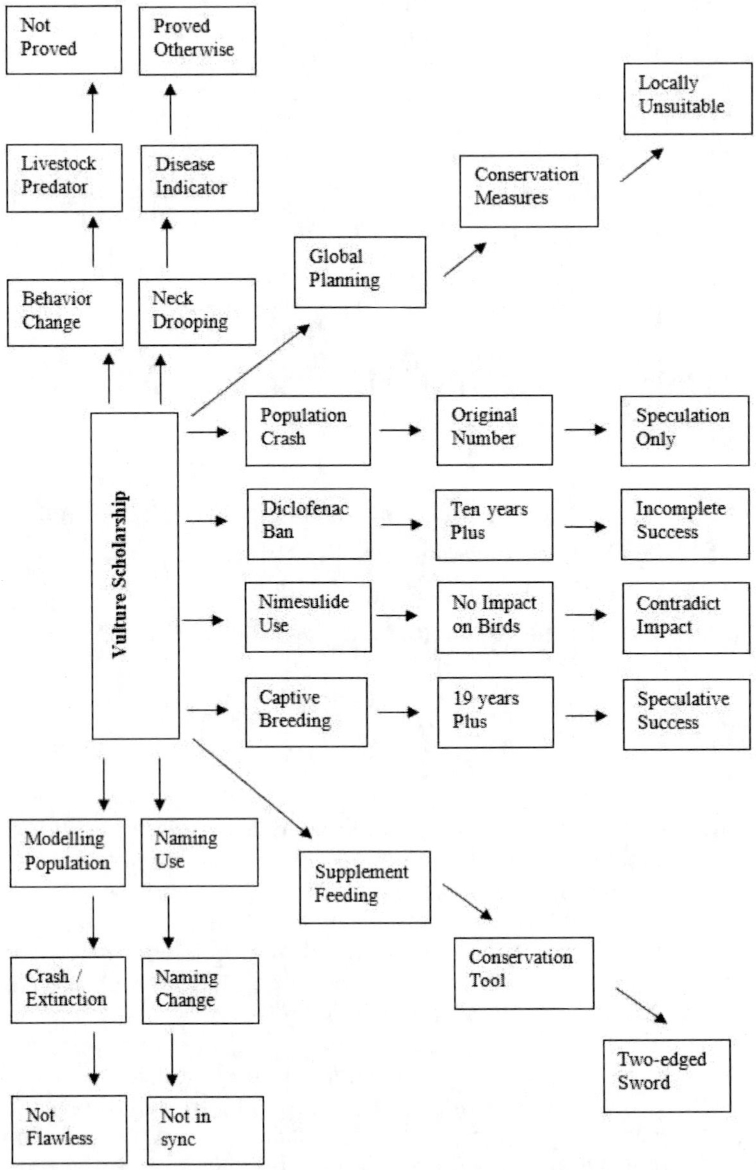

Boxes originating from source "Vulture Scholarship" are the topics selected and subsequent ones on arrows with direction explain conclusive status in literature.

Figure 1. Chart showing knowledge base on vultures of this text.

INTRODUCTION

There are 16 Old World (Accipitrid) and seven New World (Cathartid) vulture species, all of which may interact with people in their foraging activities. Currently and in recent history, the interactions of vultures with people and their life spaces have created negative consequences, which may supersede the positive roles of vultures such as their cultural value and ecological benefits. Negative consequences, real or imagined, include livestock predation, disease spread, impacts on big game poachers (as circling vultures may reveal carcass locations) and environmental pollution through defecation (Hertel 1994; Moleon et al. 2014; Campbell 2015; Duriez et al. 2019; Ballejo et al. 2020; Margalida and Donazar 2020). However, the crucial ecological roles of vultures, not to be underestimated, include carrion removal (as full time, obligate scavengers, this is superior to the role of the part time facultative scavengers), which may reduce the rate of transmission of infectious diseases (Ogada et al. 2012). This function may contribute to the reduction of pests such as feral dogs, which may be outcompeted and dissuaded by the vultures' superior foraging abilities (Moleon et al. 2014). There is evidence that the relative or total absence of vultures may be related to elevated populations of dogs and rats, the primary reservoirs for rabies and bubonic plague respectively, with serious health risks for people (Pain et al. 2003; Markandya et al. 2008). Moreover, rat and dog bites themselves are a threat to humans (Markandya et al. 2008).

Supplanting the natural services of carcass clearing by vultures, with artificial methods may have a heavy economic and environmental impacts, as natural ecosystem processes may be disrupted. This is evident from a study of mad cow disease regulation policy in Spain, where a crisis resulted (Morales-Reyes et. al. 2015). As per their findings, carcass collection and transport to authorized plants in Spain led to annual emissions of 77,344 metric tons of CO_2 eq. to the atmosphere and payments of about $50 million to insurance companies. Margalida and Colomer (2012) recorded that there would have been a minimum annual saving of around one and a half million euros in the EU.

The cultural importance of vultures is also undeniable, as they form an integral part of the cultural structures of many ancient civilizations (Campbell 2015). People, mostly in African countries, are taking direct benefits by selling and using the vulture body parts as faith-based medical cure to different ailments and witchcraft protection and consumption (McKeen et al. 2013; Gbogbo et al. 2016; Boakye et al. 2020). As reviewed by Campbell (2015), vultures have long been associated with spiritual powers, the afterlife, and heavenly contact. Vultures along with the sun are carved on the pillars of one of the oldest temples in Turkey, in ninth millennium BCE, indicating a heavenly link. In ancient Egyptian mythology, they were associated with goddesses, Nekhbet, and Mut. In Hindu mythology vulture is revered as a demigod named Jatayu. Vultures appear in Greek mythology also in the form of Prometheus and Zeus story (an eagle later replaced by a vulture in literature; Campbell 2015). 'Dokhimenishini,' the ancient practice of the Parsi cadaver being consumed by vultures, is till recently being followed by the descendants of the followers of Zoroastrianism. A similar culture of sky burials is followed by Tibetan Buddhists and Mongolians under a belief that the human soul is released and carried to heaven by vultures. Romans also regarded vultures in their divination. The new world vultures also had a strong relationship with people in prehistoric and pre-Colombian past.

There is a vast literature, with supporting databases on the roles of vultures in various fields, such as demography, animal behaviour, reproduction, threats, etc. Still, there are gaps between qualitative and quantitative research, large-scale and small-scale studies, incomplete interpretation and perception, proper model parametrization, etc. This manuscript is an attempt to integrate a few scholarly works which otherwise may remain as independent pieces and impart incomplete or misleading information.

ALTERED BEHAVIOUR AND FAKE NEWS

Reduced food availability due to various reasons including a decline in the ungulate population and provision of supplementary feeding through the establishment of Vulture Restaurants has induced behavioural changes in vulture species. The Eurasian Griffon vulture is one of the more studied birds in this regard in European conditions. Zuberogoitia et al. (2010) demonstrated that food availability and their state of hunger, guide the ability of vultures to evaluate the predation risk and can cause a change in behaviour from natural caution (shyness) towards more tolerant (fearless). The Eurasian Griffon, a conventionally carrion-eating vulture, may exploit lower quality food sources like a predictable garbage dump (Martinez-Abrain et al. 2012). Camina & Yosef (2012) observed that local adults with fidelity to their breeding colonies remain in the area while the non-breeding sub-adults and floaters stray in search of food. Another interesting example of behavioural change was reported by Margalida et al. (2011), where due to changes in European sanitary policy, there was increased food competition among *Gyps fulvus*, and increased incidents of vultures killing livestock in France and Spain. Thus, they were perceived as cattle predators, exacerbating human-wild animal conflicts.

However, Margalida and Donazar (2020) in an article in "Nature Sustainability" titled "Fake news and vultures" emphasized that the media overdid this theme, despite the absence of scientific information from any peer-reviewed scientific literature, and this facilitated a growing negative perception of vultures as predators. Thus, the age-old relationship between farmers and avian scavengers was modified by the growing perception that vultures shared the role of large carnivores as livestock killers. For example, in a published research article on some vultures (the Black vulture *Coragyps atratus*, Turkey vulture *Cathartes aura* and Andean condor *Vultur gryphus*, and others) and interactions with livestock (sheep), the perceptions of farmers concerning the actual incidence of conflict were highly exaggerated (Ballejo et al. 2020). This view largely emerged from the magnification of exceptional incidents based on particular social and cultural attitudes (Dickman 2010). Duriez et al. (2019), in another published research material, explained the perceived bias in the vulture-livestock conflict with the hypothesis of human knowledge paucity rather than the hypothesis of a recent change in vulture feeding behaviour. Still, the resulting human-vulture conflicts included the deliberate poisoning of severely threatened vulture species. Currently, the illegal use of poison baits continues and the human–vulture conflict could increase non-natural mortality effects on other species (Campbell 2015; Pacheco et al. 2020; Zuluaga et al. 2020). However, dissenting views have emerged on this topic. For example, Lambartucci et al. (2021), present a detailed and comprehensive analysis and argue that stakeholders such as the media, politicians, farmers, and scientists should not exacerbate the issues, through drama or biases, given the highly detrimental effect on the conservation of vultures.

DISEASE SYMPTOM OR ADAPTATION?

Vultures face health challenges per their carrion eating habits (Campbell, 2015). Carcass-derived microbes may pose serious pathogenic risks to non-scavenging species. However, different potential microbiome-mediated defence mechanisms account for the vulture's ability to tolerate and reduce the health-risk potential that a carcass represents as it passes through its digestive system (Mendoza et al. 2018). For example, Blumstein et al. (2017) examined several hypotheses regarding the adaptation of vultures for possible defence against microbial infections, with variable assessments. Obligate scavengers, such as vultures, have more natural antibodies against botulinum toxin than facultative scavengers (Ohishi et al. 1979). Vultures and some other scavengers have an especially low gastric pH that kills pathogens ingested from carrion (Houston and Cooper 1975). A specialized facial and gut microbiome may prevent the population growth of pathogens. These microbial communities may be an adaptation to avoid diet-induced illness (Roggenbuck et al. 2004). Some of the hypotheses like, avoidance of rotten food, bald head as an adaptation to cleanliness, acidic sterilization, rapid ingestion to avoid spore formation, etc. (Stone 1993; Houston and Cooper 1975; Kushwaha et al. 2009) did not stand the scrutiny. Nevertheless, vultures are reported to suffer from diseases, such as the Avian pox in Turkey Vulture (Mora-Carreno et al. 2019), Aspergillosis in captive Cape Vulture (Chege et al. 2013), which produce signs such as prominent nodular masses on its head, pneumonia-like symptoms, respectively. However, "Neck drooping syndrome" reported by Cunningham et al. (2003) in declining Indian vultures (*Gyps bengalensis* and *Gyps indicus*) during the 1990s was associated with an infectious, probably viral, aetiology. This was linked to an earlier report by Prakash (1999) about the vulture decline in Keoladeo National Park, Rajasthan which observed "doze off with neck limp and hanging," an unusual posture adopted by them before death which continued/progressed over 30 days. This was based on a sample of five vultures and measures taken to ensure continuity of individual observation were not described, rather circumstantial evidence was ascribed. Nevertheless, in the absence of previous reports of this posture in Indian vultures and despite healthy birds also adopting this posture, it was assumed to be a novel behaviour, a prelude to death, and was used as a primary indicator of ill health (Cunningham et al. 2003, Pain et al. 2003, Prakash et al. 2003).

Gilbert et al. (2007) worked on two hypotheses, perhaps due to the gaps in the work of Prakash (1999): whether neck drooping is (i) temperature-dependent and (ii) indicative of poor health in *Gyps bengalensis* and a prelude to death. They published the results under a title "Neck-drooping Posture in Oriental White-Backed Vultures (*Gyps bengalensis*): An Unsuccessful Predictor of Mortality and Its Probable Role in Thermoregulation" and recorded that "Neck-drooping vultures were observed significantly more frequently with their back toward the sun with their head in their shade than birds that were not neck drooping, and vultures that were not neck-drooping were observed more frequently facing the sun than that neck-drooping. Together, these observations strongly suggest that the neck-drooping posture has a role in thermoregulation. In contrast to the highly seasonal pattern of neck-drooping, the mortality of vultures occurred in all months of the year. This finding indicates that neck-drooping has low specificity and sensitivity as an indicator of poor health and impending death in vultures." Yet another report by Watson et al. (2007) suggests that neck drooping posture and behaviour may serve more than one function. They hypothesized, although based on qualitative research, that neck-drooping is a posture that *Gyps bengalensis* adopts to pose as sick when approached by humans and unable to flee.

NSAID Diclofenac Ban Impact: The Actual Story

The decline in the population of Indian vultures between 1990 and 2010 was reported and referred in several works due to the transfer of a non-steroidal anti-inflammatory drug (NSAID), diclofenac, from cattle carcass to vulture feed (Prakash et al. 2007; Prakash et al. 2012; Prakash et al. 2017). Considering the impact and serious consequences on ecosystem services provided by vultures, the use of this drug was banned in the Indian sub-continent by the respective governments of India, Pakistan, Bangladesh, and Nepal (Pain et al. 2008; Chaudhry et al. 2012; Paudel et al. 2016; Prakash et al. 2012, 2017, etc.). The government of India banned the use of the drug in May 2006 and the manufacture, sale, or use for the veterinary purpose of the same was declared an offense in 2008 (Taggart et al. 2007; DeVault et al. 2016; Richards et al. 2017). Though its sale and use were banned in India, the injectable human formulation remained widely available for unrestricted purchase (Cuthbert et al. 2011; Saini et al. 2012). Further in July 2015, multi-dose vials of this drug for human use were also banned (GSR 558(E) dated 17 July 2015) and later confirmed by the court in November 2017 (Laborate Pharmaceutical India Ltd. vs Union of India; https://indiankanoon.org/doc/194709922/). While the environmentalist lobby was pressing hard to save vultures from the NSAID menace (Chaudhry et al. 2012) the pharmaceutical lobby was not willing to concur for economic reasons and tried to persist with the drug directly and indirectly. Therefore, drug remained available in the market even after 11 years of the ban in India (Galligan et al. 2014; Prakash et al. 2017) and possibly continued at least until February 2020 (Jha et al. 2021).

Prakash et al. (2017) considered the veterinary use of diclofenac as the main and possibly the sole factor for the vulture population decline. This view was supported by the work presented by Oaks et al. (2004), Green et al. (2004), Shultz et al. (2004), etc. However, there are cases at a regional level where other causal agents were considered responsible for the loss of vultures. Abbas et al. (2013) hypothesized against the popular diclofenac theory and proposed that even after the ban of diclofenac in Pakistan, the population decrease continued. This could be due to the continued stealthy use of illegal diclofenac as explained above in India, but the author has not found such reports. Instead, Abbas et al. (2013) postulated an alternate hypothesis of food limitation based on a crashing decrease in carcass availability. Jha (2015) has also reported scarcity of food in the north Indian state of Uttar Pradesh and speculated upon its impact on vulture survival. Another finding in Bundelkhand, a north-central region of India, made suggestions on similar lines; as there is no animal husbandry in stray cattle (*Anna Pratha* of cattle rearing) and rare diclofenac deaths (Ajit Dixit, veterinarian; Amita Kanaujia, Zoologist *Pers. Comm.*). Baral et al. (2005) proposed nesting and roosting habitat as a limiting factor in vulture deaths in Rampur valley of Nepal. The sparse kapoks or the nesting trees of commercial value were extracted by the Community Forest User Group to collect revenue. These narratives suggest that while the diclofenac could be a major reason for Asian vulture deaths, a composite effect of other reasons cannot be ruled out.

Other than diclofenac, another not so popular yet damaging NSAID is nimesulide (Cuthbert et al. 2016). Reddy et al. (2006) compared the toxicity of diclofenac to domestic fowl (*Gallus domesticus*) with that of nimesulide and concluded that nimesulide and the vehicle (propylene glycol) used in the nimesulide injection preparation are completely safe for birds at the tested dose levels. It is not clear why the domestic fowl result was extrapolated to the whole group of birds, but Cuthbert et al. (2007) criticized Reddy et al. based on a threadbare analysis of a research publication "Comparative toxicity studies in birds using nimesulide and diclofenac sodium." Cuthbert et al. (2007) rejected the application of the findings of Reddy et al. (2006) to other species beyond domestic fowl and considered their conclusion to be erroneous and potentially damaging for the conservation of *Gyps* vultures and other scavenging birds.

Think Globally, Act Locally

"Think globally, act locally" is a popular global health slogan and imperative that encourages people to consider the health of the entire planet while taking actions in their cities and communities (https://www-newswise-com.cdn.ampproject.org/v/s/www.news wise.com/articles). This hypothesis appears to be equally applicable to wildlife/biodiversity projects which have a global presence and local impact. Vulture conservation and ecosystem services availability seems plausible to be discussed along these lines. Santangeli et al. (2019) developed a critical conservation priority area of the world related to Old World vultures based on information about the distribution of all 15 vulture species found in Europe, Asia, and Africa with their threats along with detailed expert knowledge on threat intensity. However, Efrat et al. (2020), while appreciating the work of Santangeli et al. (2019), in their comments published in "Conservation Biology" draw attention towards the improvement of the model. Such a generalized large-scale prediction excluded relevant data for smaller-scale conservation efforts.

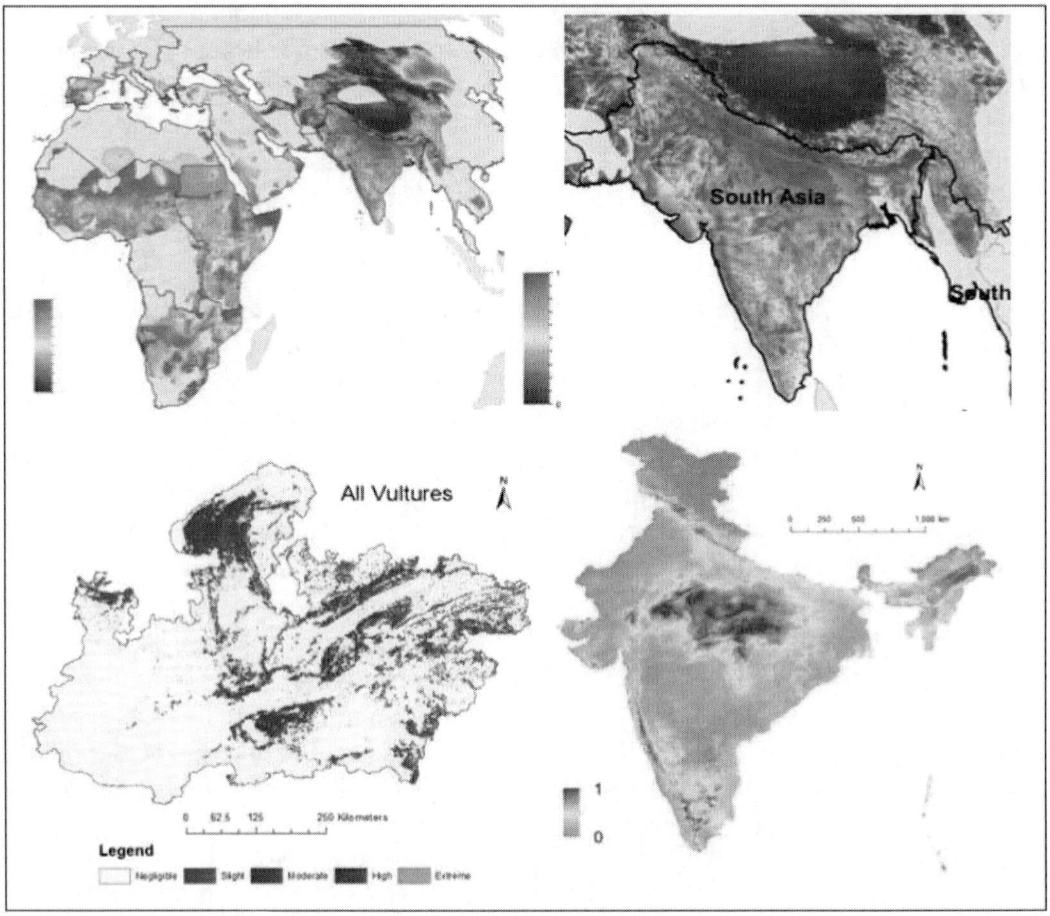

Note yellow and orange areas (above 0.4) and light or dark green area (below 0.4) in the two maps top and bottom right. They are high priority / suitability and low priority / unsuitability areas, respectively. Maps on the top have been extracted from Santangeli et al. (2019). Above maps are not on the same scale and projection. However, perceptive difference in the quantum of orange and yellow, and green areas are too high (45% vs 95%), right top and bottom maps. Map on the bottom left (Jha and Jha 2020) is clipped Central India showing MaxEnt results at state or local level. Yucca yellow area (78% of the state) is unsuitable habitat. This may be compared with suitability area at India level in the bottom right.

Figure 2. Vulture conservation priority and habitat suitability areas from Afro-Eurasian to Indian and further localised Central India.

Quoting several examples, Efrat et al. (2020) justified their claim and finally concluded that "the stakeholders and decision-makers who wish to conserve their local vulture populations to consider not only the results of large-scaled generalized models but also data that are available for their area." Our observation also adds to Efrat et al.'s hypothesis to the extent that situation projected in Santangeli et al.'s does not conform fully to local conditions. For example, the map above shows more than 95% area of India as a high priority conservation zone (Figure 2).

This map also shows vulture conservation priority areas in Africa, Europe, and Asia presented in their publication records (Efrat et al. (2020). Assuming suitability of habitat as a function of conservation priority for vulture our observation (unpublished data) does not match since only 45% area falls under the suitable habitat category (Figure 2). This observation is based on MaxEnt modeling with around 2500 presence records of eight species scattered almost throughout the country and non-correlated bioclimatic variables. Further localized suitable habitat study in Central India tells a different story, only 22% area of Madhya Pradesh (Central India) is suitable for vultures (Jha & Jha 2020).

NAME USE CONFUSION

Name changes have been observed in different vulture species over some time for the sake of phenotypic accuracy and other taxonomic reasons. For example, the Oriental (Asian/Indian) White-backed Vulture was segregated from the White-backed Vulture and later changed to White-rumped Vulture. White-backed Vulture (*Gyps africanus* Salvadori 1865) is now considered the African while White-rumped Vulture (*Gyps bengalensis* Gmelin, JF, 1788) is an Asian native. The name White-rumped Vulture was recognized no later than the year 2000 (Clement's checklist of Birds of the world, 5th edition), but some taxonomically less concerned authors still name this species as the Oriental White-backed vulture (Watson 2007), White-backed Vulture (Gadhvi and Dodiya 2006; Kushwaha and Kanaujia 2009, 2011) and Indian White-backed Vulture (Thakur and Narang 2012; Thakur and Kataria 2012). Similarly, the Red-headed Vulture (*Sarcogyps calvus* Scopoli 1786, resident of the Indian subcontinent) also known as the Asian King Vulture is confused with King Vulture (*Sarcoramphus papa* Linnaeus 1758, New world vulture).

For example, Kushwaha and Kanaujia (2009, 2011) and Chhangani and Mohnot (2008) have opted for King Vulture while addressing *Sarcogyps calvus*. The use of the English name 'Black Vulture' without the suffix identifier creates confusion with other species with the same English name. For example, Eurasian Black Vulture popularly known as Cinereous Vulture *Aegypius monachus* Linnaeus 1766, is an Old World vulture while the American Black Vulture *Coragyps atratus* Bechstein 1793 is a New World vulture.

Some vulture species ranging though the Indian subcontinent are presented in the following Figures 3-5.

Red-headed Vulture (*Sarcogyps calvus* Scopoli 1786), a critically endangered species on the left and Egyptian Vulture (*Nephron percnopterus* Linnaeus 1758), an endangered species on the right. Photo credit: Sanjay Shukla IFS.

Figure 3. Resident vultures of Indian subcontinent.

White-backed Vulture (*Gyps bengalensis* Gmelin 1788), a critically endangered species on the left and Indian Vulture (*Gyps indicus* Scopoli 1786), another critically endangered species on the right. Photo credit: Sanjay Shukla IFS (left) and Radhika Jha (right).

Figure 4. Resident vultures of the Indian subcontinent.

Himalayan Griffon Vulture (*Gyps himalayensis* Hume 1869), a threatened species on the left and Cinereous Vulture (*Aegypius monachus* Linnaeus 1766), a near threatened species on the right. Photo credit: Radhika Jha (left) BP Sirsiya (right).

Figure 5. Migratory vultures spending winter in Indian subcontinent.

Habitat Improvement Versus Captive Breeding

Prakash (1999) projected the certain extinction of some vulture species and recommended captive breeding of *Gyps* species to save them from imminent extinction, although he was not certain about the cause of the crash in the vulture population. His team further strengthened the claim of a massive decline in *Gyps bengalensis* and *G. indicus* (Prakash et al. 2003). A couple of years later his team published evidence to support that diclofenac caused catastrophic vulture population decline. Another research published by them re-emphasized that the population of the three species of vultures continued to decline at an alarming rate. The numbers of *Gyps bengalensis* declined by 99.9%. The equivalent decline of the combined total of *Gyps indicus* and G. *tenuirostris* was 96.8% (Prakash et al. 2007). After a gap of around five years, it was claimed that the populations of all three species of vulture remained at a low level, but the decline had slowed and may even have reversed for *G. bengalensis*, both in India and Nepal (Prakash et al. 2012). The reason assigned to this change in population trend was the ban of diclofenac use in cattle which is toxic to vultures. However, the diclofenac ban has been shown to have been ineffective by a few studies (Saini et al. 2012; Cuthbert et al. 2016; Jha et al. 2021). Various developments (research reports and field observation) have evidenced that the vulture population did not face extinction and diclofenac was not the sole cause. Many field practitioners believed that habitat improvement is playing a major role in recovery (CWLW, MP pers. comm.).

Captive breeding is a useful tool for increasing the population of a species in controlled conditions and releasing the bred individuals via a reintroduction program.

One such flagship program was started in 2001 at Pinjore, Haryana, and is continuing after several years with limited success. The Haryana Forest Department (http://haryanaforest.gov.in/en-us/Wild-Life/Protected-Area/Vulture-Conservation-and-Breeding-Centre-Pinjore; accessed on 19/06/2020) claims that "the Jatayu Conservation Breeding Centre (JCBC) is a joint project of the Haryana Forest Department and the Bombay Natural History Society (BNHS). It is a collaborative initiative to save the three species of vultures, White-backed, Long-billed, and Slender-billed, from looming extinction." However, even in approximately two decades, this objective has not been achieved as none of these vultures have been reintroduced in the wild. A Himalayan griffon vulture *Gyps himalayensis* was released in 2016 with some extensive media reporting, which was a success (https://www.indiatoday.in/india/haryana/story/asias-first-vulture-re-introduction-programe-launched-in-haryana-12287-2016-06-03; accessed on 03/07/2020). The question remains as to why the vulture species facing acute crises have not been able to become a part of reintroduction? Another question is that should we rely on nature's program of an automatic increase in population (Prakash et al. 2017) or continue with exorbitantly expensive programs of captive breeding with little or no results over a long period? Nevertheless, the vulture breeding program continues to gain popularity in India (https://www.thehindu.com/news/national/other-states/up-to-get-its-first-conservation-centre-for-vultures/article30119522.ece; accessed on 03/07/2020) under the guidance of BNHS.

POPULATION CRASH

Some noteworthy series of reports on this topic appeared in several research publications anchored by the scientists of BNHS. The vulture population crash in the Indian subcontinent is understood to have occurred between 1990 and 2010. One of the earlier reports is Prakash (1999) which documented a localized (29 km^2 Keoladeo National Park, Rajasthan, India) sharp reduction in population during 1985-88 and 1996-99 in White-rumped Vulture and Long-billed Vulture to the tune of 96% and 97%, respectively. Another report covering the whole of India appeared in Prakash et al. (2003) claiming a massive decline in the populations of both the species which was related to all parts of the country and exceeded 92% overall. Further, Prakash et al. (2007) results indicated that the population of White-rumped Vulture declined by 99.9%. The equivalent decline in the combined total of *Gyps indicus* and *G. tenuirostris* (a species separated from *G. indicus*) was 96.8%. An interesting feature of this report is the precision to the decimal point which means only one White-rumped Vulture survived out of 1000 and just 32 Long-billed (Slender-billed Vulture inclusive) Vultures out of 10000. Cuthbert et al. (2006) in Galligan et al. (2014) reported an 80% decline in Egyptian vulture and 91% in Red-headed Vulture. Yet another publication, Prakash et al. (2012) working beyond India reported that populations of all the three species of vultures (*G. bengalensis, G. indicus* and *G. tenuirostris*) remained at a low level, but the decline had slowed and may even have reversed for *G. bengalensis*, both in India and Nepal. They associated the diclofenac ban hypothesis with this result, although with an 'if and but' proposition. However, a slower decline mark has been set in Prakash et al. (2017) as 2006 in the case of White-rumped and 1990s in the case of Long-billed vultures. The trend in Slender-billed explained uncertainly due to a smaller population of this species.

The decline of these vulture species remains unarguable but all questions regarding the reduction should converge at one point as to "how much and from how many" in quantitative assessment since percentage expression is a simple indicator and could be misleading for a large area if it is derived from a small number of smaller areas. Larger area sampling could cover this part but again the question arises as to whether those samples are representative or not. Modeling projection has been a good answer but again it has limitations as biological functions are very difficult to parameterize to perfection.

Population Estimation

Vulture abundance records of select areas in India can be found in a few workshops and other reports (Galushin 2001; Prakash 2001; Virani et al. 2001). Based on a sample survey of protected areas (PA), their adjoining areas (AA), and connecting highways (HW) Prakash (2001) presented the proportion of *Gyps bengalensis* and *Gyps indicus* in the population as 63% and 37%. Other species were negligible. Distribution in the landscapes was in the following order: PA (62%) > HW (27%) > AA (11%). The abundance of White-rumped Vulture was unit individual 5.2 km^{-1} (PA), 13.6 km^{-1} (AA), and 22.2 km^{-1} (HW). Similarly, the figures for Long-billed Vulture were 6.48 km^{-1}, 21.47 km^{-1}, and 91.59 km^{-1}. Another report of Galushin (2001) provided raptor assessment including vultures in select urban and rural areas. The average urban density of White-rumped Vulture and Egyptian Vulture around 1970 was 5.4 km^{-2} and 1.0 km^{-2}, respectively. The average rural (agricultural sample) density of White-rumped Vulture was 2.8 km^{-2}. A small number of Egyptian Vultures was also recorded.

There are few population assessments in absolute (number) terms on a larger scale (state or country basis) in India in any of the prominent publications on the subject during or before the crisis period, though qualitative expressions like "a few thousand of this species, possibly up to the low tens of thousands," "tens of millions," "once numbering in the millions," "once abundant" can be found in "Prakash et al. 2017," "Richards et al. 2017," "Baral et al. 2013," "Gadhvi and Dodia 2006," respectively. These appear to be heuristic guesses which may always be arguable and misleading for want of a quantitative assessment. This makes it difficult to compare the actual status of the population if reference data is not available. However, a few state-level reports are giving quantitative assessments of the post-crash period. Thakur and Narang (2012) reported around 150 White-rumped vultures in Himachal Pradesh. Chhangani (2009) recorded 5080 temporally distributed vultures (1086 Long-billed, 325 White-rumped, 84 Red-headed, 2418 Egyptian, and 217 Cinereous vultures, and 240 Himalayan and 643 Eurasian griffon vultures) over four years (2004-2007) in Rajasthan. Jha (2015) recorded 2084 vultures (932 Egyptian, 516 Slender-billed, 341 Long-billed, 209 White-rumped, 16 Red-headed, and 14 Himalayan Griffon vultures) in Uttar Pradesh. Kamboj et al. (2016) reported 999 vultures of four species (458 White-rumped, 385 Long-billed, 132 Egyptian, and 34 Red-headed vultures) in Gujarat. As per Jha (2017), Madhya Pradesh has 7028 vultures of seven species (3351 Long-billed, 1874 Egyptian, 1522 White-rumped, 93 Red-headed, 53 Cinereous, 95 Eurasian, and 42 Himalayan griffon vultures). In a recent report of Jhala et al. (2017) out of 6828 reported vultures (227 Griffon, 1424 Egyptian, 2346 White-rumped, 2830 Long-billed, and one Slender-billed vulture), 92% were in the Central Indian landscape. Botha et al. (2017) have recorded the global population range of different Old World vultures in a Conservation Action Plan as follows: Bearded Vulture (2000-10000), Egyptian Vulture (18000-57000), Red-headed Vulture (3500-15000), Himalayan Griffon (66000-334000), White-rumped Vulture (8000), Indian Vulture (12000), Slender-billed Vulture (1500-3750), Griffon Vulture (80000-120000) and Cinereous Vulture (15600-21000).

Ganguly and Mukhopadhyay (2013) gave an account of the vulture population in India. They recorded in the abstract of their paper that India had 40 million vultures during the 1980s which reduced to 60,000 around 2013. Unfortunately, there is no mention of this fact in the text or how these numbers were derived. Lack of any reference to authors makes the figures unauthentic and imaginary. Wikipedia mentions that as recently as the 1980s (https://en.wikipedia. org/wiki/Indian_vulture_crisis#CITEREFGreenwood1938; accessed on 19/06/2020) there were up to 80 million white-rumped vultures (*Gyps bengalensis*) in India, but today the population numbers only several thousand. Again, for want of any reference, and due to the existence of nine vulture species, this figure about only one resident species pushes us towards skepticism; as if India had only vultures all around. Meir Rinde, a freelance writer (September 3, 2019; https://www.science history.org/distillations/poison-pill-the-mysterious-die-off-of-indias-vultures; accessed on 19/06/2020), claims that in the early 1990s at least 40 million vultures, and possibly many more, lived in India; by 2015 the combined number of Indian, white-backed, and slender-billed species had dwindled to perhaps 19,000 in the whole country. Again, this data does not have any listed source to prove authenticity. Richards et al. (2017) have recorded the vulture population in qualitative terms "By the 1980s, vulture presence was prodigious—with some species estimated to number in the tens of millions." The reference quoted in this publication is a photograph showing vultures in a plantation and carcass dumping site landscape. Although this depicts a large wake of vultures, it cannot represent tens of millions unless the expression is poetic or highly exaggerated. We tried to develop a crude model using this photograph, based on some assumptions. This could reach the figure 6 million individuals. Galushin's model suggests the vulture population of 1970's to be around 8 million (Box 1).

Extinction Prediction and Modelling

There is a popular saying "if you torture the data long enough, it will confess to anything," as stated by a Nobel Laureate of Economics Ronald H. Coase (Fleming 2010). A simple implication of this expression is that one can prove the precision of one's results to whatever extent necessary by statistical tools. At the same time, it is an accepted truism about the statistical models that "all models are wrong but some are useful" acclaimed by famous statistician GEP Box (Singh et al. 2015; Efrat et al. 2020). This takes us towards introspection since scientific models always fall short of the complexities of reality and prompt us to balance the results of two contrasting yet connected phenomena through data torturing and model fallacy. In this context, the precise results (99.9% and 96.8%) of population decline reported by Prakash et al. (2007) need to be seen along with caution mentioned in the paper that crude estimates of the absolute numbers of vultures remaining in northern India are tentative. This necessitates the need for re-examination of uncertain data and prediction connected to crude modeling and tortured results. Another strong prediction in this publication is non-comprehensible that "all three species could be down to a few hundred birds or less across the whole country, and thus functionally extinct, in less than a decade." This is not corroborated with the fact on the date, after a decade, is that the population trend is reversing. This positivity may be argued with diclofenac ban impact which in itself is strongly questionable since practically there is no effective ban even after fourteen years. However, other latent factors could have been responsible for population change, as also suggested by Choresh et al. (2018) working on another *Gyps* species (*Gyps fulvus*) in Israel. Poisoning, thought to be a root cause, might not be the only factor influencing extinction. In addition to breeding failure, some of the known mortality causes of griffons in Israel are poisoning (41%), electrocution (9%), and poaching (6%), and unknown causes comprise 30% the mortality events.

Prakash et al. (2007) is a very popular work as it has drawn a research interest score 73.2 and 129 citations on Research Gate, a scientific and social network (https://www.researchgate.net/publication/228495035_Recent_changes_in_populations_of_resident_Gyps_vultures_in_India/citations, accessed 08/07/2020). However, most of the papers have quoted 99.9% and 97.2% decline, or population crash, or diclofenac as the main causal agent (Adawaren et al. 2018; Campbell 2018; Choresh et al. 2018; Das et al. 2018; Jha 2018; Margalida and Ogada 2018; Movalli et al. 2018; Phillip et al. 2018; Saengtienchai et al. 2018; Samson et al. 2018; Jha and Jha 2020; Jha et al. 2020; Narasimhaiah et al. 2020; etc.). Probably none of them have discussed the "if" part of the results (estimation is crude, projected extinction in a decade, diclofenac ban, etc.). The point of concern here is that there is a fast transmission of incomplete knowledge among scholars.

Box 1. Photograph adopted from Richards et al. (2017) depicting vultures in the foreground assembled at a feed dump, on the trees and some soaring in the sky

A careful count of vultures in this photo frame returns the number <260. Counting omissions and doubling the frame makes it maximum 700. Such dumping sites 200 per states, for 20 north and central Indian states brings the number to 2,800,000. If we double the number (covering for the same number of individuals staying at roosts or nests), it becomes 5,400, 000 or 6 million vultures. This result is far below the findings of Richard et al. (2006), which mention tens of millions of some species.
{Another model suggests 8 million vultures during 1970s in India (Galushin's average density 3.0 km^{-2} and 70% of India had vulture distribution)}
Some assumptions:
Vultures are gregarious feeders and travel up to 100 km for food. They change their behaviour on large supplementary feeding sites with regular food availability. Only north Indian states have good number of vultures. Urban areas have higher vulture density than rural landscapes.

Prakash et al. (2017) wrote another significant article on the vulture population and summarised their findings as "Populations of all three species of vulture (*Gyps bengalensis, G. indicus* and *G. tenuirostris*) remained at a low level. The previously rapid decline of White-rumped Vulture has slowed and may have reversed since the ban on veterinary use of diclofenac in India in 2006. A few thousand of this species, possibly up to the low tens of thousands, remained in India in 2015. The population of Indian Vulture continued to decline, though probably at a much slower rate than in the 1990s." Again, unquestionable statistics have been done here to achieve regression modeling results. However, population size is subject to several caveats because of limitations in the data available. This raises the question if the model is not parameterized, properly to what extent can the results be dependable. This attracts applicability of a phenomenon GIGO, a very popular acronym of olden time which stands for Gold in Gold out, a positive expression of cause and effect. This has an analogy with modeling also since the more accurate variables you enter into the algorithm the better predictions you get. Contrary to this, the more the assumptions the more chances of imprecise results. Kane et al. (2014) also noted that any model is only as good as the data used to parameterize it. Consequently, basic and up to date, biological data are of utmost importance if we are to derive accurate predictions.

ARTIFICIAL FEEDING: BOON OR BANE ?

Vultures are among the most threatened avian guilds in the world (Sekercioglu 2006; IUCN 2010; Mateo-Tomas and Olea 2010b). The population of some of the species in different parts of the world has even crashed which warranted the immediate need for conservation. Supplementary feeding in the last fifty years has become one of the most common techniques used to alleviate threats to scavenger species and support vulture colonies and run the reintroduction program based on the quality and availability of food resources (Moreno-Opo et al. 2015; Schabo et al. 2016; Cortes-Avizanda et al. 2016). However, the use of supplementary feeding to promote the recovery of vultures has been questioned by both conservationists and wildlife managers due to the potential negative effects that may be produced (Genero et al. 2020) due to inbuilt weaknesses and inadvertent faulty practices.

The precarious conservation status of vultures is partly due to the populations having developed a strong reliance on ever-changing human activities (e.g., livestock, supplementary feeding, hunting, poisoning, etc.; Donazar et al. 2009; Mateo-Tomas and Olea 2010b; Olea and Mateo-Tomas 2009). Nevertheless, Margalida et al. (2007) suggested that the regulation of human disturbance (using the protection of nesting sectors) and maintenance of resources (e.g., by revitalizing extensive livestock management and leaving carcasses in the field as occurred until recently) may be fundamental for the maintenance and/or growth of their populations. Although Oro et al. (2008) had earlier concluded that supplementary feeding is not effective to save bearded vultures from an expected population decline, they delay population extinction and can be a useful tool for prolonging population viability while combating illegal and indirect poisoning. A recent study by Schabo et al. (2016) found a significantly positive effect of supplementary food during the nest-building stage on the number of breeding pairs.

Supplementary feeding is helpful not only in population enhancement but in behavioural studies as well. Moreno-Opo et al. (2020) studied the behavioural organization and the relationship of hierarchies according to feeding behaviour and prey selection, by comparing with other scavenger guilds and found that larger size and older age confer competitive advantage: dominance hierarchy within the European vulture guild. They showed a general dominance gradient of body size (larger species to the smaller one: CV > GV > BV > EV), and age: (from the adult age class to subadults and juveniles) as the outcome of agonistic encounters between European vultures. Supplementary feeding also helped in analyzing food preferences by Moreno-Opo et al. (2015) in Eurasian Griffon, Egyptian Vulture, Bearded Vulture, and Cinereous Vulture. The carrion features (format, scattering, prey species, biomass, and items) influenced differential selection between species and age-classes. At a species level, large inputs of un-scattered carrion increased the abundance of actively feeding griffon vultures. The ratio of the abundance of Bearded Vulture, Egyptian Vulture, and Cinereous Vulture to Griffon vulture was favoured when less biomass was supplied and when the food provided was not presented as whole carcasses. Thus, using medium-size ungulates (i.e., sheep and goats) presented as small, abundant, and scattered pieces was favoured for consumption by the most endangered species. It was observed by Moreno-Opo et al. (2010) that Cinereous Vultures prefer individual, medium-sized muscular pieces and small peripheral scraps of meat and tendon.

However, supplementary feeding is not a panacea for the vulture crises since different studies have shown negative consequences for target populations as well as for non-target species (Cortes-Avizanda et al. 2009). Yarnell et al. (2014) presented evidence that providing supplementary carrion for vultures stimulated an increase in local abundance of two species of mammalian carnivores, the brown hyena and the black-backed jackal which are opportunistic scavengers and interact competitively with each other and with vultures also. From a management point of view, a study shows that supplementary feeding can have detrimental effects on fecundity (Margalida 2017). In recent years, many studies have speculated on the likelihood that avian scavengers may be infected by feeding on pig carcasses at supplementary feeding stations (SFS) from intensive livestock. The current study indicates that free-living griffon vultures and pig farms that provide the carcasses at SFS share several zoonotic Salmonella strains (Marin et al. 2018). It was found that supplementary feeding appears to be discouraging population expansion outside the Pyrenees and has been related to habitat saturation processes which are triggering negative population effects such as a reduction in productivity and changes in mating behaviour (Carrete et al. 2006; Margalida et al. 2008; Oro et al. 2008). Preliminary results of Margalida (2010) suggested that the provision of supplementary food during the chick-rearing period aimed at increasing breeding success does not seem to be effective. Another study of Gomez-Ramirez (2018) revealed that excessive use of livestock and deficient management of the carcasses can lead to adverse effects in the scavengers that ingest them, especially in supplementary feeding. The supplementary feeding is a special case of a carcass site where facultative scavengers concentrate permanently because of the constant food supply. There, the probability of predation risk could be persistent over time, having a stronger effect on the population dynamics of prey species (Cortes-Avizanda et al. 2009). The management of supplementary food through a network of numerous small sites does not substantially disrupt the natural foraging behaviour of vultures, but large, frequently replenished sites tend to artificially increase their level of routine (Fluhr et al. 2017). On the safety side, the risk of chemical pollutant and pathogen transfer cannot be ruled out in the case of a supplementary feeding mechanism (Casas-Diaz et al. 2016; Cortes-Avizanda et al. 2016). Therefore, secondary feeding in Vulture Restaurants (Figure 6) may be characterized as a necessary evil or a double-edged sword (Cortes-Avizanda et al. 2016).

Provisioning of carcass and facilitation by skin removal on the left and a group of vultures feeding on carcass at a restaurant on the right. Photo courtesy: Sujoy Banerjee IFS.

Figure 6. Newly established operational Vulture Restaurant in *Terai* area of Uttar Pradesh, North India.

References

Abbas, F., Rooney, T. P., Haider, J. & Mian, A. 2013. Food limitation as a potentially emerging contributor to the Asian vulture crisis. *The Journal of Animal & Plant Sciences*, 23(6): 1758-1760.

Adawaren, E. O., Mukandiwa, L., Njoya, E. M., Bekker, L., Duncan, N. & Naidoo V. 2018. The use of liver slices from the Cape vulture (*Gyps coprotheres*) to better understand the role of liver toxicity of non-steroidal anti-inflammatory drugs (NSAIDs) in vultures. *Environmental Toxicology and Pharmacology*, 62: 147-155. https://doi.org/10.1016/j.etap.2018.07.001.

Ballejo, F., Plaza, P. I. & Lambertucci, S. A. 2020. The conflict between scavenging birds and farmers: Field observations do not support people's perceptions. *Biological Conservation*, 248 (2020) 108627. https://doi.org/10.1016/j.biocon.2020.108627.

Baral, N., Nagy, C., Crain, B. J. & Gautam, R. 2013. Population viability analysis of Critically Endangered white-rumped vultures *Gyps bengalensis*. *Endangered Species Research*, 21: 65–76. https://doi.org/10.3354/esr00511.

Baral, N., Gautam, R. & Tamang, B. 2005. Population status and breeding ecology of White-rumped Vulture *Gyps bengalensis* in Rampur Valley, Nepal. *Forktail*, 21: 87–91.

Blumstein, D. T., Rangchi, T. N., Briggs, T., De Andrade, F. S. & Natterson-Horowtiz, B. 2017. A Systematic Review of Carrion Eaters' Adaptations to Avoid Sickness. *Journal of Wildlife Diseases*, 53(3): 1-5. https://doi.org/10.7589/2016-07-162.

Boakye, M. K., Wiafe, E. D. & Ziekah, M. Y. 2019. *Ethnomedicinal use of vultures by traditional medicinal practitioners in Ghana*. Ostrich, June 2019. https://doi.org/10.2989/00306525.2019.1578834.

Botha, A. J., Andevski, J., Bowden, C. G. R., Gudka, M., Safford, R. J., Tavares, J. & Williams, N. P. 2017. *Multi---species Action Plan to Conserve African---Eurasian Vultures. CMS Raptors MOU Technical Publication No. 4.* CMS Technical Series No. 33. Coordinating Unit of the CMS Raptors MOU, Abu Dhabi, United Arab Emirates.

Camina, A. & Yosef, R. 2012. Effect of European Union BSE-related enactments on fledgling Eurasian Griffons *Gyps fulvus*. *Acta Ornithologica*, 47: 101–109. https://doi.org/10.3161/000164512X662205.

Campbell, M. O. 2018. GPS, animal tracking and conservation biology. In Campbell. M. O. Ed. *Geomatics and Conservation Biology*. NY: New York, USA, Nova Science Publishers, Inc.

Carrete, M., Donazar, J. A. & Margalida, A. 2006. Density-dependent productivity depression in Pyrenean bearded vultures: implications for conservation. *Ecological Applications*, 15:1674–1682.

Casas-Diaz, E., Cristofol, C., Cuenca, R., Agustin, S., Carneiro, M., Marco, I., Lavin, S. & Margalida, A. 2016. Determination of antibiotic residues in plasma of two populations of Eurasian griffon vulture (*Gyps fulvus*) in Spain. *Science of the Total Environment*, 557:620–626.

Chaudhry, M. J. I., Ogada, D. L., Malik, R. N., Virani, M. Z. & Giovanni, M. D. 2012. First evidence that populations of the critically endangered Long-billed Vulture *Gyps indicus* in Pakistan have increased following the ban of the toxic veterinary drug diclofenac in south Asia, *Bird Conservation International*, 22:389–397. https://doi. org/10.1017/S0959270912000445.

Chege, S., Howlett, J., Al Qassimi, M., Toosy, A., Kinne, J. & Obanda, V. 2013. Opportunistic infection of *Aspergillus* and bacteria in captive Cape vultures (*Gyps coprotheres*). *Asian Pacific Journal of Tropical Biomedicine*, 3(5): 401-406. https://doi.org/10.1016/S2221-1691(13)60084-2.

Chhangani, A. K. & Mohnot, S. M. 2008. Demography of migratory vultures in and around Jodhpur, India. *Vulture News*, 58:23-38.

Choresh, Y., Burg, D., Trop, T., Izhaki, I. & Malkinson, D. 2018. Long-Term Griffon Vulture Population Dynamics at Gamla Nature Reserve. *The Journal of Wildlife Management,* https://doi.org/10.1002/jwmg.21566.

Cortes-Avizanda, A., Blanco, G., DeVault, T.L., Markandaya, A., Virani, M. Z., Brandt. J. & Donazar, J. A. 2016. Supplementary feeding and endangered avian scavengers: benefits, caveats and controversies. *Frontiers in Ecology and the Environment*, 14(4): 191-199. https://doi.org/10.1002/fee.1257.

Cunningham, A. A., Prakash, V., Pain, D., Ghalsasi, G. R., Wells, G. A. H., Kolte, G. N., Nighot, P., Goudar, M. S., Kshirsagar, S. & Rahmani, A. 2003. Indian vultures: victims of an infectious disease epidemic? *Animal Conservation* 6: 189–197. https://doi.org/10.1017/S136794300300324X.

Cuthbert, R., Green, R.E., Ranade, S., Saravanan, S.S., Pain, D.J., Cunningham, A.A. & Prakash, V., 2006. Rapid population declines of Egyptian Vulture Neophron percnopterus and Red-headed Vulture Sarcogyps calvus in India. *Animal Conservation* 9: 349–354.

Cuthbert, R., Pain, D. J., Green, R. E., Swan, G. & Swarup, D. 2007. Comparative toxicity studies of NSAIDs in birds: A criticism of Reddy et al. *Environmental Toxicology and Pharmacology*, 23 254–25.

Cuthbert, R. J., Taggart, M. A., Saini, M., Sharma, A., Das, A., Kulkarni, M. D., Deori, P., Ranade, S., Shiringarpure, R. N, Galligan T. N., & Green R. E. 2016. Continuing mortality of vultures in India associated with illegal veterinary use of diclofenac and a potential threat from nimesulide. *Oryx*, 50(1): 104-112. https://doi.org/10.1017/S003060531500037X.

Cuthbert, R. J., Dave, R., Chakraborty, S. S., Kumar, S., Prakash. S., Ranade, S. P. & Prakash, V. 2011. Assessing the ongoing threat from veterinary nonsteroidal anti-inflammatory drugs to Critically Endangered Gyps vultures in India. *Oryx*, 45: 420–426.

Cuthbert, R. J., Green, R. E., Ranade, S., Saravanan, S., Pain, D., Prakash, V. & Cunningham, A. A. 2006. Rapid population declines of Egyptian Vulture (*Neophron percnopterus*) and Red-headed Vulture (*Sarcogyps calvus*) in India. *Animal Conservation*, 9: 349 – 354.

Das, S., Chockalingam, J., Mondal, S. & Sharma R. 2018. Geospatial Modelling of Human-Elephant Conflicts in Dalma Wildlife Santuary and its Surroundings in India. In Campbell M. O. (Ed), *Geomatics and Conservation Biology*, New York, NY USA, Nova Science Publishers, Inc.

DeVault, T. L., Beasley J. C., Olson Z. H., Moleon M., Carrete M., Margalida A., & Sanchez-Zapata, J. A. 2016. Ecosystem Services Provided by Avian Scavengers. In Şekercioglu, C. H., Wenny, D. G. & Whelan, C. J. (Ed), *Why Birds Matter: Avian Ecological Function and Ecosystem Services*. Chicago: University of Chicago Press.

Dickman, A. J. 2010. Complexities of conflict: the importance of considering social factors for effectively resolving human-wildlife conflict. *Animal Conservation*, 13: 458–466. https://doi.org/10.1111/j.1469-1795.2010.00368.x.

Donazar, J. A., Margalida, A., Carrete, M. & Sanchez-Zapata, J. A. 2009. Too sanitary for vultures. *Science*, 326(5953): 664.

Duriez, O., Descaves, S., Gallais, R., Neouze, R., Fluhr, J. & Decante, F. 2019. Vultures attacking livestock: a problem of vulture behavioural change or farmers' perception? *Bird Conservation International*, 1-17. https://doi.org/10.1017/S0959270918000345.

Efrat, R., Hatzofe, O. & Berger-Tal, O. 2020. Translating large-scale prioritization models for vultures to local-scale decision making. *Conservation Biology*, Volume 0, No. 0, 1–3. https://doi.org/10.1111/cobi.13557.

Fleming, T. R. 2010. Clinical Trials: Discerning Hype from Substance. *Annals of Internal Medicine*, 153(6): 400–406. https://doi.org/10.7326/0003-4819-153-6-201009210-00008.

Gadhvi, I. R. & Dodia P. P. 2006. *Indian White-backed Vultures Gyps bengalensis nesting in Mahuva*, Bhavnagar district, Gujarat, India. Indian Birds, 2(2):36.

Galligan, T. H., Amano, T., Prakash, V. M., Kulkarni, M., Shringarpure, R., Prakash, N., Ranade, S., Green, R. E. & Cuthbert, R. J. 2014. Have population declines in Egyptian Vulture and Redheaded Vulture in India slowed since the 2006 ban on veterinary diclofenac? *Bird Conservation International / FirstView Article* / April 2014, pp 1–10. https://doi.org/10.1017/S0959270913000580.

Galushin, V. M. 2001. Population of vultures and other raptors in Delhi and neighbouring areas from 1970s to 1990s. In Katzner T. & Pary-Jones J. (Eds), *Reports from workshop on Indian Gyps vultures, 4th Eurasian Congress on raptors.* Sevilla, Spain.

Gbogbo, F., Roberts, J. S. T. & Awotwe-Pratt, V. 2016. Some Important Observations on the Populations of Hooded Vultures *Necrosyrtes monachus* in Urban Ghana. *International Journal of Zoology*, 2016, ID 7946172. http://dx.doi. org/10. 1155/2016/7946172.

Genero, F., Franchini, M., Fanin, Y., & Filacorda, S. 2020. Spatial ecology of non-breeding Eurasian Griffon Vultures *Gyps fulvus* in relation to natural and artificial food availability. *Bird Study*, https://doi.org/10.1080/00063657.2020.1734534.

Gilbert, M., Watson, R. T., Virani, M. Z., Oaks J. L., Ahmed, S., Chaudhry, M. J. I., Arshad, M., Mahmood, S., Ali, A. & Khan, A. A. 2007. Neck-Drooping Posture in Oriental White-Backed Vultures (*Gyps bengalensis*): An Unsuccessful Predictor of Mortality and its Probable Role in Thermoregulation. *Journal of Raptor Research*, 41(1): 35–40.

Gomez-Ramirez, P., Jiminez-Montalban, P. J., Delgado, D., Martinez-Lopez, E., Maria-Mojica, P., Godino, A. & Garcia-Fernandez, A. J. 2018. Development of a QuEChERS method for simultaneous analysis of antibiotics in carcasses for supplementary feeding of endangered vultures. *Science of the Total Environment*, 626: 319-327. https://doi.org/10.1016/j.scitotenv.2018.01.060.

Green, R. E., Newton, I., Shultz, S., Cunningham, A. A., Gilbert, G., Pain, D. J. & Prakash, V. 2004. Diclofenac poisoning as a cause of vulture population declines across the Indian subcontinent. *Journal of Applied Ecology*, 41: 793–800.

Hertel, F. 1994. Diversity in body-size and feeding morphology within past and present vulture assemblages. *Ecology*, 75: 1074-1084.

Houston, D. C. & Cooper, J. E. 1975. The digestive tract of the white back griffon vulture and its role in disease transmission among wild ungulates. *Journal Wildlife Diseases*, 11:306–313.

IUCN (2010) *IUCN Red List of Threatened Species*. Version 2010. 1., http://www. iucnredlist.org.

Jha, K. K. 2015. Distribution of vultures in Uttar Pradesh, India. *Journal of Threatened Taxa*, 7(1): 6750-6763.

Jha, K. K. 2018. Mapping and management of vultures in an Indian stronghold. In Campbell, M. O. (Ed), *Geomatics and Conservation Biology*, New York, NY: USA, Nova Science Publishers Inc.

Jha, K. K., Campbell, M. O. &, Jha R. 2020. Vultures, their population status and some ecological aspects in an Indian stronghold. *Notulae Scientia Biologicae*, 12(1): 124-142. https://doi.org/10.15835/nsb12110547.

Jha, K. K. & Jha, R. 2020. Habitat suitability mapping for migratory and resident vultures: A case of Indian stronghold and species distribution model. *Journal of Wildlife and Biodiversity*, 4(3): 91-111. https://doi.org/10.22120/jwb.2020.120246.1111.

Jha, K.K., Jha, R. & Campbell, M.O. 2021.The Distribution, Nesting Habits and Status of Threatened Vulture Species in Protected Areas of Central India. *Ecological Questions* 32(2021)2. http://dx.doi.org/10.12775/EQ.2021.020.

Kane, A., Jackson, A. L., Monadjem, A., Colomer, M. A. & Margalida, A. 2014. Carrion ecology modelling for vulture conservation: are vulture restaurants needed to sustain the densest breeding population of the African white-backed vulture? *Animal Conservation*. https://doi.org/10.1111/acv.12169.

Khatri, P. C. 2013. Home range use of winter migratory vultures in and around Jorbeer, Bikaner (Rajasthan) India. *Bioscience Discovery*, 4(1): 96-99.

Kushwaha, S. & Kanaujia, A. 2011. Observations on Migratory and Residential vultures in Jorbeer, Rajasthan, India. Asian Journal of Experimental. *Biology and Science* 2(3): 404-411.

Kushwaha, S., Kanaujia, A. & Basu, D. 2009. Nature's scavengers: Vultures. *Research Environment Life Science*, 2:185–192.

Lambertucci, S.A., Margalida, A., Speziale, K.L., Amar, A., Ballejo, A., Bildstein, K.L., Blanco, G., Botha, A.J., Bowden, C.G.R., Cortés-Avizanda, A., Duriez, O., Green, R.E., Hiraldo, F., Ogada, D., Plaza, P., Sánchez-Zapata, J.A., Santangeli, A., Selva, N., Spiegel, O. & Donázar, J.A. 2021. Presumed killers? Vultures, stakeholders, misperceptions, and fake news. *Conservation Science and Practice*. 2021;e415,1-10. https://doi.org/10.1111/csp2.415.

Margalida, A. 2008. Bearded vultures (*Gypaetus barbatus*) prefer fatty bones. *Behavioural Ecology and Sociobiology*, 63:187–193.

Margalida, A. 2010. Supplementary feeding during the chick-rearing period is ineffective in increasing the breeding success in the bearded vulture (*Gypaetus barbatus*). *European Journal of Wildlife Research*, 56: 673–678.

Margalida, A., Donazar, J. A., Bustamante, J., Hernandez, F., & Romero-Pujante, M. 2008. Application of a predictive model to detect long-term changes in nest-site selection in the bearded vultures: conservation in relation to territory shrinkage. *Ibis*, 150: 242–249.

Margalida A. & Ogada D. 2018. Old World Vultures in a Changing Environment. In: Sarasola J., Grande J., Negro J. (Eds) *Birds of Prey*. Springer, Cham. https://doi.org/10.1007/978-3-319-73745-4_19.

Margalida, A. & Colomer, M. A. 2012. Modelling the effects of sanitary policies on European vulture conservation. *Scientific Reports*, 2:753.

Margalida, A. & Donazar, J. A. 2020. Fake news and vultures. *Nature Sustainability*. 2p. https://doi.org/10.1038/s41893-020-0534-5.

Margalida, A. 2017. Importance of Long-Term Studies to Conservation Practice: The Case of the Bearded Vulture in the Pyrenees. In Catalan, J. (Ed.), High Mountain Conservation in a Changing World, *Advances in Global Change Research*, 343-383. doi:10.1007/978-3-319-55982-7_15.

Margalida, A., Campion, D. & Donazar, J. A. 2011. European vultures' altered behaviour. *Nature*, 480:457.

Margalida, A., Garcia, D. & Cortes-Avizanda, A. 2007. Factors influencing the breeding density of Bearded Vultures, Egyptian Vultures and Eurasian Griffon Vultures in Catalonia (NE Spain): management implications. *Animal Biodiversity and Conservation,* 30(2): 189-200.

Marin, C., Torres, C., Marco-Jiminez, F., Cerda-Cuellar, M., Sevilla, S., Ayats, T. & Vega, S. 2018. Supplementary feeding stations for conservation of vultures could be an important source of monophasic *Salmonella typhimurium. Science of the Total Environment*, 636: 449-455. https://doi.org/10.1016/j.scitotenv.2018.04.310.

Markandya, A., Taylor, T., Longo, A., Murty, M. N., Murty, S. & Dhavala, K. 2008. Counting the cost of vulture decline: An appraisal of the human health and other benefits of vultures in India. *Ecological Economics*, 67:194–204.

Martinez-Abrain, A., Tavecchia, G., Regan, H. M., Jimenez, J., Surroca, M. & Oro, D. 2012. Effects of wind farms and food scarcity on a large scavenging bird species following an epidemic of bovine spongiform encephalopathy. *Journal of Applied Ecology*, 49:109–117. https://doi.org/10.1111/j.1365-2664.2011.02080.x.

Mateo-Tomas, P. & Olea, P.P. 2010a. Anticipating Knowledge to Inform Species Management: Predicting Spatially Explicit Habitat Suitability of a Colonial Vulture Spreading Its Range. *PLoS ONE* 5(8): e12374. doi:10.1371/journal.pone. 00123 74.

Mateo-Tomas, P. & Olea, P. P. 2010b. When hunting benefits raptors: a case study of game species and vultures. *European Journal of Wildlife Research*, 56: 519–528.

McKean, S., Mander, M., Diederichs, N., Ntuli, L., Mavundla, K., Williams, V. & Wakelin, J. 2013. The impact of traditional use on vultures in South Africa. *Vulture News*, 65: 15-36. http://dx.doi.org/10.4314/vulnew.v65i1.2.

Mendoza, M. L. Z., Roggenbuck, M., Vargas, K. M., Hansen, L. H., Brunak, S., Gilbert, M. T. P. & Sicheritz-Ponten, T. 2018. Protective role of the vulture facial skin and gut microbiomes aid adaptation to scavenging. *Acta Veterinaria Scandinavica* (2018), 60:61. https://doi.org/10.1186/s13028-018-0415-3.

Moleon, M., Sanchez-Zapata, J. A., Margalida, A., Carrete, M., Donazar, J. A., and Owen-Smith, N. 2014. Humans and scavengers: The evolution of interactions and ecosystem services. *BioScience*, 64: 394–403.

Mora-Carreño, M., Guerra-Correa, C., Moroni, D. & Paredes, E. 2019. Avian pox in a turkey vulture (*Cathartes aura*) from northern Chile. *Australian Journal of Veterinary Science*, 51: 41-43.

Morales-Reyes, Z., Perez-Garcia, J. M., Moleon, M., Botella, F., Carrete, M., Lazcano, C., Moreno-Opo, R., Margalida, A., Donazar, J. A. & Sanchez-Zapata, J. A. 2015. Supplanting ecosystem services provided by scavengersraises greenhouse gas emissions. *Scientific Reports*, 5:7811.

Moreno-Opo, R., Margalida, A., Arredondo, A., Guil, F., Martın, M., Higuero, R., Soria, C. & Guzman J. 2010. Factors influencing the presence of the cinereous vulture Aegypius monachus at carcasses: food preferences and implications for the management of supplementary feeding sites. *Wildlife Biology*, 16(1): 25-34. https://doi.org/10.2981/09-037.

Moreno-Opo, R., Trujillano, A. & Margalida, A. 2020. Larger size and older age confer competitive advantage: dominance hierarchy within European vulture guild. *Scientific Reports*, 10: 2430. https://doi.org/10.1038/s41598-020-59387-4.

Moreno-Opo, R., Trujillano, A., Arredondo, A., González, L. M. & Margalida, A. 2015. Manipulating size, amount and appearance of food inputs to optimize supplementary feeding programs for European vultures. *Biological Conservation*, 181: 27–35.

Movalli, P., Krone, O., Osborn, D. & Pain, D. 2018. Monitoring contaminants, emerging infectious diseases and environmental change with raptors, and links to human health, *Bird Study*, 65: xx-xx. https://doi.org/10.1080/00063657. 2018.1506735.

Narasimhaiah, M. G., James, A. & Kalaburgi R. A. 2020. Eco-pharmacology: awareness and practice of drug disposal among medical students. *International Journal of Basic & Clinical Pharmacology*, 9 (7): 1007-1010.

Oaks, J. L., Gilbert, M., Virani, M. Z., Watson, R. T., Meteyer, C. U., Rideout, B. A., Shivaprasad, H. L., Ahmed, S., Chaudry, M. J. I., Arshad, M., Mahmood, S., Ali, A. & Khan, A. A. 2004. Diclofenac residues as the cause of population decline of vultures in Pakistan. *Nature*, 427: 630–633.

Ogada, D. L., Torchin, M. E., Kinnaird, M. F. & Ezenwa, V. O. 2012. Effects of vulture declines on facultative scavengers and potential implications for mammalian disease transmission. *Conservation Biology*, 26:453–60.

Ohishi, I., Sakaguchi, G., Riemann, H., Behymer, D. & Hurvell, B. 1979. Antibodies to Clostridium botulinum toxins in free-living birds and mammals. *Journal of Wildlife Diseases*, 15:3–9.

Olea, P. P. & Mateo-Tomas, P. 2009. The role of traditional farming practices in ecosystem conservation: the case of transhumance and vultures. *Biological Conservation*, 142: 1844–1853.

Oro, D., Genovart, M., Tavecchia, G., Fowler, M. S. & Martınez-Abraın, A. 2013. Ecological and evolutionary implications of food subsidies from humans. *Ecology Letters*, 16: 1501–1514.

Oro, D., Margalida, A., Carrete, M., Heredia, R. & Donazar, J. A. 2008. Testing the Goodness of Supplementary Feeding to Enhance Population Viability in an Endangered Vulture. *PLoS ONE* 3(12): e4084. https://doi.org/10.1371/journal. pone.0004084.

Pacheco, R. E., Jacome, N. L., Astore, V., Borghi, C. E. & Piña C. I. 2020. Pesticides: The most threat to the conservation of the Andean condor (*Vultur gryphus*). *Biological Conservation* 242 (2020) 108418. https://doi.org/10.1016/j. biocon. 2020.108418.

Pain, D. J., Bowden, C. G. R., Cunningham, A. A., Cuthbert, R., Das, D., Gilbert, M., Jakati, R. D., Jhala, Y., Khan, A. A., Naidoo, V., Oaks, J. L., Parry-Jones, J.,

Prakash, V., Rahmani, A., Ranade, S. P., Baral, H. S., Senacha, K. R., Saravanan, S., Shah, N. Swan, G., Swarup, D., Taggart, M. A., Watson, R. T., Virani, M. Z., Wolter, K. & Green, R. E. 2008. The race to prevent the extinction of South Asian Vultures. *Bird Conservation International*, 18: S30–S48.

Pain, D. J., Cunningham, A. A., Donald, P. F., Duckworth, J. W., Houston, D. C., Katzner, T., Parry-Jones, J., Poole, C., Prakash, V., Round, P. & Timmins, R. 2003. Causes and effects of temporospatial declines of *Gyps* vultures in Asia. *Conservation Biology*, 17:661–71.

Paudel, K., Bhusal, K. P., Acharya, R., Chaudhary, A., Baral, H. S., Chaudhary, I. P., Green, R. E., Cuthbert, R. J., & Galligan, T. H. 2016. Is the population trend of the Bearded Vulture *Gypaetus barbatus* in Upper Mustang, Nepal, shaped by diclofenac? *Forktail*, 32: 54–57.

Philip, J. M., Aravind U. K. & Aravindakumar, C. T. 2018. Use of Antibiotics in Animals and Its Possible Impacts in the Environment. In: *Handbook of Research on Social Marketing and Its Influence on Animal Origin Food Product Consumption*. IGL Publisher. https://doi.org/10.4018/978-1-5225-4757-0.ch005.

Prakash, V., Bishwakarma, M. C., Chaudhary, A., Cuthbert, R., Dave, R., Kulkarni, M., Sashi K., Paudel, K., Ranade, S., Shringarpure, R. & Green R. E. (2012). The population decline of Gyps vultures in India and Nepal has slowed since veterinary use of diclofenac was banned. *PLoS One* 7(11). https://doi.org/10.1371/journal.pone.0049118.

Prakash, V. 1999. Status of Vultures in Keoladeo National Park, Bharatpur, Rajasthan, with Special Reference to Population Crash in *Gyps* Species. *The journal of the Bombay Natural History Society*, 96: 365-376.

Prakash, V. 2001. Status and distribution of vultures in India with special reference to population crash in *Gyps* species. In Katzner T. & Pary-Jones J. (Eds), *Reports from workshop on Indian Gyps vultures, 4th Eurasian Congress on raptors*. Sevilla, Spain.

Prakash, V., Galligan, T. H., Chakraborty, S. S., Dave, R., Kulkarni, M. D., Prakash, N., Shringarpure, R. N., Ranade, S. P. & Green, R. E. 2017, Recent changes in populations of Critically Endangered Gyps vultures in India. *Bird Conservation International*, 29(1): 55-70. https://doi.org/10.1017/S0959270917000545.

Prakash, V., Green, R. E., Pain, D. J., Ranade, S. P., Saravanan, S., Prakash, N., Venkitachalam, R., Cuthbert, R., Rahmani, A. R. & Cunningham, A. A. 2007. Recent changes in populations of resident *Gyps* vultures in India. *Journal of the Bombay Natural History Society*, 104(2): 129-135. https://doi.org/10.1017/S0959270917000545.

Prakash, V., Pain, D. J., Cunningham, A. A., Donald, P. F., Prakash, N., Verma, A., Gargi, R., Sivakumar, S. & Rahmani A. R. (2003). Catastrophic collapse of Indian White Backed (*Gyps bengalensis*) and Long Billed (*Gyps indicus*) Vultures population. *Biological Conservation*, 109(3): 381-390.

Reddy, N. C. P., Anjaneyulu, Y., Sivasankari B. & Rao K. A. 2006. Comparative toxicity studies in birds using nimesulide and diclofenac sodium. *Environmental Toxicology and Pharmacology*, 22: 142–147.

Richards, N. L., Gilbert, M., Taggart, M. & Naidoo, V. 2017. A Cautionary Tale: Diclofenac and Its Profound Impact on Vultures. *Encyclopedia of the Anthropocene pp 1-10.* http://dx.doi.org/10.1016/B978-0-12-409548-9.09990-5.

Roggenbuck, M., Schnell, I. B., Blom, N., Baelum, J., Bertelsen, M. F., Sicheritz-Ponten, T., Sorensen, S. J., Gilbert, M. T., Graves, G. R. & Hansen, L. H. 2014. The microbiome of New World vultures. *Nature Communications*, 5:5498.

Saengtienchai, A., Ikenaka, Y., Kawata, M., Kawai, Y., Takeda, K., Kondo, T., Bortey-Sam, N., Nakayama, S. M. M., Mizukawa, H. & Ishizuka M. 2018. Comparison of xenobiotic metabolism in phase I oxidation and phase II conjugation between rats and bird species. *Comparative Biochemistry and Physiology Part C Toxicology & Pharmacology*, 214: 28-35. https://doi.org/10. 1016/j.cbpc.2018.08.007.

Saini, M., Taggart, M.A., Knopp, D., Upreti, S., Swarup, D., Das, A., Gupta, P. K., Niessner, R., Prakash, V., Mateo, R. & Cuthbert, R. J., 2012. Detecting diclofenac in livestock carcasses in India with an ELISA: A tool to prevent widespread vulture poisoning, *Environmental Pollution* 160: 11-16.

Samson, A., Ramakrishnan, B., Kannan, G., Veeramani, A. & Ramasubramanian, S. 2018. Conservation Threats on Critically Endangered *Gyps* Vultures in the Tamil Nadu Part of the Nilgiri Biosphere Reserve, Southern India. In Sivaperuman, C. &Venkataraman K. (Eds). *Indian Hotspots Vertebrate Faunal Diversity, Conservation and Management Volume 2*. Springer Nature Singapore Pte Ltd.

Santangeli, A., Girardello, M., Buechley, E., Botha, A., Minin, E. D. & Moilanen, A. 2019. Priority areas for conservation of Old World vultures. *Conservation Biology, Volume 0*, No. 0, 1–10. https://doi.org/10.1111/cobi.13282.

Schabo, D. G., Heuner, S., Neethling, M. V., Rosner, S., Uys, R. & Farwig, N. 2016. Long-term data indicates that supplementary food enhances the number of breeding pairs in a Cape Vulture *Gyps coprotheres* colony. *Birdlife International*, 140-152. https://doi.org/10.1017/S0959270915000350.

Sekercioglu, C. H. 2006. Ecological significance of bird populations. *Handbook of the Birds of the World* 11: 15–51.

Shultz, S., Baral, H. S., Charman, S., Cunningham, A. A., Das, D., Ghalsasi, G. R., Goudar, M. S., Green, R. E., Jones, A., Nighot, P., Pain, D. J. & Prakash, V. 2004. Diclofenac poisoning is widespread in declining vulture populations across the Indian subcontinent. *P. Roy. Soc. Lond. B (Suppl.)* 271: S458–S460.

Singh, G. G., McKechnie, I., Braje, T. J. & Campbell, B. 2015. All models are wrong but some are useful: A response to Campbell's comment on estimating *Mytilus californianus* shell size. *Journal of Archaeological Science* 63: 160e163.

Stone LM. 1993. *Vultures*. Minneapolis, Minnesota, USA Carolrhoda Books.

Taggart, M. A., Senacha, K. R., Green, R. E., Jhala, Y. V., Raghavan, B., Rahmani, A. R., Cuthbert, R., Pain, D. J., & Meharg, A. A., 2007. *Diclofenac residues in carcasses of domestic ungulates available to vultures in India. Environment International*, 33:759-765.

Thakur, M. L. & Kataria, R. C. 2012. Breeding records and nest site preference of Indian White-backed vulture in Kangra valley of Himachal Pradesh, India. *International Journal of Science and Nature*, 3(2): 350-353.

Thakur, M. L. & Narang, S. K. 2012. Population status and habitat-use pattern of Indian white-backed vulture (*Gyps bengalensis)* in Himachal Pradesh, India. *Journal of Ecology and the Natural Environment*, 4(7):173-180.

Virani, M., Gilbert, M. & Watson R. 2001. Asian vulture crisis project: Field results from Pakistan and Nepal for the 2000-2001 field season. In Katzner T. & Pary-Jones J. (Eds), *Reports from workshop on Indian Gyps vultures. 4th Eurasian Congress on raptors.* Sevilla, Spain.

Watson, RT., Gilbert, M. & Virani, M. 2008. Neck-Drooping Posture of Oriental White-Backed Vultures (*Gyps bengalensis*) In Close Proximity to Human Observers. *Journal of Raptor Research*, 42(1):66–67.

Yarnell, R. W., Phipps, W. L., Dell, S., MacTavish, L. M. & Scot, D. M. 2014. Evidence that vulture restaurants increase the local abundance of mammalian carnivores in South Africa. *African Journal of Ecology*, 53(3): 287-294. https://doi. org/10.1111/ aje.12178.

Zuberogoitia, I., Martínez, J. E., Margalida, A., Gómez, I., Azkona, A. & Martínez, J. A. 2010. Reduced food availability induces behavioural changes in Griffon Vulture *Gyps fulvus. Ornis Fennica*, 87:52–60.

Zuluaga, S., Salom, A., Vargas, F. H., Coulson, J. O., Kohn, S. & Grande, J. M. 2020. Acknowledging Andean Condor predation on livestock, a first step in addressing the human-condor conflict: A commentary to Estrada Pacheco et al. (2020). *Biological Conservation*, 247 (2020) 108618. https://doi.org/10.1016/j. biocon.2020.108618.

In: Critical Research Techniques in Animal …
Editors: Michael O'Neal Campbell et al.
ISBN: 978-1-53619-846-1

Chapter 3

BROWN WORLD OF VULTURES: A CASE OF VULTURE RESTAURANT

Kaushalendra Kumar Jha[1,*] and Radhika Jha[2]

[1]Indian Institute of Forest Management, Bhopal, India
[2]Zoology Department, Lucknow University, Lucknow, India

ABSTRACT

Vultures are unique organisms and one of the most important components of an ecosystem. Old and New World vultures occupy at least five continents and provide ecosystem services as obligate scavengers. Most vulture species are endangered or threatened throughout the world as they depend on the brown world comprising the dead diversity of mammals, primarily ungulates which are themselves dwindling. Even available carcasses on traditional or modern feed stations may be infested with contaminants or competitors. Although supplementary feeding stations are known to augment vulture populations, they have many side effects sabotaging conservation efforts in terms of reduced fecundity, changed social behaviour and even life threats. The concept of mapping Vulture Restaurants (VR) has been developed and strength, threat, opportunity, weakness and suitability class maps of VR are suggested. However, strengthened management based on research results may reduce the gap in existing scholarship and achieve the objective of protecting vultures for developed and efficient services.

* Corresponding Author's E-mail: jhakk1959@gmail.com.

INTRODUCTION

Many populations of raptors around the world, including vultures, have substantially declined or currently exist at levels that merit population augmentation. This is defined as increasing the population of a species whose numbers have reduced considerably (Barclay, 1987; Stattersfield & Capper, 2000; Negro et al., 2007). Vultures are long-lived scavenging raptors which depend on the availability of medium and large carcasses. The shortage of safe food resources, among others, is the main causative factors leading to the decline of scavengers in Europe, Africa and Asia over the last hundred years (Cramp & Simmons, 1980; Sarrazin et al., 1994; Carrete et al., 2009 & 2011; Olea & Mateo-Tomás, 2009; Tellería, 2009; García- Ripollés & López-López, 2011; Zuberogoitia et al., 2013).

Avian scavengers are part of the detrital food web of ecosystems. They provide the important ecological service of recycling carrion biomass to prevent the accumulation of dead biomass, thereby contributing to waste removal, disease regulation, and nutrient cycling (DeVault et al., 2003; Margalida & Colomer, 2012). Supplementary food provisioning to vultures is only effective when tailored to the feeding ecology of all species involved, particularly because differences in diet and behavioural dominance lead to unequal benefits between species and age classes (Moreno-Opo et al., 2015). Carcass or carrion ecology and its role in the feeding of wild scavengers is a field of study that has gained special relevance in recent years (Barton et al., 2013; Oro et al., 2013; Fielding et al., 2014), particularly from the perspective of conservation management (Moreno Opo et al., 2015). The carcass or animal detritus which provides nutrients to obligatory scavengers like vultures is known as the brown world.

In many places, traditional sources of food for many populations of vultures and other scavengers have declined dramatically in the last 100 years. Wild ungulates that once provided food for vultures in western U.S., Africa, and Asia, are now absent or severely diminished in many places (Mundy et al., 1992; Negro et al., 2007) even in protected areas (Campbell, 2015). Therefore, the recent management strategy of the brown world has developed into Vulture Restaurants (VR) or vulture feeding stations to augment the shrinking population of vultures in almost all the vulture inhabiting continents. However, VR strategy needs careful attention as it has plenty of side effects due to inadvertent management errors. This review, based on the screening of relevant literature from reputed journals, books and print media, covers different VR related topics such as, vulture feeding habits, trophic sources, nomenclature issues, needs of feeding stations, risks and threats, mapping and methodologies used for this review.

Vultures (Table 1) are a strongly disliked and highly ignored group of birds due to their food habits which are perceived to be unclean and related to the dead and people's ignorance about the ecosystem services they provide. As specialised obligate scavengers, vultures eat dead animals, even human beings, providing valuable ecosystem services by contributing to nutrient cycling and potentially limiting the spread of disease among humans and animals (DeVault et al., 2003; Sekercioglu, 2006; Yarnell et al., 2014). Vultures occur on five continents; Africa, Asia, Europe, and North and South America but interspecies differences are minimal. There are two main groups of vultures; the Accipitrid or Old-World vultures, which are classified with the predatory birds of the order *Falconiformes*; and the *Cathartid* or New World vultures of the Americas, which are closely or distantly related to the storks of the order *Ciconiiformes* (Bonaparte, 1854).

Among each of these two large groupings, there are three main size grades; the huge *Cathartid* condors and the larger *Accipitrid* Griffons and Cinereous vultures; the medium size *Cathartid* King Vulture and the *Accipitrid* Indian, Slender-billed, White-backed and White-rumped vultures; and finally, the smaller *Cathartid* Turkey and Black vultures and *Accipitrid* Egyptian and Hooded Vultures (Campbell in Jha, 2017) (Table 1). Ecologically, they can also be categorised broadly as forest, savanna, desert, and mountain vultures as these regions have different biomes with varying animals and plants playing important roles for vultures providing food (Campbell, 2015) beyond SFS as it is supplementary and in some cases post SFS care e.g., closure of SFS after rehabilitation or population augmentation and demise of SFS due to sanitary regulation in Europe (Donazar et al., 2010).

Del Hoyo et al., (1994), Botha et al., (2017) and others have studied the feeding habits of different Old-World vultures. For example, the Bearded Vulture consumes prey remains left by predators or other scavengers. Its diet consists of 70% bones, 25% of soft tissue and 5% of skin (Hiraldo et al., 1979).

Table 1a. Cathartid (New World) Vulture species ranked in approximate order of size

Species (English name)	Scientific name	Length (cm)	Wingspan (m)	Weight (kg)
Andean Condor	*Vultur gryphus* (Linnaeus, 1758)	100-130	2.7-3.2	8-15
California Condor	*Gymnogyps californianus* (Shaw, 1797)	109-140	2.5-3	7-14.1
King Vulture	*Sarcoramphus papa* (Linnaeus, 1758)	67-81	1.2-2	2.7–4.5
Greater Yellow-headed Vulture	*Cathartes melambrotus* (Wetmore, 1964)	64-75	1.6–1.8	1.7
Turkey Vulture	*Cathartes aura (Linnaeus, 1758)*	62-81	1.6–1.8	1–2.3
Yellow-headed Vulture	*Cathartes burrovianus (Cassin, 1845)*	53-66	1.5–1.7	1–1.6
Black Vulture	*Coragyps atratus (Bechstein, 1793)*	56-74	1.3–1.7	1.2–1.9

Source: Campbell (2015).

Table 1b. Accipitrid (Old World) Vulture species ranked in approximate order of size

Species (English name)	Scientific name	Length (cm)	Wingspan (m)	Weight (kg)
Cinereous Vulture	*Aegypius monachus* (Linnaeus, 1766)	98–120	2.5-3.1	6.3-14
Himalayan Vulture	*Gyps himalayensis* (Hume, 1869)	109 to 115	2.6-3.1	8-12
Lappet-faced Vulture	*Torgos tracheliotos* (Forster, 1791)	78–115	2.5-2.9	4.4.-13.6
Griffon Vulture	*Gyps fulvus* (Hablitz, 1783)	93-122	2.3–2.8	6.2–11.3
Bearded Vulture	*Gypaetus barbatus* (Linnaeus, 1758)	94-125	2.3–2.8	4.5–7.8
Cape Griffon Vulture	*Gyps coprotheres* (Forster, 1798)	96-115	2.3–2.6	7-11
Ruppell's Griffon Vulture	*Gyps rueppelli* (Alfred Brehm, 1852)	85-103	2.3–2.6	6.4-9
White-headed Vulture	*Trigonoceps occipitalis* (Burchell, 1824)	72-85	2.1–2.3	4–4.7
Red-headed Vulture	*Sarcogyps calvus* (Scopoli, 1786)	76-86	2–2.6	3.5–6.3
Indian Vulture	*Gyps indicus* (Scopoli, 1786)	80-103	2–2.4	5.5–6.3
Slender-billed Vulture	*Gyps tenuirostris* (Gray, 1844)	80-103	2–2.4	5.5–6.3
White-backed Vulture	*Gyps africanus* (Salvadori, 1865)	78-98	2–2.3	4.2–7.2
White-rumped Vulture	*Gyps bengalensis* (Gmelin, 1788)	75-93	1.8–2.6	3.5–7.5
Hooded Vulture	*Necrosyrtes monachus* (Temminck, 1823)	62-72	1.6–1.7	1.2–2.6
Egyptian Vulture	*Neophron percnopterus* (Linnaeus, 1758)	47-65	1.6–1.7	1.9–2.4
Palm-nut Vulture	*Gypohierax angolensis* (Gmelin, 1788)	60	1.5	1.2–1.5

Source: Campbell (2015).

The Egyptian Vulture forages in lowland and montane regions over open, often arid, country. Being an opportunistic scavenger, it also scavenges at human settlements. Its broad diet includes carrion of livestock, domestic chicken, tortoise, organic waste, insects, young vertebrates, eggs and faeces (Margalida et al., 2012; Dobrev et al., 2015 & 2016). The White-headed Vulture is often the first vulture species to arrive at a carcass. It is known to sometimes snatch food from other vulture species, consuming it nearby (Mundy et al., 1992). The Hooded Vulture feeds on carrion in the countryside, but in urban areas it congregates at slaughterhouse disposal sites and rubbish dumps.

The Himalayan griffon soars over large areas often with other vultures in search of carcasses. Small numbers attend carcasses which can be consumed rapidly. It is dominant over other vulture species except Black Vulture. White-rumped Vulture feeds exclusively on carrion and often associates with other vulture species when scavenging at rubbish dumps and slaughterhouses. White-backed Vulture feeds on carrion and bone fragments of larger carcasses, mainly soft muscle and organ tissue. Indian Vultures are generally heavily dependent on animal (especially cattle) and human remains. The Slender-billed Vulture, previously considered a sub-species of Indian vultures, is known to adapt well to supplementary food provided at vulture restaurants. Cape Vulture specialises in larger carcasses, mainly soft muscle and organ tissue. Ruppell's Vulture feeds on both carrion and bone fragments of larger carcasses and rarely comes down to small carrion. It mainly eats soft muscle and organ tissue. It follows other vultures and migrant game or stock herds to locate much of its food (Del Hoyo et al., 1994). Eurasian Griffon feeds almost exclusively on carrion of medium-sized and large domestic and wild animals. Cinereous Vulture's diet consists mainly of carrion from medium-sized or large mammal carcass, although snakes and insects have also been recorded as their food items.

The Brown World: Trophic Sources for Vultures

The world of living organisms in an ecosystem is considered as "green world" wherein producers, consumers and decomposers play an important role in energy transfer. In contrast to this, there exists a "brown world" which comprises of organic matter or organic detritus (Swift et al., 1979; Moore et al., 2004). Detritus is defined as dead organic matter of different origin (generally dead plants and animals) and quality in terms of chemical composition and assimilation efficiency by consumers. High-quality detritus is produced from carrion (Olea et al., 2019), a source of energy and nutrients to living organisms which are scavengers and predators (Schoenly & Reid, 1983; Braack, 1987; Schoenly, 1991; Parmenter & MacMahon, 2009; Wilson & Wolkovich, 2011; Barton et al., 2013) in most food webs (Moore et al., 2004). A vulture-centric "brown world" representing partial food web is depicted in Figure 1, a synthesis of various studies (Moore et al., 2004, Barton et al., 2013, Pereira & Novarro, 2015; Selva & Cortes-Avizanda, 2009).

Source of the brown world for vultures is mainly the carrion available in a particular area. Carrion is defined as a pulsed food resource of unpredictable occurrence in space and time that offers a high nutritive biomass but is not globally widespread across all habitats and territories. This can be considered free because it does not require a large physical investment derived from predation (Ostfeld & Keesing, 2000; DeVault et al., 2003; Selva & Fortuna, 2007; Barton et al., 2013; Moleón et al., 2014, Margalida & Ogada, 2018). Vultures, which evolved as obligate scavengers, prefer "freshly dead" carrion but they can consume decaying carcass also. Mammals serve as the predominant source of carrion, but carrion may also refer dead birds, reptiles, amphibians, and invertebrates. Vultures may share carrion sources with other scavengers, including groups of the same or different vulture species and facultative (temporary) scavengers such as dogs, common ravens, red kites, golden eagles etc. Some competitors, such as brown hyena, black-backed jackal, feral domestic dogs, baboons, bush-pigs etc. (Piper, 2004; Yarnell et al., 2014) may out-compete vultures for the same carcass.

In general, the Eurasian Griffon Vulture (*Gyps fulvus*, (Hablitz, 1783), the Egyptian Vulture (*Neophron percnopterus*, Linnaeus, 1758) and the Cinereous Vulture (*Aegypius monachus*, (Linnaeus, 1766) are considered as meat eaters while the bearded vulture (*Gypaetus barbatus,* Linnaeus, 1758), is a specialized bone-eater (Margalida & Colomer, 2012). Cramp & Simmons (1980) and del Hoyo et al., (1994) reported that the Egyptian Vulture has the ability to use a wide range of food sources in different parts of the species' range. Mammals are the most common group of vertebrates in the vultures' diet composition followed by reptiles and birds (Dobrev et al., 2015).

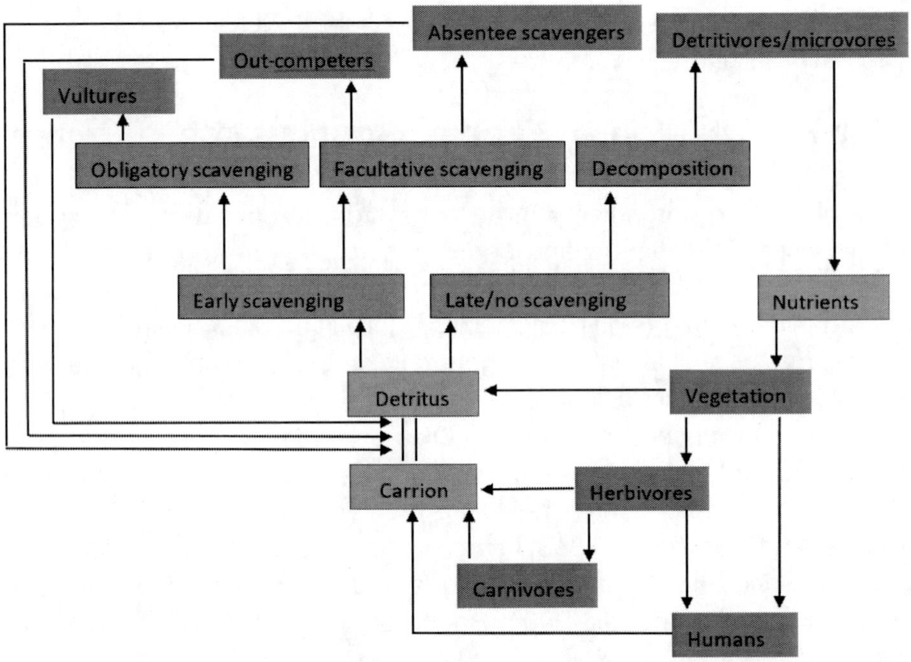

Brown, green and blue boxes represent living, dead and process components, respectively. Arrows are unidirectional connecters. Adopted and modified from Moore et al., (2004), Barton et al., (2013), Pereira & Novarro (2015), Selva & Cortes-Avizanda, 2009).

Figure 1. Diagrammatic representation of a conceptual model of "Vulture Trophic World."

However, the food of roosting Egyptian Vultures varies between European rabbits (*Oryctolagus cuniculus*, Linnaeus, 1758), pigs (*Sus scrofa domesticus*, Erxleben, 1777), sheep (*Ovis aries*, Linnaeus, 1758), and domestic fowl (*Gallus gallus domesticus*, Linnaeus, 1758) (Cebalos & Donazar, 1990). Margalida et al., (2011b) observed that Pyrenean chamois (*Rupicapra pyrenaica*, Bonaparte, 1845) red deer

(*Cervus elaphus*, Linnaeus, 1758) wild boar (*Sus scrofa*, Linnaeus, 1758) fallow deer (*Dama dama*, Linnaeus, 1758) mouflon (*Ovis orientalis*, Linnaeus, 1758) and roe deer *Capreolus capreolus*, Linnaeus, 1758) are the wild ungulates providing food to different vultures (Eurasian Griffon vulture, Cinereous Vulture, Egyptian Vulture, and Bearded Vulture) in the Pyrenees and pre-Pyrenees. African white-backed vultures (*Gyps africanus* Salvadori, 1865) used to eat elephants (*Loxodonta africana*, Blumenbach, 1797) in Zimbabwe (Mundy et al., 1992; Monadjem & Garcelon, 2005) and impala (*Aepyceros melampus*, Lichtenstein, 1812), blue wildebeest (*Connochaetes taurinus*, Burchell, 1823), zebra (*Equus quagga burchellii*, Gray, 1824) and 21 other species in Swaziland (Kane et al., 2014). The diet of Cinereous Vulture consists mainly of carrion from medium-sized or large mammal carcasses, although snakes and insects too have been recorded as food items (Botha et al., 2017).

Vultures' brown world exists in nature in the form of wild as well as domestic ungulates supplying unpredictable flows of food as carcasses to the obligate scavengers along with some other consumers. However, anthropogenic sources (animal husbandry) also supply nutrients to vultures. These are somewhat predictable sources, human arranged carcasses and products of livestock at predefined or established places for the specific goal of conservation, i.e., saving endangered species by augmenting the population, enhancing breeding efficiency etc. Such places are named variously in conservation ecology as Vulture Restaurants, Supplementary Feeding Sites, Muladares etc. (Campbell, 2015).

What Is There in the Name

Many senior citizens from the northern part of the Indian subcontinent, while recalling their childhood days, may remember the barren areas at the outskirts of villages known as *Khalar,* where cattle carcasses were dumped after their skin was removed and eventually turned into leather. The exposed carrion was left to be consumed by the waiting vultures. Such dumping sites appear comparable to the "*Muladares*" in Spanish which could be defined as traditional sites where livestock carrion has been intentionally discarded for elimination by scavenger species for centuries (Margalida et al., 2007; Martinez-Abrain et al., 2012). The English equivalent of "*Muladares*," popular in the scientific world, is "Vulture Restaurant (VR)," especially so in non-English speaking countries where vultures are a common sight. The phrase refers to a vulture related eating place created by human beings. With explanation the term makes sense, but it can also be confusing to a layperson. To the uninitiated it could either be the name of a restaurant where vulture is served as a meal or a place where meals are served to vultures. With this confusing understanding of the phrase, one could even argue that it is a misnomer.

Since VR is not dictionary defined, to further analyse this confusing term, the dictionary definition could be consulted separately (vulture and restaurant) which make things a little complicated. A search on dictionary.com defined vultures as "any of several large, primarily carrion-eating Old World birds of prey of the family Accipitridae, often having a naked head and less powerful feet than those of the related hawks and eagles" and "any of several superficially similar New World birds of the family Cathartidae, as the turkey vulture." Additionally, it defined a restaurant as "an establishment where meals are served to customers" or "a commercial establishment where meals are prepared and served to customers." However, no VR referred to in the literature is a commercial establishment, nor are vultures the customers. Different from the norms in restaurants for humans, in vulture restaurants the vultures are served only domesticated food (cattle reared by farmers) not exotic or wild (ungulates of wild origin or overseas animals). Somehow VR does not conform a sound description of the *Muladares*.

Nevertheless, the meaning defined in scientific literature clears the mist. As described by Piper (2004), a VR is a source of animal protein in fairly large parcels, often placed with some regularity. Mundy et al., (1992) defined it as a site at which supplementary food, often in the form of dead livestock, is placed at regular or frequent intervals. Moreno-Opo et al., (2010) further defined a VR as an alternative food source, which included the traditional *Muladares,* a place where livestock carcasses were traditionally dumped so that scavenger birds could eat and get rid of them.

Although, conservationists created VR in the late 1960s (first VR was opened in 1966; Piper, 2004; Amar et al., 2018), many research publications have referred to the same, more recently, as supplementary feeding sites or stations (SFS) meant to increase the availability of food resources in southern Europe and southern Africa (Bijleveld, 1974; Cortes-Avizanda et al., 2016, Marin et al., 2018; Brink et al., 2020; Moreno-Opo et al., 2020) except one (Oro et al., 2008) calling it as artificial feeding site. Since SFS does not attract any connotation related conflict, it appears to be a better option along with the continued use of VR (Margalida, 2017; Amar et al., 2018; Blanco et al., 2019). These terms have been used in this text interchangeably and frequently. Nevertheless, after going through a considerable amount of literature and considering the objectives of and practices at VR, the authors feel that "Vultures' Supplementary Feeding Station" should be more appropriate nomenclature.

SUPPLEMENTARY FEEDING SITES

Supplementary feeding sites/stations, VR or food bonanzas are a key conservation intervention for vultures in Europe, the United States of America and southern Africa (Gilbert et al., 2007; Cortés-Avizanda et al., 2010; Schabo et al., 2016). The main aim of SFS is to provide a regular supply of poison-free food to discourage vultures from feeding at uncontrolled sources (Schabo et al., 2016) frequently used for long-term population monitoring. Depending on the objectives of management, SFS could be of two types: (i) Population conservation SFS and (ii) Breeding improvement SFS (Ferrer et al., 2017). The former is aimed at feeding large numbers of birds in one place (Cortés-Avizanda et al., 2016). At such communal feeding stations, food provision usually takes place year-round, and over many successive years. Such an arrangement is aimed mainly to increase the overall population size. At the second type of SFS, feeding is targeted at a particular territorial pair within a population for some part of the year where food is placed close to the nest every day to reduce nestling deaths and increase clutch size. This technique has shown success in poor quality areas (less suitable habitat, low food availability, high human disturbance, etc.) where the pairs produced bigger broods.

Top left: Carrion consumption by vultures at Indian equivalent of *Muladares* or Vulture Restaurant with irregular supply of carcass in the vicinity of tropical moist deciduous forest. Top right: Carcass eating by the aggressive dogs in semiarid agriculturally dominant area with frequent disposal of carcass where vulture visitation is uncommon. A hypothesis is that a vulture restaurant should only be started in areas where vultures are indeed seen from time to time (Dasa, 2014) is truly reflected in these two locations/photographs. Bottom: community run fenced area or formal Vulture Restaurant which remain useful for initial 2 – 3 years then later left unused due to financial difficulty.

Figure 2. Community land used for disposal of dead cattle outside the village in Uttar Pradesh, north India.

Supplementary Feeding Stations are created for different purposes, including providing micronutrients, restoring declining vulture populations, managing scavenger reintroductions and reducing ingestion of toxic compounds like diclofenac (Indian sub-continent) and lead (North America) (Campbell, 2015; Cortes-Avizanda et al., 2016). Supplementary feeding is a common practice to raise reproductive output in raptors and other species e.g., California Condor (*Gymnogyps californianus*, Shaw, 1797), sparrow-hawk (*Accipiter nisus*, Linnaeus, 1758), common kestrel (*Falco tinnunculus*, Linnaeus, 1758), Spanish imperial eagle (*Aquila adalberti*, C. L. Brehm, 1861), Bearded Vulture etc. This practice is cost effective (8X) and time efficient when compared to captive breeding programs (Ferrer et al., 2017; Marin et al., 2018). Supplementary Feeding Stations can also be used to study unusual or different from normal behaviour (e.g., polyandrous trio formation in *Gypateus barbatus,* Margalida, 2017), survey types and number of vultures in an area (Hla et al., 2011) etc.

Box 1. Regulation and VR crisis in Spain/Europe

Spain is the home of 95% of European vultures (Morales-Rayes et al., 2015) with conservation priorities. There was a legal arrangement for the disposal of cattle carcass by burial during the second half of the twentieth century. But the traditional practices of open disposal at *Muladares* or throwing in the field, in turn providing food to vultures, continued in violation of the provision due to inconvenient and uneconomical way of disposal. Due to the Bovine Spongiform Encephalopathy (BSE) outbreak, this provision was further tightened at the beginning of twenty first century as the farmers were made to transport the cattle to the disposal plant and get them incinerated instead of local burial or dumping at *Muladares*. This provision was not only inconvenient and costly to the farmers but attracted criticism from environmentalists due also to high carbon foot printing. The lowest amount of carbon emitted per carcass per year was 0.16 kg to 0.63 kg of CO_2 (Dupont et al., 2012). The feeding station scenario produced no carbon emission at all.

Morales-Rayes et al. (2015) evaluated the impact of transportation and incineration vis a vis traditional disposal of carcass in Spain and concluded that supplanting the natural removal of dead livestock by scavengers with carcass collection and transport to intermediate and processing plants meant the emission of 77,344 metric tons of CO_2 eq. to the atmosphere per year, in addition to annual payments of ca. $50 million to insurance companies. Thus, replacing the ecosystem services provided by scavengers has not only conservation costs, but also important and unnecessary environmental and economic costs. However, for conservation of carrion eating birds in some EU countries like, Spain, France, Portugal, Italy and Greece the carcass incinerating provision was modified and negative BSE tested carcasses were to be supplied as vulture food. This provision created high discomfort among the shepherds as they were supposed to guard the carcass during the 48 hour period of BSE testing (Tella, 2001; Camina, 2004; Camina & Yosef 2012; Margalida et al., 2012).

However, the BSE crisis affected VR management severely, as almost all of them were closed overnight (Margalida & Ogada 2018). This resulted in scarcity and low- quality supply of food for vultures forcing them to change their territorial, guild, and foraging behaviours (Zuberogoitia et al., 2010; Camina & Yosef, 2012; Martinez-Abrain et al., 2012). The post BSE impact also resulted in an annual decrease in breeding success, reduced population growth, increased mortality of young age classes, changes in the diet, of Eurasian Griffon vulture and a reduction in egg quality of Bearded Vulture (Donázar et al., 2009a,b & 2010; Zuberogoitia et al., 2010; Margalida et al., 2011a, 2014a & b; Hernández et al., 2018).

Box 2. VR strategy in India/Subcontinent

The vulture crisis in the Indian subcontinent was reported with the threat of extinction of indigenous *Gyps* species. The reason of population crash was primarily assigned to diclofenac use in cattle. A serious step was taken by the government authorities to ban the use of veterinary formulations of the drug in order to protect vultures. But the abuse continued stealthily under the garb of untenable legal remedies against ban. Surprisingly, the establishment of VR which was popular in Africa and Europe, in particular Spain and South Africa, did not pickup in the subcontinent. An attempt is made in the following paragraphs to compile the status of VR in brief in India. There is scanty information on VR in South Asia in peer reviewed journals. The intention was not to do a compendium search, the authors sifted through roughly a hundred web pages on google search using the key phrase "vulture restaurants in India" to get a fair idea of the establishment of VR in numbers. This was actually low considering widely spread and considerable number of vulture species and individuals.

A vast country with a significant population of vultures, at least in the central and northern states of the country where indigenous vultures are in critically endangered to endangered category, has adopted VR, one of the most accepted strategies for vulture conservation in the world, with a half-hearted attempt. Establishment of VR, a fenced area with wooden machan; pucca platform; shallow trough or just plain field in the forest or outside a village on community land, in India picked up around the second decade of the 2000s and the first VR was established by the Forest Department of Maharashtra in 2011 in Phansad Wildlife Sanctuary (Doreen Fiedler 2015; Aisling Maria Cornin 2015). Then onwards VRs were opened in limited numbers in vulture areas of different states like, Maharashtra (Mallikarjuna 2012), Punjab (Dharmendra Rataul 2011), Himachal Pradesh (Archana Phull 2020), Uttar Pradesh (Sonika Kushwaha 2014), as per the staggered information available in the e-print media. However, VRs (not more than 30 in the country) in general have not been able to serve the purpose of feeding vultures for long term since the operations were closed after a few years due to the lack of continued flow of funds from the government (Gautam Dheer, 2015; Yudhvir Rana 2016; Sumita Sarkar 2017), but it has definitely propagated knowledge about the importance of vultures and their conservation.

Pakistan also opened VR, one each in Sindh and Punjab province, to protect vultures (Gilbert, 2007; Mariya Karimjee 2013). The Nepalese government established several VR with an aim to preserve and rebuild the species as well (DeCandido et al., 2012; Alexander Lerche 2017). They are also building a vulture safe zone (Galligan, 2019) along with Bangladesh, an area of 100 km radius free of any environmental threat, (Alam et al., 2016; Warren Cornwall 2018, www.sciencemag.org) which may enhance the utility of VR in the region, should it continue, unlike in India.

Supplementary feeding presents both economic advantages for farmers and ecological advantages for the society, therefore, it should be promoted in the regions where vultures are present instead of eliminating carcasses by using polluting industrial quartering plants (Cortés-Avizanda et al., 2016; Dupont et al., 2011; Morales-Reyes et al., 2015). The development and management of SFS should, however, aim at limiting the risk of routine behaviour and complete dependence of vultures on SFS (Fluhr et al., 2017). The high availability of food resources associated with supplementary feeding points allows a great number of non-breeding bearded vultures

(Margalida, 2017) which otherwise would be occupying different areas (Brown, 1997; Xirouchakis & Nikolakalis 2002; Hirzel et al., 2004; Margalida et al., 2011a) - to coexist within the spatial distribution of the breeding population. Thus, high concentrations of floaters around breeding territories - as happens near supplementary feeding points (Sesé et al., 2005; Margalida et al., 2011b) - may increase the time being spent in agonistic encounters and, therefore, reduce breeding success. Vulture Restaurants in Africa and/or Supplementary Feeding stations in Europe are the well-practiced long-term institution but in Asia, particularly in Indian subcontinent it is a recent and under-developed activity (Figure 2). Two case studies regarding status of VR/SFS and crisis impact in Spain (Europe) and India (Sub-continent) are presented in Box 1 and 2.

The high availability of food resources associated with supplementary feeding points allows a great number of non-breeding bearded vultures (Margalida, 2017) which otherwise would be occupying different areas (Brown, 1997; Xirouchakis & Nikolakalis 2002; Hirzel et al., 2004; Margalida et al., 2011a) - to coexist within the spatial distribution of the breeding population. Thus, high concentrations of floaters around breeding territories - as happens near supplementary feeding points (Sesé et al., 2005; Margalida et al., 2011b) - may increase the time being spent in agonistic encounters and, therefore, reduce breeding success. Vulture Restaurants in Africa and/or SFS in Europe are well-practiced long-term institutions but in Asia, particularly in the Indian subcontinent, it is a recent and under developed activity.

WHAT IS THE NEED OF VR

Fluhr et al., (2017) state that supplementary feeding is widely used as a conservation tool for many species of birds (Fasciolo et al., 2016; González et al., 2006; Robb et al., 2008) and mammals (López-Bao et al., 2010) whose productivities or individual survival is limited by food scarcity or toxicity. Vulture Restaurants, in particular, are used worldwide to provide threatened vultures with a source of supplementary carrion free from anthropogenic contaminants such as poisons and veterinary drugs (Yarnell et al., 2014). It is helpful when the prey population crashes, perhaps as a result of sudden disease or poisoning (Ferrer et al., 2017).

Supplementary feeding stations are essential at the first stages of conservation programs like reintroductions, where it is necessary to provide released animals with safe food, until they settle and learn how to feed by themselves in the environment (Ewen et al., 2015; Fluhr et al., 2017). The second case of supplementary feeding would be extensive grazing and scarce food availability to help vultures increase breeding success and prevent non–natural mortality due to direct poisoning (Margalida & Cortes-Avizanda, 2007). Productivity of poor territories could also be improved by supplementary feeding techniques if the reproductive rate potential of the birds is high and food availability is the limiting factor. Ferrer et al., (2017), on the basis of their research on two target species (Spanish Imperial eagle *Aquila adalberti* C. L. Brehm, 1861 and Bearded Vulture) in food scarce areas suggested that it can be applied to other species with a similar life history. Supplemental feeding has also been used to provide a contaminant-free food resource in areas where poisoning is suspected (Terrasse, 1985), and as a way to supplement essential nutrients lacking in depauperate natural food resources (Friedman & Mundy, 1983).

Vulture restaurants can be used as field laboratories to conduct studies related to carrion and scavengers. One of the latest examples is Moreno-Opo et al., (2020)'s study at SFS in the Pyrenean mountains and the Mediterranean basin on behavioural aspects of European vulture regarding carrion consumption. They studied the vulture behavioural organization and the relationship of hierarchies according to the feeding behaviour and prey selection, by comparing with other scavenger guilds. They found a general dominance gradient of body size (larger species to the smaller ones Cinereous Vulture > Griffon Vulture > Bearded Vulture > Egyptian Vulture), and age (from the adult age class to subadults and juveniles) as the outcome of agonistic encounters between European vultures.

VR: A Necessary Evil?

In light of decreasing biodiversity worldwide (Tickner et al., 2020) and shrinking population of vultures in Africa, Asia and Europe (Ogada et al., 2011; Ogada et al., 2016), immediate measure of recovery and maintenance has become high priority. Considering food shortage as one of the reasons, worldwide effort and investment have been devoted to creating SFS in an attempt to reverse or reduce the observed decline of large avian scavengers (Cortes-Avizanda et al., 2016). Such shortage was caused due to severely diminishing wild ungulates in western America, Africa and Asia (Mundy et al., 1992), changes in livestock management and traditional stock raising practices (Negro et al., 2007), incineration policy of dead livestock due to madcow disease in Europe (Tella, 2001), intentional poisoning of carnivorous carcass in Africa (Ogada, 2014).

Providing supplementary carrion for vultures has varied impact on them depending on density and size of SFS as well as frequency and quality of carcass supply. The primary advantage is recovery and maintenance of endangered (or starving) vultures. However, VR exerts undesirable impact within target populations and on non-target species and communities (Cortes-Avizanda et al., 2016), although Ferrer et al., (2017) disagree to this to some extent. Nevertheless, there are several disadvantages (risks) of VR management right from loss of vulture population and local prey biodiversity to transmission of contaminants (Figure 3), even in human being (Gomez-Ramirez et al., 2018). Therefore, VR or SFS could possibly be considered as a double-edged sword (Cortes-Avizanda et al., 2016).

Feeding provision at SFS increases survival of individual vultures and helps maintain neighbouring breeding population giving immediate conservation benefits (example, *Gypaetus barbatus* and *Neophron percnopterus* in the Pyrenees and south France; Cortes-Avizanda et al., 2016). Proximity to predictable feeding sites favours the establishment of communal roosts where immature birds gather; the creation of SFS is therefore often considered a useful management tool for attracting individuals that will disperse to isolated and declining populations (Donázar et al., 2009a; Lieury et al., 2015). These concentrations increase the probability of long-term territory occupancy, enhancing population viability (Grande et al., 2009b). However, such arrangement is known to affect fecundity negatively in spite of increased preadult survival (Margalida, 2017) as a consequence of demographic pressures (Carrete et al., 2006a & b). The aggregation of non-breeding birds at SFS might contribute to reduced reproductive success in breeding birds with territories located near the feeding stations, perhaps due to increased interactions between breeding adults and non-breeding birds, and to crowding of breeding populations (Carrete et al., 2006a & b). Another impact of management of supplementary food through a network of numerous small SFS, as suggested by Fluhr et al., (2017), does not show substantial disruption in the natural foraging behaviour of vultures, but large, frequently replenished SFS tend to artificially increase their level of routine. Therefore, Fluhr et al., (2017) recommended to the managers to preferentially rely upon a system of dilution of carcasses across the environment to protect opportunistic or normal foraging behaviour (Houston, 2006) typical of wild vultures. Martinez-Abrain et al., (2012) also recommended the establishment of scattered, low-value food sources to replicate historical conditions and to avoid the problems (problematic wind turbines, or other anthropogenic sources of mortality) associated with high concentrations of individuals in one place.

The supply of supplementary carrion for vultures is known to affect non-target species as well (Yarnell et al., 2014). It stimulates an increase in local abundance of two species of mammalian carnivores, the brown hyaena (*Hyaena brunnea*) and the black-backed jackal (*Canis mesomelas*). Of course, from wildlife conservation point of view this may be considered as a positive impact but the target species, that is vulture, faces competition for carrion from these carnivores which is against the VR objective. On the other hand, predation pressure on another set of non-target species (for example, ground nesting birds, as choosing the location for vulture restaurants) may be the key not only for scavengers but also for the conservation of the species living nearby. Aggregations of facultative scavengers around carcasses can increase predation pressure on surrounding passerine nests. Cortes-Avizanda et al., (2009) found that nearby nests/eggs were destroyed/eaten by facultative scavengers and the nest predation risk increased when a carcass appeared. This happens because facultative scavengers are also facultative predators and vice versa. Additionally, such scavengers may act as predators of vultures (Schabo et al., 2016) too.

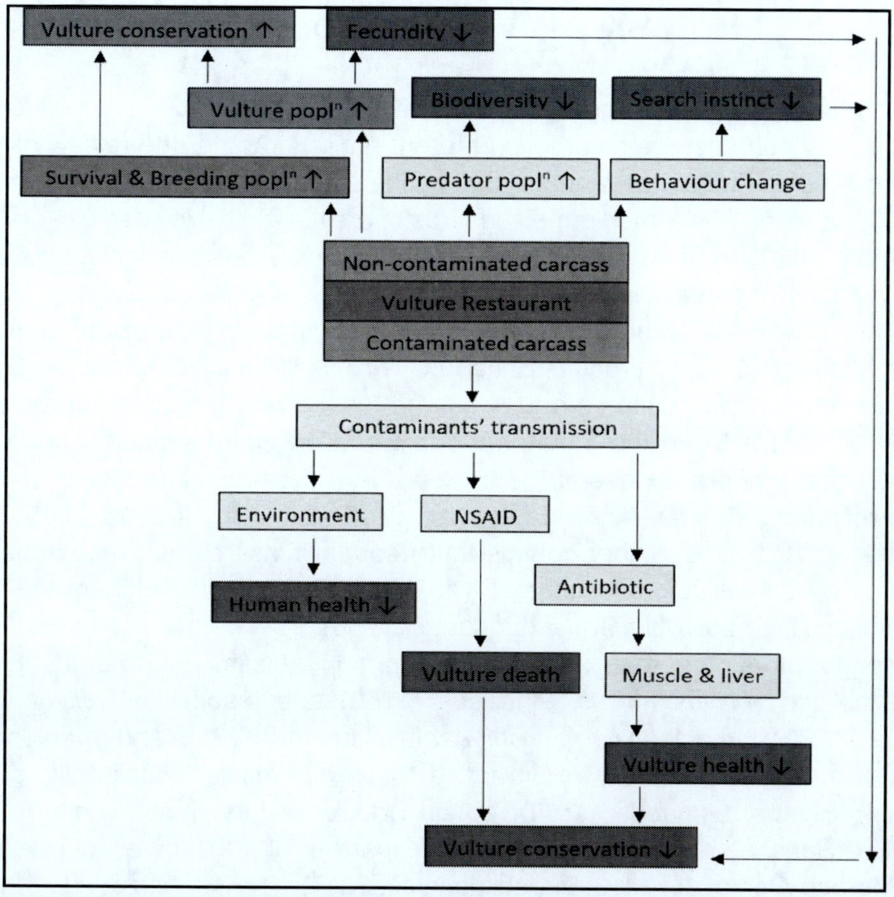

Boxes are sources (blue), positive impacts (green), negative impacts (red) and intermediate link (yellow). Arrows are the connecters. Up and down arrows in the boxes indicate increase and decrease, respectively. Negative impacts weigh heavier due to higher chance factor in their favor.

Figure 3. Diagrammatic representation of contrasting impacts (advantages and side effects/risks) of Supplementary Feeding Sites or Vulture Restaurants.

Predictable human-provided food sources could increase the survival of individuals that would otherwise disappear as a result of natural selective processes, favouring some kind of artificial selection (Blanco, 2006; Ferrer et al., 2017). Additionally, González et al., (2006) and Ferrer et al., (2014) claimed that SFS have probably aided the reproduction of facultative and specialist scavengers, but other authors did not find a similar relationship (Blanco, 2006; Oro et al., 2008; Margalida, 2010; Cortes-Avizanda et al., 2016). Since time and space of dumping food at SFS becomes predictable, the dominant scavenger (*Gyps fulvus*) species prospers at the cost of others (Arrondo et al., 2015) which can be countered by randomizing the time of carcass disposal.

How Safe Is VR from Competitors

Facultative scavengers, generally the vultures' competitors for natural or human provided food sources, are diverse in different regions. As per the review of Selva et al., (2019), the African savanna has corvids (e.g., White-necked raven *Corvus albicollis*, Latham, 1790), medium-sized birds of prey (e.g., Tawny eagle, *Aquila rapax,* Temminck, 1828), mammalian carnivores (e.g., side-striped jackal *Canis adustus*, Sundevall, 1847), and apex predators such as hyaenas (*Hyaena*, Brisson, 1762) and large cats (*Panthera* Oken, 1816). In areas with resident herbivore populations, spotted hyaenas (*Crocuta crocuta*, Erxleben, 1777) and lions (*Panthera leo*, Linnaeus, 1758) can outcompete vultures and consume most of the ungulate carrion biomass (Moleón et al., 2015). In the Mediterranean Eurasian Griffon vultures and common ravens (*Corvus corax*, Linnaeus, 1758) are the most habitual species scavenging carrion resources (Cortés-Avizanda et al., 2010; Mateo-Tomás et al., 2015; Moreno-Opo et al., 2015, 2016). Other corvids that frequently visit carrion resources are the carrion crow (*Corvus corone*, Linnaeus, 1758) magpie (*Pica pica*, Linnaeus, 1758) and Eurasian jay (*Garrulus glandarius*, (Linnaeus, 1758).

Among raptors other than vultures, the common buzzard (*Buteo buteo*, Linnaeus, 1758), the red (*Milvus milvus*, Linnaeus, 1758) and black kite (*Milvus migrans*, Boddaert, 1783), the golden (*Aquila chrysaetos*, Linnaeus, 1758) and Spanish imperial eagles (*Aquila adalberti*, C. L. Brehm, 1861) and, more occasionally goshawks (*Accipiter gentilis*, Linnaeus, 1758), obtain additional food resources from carrion subsidies (Sánchez et al., 2009; Sánchez-Zapata et al., 2010; Mateo-Tomás et al., 2015; Moreno-Opo et al., 2016; Margalida et al., 2017). The Andean condor, the turkey vulture and the black vulture are the species of obligate scavengers in South American steppes. Other bird species facultatively consuming carcasses are the southern caracaras (*Caracara plancus*, Miller, 1777) and chimango (*Milvago chimango*, Vieillot, 1816) and the black-chested buzzard eagle (*Geranoaetus melanoleucus*, Vieillot, 1819) (Travaini et al., 1998; Lambertucci et al., 2009). Carcasses of domestic and wild (red deer *Cervus elaphus*, the European hare (*Lepus europaeus*, Pallas, 1778), guanacos (*Lama guanicoe*, Müller, 1776), and rheas (*Rhea pennata*, d'Orbigny, 1834) animals are also consumed by mammals (pumas *Puma concolor*, Linnaeus, 1771), foxes (*Lycalopex*, Burmeister 1854), skunks (*Conepatus*, Gray, 1837) and armadillos (*Zaedyus pichiy*, Desmarest, 1804), and also by passerines and lizards (Travaini et al., 1998; Elbroch & Wittmer, 2012).

All the scavengers in different regions where vultures are found, some of which are mentioned above, are potential competitors of vulture for the carcass but there are selected studies available which provide information about competition at VR. For example, in the case of specific SFS for Bearded Vulture in the proximity of breeding nests in the Spanish Pyrenees, other species like the griffon vulture, common raven, red kite, golden eagle and Egyptian Vulture were also observed eating the remains at SFS (Margalida, 2010). A feeding site originally devised for supplementing a breeding population of Cinereous Vulture in southern Spain, for example, attracted a group of about 1,000 young griffon vultures, which were not present in large numbers earlier (Negro et al., 2007). Jha (2015) and Jha et al., (2020) reported that disposal of dead animals was mainly done by removing carcasses to village outskirts (north Indian equivalent of *Muladares*), where dogs (*Canis lupus familiaris*, Linnaeus, 1758), crows, egrets and pigs compete with vultures. Therefore, Negro et al., (2007) also suggested that when choosing a location for a vulture restaurant, several factors should be considered for example, first and foremost, the presence of opportunistic mammalian scavengers, including foxes, wolves (*Canis lupus*, Linnaeus, 1758), and dogs which may visit the restaurant as well.

How Secured Is Carrion from Contaminants

Environmental contaminants are the most serious threat to avian scavengers (Buechley & Şekercioğlu, 2016; Pauli et al., 2018). Vultures being at the top of the food chain, can be considered at risk of exposure and accumulation of environmental contaminants such as heavy metals and pesticides (García-Fernández et al., 2005; Gómara et al., 2004). Their exposure to veterinary pharmaceuticals has become of increasing concern after the diclofenac crisis of *Gyps* vultures in Asia feeding on medicated cows (Cuthbert et al., 2011; Taggart et al., 2009) and the looming threat in Spain due to high availability of this drug (Campbell, 2015). Other veterinary drugs harmful to vultures are barbiturics, ketoprofen, flunixin, corticosteroids, endocrine disruptors, antibiotics like penicillin, tetracycline, oxytetracycline and sulphonamide etc. (Gomez-Ramirez et al., 2018). Vultures are prone to harmful bacteria; *Salmonella*, was detected in the Griffon Vulture (Marin, et al., 2018) and the Egyptian Vulture (Blanco, 2018), entering through dead pigs discarded at SFS.

Such problems are associated with *Muladares* or vulture restaurants mainly from stabled livestock carrion, as a consequence of the residues of antimicrobial agents and other drugs which may be consumed by vultures and cause death or diseases (Margalida et al., 2007). If the carcasses to be dumped at the feeding site have been treated with excessive antibiotics, chronic exposure to it can lead to adverse effects on the scavengers. Therefore, disposal of medicated animals in SFS should be regulated to minimise the risk for scavenger birds (Gomez-Ramirez et al., 2018). However, the unfortunate part is sociocultural compulsions which make the regulation ineffective and defeat the purpose. Two important examples are eye-opening in two differently developed countries. (i) Diclofenac ban in Indian subcontinent: In spite of the ban on production of the veterinary drug by the governments in India and Nepal, the sale and use of diclofenac drug continued, posing impending threat to vultures (Adhikari & Bhattarai, 2017; Cuthbert et al., 2016) and (ii) Removal of carcasses in an attempt to prevent possible outbreak of BSE (bovine spongiform encephalopathy) in Spain: Instead of the burial of carcass with lime cover, farmers adopted a cheaper and easier method of livestock carcass disposal at *Muladares* in Griffon foraging area (Camina, 2004b). Possibly this was done because the regulation proved to be unworkable, both from the economical and practical point of view (Camina, 2004a; Camina & Montelio, 2006).

Since different drugs act in different tissues and vultures have preferences for carcass parts, different species of vultures are affected differently by the same drug. Antibiotics are found more in muscle and the kidney than in the liver, therefore, there are chances that they will affect those species which eat muscles and cartilage (Cinereous Vulture) more than those that eat viscera (Eurasian Griffon Vulture). Several studies have demonstrated that Eurasian Griffon Vultures are indeed accessing antibiotic residues in livestock carcasses provided at feeding stations (Casas-Díaz et al., 2016).

It is a normal and inevitable practice also to treat sick domestic animals with different veterinary formulations all over the world. Many such formulations are harmful for vultures leading to death after consumption of contaminated carcasses. Use of Diclofenac is very common in the Indian subcontinent and it was reported fatal for vultures eating on Diclofenac administered dead cattle (Campbell, 2015). Despite the ban on the production of such veterinary drugs by the governments in India, Pakistan and Nepal its sale continued unabated, posing a threat to vultures (Adhikari & Bhattarai, 2017; Jha et al. 2021).

The African vultures are often unintended victims of poisoning incidents (Ogada et al., 2016), in which carcasses are baited with highly toxic agricultural pesticides to kill carnivores such as lions, hyenas, and jackals (Ogada, 2014), or to control feral dog populations (Abebe, 2013). Although this happens generally in the forest or savannah or countryside, occasional incidents at VR near wildlife area cannot be ruled out.

Disposal of carcasses, offal, urine, faeces, and unused products can contaminate the environment with veterinary pharmaceuticals which in turn can make their way to humans as well, especially antibiotics and their metabolites which can cause antibiotic resistance, predisposition to fungal infections, allergic reactions or disruption of microbiome in humans and wildlife (Gomez-Ramirez et al., 2018).

MAPPING AND MANAGEMENT OF VR

Management of any activity or project is preceded by measurement or gathering knowledge about it which may be reflected in the inventory of resources and limitations. In business terms these could be categorised into strength, weaknesses, opportunities and threats. By using geomatics tools these features of wildlife related projects could be recorded in the form of maps. This is a highly useful document which imparts quick geospatial information for management. In the case of VR, different factors could be mapped in relation to the area and categorised into Strength, Threat, Opportunity, and Weakness Maps. A strength map may have features like good availability of carcass providing ungulates in the area or ungulate density etc. A threat map may have distribution of power structures, windmill presence etc. in the area. An opportunity map may have information about *Muladares*, the presence of livestock farms etc. A weakness map may have the presence of the population of carrion competitors, absence of waterbody etc. in the area. Finally, after combining these maps, a suitability class map could be generated. Inventory of knowledge and mapping strategy are discussed in the following paragraphs.

Management Requirements

Since the late 1960s, VR were opened primarily for supplying feed to vultures as conservation efforts in different countries, for example, Bulgaria (Dobrev et al., 2015), Cambodia (Hille et al., 2016), France (Fluhr et al., 2017), Indian subcontinent (Gilbert et al., 2007), Myanmar (Hla et al., 2011), Namibia (Bamford et al., 2007), Nepal (Adhikari & Bhattarai, 2017), Portugal (Gomez-Ramirez et al., 2018), South Africa (Schabo et al., 2016; Brink et al., 2020), Spain (Lopez-Lopez et al., 2014; Arrondo et al., 2015), Swaziland (Kane et al., 2014) etc. Various types of studies were conducted (impact on population, breeding success etc.) but no comprehensive attempt was made to characterise the feeding site or to develop the index of site suitability based on different variables.

Generally, feeding sites must be located in areas that maximize the attendance of the target species and benefit them. To achieve this, many different parameters should be considered and evaluated beforehand. These include variables such as physiographic characteristics of the area, distance to the breeding territories, dangerous infrastructure, biology and behaviour of target species, population of other scavengers, carcass contamination, anthropogenic disturbances etc. (Piper, 2004; Bounas, 2015). Since research results on the topics are limited and species specific, the values could be taken for other nearest species also.

There are several variables affecting the establishment of VR positively or negatively. For example, the creation of feeding stations favours immature or non-territorial adult birds or floaters for survival and even leads to re-colonization of abandoned territories (Benitez et al., 2009; Grande et al., 2009a). Negro et al., (2007) suggested that several things should be kept in mind while choosing a location for a vulture restaurant. The presence of opportunistic mammalian scavengers, including foxes, wolves and dogs in the area which may visit the restaurant and eat the feed. Yarnell et al., (2014) working on VR and scavengers, evidenced that providing supplementary carrion for vultures stimulated an increase in local abundance of two species of mammalian carnivores, the brown hyaena (*Hyaena brunnea*, Thunberg, 1820) and the black-backed jackal (*Canis mesomelas*, Schreber, 1775). Fencing is essential for a supplementary feeding station in order to avoid mammal scavengers to approach and consume the remains intended for vultures. Although there is an occasional report of vulture injury after getting trapped during take-off (Piper, 2004), Arrondo et al., (2015) reported practicing of vertically open and peripherally fenced VR to keep carnivores away. However, in most regions in Spain, SFS are fenced off, located far from watercourses and situated in open areas.

Another variable is slope. The attendance of vultures on VR is related to the slope, over a radius of 1 km in the vicinity of the feeding site. Feeding stations placed in rugged areas received more visits from Bearded Vultures, whereas Griffon and Egyptian Vultures used places located in flat areas (Cortes-Avizanda et al., 2010). Adhikari & Bhattarai (2018) found negative relationship between species and distance to VR which meant that the species prefer to stay near the VR because of the regular availability of food/carcass. Trees are another variable. Trees that could provide roosting or perching (preferably dead ones) for birds, should be present in a radius of 3 km from the feeding site (Donazar et al., 1996). Taller trees of *Bombax ceiba* were preferred in the vicinity of VR for perching of White-rumped Vulture, Himalayan Griffon and Egyptian Vulture (Adhikari & Bhattarai, 2018). Targeting is another issue. Supplementary feeding needs to be targeted for greatest efficiency, at poorer territories in which the reproductive rate has the potential to be raised by provision of extra food (Ferrer et al., 2017). For example, there could be a worsening of trophic conditions which caused a decrease of individuals in population and pairs engaged in breeding. Supplementary feeding of Egyptian Vultures enabled stabilization of the number of pairs engaged in breeding in the population (Karimov & Mamadov, 2019).

Wind farms are another problem. Wind farms are the preferred option for alternate energy generation due to the reduction of air pollution and fossil fuel consumption. However, they contribute to raptor population reduction (Leddy et al., 1999; Laiolo & Tella, 2006; Kuvlesky et al., 2007; Garvin et al., 2011). Presence of power lines and wind farms are generally dangerous for flying birds with risk of death by electrocution or collision. Therefore, potential feeding sites should be established at least 2 km away from such infrastructure (Bounas et al., 2015). In general, a buffer of 10 km radius should be maintained around a breeding colony of Bearded Vulture (Kruger et al., 2015). Traditional dump sites that are conventionally used by farmers to dispose of carrion and are visited by scavengers should be favoured during the selection process. However, it has the disadvantage of inviting companion scavengers like, birds and mammals. An existent road provides easy access for supplying the carcass to the feeding site. At the same time frequent and regular traffic of local stakeholders, could create disturbance for the feeding birds. However, this can either be regulated or avoided while running a feeding site.

Carcass size is another important variable. Carcass-size manipulation may be a valuable tool for directing supplementary feeding towards species of interest. Cortes-Avizanda et al., (2010) reported that griffon vultures do not prefer feeding stations supplied with large carcasses, but Egyptian Vultures do. Another example of higher dietary plasticity is Griffon and Cinereous Vulture where the former is mostly dependent on livestock carcasses, while the diet of the latter includes the remains of small- and medium-sized vertebrates such as European Rabbits *Oryctolagus cuniculus* (Arrondo et al., 2018; Genro et al., 2020). The presence of facultative scavengers also presents an issue.

The concentration of facultative scavengers around the SFS could increase the probability of disturbance or even nest predation (CortesAvizanda et al., 2009). Thus, the location should be selected at least 2 km away from an active nest in the case of Egyptian Vulture (Bounas et al., 2015).

Pesticides have presented problems as well. The most widely used pesticides affecting vulture populations, and associated with deliberate poisoning, are carbamates and organophosphorus compounds. If poisoning with pesticides is not stopped, this threatened avian group could go extinct very soon (Plaza et al., 2019). Pesticide poisoning is currently the greatest threat to the Andean Condor, and it affects adult condors more than immature ones. Intoxication or poisoning mortality (13%) as compared to natural death rate (5%) has severe consequences in the Andean Condor (Temple & Wallace, 1989; Pacheco et al., 2020). The most commonly used poison is Carbofuran, and to a lesser extent Paration, which has reached alarming levels that could lead to extinction (Pacheco et al., 2020). Poisoned carcasses that are not eliminated from the natural environment would have a drastic effect on the natural population of African vultures even leading them to extinction (Murn & Botha, 2017). Bhusal (2018) and others (Mukherjee et al., 2014; Alam et al., 2016 etc.) have proposed a vulture safe zone of a 100 km radius where there should be no use of direct or indirect poison.

The impacts of feedings stations on foraging behavior present another issue. The findings of Kane et al., (2016) suggest that the use of SFS does not impact natural foraging behaviour of *Gyps coprotheres* significantly. But reduced food availability changed the foraging habit of other vultures. Many individuals of Griffon Vulture *Gyps fulvus* exploit a garbage dump, another predictable but lower-quality food source not exploited conventionally (Martinez-Abrain et al., 2012). Food scarcity also induced behavioural changes in Griffon Vulture. Zuberogoitia et al., (2010) demonstrated that food availability and their state of hunger guide the ability of vultures to evaluate the predation risk change in behaviour from natural caution (shyness) towards more tolerant (fearless). However, Camina & Yosef (2012) observed that the local adults with fidelity to their breeding colonies remain in the area while the non-breeding sub-adults and floaters strayed in search of food. Another interesting example of behavioural change was reported by Margalida et al., (2011c) that due to a change in European sanitary policy, increased competition among *Gyps fulvus* for food took them to killing livestock in France and Spain. The illegal trade in vulture body parts for use in traditional medicine is a significant threat (McKean et al., 2013; Saidu & Buij, 2013; Ogada et al., 2016). There are fears that if SFS locations are freely accessible, they could be misused by poachers for vulture harvesting, or that provided carcasses would be taken for human consumption (Brink et al., 2020).

Management Guidelines

It is evident that VR establishment automatically brings in side effects which are detrimental to vulture conservation, if not managed scientifically. Therefore, to ameliorate inadvertent damage in establishing a VR to the already endangered population or even the least concerned ones, management actions should be directed towards the creation of a chain of feeding stations supplied sporadically with food, which would avoid the concentration of resources in just a few places (Deygout et al., 2009; Cortes-Avizanda et al., 2010). This needs an approach of focus, control and monitoring (Cortes-Avizanda et al., 2016), detailed in following paragraphs.

The focus of VR establishment should be at suitable and variable places based on the distribution and movement of less abundant and less dominant scavengers. Also important is the creation of multiple small and dispersed VR, rather than few and large ones. Another point of focus should be requirement based spatiotemporal distribution/availability of carrion for the target species. If required, VR may also be utilised for threatened facultative scavengers instead of abundant and dominant vultures by provisioning small carcass or small pieces of carrion remains.

It is well established that veterinary pharmaceuticals, heavy metals and livestock pathogens get transferred to vultures through carcasses which may cause irreparable damage to vultures, and sometimes even death. Lapses in such carrion supply must be strictly regulated. Quality control, especially in third world countries facing population crash, should be of the highest order since recovery of truncated populations is very difficult owing to slow breeding and low survival rate in vultures. Supply of swine, poultry and viscera and offal must be avoided to improve food quality and to reduce chances of contamination. Implementation of strict monitoring of schemes for supplied food and scavenger use throughout the year must be ensured, and eventually adaptation of carrion provisioning based on scientifically rigorous and adaptive approaches (Cortes-Avizanda et al., 2016).

VR Suitability Mapping

Vulture Restaurants or feeding stations are created, generally, for endangered resident species but its advantage can be drawn by migratory and non-targeted species also. However, to maximise the utility of such stations several independent maps (Strength, Threat, Opportunity, Weakness) and then a combined map (Suitability) could be prepared. Such records may become the guideline for site selection. Suitable area map of VR having features like, high domestic ungulate population, very low-density forest or open area, anthropogenic disturbance free locality etc. could be very useful information to be used for site selection. On the other hand, threat map of VR having features like, dangerous structures in the area, high density carnivore area etc. would be an important document to be prepared in order to avoid such areas while selecting the site.

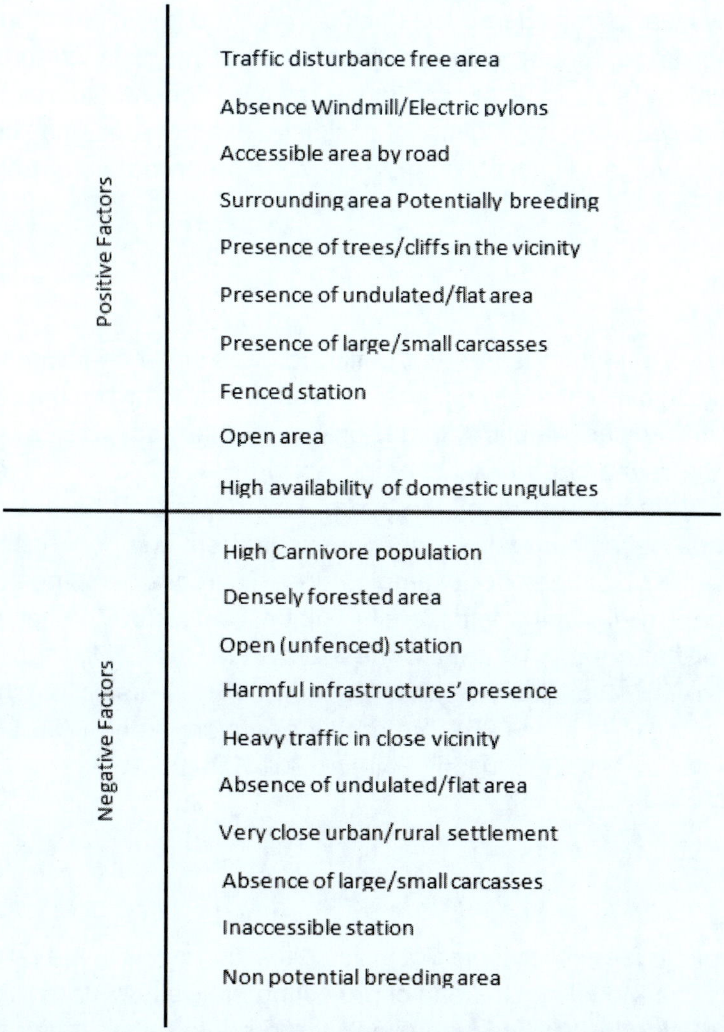

These factors could be used as variables in determining suitability of an area based on index or modeling. Factors closer to the axes heuristically appear more important or valuable. They may be tested statistically to arrive at precise order/value.

Figure 4. Presentation of factors influencing establishment of Vulture Restaurant.

Any VR site has biophysical and environmental features, a combination of which may make it a good or bad site. However, a set of suitable natural and anthropogenic features or variables (discussed in management requirements section above and summarised in Figure 4) could be identified and quantified along with climatic factors (rainfall, temperature, etc.). Using such a set of attributes of a known area and putting them in an appropriate model, new suitable areas could be identified. This would be helpful in expanding the scope of appropriate places to be used for VR establishment. Similar attempts, as Camina (2006), could be made or better models could be developed by using more variables. Camina used GLM models considering the amount of food available at the feeding site, numbers and ages of the vultures that are present, distances to colonies, presence of other species and climatic variables.

Camina's analysis also considered livestock density and forest cover on the estimated colony foraging range. Finally, number of immatures, the distance to the colony and the average colony size in a 15 km radius around the *Muladares* were the only considered variables. After having model assessed a larger area, may be country level or at least state level, a VR suitability map could be prepared for future planning use.

VR Blueprint

Although lot of research is needed to fill the gaps while deciding the structure of VR, it is possible to hypothetically ascertain VR space/structure on the basis of existing information with an assumption that non-researched species will have similar behaviour as the researched ones exemplified in the text. An effective space of VR may vary from a fraction of a km^2 to several km^2 depending on the requirement of surroundings and behaviour of the scavengers. Different components of VR structure may be represented by zones, for example, fenced carcass dumping zone or nucleus (0.003 km^2 with 1m high perimeter fence; Gilbert et al., 2007), specific topography zone (sloped or flat area, 3.14 km^2; Cortes-Avizanda et al., 2010), no-nest and no-infrastructure zone (12.57 km^2; Bounas et al., 2015), sporadic perching tree zone (28.26 km^2; Donazar et al., 1996), preferably no-colony zone (706.5 km^2; Camina, 2006), poison free zone (>30000 km^2; Bhusal, 2014) etc.

VR-Pixel

It is plausible to identify and predict area suitability for VR in the region on the line of Ecological Niche Modelling. Instead of projecting the well-defined niche of a species (by naturally existing biophysical variables), the predetermined index-based value of suitability for VR calculated by the weighted value of different positive and negative factors could be located in the potential region. Such index value could be used in VR-Pixel (VR space equivalent of a pixel) which could be of predetermined size based on VR blueprint and treated as cells of fishnet in ArcGIS software (for the sake of area contiguity and complete assessment of the region, unit area should be taken as a square instead of a circle). Range value of VR-Pixel may determine the suitability, like SDM index in MaxEnt. Thus, VR suitability mapping could be done in a region and used as a guide for establishing VR as and when required. The "Shelter site" of 10 km^2 for vultures as proposed by Jha et al., (2020) could be taken as VR-Pixel characterization and "Foraging site" of 250 km^2 be treated as VR zone. Depending on the abundance of vultures in the colony/shelter, the number of VR should be decided in a VR zone. This concept may be illustrated through an example of Central India – Madhya Pradesh (Figure 5). Since vultures can move up to 100 km for foraging, VR could be established in a grid or VR suitable zone without vulture sites but closer to the colony.

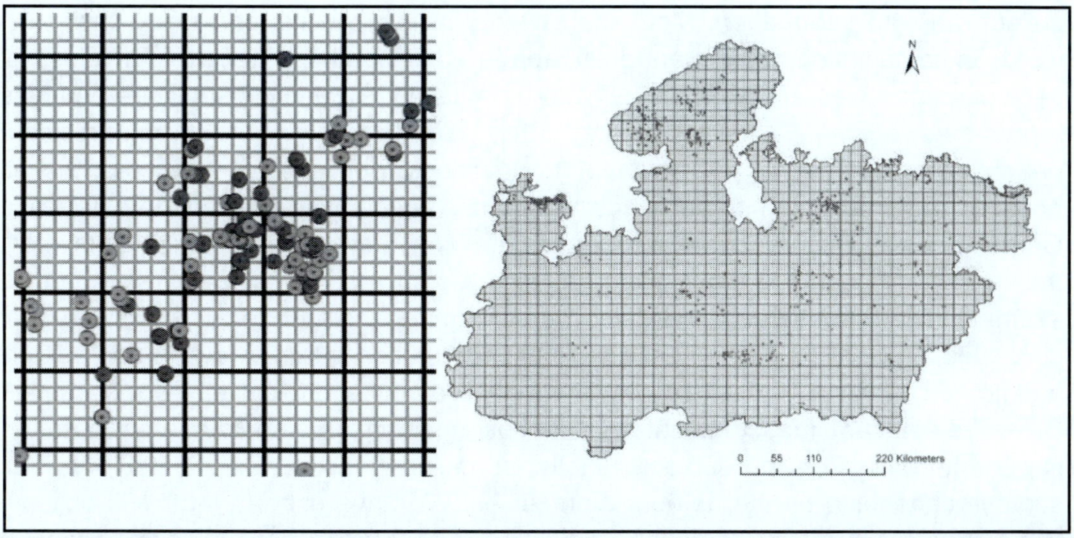

Vulture locations are red and green circles. Subgrids or shelter area may represent VR. Pixel and Grids or foraging areas may be treated as VR Zones. This illustration is adopted from Jha et al., (2020).

Figure 5. Central India - Madhya Pradesh on right and zoomed part on left divided into grids (black squares of 250 km^2) and subgrids (grey squares of 10 km^2).

Conclusion

From the narrative in the text, it is evident that VR establishment is highly useful for saving endangered vultures but the selection of the site for VR for effective management is crucial to minimise VR risks from competitors and contaminants. It is highly recommended that VR suitability zone mapping using geomatics, along with strength, threat, opportunity and weakness maps, must be ensured on suggested or modified lines. Simultaneously, more and more research should be done on different species to have generalised or species-specific recommendation for the use of VR with minimised or preferably no negative impacts, while aiming for the absolute conservation of this most important and free of cost service provider to people globally.

Review Methodology

This review is not claimed to be exhaustive. It is based on previously published research articles, book chapters, print media stories and field observation until March 2020. There are some methodologies for research reviewing available in the literature as per the need and scope of the subject (e.g., Larsen and Olsen, 2007; Smith-Hall et al., 2012 etc.). For the present chapter, certain keywords were decided (Vulture/Scavenger/Raptor species, Vulture restaurant, Vulture feeding sites/stations, Vulture feeding habit, *Muladares*) and run through Google's search engine in order to collect required information. More than twenty million results, containing these key words in combination or independently, appeared. From these websites, articles with subject matter relevant to the review were shortlisted by rejecting duplicate and least representative sites. The results were further narrowed down to research articles in peer reviewed journals as well as chapters from books published by reputed publishing houses. After scanning through the abstracts, academic research networks like, Google Scholar, Researchgate, Academia edu, Mendeley, etc. were used to download the reprints for detailed study. Articles from e-print media with good standing, for example, The Hindu, Deccan Herald, Times of India, Indian Express etc., were also consulted to incorporate the latest on-ground information in the review. For some facts and figures quoted in this text, government publications and websites were consulted. Cross references, mostly about books/book chapters, were also referred to get important content on the subject. Finally, some information was collated with field surveys conducted by the authors themselves. We have used in total 156 citations contributed by 607 authors dating from 1974 to 2020 (114 peer reviewed research papers, 31 non-peer reviewed book/book chapters/conference proceedings/official publication and 11 newspaper-based web stories) in the text and listed them in the reference section.

References

Adhikari, J. N. & Bhattarai B. P. 2017. Can Vulture Restaurant Protect Critically Endangered Species of Vultures in Lowland Nepal? *International Journal of Science and Research*, 6 (11): 1717-1723. https://doi.org/10.21275/ART20178408.

Aisling, Maria Cornin. 2020. *India Opens First-Ever Vultures-Only Restaurant in an Attempt to Save Them from Extinction.* Retrieved from https://www.onegreenplanet.org/news/india-opens-its-first-restaurant-for-vultures-to-save-them-from-extinction/.

Alam, A. B. M. S., Ahmed, S. & Rahman, S. 2016. *Vultures and Vulture Safe Zones of Bangladesh. IUCN, International Union for Conservation of Nature*, Bangladesh Country Office, Dhaka, Bangladesh.

Alexander, Lerche. 2017. *The vulture restaurant.* 23 DEC 2017. Retrieved from https://www.aljazeera.com/indepth/inpictures/vulture-restaurant-171221084051492.html.

Amar, A., Buij, R., Suri, J., Sumasgutner, P. & Virani, M. Z. 2018. Conservation and Ecology of African Raptors. In: Sarasola, J. H., Grande, J. M. & Negro, J. J. et al., (Eds.), *Birds of Prey*, Springer International Publishing AG, part of Springer Nature 2018. https://doi.org/10.1007/978-3-319-73745-4_18.

Archana, Phull. 2020. *Restaurant' for vultures in HP wild life sanctuary.* Retrieved from https://www.thestatesman.com/cities/shimla/restaurant-for-vultures-in-hp-wild-life-sanctuary-1502856301.html.

Arrondo, E., Cortes-Avizanda A. & Donazar, J. A. 2015. Temporally unpredictable supplementary feeding may benefit endangered scavengers. *Ibis*, 157(3): 648-651. https://doi.org/10.1111/ibi.12267.

Arrondo, E., Moleon, M., Cortes-Avizanda, A., Jimenez, J., Beja, P., Sanchez-Zapada, J. A. & Donazar, J. A. 2018. Invisible barriers: differential sanitary regulations constrain vulture movements across country borders. *Biological Conservation*, 219: 46–52. https://doi.org/10.1080/00063657.2020.1734534.

Barclay, J. H. 1987. Augmenting wild populations. In Pendleton, B. A. G., Millsap, B. A., Cline, K. W. & Bird, D. M. (Eds.), *Raptor management techniques manual*. 239–247. National Wildlife Federation, Washington, DC U.S.A.

Barton, P. S., Cunningham, S. A., Lindenmayer, D. B. & Manning, A. D. 2013. The role of carrion in maintaining biodiversity and ecological processes in terrestrial ecosystems. *Oecologia*, 171:761–772. https://doi.org/10.1007/s00442-012-2460-3.

Benitez, J. R., Cortes-Avizanda, A., Avila, E. & Garcia R. 2009. Effects of the creation of a vulture restaurant for the conservation of an Egyptian vulture Neophron percnopterus population in Andalucia (Southern Spain). In Donazar J. A., Margalida A., Campión D. (Eds). *Vultures, feeding stations and sanitary legislation: a conflict and their consequences from the perspective of conservation biology*. 276-291. Munibe 29 (Suppl.) Sociedad de Ciencias Aranzadi. Donostia.

Bhusal, K. P. 2018. Vulture Safe Zone: a landscape level approach to save the threatened vultures in Nepal. *The Himalayan Naturalist*, 1(1): 25-26.

Bijleveld, M. 1974. *Birds of Prey in Europe*. London, UK: Macmillan.

Blanco, G. 2018. Supplementary feeding as a source of multiresistant *Salmonella* in endangered Egyptian vultures. *Transboundary and Emerging Diseases*, 2018: 1–11. https://doi.org/10.1111/tbed.12806.

Blanco, G., Cortes-Avizanda, A., Frias, O., Arrondo, E., Donazar, J. A. 2019. Livestock farming practices modulate vulture diet-disease Interactions. *Global Ecology and Conservation*, 17(2019): e00518. https://doi.org/10.1016/j.gecco. 2018.e00518.

Bounas, A. 2015. Creating supplementary feeding stations for the conservation of Egyptian Vulture in Greece. Technical report under action C3 of the LIFE+ project "The Return of the Neophron" (LIFE10 NAT/BG/000152). *Hellenic Ornithological Society*, Athens, Greece.

Braack, L. E. O. 1987. Community dynamics of carrion-attendant arthropods in tropical African woodland. *Oecologia*, 72: 402–409. https://doi.org/10.1007/BF0 0377571.

Brink, C. W., Santangeli, A., Amar, A., Wolter, K., Tate, G., Kruger, S., Tucker, A.S. & Thomson, R. L. 2020. Quantifying the spatial distribution and trends of supplementary feeding sites in South Africa and their potential contribution to vulture energetic requirements. *Animal Conservation*. https://doi.org/10.1111/acv.12561.

Brown, C.J. 1997. Population dynamics of the bearded vulture *Gypaetus barbatus* in southern Africa. *African Journal of Ecology*, 35: 53–63. https://doi.org/10.1111/j.1365-2028.1997.048-89048.x.

Buechley, E. R. & Sekercioglu, C. H. 2016. The avian scavenger crisis: looming extinctions, trophic cascades, and loss of critical ecosystem functions. *Biological Conservation*, 198: 220–228.

Camina, A. 2004a. Consequences of the Bovine Spongiform Encephalopathy (BSE) on breeding success and food availability of Spanish Vulture populations. In Meyburg B. U., Chancellor R. D. (Eds.), Raptors Worldwide. 27–44. *MME-World Working group on Birds of Prey*. Hungary.

Camina, A. 2004b. Griffon Vulture *Gyps fulvus* monitoring in Spain: current research and conservation projects. In Chancellor, R. D. & B.-U. Meyburg (Eds.), Raptors Worldwide. 45-66. *MME-World Working group on Birds of Prey*. Hungary.

Camina, A. 2006. Food Exploitation by Griffon Vultures: The Effect of Vulture Restaurants in Spain. In Houston, D. C. & S. E. Piper (Eds.), *Proceedings of the International Conference on Conservation and Management of Vulture Populations, 2006*. 14-16 November 2005, Thessaloniki, Greece. Natural History Museum of Crete & WWF Greece.

Camina, A. & Montelio, E. 2006. Griffon Vulture *Gyps fulvus* food shortages in the Ebro Valley (NE Spain) caused by regulations against Bovine Spongiform Encephalopathy (BSE). *Acta Ornithologica*, 41: 07–13. https://doi.org/10.3161/000164506777834679.

Camina, A. & Yosef, R. 2012. Effect of European Union BSE-related enactments on fledgling Eurasian Griffons *Gyps fulvus*. *Acta Ornithologica*, 47: 101–109. https://doi.org/10.3161/000164512X662205.

Campbell, MO. 2015. *Vultures their Evolution, Ecology and Conservation*. Boca Ratoon, FL: CRC Press.

Carrete, M., Donazar, J. A. & Margalida, A. 2006a. Density–dependent productivity depression in Pyrenean Bearded Vultures: implications for conservation. *Ecological Applications*, 15: 1674-1682. https://doi.org/10.1890/1051-0761(2006)016[1674:DPDIPB]2.0.CO;2.

Carrete, M., Donazar, J. A., Margalida, A. & Bertran, J. 2006b. Linking ecology, behaviour and conservation: does habitat saturation changes mating system in bearded vultures? *Biology Letters*, 2: 624–627. https://doi.org/10.1098/rsbl.2006.0498.

Carrete, M., Sanchez-Zapata, J. A., Benítez, J. R., Lobon, J. & Donazar, J. A. 2009. Large scale risk assessment of wind-farms on population viability of a globally endangered raptor. *Biological Conservation*, 142: 2954–2961. https://doi.org/10.1016/j.biocon.2009.07.027.

Carrete, M., Sanchez-Zapata, J. A., Benitez, J. R., Lobon, M., Montoya, F. & Donazar, J. A. 2011. Mortality at wind-farms is positively related to large-scale distribution and aggregation in griffon vultures. *Biological Conservation*, 145: 102–108. https://doi.org/10.1016/j.biocon.2011.10.017.

Casas-Diaz, E., Cristofol, C., Cuenca, R., Agustin, S., Carneiro, M., Marco, I., Lavin, S. & Margalida, A. 2016. Determination of antibiotic residues in plasma of two populations of Eurasian griffon vulture (*Gyps fulvus*) in Spain. *Science of the Total Environment*, 557: 620–626. https://doi.org/10.1016/j.scitotenv.2016. 03.083.

Ceballos, O. & Donazar J. A. 1990. Roost-tree Characteristics, Food Habits and Seasonal Abundance of Roosting Egyptian Vultures in Northern Spain. *Journal of Raptor Research*, 24(1-2): 19-25. Retrieved from https://pdfs.semanticscholar.org/acb5/ae20f07ccf7e0f7243c4dca4aea23bb66b69.pdf?_ga=2.202083986.262720609.1592945489-1738787424. 159 2945489.

Cortes-Avizanda, A., Carrete, M., Serrano, D. & Donazar, J. A. 2009. Carcasses increase the probability of predation of ground-nesting birds: a caveat regarding the conservation value of vulture restaurants. *Animal Conservation*, 12: 85–88. https://doi.org/10.1111/j.1469-1795.2008.00231.x.

Cortes-Avizanda, A., Carrete, M. & Donazar, J.A. 2010. Managing supplementary feeding for avian scavengers: guidelines for optimal design using ecological criteria. *Biological Conservation*, 143:1707–1715. https://doi.org/10.1016/j.biocon.2010.04.0167.

Cortes-Avizanda, A., Blanco, G., DeVault, T. L., Markandaya, A., Virani, M. Z., Brandt. J. & Donazar, J. A. 2016. Supplementary feeding and endangered avian scavengers: benefits, caveats and controversies. *Frontiers in Ecology and the Environment*, 14(4): 191-199. https://doi.org/10.1002/fee.1257.

Cramp, S. & Simmons, K. 1980. *The birds of the western Palearctic. Vol. 2*, Oxford: Oxford University Press.

Cuthbert, R., Taggart, M. A., Prakash, V., Saini, M., Swarup, D., Upreti, S., Mateo, R., Chakraborty, S. S., Deori, P. & Green, R. E. 2011. Effectiveness of action in India to reduce exposure of Gyps vultures to the toxic veterinary drug diclofenac. *PLoS One* 6: e19069. https://doi.org/10.1371/journal.pone.0019069.

Cuthbert, R. J., Taggart, M. A., Saini, M., Sharma, A., Das, A., Kulkarni, M. D., Deori, P., Ranade, S., Shringarpure, R. N., Galligan, T. H. & Green, R. E. 2016. Continuing mortality of vultures in India associated with illegal veterinary use of diclofenac and a potential threat from nimesulide. *Oryx,* 50(1): 104-112. https://doi.org/10.1017/S003060531500037X.

Dasa, S. 2014. *Spare us some carcasses and stop eating our share, An appeal from the vultures.* Hyderabad. Soul Science University Press.

DeCandido, R., Subedi, T. & Allen, D. 2012. Jatayu: the vulture restaurants of Nepal. *BirdingASIA*, 17: 49–56.

Del Hoyo, J., Elliott, A. & Sargatal, J. 1994. *Handbook of the Birds of the World. Vol. 2.* Barcelona: Lynx Edicions.

DeVault, T. L., Rhodes, O. E. & Shivik, J. A. 2003. Scavenging by vertebrates: behavioural, ecological, and evolutionary perspectives on an important energy transfer pathway in terrestrial ecosystems. *Oikos*, 102: 225–234.

Deygout, C., Gault, A., Sarrazin, F. & Bessa-Gomes, C. 2009. Modeling the impact of feeding stations on vulture scavenging service efficiency. *Ecological Modelling*, 220: 1826–1835. https://doi.org/10.1016/j.ecolmodel.2009.04.030.

Dharmendra, Ratul. 2011. 3 *'vulture restaurants' set up to feed the rare birds.* Retrieved from http://archive.indianexpress.com/news/3--vulture-restaurants--set-up-to-feed-the-rare-birds/746917/.

Dobrev, V., Boev, Z., Oppel, S., Arkumarev, V., Dobrev, D., Kret, E., Vavylis, D., Saravia, V., T., Bounas, A. & Nikolov, S. C. 2015. Diet of the Egyptian vulture (*Neophron percnopterus*) in Bulgaria and Greece (2005-2013). Technical report under action A5 of the LIFE+ project "*The Return of the Neophron*" (LIFE10 NAT/BG/000152). BSPB, Sofia.

Donazar, J. A., Margalida, A., Campion, D. 2009a. Vultures, feeding stations and sanitary legislation: a conflict and its consequences from the perspective of Conservation Biology. Munibe 29 (Suplemento). *Sociedad de Ciencias* Aranzadi, San Sebastian.

Donazar, J. A., Margalida, A., Carrete, M. & Sanchez-Zapata, J. A. 2009b. Too sanitary for vultures. *Science*, 326: 664. https://doi.org/10.1126/science. 326_664a.

Donazar, J.A., Cortes-Avizanda, A. & Carrete, M. 2010. Dietary shifts in two vultures after the demise of supplementary feeding stations: consequences of the EU sanitary legislation. *European Journal of Wildlife Research* 56: 613–621. https://doi.org/10.1007/s10344-009-0358-0.

Doreen, F. 2015. *India gets first restaurant for vultures to save the birds from extinction.* September 08, 2015. Retrieved from https://www.firstpost.com/living/i ndia-gets-first-restaurant-for-vultures-to-save-the-birds-from-extinction-2425900.html.

Dupont, H., Mihoub, J-B., Becu, N. & Sarrazin, F. 2011. Modelling interactions between scavenger behaviour and farming practices: impacts on scavenger population and ecosystem service efficiency. *Ecological Modelling*, 222: 982–992. https://doi.org/10.1016/j.ecolmodel.2010.11.029.

Dupont, H., Mihoub, J. B., Bobbe, S. & Sarrazin, F. 2012. Modelling carcass disposal practices: implications for the management of an ecological service provided by vultures. *Journal of Applied Ecology*, 49: 404–411. https://doi.org/10.1111/j.1365-2664.2012.02111.x.

Elbroch, L. M. & Wittmer, H. U. 2012. Table scraps: inter-trophic food provisioning by pumas. *Biological Letters*, 8: 776–779. https://doi.org/10.1098/rsbl.2012.0423.

Ewen, J. G., Walker, L., Canessa, S. & Groombridge, J.J. 2015. Improving supplementary feeding in species conservation. *Conservation Biology*, 29: 341–349. https://doi.org/10.1111/cobi.12410.

Fasciolo, A., Delgado, M. D. M., Cortes, G., Soutullo, A. & Penteriani, V. (2016). Limited prospecting behaviour of juvenile eagle owls *Bubo bubo* during natal dispersal: implications for conservation. *Bird Study*, 63: 128–135. https://doi.org/10.1080/00063657.2016.1141166.

Ferrer, M., Morandini, V., Baguena, G., & Newton, I. 2017. Reintroducing endangered raptors: A case study of supplementary feeding and removal of nestlings from wild populations. *Journal of Applied Ecology*, 1(8). https://doi.org/ 10.1111/1365-2664.13014.

Fielding, D., Newey, S., Van der Wal, R., & Justin, I. R. 2014. Carcass provisioning to support scavengers, evaluating a controversial nature conservation practice. *Ambio*, 43: 810–819. https://doi.org/10.1007/s13280-013-0469-4.

Fluhr, J., Benhamou, S., Riotte-Lambert, L. & Duriez, O. 2017. Assessing the risk for an obligate scavenger to be dependent on predictable feeding sources. *Biological Conservation*, 215: 92–98. https://doi.org/10.1016/j.biocon.2017. 07.030.

Friedman, R. & Mundy, P. 1983. The use of "restaurants" for the survival of vultures in South Africa. In Wilbur S. R. & Jackson J. A. (Eds.), *Vulture biology and management.* 345–355. Berkeley, CA: University of California Press.

Galligan, T. H. 2019. Building a vulture safe zone. Vulture Bulletin, *Annual News Letter*, 8:3-4. Bird conservation Nepal.

Gautam, D. 2015. *Vulture restaurant in Punjab on brink*. DHNS April 24, 2015. Retrieved from https://www.deccanherald.com/content/473713/vulture-restaurant -punjab-brink.html.

Garcia-Fernandez, A. J., Martinez-Lopez, E., Romero, D., Maria-Mojica, P., Godino, A. & Jimenez, P. 2005. High levels of blood lead in griffon vultures (*Gyps fulvus*) from Cazorla natural park (southern Spain). *Environmental Toxicology*, 20: 459–463. https://doi.org/10.1002/tox.20132.

Garcia-Ripolles, C. & Lopez-Lopez, P. 2011. Integrating effects of supplementary feeding, poisoning, pollutant ingestion and wind farms of two vulture species in Spain using a population viability analysis. *Journal of Ornithology*, 152: 879–888. https://doi.org/10.1007/s10336-011-0671-8.

Garvin, J. C., Jennelle, C. S., Drake, D. & Grodsky, S. M. 2011. Response of raptors to a windfarm. *Journal of Applied Ecology*, 48: 199–209. https://doi.org/10. 1111/j.1365-2664.2010.01912.x.

Genero, F., Franchini, M., Fanin, Y., & Filacorda, S. 2020. Spatial ecology of non-breeding Eurasian Griffon Vultures *Gyps fulvus* in relation to natural and artificial food availability. *Bird Study*. https://doi.org/10.1080/00063657.2020.1734534.

Gilbert, M., Watson, R. T., Ahmed, S., Asim, M. & Johnson, J. A. 2007. Vulture restaurants and their role in reducing diclofenac exposure in Asian vultures. *Bird Conservation International*, 17: 63–77. https://doi.org/10.1017/S095927090 6000621.

Gomara, B., Ramos, L., Gangoso, L., Donazar, J. A. & Gonzalez, M. J. 2004. Levels of polychlorinated biphenyls and organochlorine pesticides in serum samples of Egyptian vulture (*Neophron percnopterus*) from Spain. *Chemosphere*, 55: 577–583. https://doi.org/10.1016/j.chemosphere.2003.11.034.

Gomez-Ramirez, P., Jiminez-Montalban, P. J., Delgado, D., Martinez-Lopez, E., Maria-Mojica, P., Godino, A., & Garcia-Fernandez, A.J. 2018. Development of a QuEChERS method for simultaneous analysis of antibiotics in carcasses for supplementary feeding of endangered vultures. *Science of the Total Environment*, 626: 319-327. https://doi.org/10.1016/j.scitotenv.2018.01.060.

Gonzalez, L. M., Margalida, A., Sanchez, R. & Oria, J. 2006. Supplementary feeding as an effective tool for improving breeding success in the Spanish imperial eagle (*Aquila adalberti*). *Biological Conservation*, 129: 477–486. https://doi.org/ 10.1016/j.biocon.2005.11.014.

Grande, J. M., Carrete, M., Ceballos, O., Tella, J. L. & Donazar, J. A. 2009a. Relevance of vulture restaurants for the conservation of the Egyptian Vulture Neophron percnopterus in Spain. In: Donazar J. A., Margalida A., Campión D. (Eds.), *Vultures, feeding stations and sanitary legislation: a conflict and their consequences from the perspective of conservation biology*, 254-275. Munibe 29 (Suppl.) Sociedad de Ciencias Aranzadi. Donostia.

Grande, J. M., Serrano, D., Tavecchia, G., Carrete, M., Ceballos, O., Dıaz-Delgado, R., Tella, J. L. & Donazar, J. A. 2009b. Survival in a long-lived territorial migrant: effects of life-history traits and ecological conditions in wintering and breeding areas. *Oikos,* 118: 580 - 590. https://doi.org/10.1111/j.1600-0706.2009.17218.x.

Hernandez, M., Colomer, M. A., Pizarro, M. & Margalida, A. 2018. Changes in eggshell thickness and ultrastructure in the Bearded vulture (*Gypaetus barbatus*) Pyrenean population: a long-term analysis. *Science of the Total Environment*, 624: 713–721. https://doi.org/10.1016/j.scitotenv.2017.12.150.

Hille, S. M., Korner-Nievergelt, F., Bleeker, M. & Collar, N. J. 2016. Foraging behaviour at carcasses in an Asian vulture assemblage: towards a good restaurant guide. *Bird Conservation International* 26: 263–272. https://doi.org/ 10.1017/S09 59270915000349.

Hirzel, A. H., Posse, B., Oggier, P.-A., Crettenand, Y., Glenz, C. & Arlettaz, R. 2004. Ecological requirements of reintroduced species and the implications for release policy: the case of the bearded vulture. *Journal of Applied Ecology*, 41: 1103–1116. https://doi.org/10.1111/j.0021-8901.2004.00980.x.

Hla, H., Shwe, NM., Htun, TW., Zaw, SM., Mahood, S., Eames, JC. & Pilgrim, JD. 2011. Historical and current status of vultures in Myanmar. *Bird Conservation International*, 2010: 1-12. https://doi.org/10.1017/S0959270910000560.

Houston, D. C. 2006. Reintroduction programmes for vulture species. In Houston, D. C. & Piper, S. E. (Eds.), *Proceedings of the International Conference on Conservation and Management of Vulture Populations: 87–97 Natural History Museum of Crete & WWF Greece*, Thessaloniki.

Jha, K. K. 2015. Distribution of vultures in Uttar Pradesh, India. *Journal of Threatened Taxa*, 7(1): 6750–6763; http://dx.doi.org/10.11609/JoTT.o3319.6750-63.

Jha, K. K. 2017. *Vulture Atlas Central India – MP. Indian Institute of Forest Management*, Bhopal, India.

Jha, K. K., Campbell, M. O. & Jha, R. 2020. Vultures, their population status and some ecological aspects in an Indian stronghold. *Notulae Scientia Biologicae,* 12(1): 124-142. https://doi.org/10.15835/nsb12110547.

Jha, K.K., Jha, R. & Campbell, M.O. 2021.The Distribution, Nesting Habits and Status of Threatened Vulture Species in Protected Areas of Central India. *Ecological Questions* 32(2021)2. http://dx.doi.org/10.12775/EQ.2021.020.

Kane, A., Jackson, A. L., Monadjem, A., Colomer, M. A. & Margalida, A. 2014. Carrion ecology modelling for vulture conservation: are vulture restaurants needed to sustain the densest breeding population of the African white-backed vulture? *Animal Conservation*, 18(3): 279 - 286. https://doi.org/10.1111/acv. 12169.

Kane, A., Wolter, K., Neser, W., Kotze, A., Naidoo, V. & Monadjem, A. 2016. Home range and habitat selection of Cape Vultures *Gyps coprotheres* in relation to supplementary feeding. *Bird Study*, 63(3): 387–394. http://dx.doi.org/10.1080/00063657.2016.1214105.

Karimov, T. & Mammadov, A. 2018. Impact of Trophic Conditions on Biological Parameters of Egyptian Vultures (*Neophron percnopterus*) in Azerbaijan. *Ciencia e Tecnica*, 38(11): 136-148.

Kruger, S. C., Simmons, R. E. & Amar, A. 2015. Anthropogenic activities influence the abandonment of Bearded Vulture (*Gypaetus barbatus*) territories in southern Africa. *The Condor: Ornithological Applications*, 117: 94–107. https://doi.org/10.1650/CONDOR-14-121.1.

Kuvlesky, W. P., Brennan, L. A., Morrison, M. L., Boydston, K. K., Ballard, B. M. & Bryant, F. C. 2007. Wind energy development and wildlife conservation: challenges and opportunities. *Journal of Wildlife Management*, 71: 2487–2498. https://doi.org/10.2193/2007-248.

Laiolo, P. & Tella, J. L. 2006. Fate of unproductive and unattractive habitats: recent changes in Iberian steppes and their effects on endangered avifauna. *Environmental Conservation*, 33: 223–232. https://doi.org/10.1017/S0376892906003146.

Lambertucci, S. A., Speziale, K. L., Rogers, T. E. & Morales, J. M. 2009. How do roads affect the habitat use of an assemblage of scavenging raptors? *Biodiversity and Conservation*, 18: 2063–2074. https://doi.org/10.1007/s10531-008-9573-3.

Larsen, H. O. & Olsen, C. S. 2007. Unsustainable collection and unfair trade? Uncovering and assessing assumptions regarding Central Himalayan medicinal plant conservation. *Biodiversity and Conservation*, 16:1679–1697. https://doi.org/10.1007/s10531-006-9039-4.

Leddy, K. L., Higgins, K. F. & Naugle, D. E. 1999. Effects of wind turbines on upland nesting birds in conservation reserve program grasslands. *Wilson Bulletin*, 11: 100–104.

Lieury, N., Gallardo, M., Poncho, C., Besnard, A. & Millon, A. 2015. Relative contribution of local demography and immigration in the recovery of a geographically-isolated population of the endangered Egyptian vulture. *Biological Conservation*, 191: 349–356. https://doi.org/10.1016/j.biocon.2015.07.008.

Lopez-Bao, J.V., Palomares, F., Rodríguez, A. & Delibes, M. 2010. Effects of food supplementation on home-range size, reproductive success, productivity and recruitment in a small population of Iberian lynx. *Animal Conservation*, 13: 35–42. https://doi.org/10.1111/j.1469-1795.2009.00300.x.

Lopez-Lopez, P., Garcia-Ripolles, C. & Urios, V. 2014. Food predictability determines space use of endangered vultures: implications for management of supplementary feeding. *Ecological Applications*, 24(5): 938-949. https://doi.org/ 10.1890/13-2000.1.

Mallikarjuna, G. 2012. *A restaurant for vultures*. June 06, 2012. Retrieved from https://www.thehindu.com/news/cities/Delhi/a-restaurant-for-vultures/article3497590.ece.

Margalida, A., Garcia, D., & Cortes-Avizanda, A. 2007. Factors influencing the breeding density of Bearded Vultures, Egyptian Vultures and Eurasian Griffon

Vultures in Catalonia (NE Spain): management implications. *Animal Biodiversity and Conservation,* 30(2): 189-200. http://hdl.handle.net/10261/178325.

Margalida, A. 2010. Supplementary feeding during the chick-rearing period is ineffective in increasing the breeding success in the bearded vulture (*Gypaetus barbatus*). *European Journal of Wildlife Research*, 56: 673–678. https://doi.org/10.1007/s10344-010-0366-0.

Margalida, A. 2017. Importance of Long-Term Studies to Conservation Practice: The Case of the Bearded Vulture in the Pyrenees. In Catalan, J. (Ed.), High Mountain Conservation in a Changing World, *Advances in Global Change Research*, 343-383. https://doi.org/10.1007/978-3-319-55982-7_15.

Margalida, A. & Colomer, M. A. 2012. Modelling the effects of sanitary policies on European vulture conservation. *Science Reporter* 2: 753. https://doi.org/10.1038/srep00753.

Margalida, A., Colomer, MA., Sanchez, R., Sanchez FJ., Oria J. & Gonzalez, LM. 2017. Behavioural evidence of hunting and foraging techniques by a top predator suggests the importance of scavenging for pre-adults. *Ecology and Evolution*, 7: 4192–4199. https://doi.org/10.1002/ece3.2944.

Margalida, A., Moreno-Opo, R., Arroyo, B. E. & Arredondo, A. 2011a. Reconciling the conservation of an endangered species with an economically important anthropogenic activity: interactions between cork exploitation and the cinereous vulture *Aegypius monachus* in Spain. *Animal Conservation*, 14: 167–174. https://doi.org/10.1111/j.1469-1795.2010.00412.x.

Margalida, A., Colomer, M.A. & Sanuy, D. 2011b. Can wild ungulate carcasses provide enough biomass to maintain avian scavenger populations? An empirical assessment using a bio-inspired computational model. *PLoS ONE* 6: e20248. https://doi.org/10.1371/journal.pone.0020248.

Margalida, A., Campion, D. & Donazar, J.A. 2011c. European vultures' altered behaviour. *Nature*, 480: 457. https://doi.org/10.1038/480457b.

Margalida, A. & Ogada D. 2018. Old World Vultures in a Changing Environment. In Sarsola, J. H., Grand, J. M. & Negro, J. J. (Eds.), *Birds of Prey*, 457-472. Biology and Conservation in XXI Century. Springer International Publishing, Gewerbestrasse 11, 6330 Cham, Switzerland.

Margalida, A., Campión, D. & Donazar, J. A. 2014a. Vultures vs livestock: conservation relationships in an emerging conflict between humans and wildlife. *Oryx*, 48: 172–176. https://doi.org/10.1017/S0030605312000889.

Margalida, A., Colomer, M. A. & Oro, D. 2014b. Man-induced activities modify demographic parameters in a long-lived species: effects of poisoning and health policies. *Ecological Application*, 24: 436–444. https://doi.org/10.1890/13-0414.1.

Mariya, Karimjee. 2013. *Pakistan Inside/Out: A restaurant for vultures.* Global Post. September 03, 2013. Retrieved from https://www.pri.org/stories/2013-09-03/pakistan-insideout-restaurant-vultures.

Marin, C., Torres, C., Marco-Jiminez, F., Cerda-Cuellar, M., Sevilla, S., Ayats, T. & Vega, S. 2018. Supplementary feeding stations for conservation of vultures could be an important source of monophasic *Salmonella typhimurium*. *Science of the Total Environment*, 636: 449-455. https://doi.org/10.1016/j.scitotenv.2018.04.310.

Martinez-Abrain, A., Tavecchia, G., Regan, H. M., Jimenez, J., Surroca, M. & Oro, D. 2012. Effects of wind farms and food scarcity on a large scavenging bird species following an epidemic of bovine spongiform encephalopathy. *Journal of Applied Ecology*, 49: 109–117. https://doi.org/10.1111/j.1365-2664.2011.0208.

Mateo-Tomas, P., Olea, P. P., Moleon, M., Vicente, J., Botella F., Selva, N., Vinuela, J. & Sanchez-Zapata, JA. 2015. From regional to global patterns in vertebrate scavenger communities subsidized by big game hunting. *Diversity and Distributions*, 21: 913–924. https://doi.org/10.1111/ddi.12330.

McKean, S., Mander, M., Diederichs, N., Ntuli, L., Mavundla, K., Williams, V. & Wakelin, J. 2013. The impact of traditional use on vultures in South Africa. *Vulture News*, 65: 15-36. https://doi.org/10.4314/vulnew.v65i1.2.

Moore, J. C., Berlow, E. L., Coleman, D. C., de Ruiter P. C., Dong, Q., Hastings, A., Johnson, N. C., McCann, K. S., Melville, K., Morin P. J., Nadelhoffer, K., Rosemond, A. D., Post, D. M., Sabo, J. L., Scow, K. M., Vanni M. J. & Wall D. H. 2004. Detritus, trophic dynamics and biodiversity. *Ecology Letters*, 7: 584 - 600. https://doi.org/10.1111/j.1461-0248.2004.00606.x/.

Moleon, M., Sanchez-Zapata J. A., Margalida, A., Carrete, M., Donazar, J. A. & Owen-Smith, N. 2014. Hominids and scavengers: the evolution of interactions and ecosystem services. *Bioscience*, 64(5): 394–403. https://doi.org/10.1093/biosci/biu034/.

Morales-Reyes, Z., Pérez-García, J. M., Moleón, M., Botella, F., Carrete, M., Lazcano, C., Moreno-Opo, R., Margalida, A., Donázar, J. A. & Sánchez-Zapata, J. A. 2015. Supplanting ecosystem services provided by scavengers raises greenhouse gas emissions. *Scientific Reports*, 5: 7811. https://doi.org/10. 1038/srep07811.

Moreno-Opo, R., Trujillano, A., Arredondo, A., González, L. M. & Margalida, A. 2015. Manipulating size, amount and appearance of food inputs to optimize supplementary feeding programs for European vultures. *Biological Conservation*, 181: 27–35. https://doi.org/10.1016/j.biocon.2014.10.022.

Moreno-Opo, R., Trujillano, A. & Margalida, A. 2016. Behavioral coexistence and feeding efficiency drive niche partitioning in European avian scavengers. *Behavioral Ecology*, 27: 1041–1052. https://doi.org/10.1093/beheco/arw010.

Moreno-Opo, R., Trujillano, A. & Margalida, A. 2020. Larger size and older age confer competitive advantage: dominance hierarchy within European vulture guild. *Scientific Reports*, 10: 2430. https://doi.org/10.1038/s41598-020-59387-4.

Mukherjee, A., Galligan, T. H., Prakash, V., Paudel, K., Khan, U., Prakash, S., Ranade, S., Shastri, K., Dave, R., Donald, P., Bowden, C. 2014. Vulture Safe Zones to save Gyps Vultures in South Asia. *MISTNET*, 15(3): 4-21.

Mundy, P. J., Butchart, D., Ledger, J. A. & Piper, S. E. 1992. *The vultures of Africa*. San Diego, CA: Academic Press Inc.

Negro, J. J., Sarasola, J. H. & Barclay, J. H. 2007. Augmenting wild populations and food resources. In Bird, D. & Bildstein K. (Eds.), *Raptor conservation and management techniques*. Publisher: Hancock House Publisher.

Ogada, D., Keesing, F. & Virani, M. Z. 2011. Dropping dead: Causes and consequences of vulture population declines worldwide. *Annals of the New York Academy of Sciences*, 1249(1): 57-71. https://doi.org/10.1111/j.1749-6632.2011.06293.x.

Ogada, D.L. 2014. Power of poison: pesticide poisoning of Africa's wildlife. *Annals of the New York Academy of Sciences*, 1322: 1-20. https://doi.org/10.1111/nyas.12405.

Ogada, D., Shaw, P., Beyers, R. L., Buij, R., Murn, C., Thiollay, J. M., Beale, C. M., Holdo, R. M., Pomeroy, D., Baker, N., Kruger, S. C., Botha, A., Virani, M. Z., Monadjem, A., & Sinclair, A. R. E. 2016. Another Continental Vulture Crisis: Africa's Vultures Collapsing toward Extinction. *Conservation Letters*, 9(2): 89–97. https://doi.org/10.1111/conl.12182.

Olea, P. P. & Mateo-Tomás, P. 2009. The role of traditional farming practices in ecosystems conservation: the case of transhumance and vultures. *Biological Conservation*, 142: 1844–1853. https://doi.org/10.1016/j.biocon.2009.03.024.

Olea, P. P., Mateo-Tomas, P. & Sanchez-Zapata, J. A. 2019. Introduction to the Topic of Carrion Ecology and Management. In Olea, P. P., Mateo-Tomas, P., Sanchez-Zapata, J. A. (Eds.), *Carrion Ecology and Management*, Springer Nature Switzerland AG.

Oro, D., Margalida, A., Carrete, M., Heredia, R. & Donazar, J. A. 2008. Testing the Goodness of Supplementary Feeding to Enhance Population Viability in an Endangered Vulture. *PLoS ONE*, 3(12): e4084. https://doi.org/10.1371/journal.pone.0004084.

Oro, D., Genovart, M., Tavecchia, G., Fowler, M. S. & Martınez-Abraın, A. 2013. Ecological and evolutionary implications of food subsidies from humans. *Ecology Letters*, 16: 1501–1514. https://doi.org/10.1111/ele.12187.

Ostfeld, R. S. & Keesing, F. 2000. Pulsed resources and community dynamics of consumers in terrestrial ecosystems. *Trends in Ecology & Evolution*, 15: 232–237. https://doi.org/10.1016/S0169-5347(00)01862-0.

Pacheco, R. E., Jacome, N. L., Astore, V., Borghi, C. E. & Pina, C. I. 2020. Pesticides: The most threat to the conservation of the Andean condor (*Vultur gryphus*). *Biological Conservation*, 242 (2020) 108418. https://doi.org/10.1016/j.biocon.2020.108418.

Parmenter, R. R. & MacMahon, J. A. 2009. Carrion decomposition and nutrient cycling in a semiarid shrub-steppe ecosystem. *Ecological Monograph*, 79: 637–661. https://doi.org/10.1890/08-0972.1.

Pauli, J. N., Donadio, E. & Lambertucci, S. A. 2018. The corrupted carnivore: how humans are rearranging the return of the carnivore-scavenger relationship. *Ecology*, 99: 2122–2124. https://doi.org/10.1002/ecy.2385.

Plaza, P. I., Martinez-Lopez, E. & Lambertucci, S. A. 2019. The perfect threat: Pesticides and vultures. *Science of the Total Environment*, 687: 1207-1218. https://doi.org/10.1016/j.scitotenv.2019.06.160.

Robb, G. N., McDonald, R. A., Chamberlain, D.E. & Bearhop, S. 2008. Food for thought: supplementary feeding as a driver of ecological change in avian populations. *Frontiers in Ecology and the Environment*, 6: 476–484. https://doi.org/10.1890/060152.

Saidu, Y. & Buij, R. 2013. Traditional medicine trade in vulture parts in northern Nigeria. *Vulture News*, 65: 4-14. https://doi.org/10.4314/vulnew.v65i1.1.

Sánchez, R., Margalida, A., González, L. M. & Oria J. 2009. Temporal and spatial differences in the feeding ecology of the Spanish imperial eagle *Aquila adalberti* during the non-breeding season: effects of the rabbit population crash. *Acta Ornithologica*, 44: 53–58. https://doi.org/10.3161/000164509X464885.

Sanchez-Zapata, J. A., Eguia, S., Blazquez, M. Moleon, M. & Botella, F. 2010. Unexpected role of ungulate carcasses in the diet of golden eagles *Aquila chrysaetos* in Mediterranean mountains. *Bird Study*, 57: 352–360. https://doi.org/10.1080/00063651003674946.

Sarrazin, F., Bagnolini, C., Pinna, J. L., Danchin, E., & Clobert, J. 1994. High survival estimates of griffon vultures (*Gyps fulvus fulvus*) in a reintroduced population. *Auk*, 111: 853–862.

Schabo, D. G., Heuner, S., Neethling, M. V., Rosner, S., Uys, R. & Farwig, N. 2016. Long-term data indicates that supplementary food enhances the number of breeding pairs in a Cape Vulture *Gyps coprotheres* colony. *Birdlife International*, 140-152. https://doi.org/10.1017/S0959270915000350.

Schoenly, K. 1991. Food web structure in dung and carrion arthropod assemblages, null models and monte-carlo simulation: applications to medical veterinary entomology. *Journal of Agriculture Entomology*, 8: 227–249. Retreived from http://scentsoc.org/Volumes/JAE/v8/4/00084227.pdf.

Schoenly, K. & Reid, W. 1987. Dynamics of heterotrophic succession in carrion arthropod assemblages: discrete series or a continuum of change? *Oecologia*, 73: 192–202. https://doi.org/10.1007/BF00377507.

Sekercioglu, C. H. 2006. Increasing awareness of avian ecological function. *Trends in Ecology & Evolution*, 21: 464–471. https://doi.org/10.1016/j.tree.2006.05.007.

Selva, N. & Fortuna, M. A. 2007. The nested structure of a scavenger community. *Proceedings of Royal Society of London*, B 274: 1101–1108. https://doi.org/10.1098/rspb.2006.0232.

Selva, N., Moleon, M., Sebastian-Gonzalez, E., DeVault, T. L., Quaggiotto, M. M., Bailey, D. M., Lambertucci, S. A., & Margalida, A. 2019. Vertebrate Scavenging Communities. In Olea, P. P., Mateo-Tomas, P., Sanchez-Zapata, J. A. (Eds.), *Carrion Ecology and Management*, Springer Nature Switzerland AG 2019.

Sese, J. A., Antor, R. J., Alcantara, M., Ascaso, J. C. & Gil, J. A. 2005. La alimentacion suplementaria en el quebrantahuesos: estudio de un comedero del Pirineo occidental aragones [Supplementary feeding in the Bearded Vulture: a feeding study of the western Pyrenees Aragon]. In Margalida, A. & Heredia, R. (Ed.), Biología de la conservación del Quebrantahuesos *Gypaetus barbatus* en España [*Conservation biology of the Bearded Vulture Gypaetus barbatus in Spain*], 279–304. Madrid, Organismo Autónomo Parques Nacionales.

Smith-Hall, C., Larsen, H. O. & Pouliot, M. 2012. People, plants and health: a conceptual framework for assessing changes in medicinal plant consumption.

Journal of Ethnobiology and Ethnomedicine, 2012, 8: 43. https://doi.org/10.1186/1746-4269-8-43.

Stattersfield, A. J. & Capper, D. R. 2000. Threatened birds of the world. *BirdLife International*, Cambridge, United Kingdom.

Sumita, Sarkar. 2017. *Vultures no longer visit conservation 'restaurant' at Harsul*. TNN, 31/10/2017. Retrieved from http://timesofindia.indiatimes.com/article show/61364638.cms?utm_source=contentofinterest&utm_medium=text&utm_campaign=cppst.

Swift, M. J., Heal, O. W. & Anderson, J. M. 1979. *Decomposition in terrestrial ecosystems*. Berkeley, CA: University of California Press.

Taggart, M. A., Senacha, K. R., Green, R. E., Cuthbert, R., Jhala, Y. V., Meharg, A. A., Mateo, R. & Pain, D. J. 2009. Analysis of nine NSAIDs in ungulate tissues available to critically endangered vultures in India. *Environmental Science & Technology*, 43: 4561–4566. https://doi.org/10.1021/es9002026.

Tella, J. L. 2001. Action is needed now, or BSE crisis could wipe out endangered birds of prey. *Nature*, 410: 408. https://doi.org/10.1038/35068717.

Telleria, J. L. 2009. Overlap between wind power plants and griffon vultures Gyps fulvus in Spain. *Bird Study*, 57: 352–360. https://doi.org/10.1080/00063650902792056.

Terrasse, J. F. 1985. The effects of artificial feeding on Griffon, Bearded and Egyptian vultures in the Pyrenees. In Newton, I. & Chancellor, R. D. (Eds.), Conservation studies on raptors. 429–430. *International Council for Bird Preservation Tech. Publ. 5*. Cambridge, United Kingdom.

Tickner, D., Opperman, J. J., Abell, R., Acreman, M., Arthington, A. H., Bunn, S. E., Cooke, S. J., Dalton, J., Darwall, W., Edwards, G., Harrison, I., Hughes, K., Jones, T., Leclère, D., Lynch, A. J., Leonard, P., McClain, M. E., Muruven, D., Olden, J. D., Ormerod, S. J., Robinson, J., Tharme, R. E., Thieme, M., Tockner, K., Wright, M. & Young, L. 2020. Bending the Curve of Global Freshwater Biodiversity Loss: An Emergency Recovery Plan. *BioScience*, 70(4): 330–342. https://doi.org/10.1093/biosci/biaa002.

Travaini, A., Donazar, J. A., Rodriguez, A. Ceballos, O., Funes, M., Delibes, M. & Hiraldo, F. 1998. Use of European hare (*Lepus europaeus*) carcasses by an avian scavenging assemblage in Patagonia. *Journal of Zoology*, 246: 175–181.

Warren, Cornwall. 2018. *Vulture safe zones' aim to rescue a vital but unloved scavenger*. Mar. 9, 2018. Retrieved from https://www.sciencemag.org/news/2018/03/vulture-safe-zones-aim-rescue-vital-unloved-scavenger.

Wilson, E. E. & Wolkovich, E. M. 2011. Scavenging: how carnivores and carrion structure communities. *Trends in Ecology & Evolution*, 26: 129–135. https://doi.org/10.1016/j.tree.2010.12.011.

Xirouchakis, S. & Nikolakakis, M. 2002. Conservation implications of the temporal and spatial distribution of bearded vulture *Gypaetus barbatus* in Crete. *Bird Conservation International*, 12: 269–280. https://doi.org/10.1017/S0959270902002174.

Yarnell, R. W., Phipps, W. L., Dell, S., MacTavish, L. M. & Scot, D. M. 2014. Evidence that vulture restaurants increase the local abundance of mammalian carnivores in South Africa. *African Journal of Ecology*, 53(3): 287-294. https://doi.org/10.1111/aje.12178.

Yudhvir, Rana. 2016. *With no funds, last 'vulture restaurant' shuts down in Pathankot village.* TNN Mar 2, 2016. Retrieved from http://timesofindia.indiatimes.com/articleshow/51216528.cms?utm_source=contentofinterest&utm_medium=text&utm_campaign=cppst.

Zuberogoitia, I., Martinez, J. E., Margalida, A., Gomez, I., Azkona, A. & Martínez, J. A. 2010. Reduced food availability induces behavioural changes in Griffon Vulture *Gyps fulvus*. *Ornis Fennica*, 87: 52–60. Retrieved from http://www.globalraptors.org/Grin/researchers/uploads/521/reduced_food_2010.pdf.

In: Critical Research Techniques in Animal ...
Editors: Michael O'Neal Campbell et al. ISBN: 978-1-53619-846-1

Chapter 4

ENVIRONMENTAL CHEMISTRY AND THE CONSERVATION BIOLOGY OF RAPTORS AND WATERBIRDS

Michael O'Neal Campbell*
Simon Fraser University, British Columbia, Canada

ABSTRACT

The pollution chemistry of the environment, especially chemical element and compound accumulation at the upper levels of the food chain is a fundamental aspect of the study of the conservation biology of raptors (birds of prey) and waterbirds, with increasing relevance given the largely human-induced pollution of ecosystems and food chains. Crucially, chemical element and compound intrusion is a factor behind local species extinction and the mitigation of species recovery efforts. Chemical pollutants are variously defined, but generally they are elements or compounds that may negatively affect the physiology and health of animals. The key points concern the contributory factors of industrial, agricultural and urban technology, and related human activities, with immediate or potential hazards to animals, plants and people, and animal body threshold levels for chemical pollutants. Such chemical elements and compounds include mercury, lead, copper, arsenic, chromium, copper, rodenticides and insecticides, nonsteroidal anti-inflammatory drugs (NSAIDs) and polychlorinated biphenyls -PCBs. Raptors and waterbirds are particularly susceptible to chemical poisoning, as they eat animals that have ingested such poisons from infected materials, with accumulations at the upper levels of the food chain. This chapter takes a snapshot of the vast area of environmental chemical pollution and examines the factors for such events, the impacts and the ameliorative actions that have been attempted as part of conservation management. Raptors and waterbirds are taken as examples, due to their food chain status and the primary impact of pollutants on small mammals and fish, and carcasses of these and larger mammals. The

* Corresponding Author's E-mail: ecol55@hotmail.com.

methodologies employed in the studies are also examined. It is argued that the main culprits for such events, such as agricultural and industrial development, have been increasingly managed by technological advancements and increased ecological knowledge and more competent management. However, the increasing intensity of agricultural and industrial development, compromises the benefits derived from such efforts. Increasingly, fields of knowledge such veterinary science and environmental management claim the interface between environmental chemistry and conservation biology. This study contributes to ornithological, chemical, ecological, social and policy studies.

INTRODUCTION

Chemical poisoning is a major determinant of avian health, distribution and numbers globally (Campbell, 2015). Additionally, with the outreach of relevant research, "public interest in environmental contamination has increased in recent years and therefore monitoring, evaluation, management, and remediation of ecological damage are required" (Nighat et al. (2013, 488). Birds of prey (raptors) and waterbirds (sea and river denizens) are particularly susceptible to such possibilities, as they sit at the apex of the food chain, hence any chemicals in the environment may be transmitted up the food chain through plants and levels of prey animals (including rodents, small birds, fish, and carcasses of these and larger animals) cross (Cross, 2009; Campbell, 2015, 2018). Raptors and waterbirds (Table 1) are major also players in global ecosystems and consequently their extinction or population reduction may jeopardize ecological stability (Campbell, 2018). Here, raptors are defined as birds of the Orders *Accipitriformes*, Vieillot, 1816 (formerly *Falconiformes*, including diurnal hawks, eagles, falcons and vultures), *Cathartiformes*, Coues, 1884 (New World vultures) and *Strigiformes* Wagler, 1830 (nocturnal owls) (Table 1). Waterbirds are a wider, more diverse grouping including the orders *Charadriiformes* Huxley, 1867 (gulls and waders), *Anseriformes* Wagler, 1831 (swans, geese and ducks), *Podicipediformes* Fürbringer, 1888 (grebes), *Gaviiformes* Wetmore & Miller, 1926 (loons or divers), *Ciconiiformes* Bonaparte, 1854 (storks), *Pelecaniformes* Sharpe, 1891 (pelicans, herons), *Suliformes,* Sharpe, 1891 (gannets, boobies, and possibly cormorants and shags) and *Gruiformes*, Bonaparte, 1854. Commoner families of these Orders are listed in Table 1.

Raptors and waterbirds share similar predatory activities, may prey on similarly sized prey species and are principal obligate (full-time) and facultative (part-time) scavengers in marine, riverine, vegetated and urban areas (Campbell, 2015). The concentration or mere existence of pollutants in the environments of raptors and waterbirds (i.e., plants, water, and prey species such as small fish, small mammals such as rodents, and small birds such as passerines and pigeons) begins direct and indirect paths to avian toxicity with dire consequences for raptor and waterbird pollutant levels. For example, for waterbirds, "heavy metals such as mercury and cadmium are of particular concern because of their high toxicity. Both occur naturally in seawater, and locally increased levels can arise through natural causes as well as through human-induced contamination. "Fish-eating seabirds contain higher levels of mercury in their livers" than other birds (Parslow & Jefferies, 1973). For raptors, such pollutants are ingested by their prey, usually passerines, pigeons and small rodents (Campbell, 2015). These chemicals may be ingested and accumulate in the livers, kidneys, gastrointestinal (digestive) tract, nervous system, eggs, bones and even feathers of affected birds, resulting in illness, incapacity and even death, depending on the level of accumulation (Frantz et al., 2012, 23). Heavy metal exposure may affect bird reproduction and migration, in addition to hormone and respiratory systems (Scheuhammer, 1987; Haschek et al., 2013; Williams et al., 2014; Kitowski et al., 2016). These threshold levels may vary according to the avian species, the body organ affected, and the toxicity of the chemical considered, but there may be generalizable levels (Campbell, 2015). Heavy metals (a metal of relatively high density or atomic weight), such as lead, iron and zinc, and chemicals manufactured for agricultural and industrial use, such as anti-inflammatory drugs (NSAIDs), polychlorinated biphenyls (PCBs), insecticides, rodenticides and insecticides are currently the commonest forms of chemical poisoning for birds (Campbell, 2018).

Table 1. Classification of common raptors and waterbirds

Orders	**Families**	**English names**
Birds of Prey or Raptors		
Accipitriformes	*Accipitridae*, Vieillot, 1816	Eagles, hawks, kites, vultures
	Pandionidae, Sclater & Salvin, 1873	Ospreys
Cathartiformes	*Cathartidae*, Lafresnaye, 1839	New World vultures
	Sagittariidae, Finsch & Hartlaub, 1870	Secretary birds
	Falconidae, Leach, 1820	Falcons, caracaras
Strigiformes	*Strigidae*, Leach, 1820	Typical owls
	Tytonidae, Ridgway, 1914	Barn owls
Waterbirds		
Charadriiformes	*Laridae*, Rafinesque, 1815	Gulls, terns
	Charadriidae, Leach, 1820	Plovers, lapwings
Anseriformes	*Anatidae*, Leach, 1820	Ducks, geese, swans
Ciconiiformes	*Ciconiidae*, J. E. Gray, 1840	Storks
Pelecaniformes	*Pelecanidae*, Rafinesque, 1815	Pelicans
	Ardeidae, Leach, 1820	Herons, egrets, bitterns
Suliformes	*Sulidae*, Reichenbach, 1849	Gannets, boobies
	Phalacrocoracidae, Reichenbach, 1850	Cormorants, shags
Gaviiformes	*Gaviidae*, Coues, 1903	Loons
Coraciiformes	*Alcedinidae*, Rafinesque, 1815	Kingfishers
Gruiformes	*Rallidae*, Rafinesque, 1815	Rails, crakes, coots, moorhens

Note that some classifications place the New World vultures in *Accipitriformes.* The waterbirds have nearly 20 families, so the commoner families are presented.

Further problems for bird physiology and behavior may occur (Campbell, 2015). For example, for some species including the passerine prey of raptors, certain heavy metals may contribute to declines in plumage intensity (especially copper, zinc, nickel and lead) (Eeva et al., 1998), fledging success (especially mercury, lead and cadium) (Nam & Lee, 2006; Evers et al., 2008), genetic diversity (especially copper, zinc cadium and lead) (Eeva et al., 2006), and humoral immune responsiveness (especially lead and cadium) (Snoeijs et al., 2004). There may also be "aberrant incubation behaviour, lethargy and wing area asymmetry" (Frantz et al. (2012, 23) largely due to mercury poison (Evers et al., 2008). There are also problems connected to the process of bone mineralization, mostly from lead (Gangoso et al., 2009; Frantz et al., 2012). Here bone mineralization refers to "the process by which the organic bone matrix becomes filled with calcium phosphate nanocrystals" (Kuhn, 2001, 50). This is a key process in bone formation, which is "accomplished by coordinated multicellular activity," basically "a cell-mediated process involving the specific, highly ordered deposition of a unique form of calcium phosphate confined to precise locations within the organic matrix... essential for bone to meet its multifunctional property requirements" (Kuhn, 2001, 50).

Studies of such complex phenomena are dispersed and ongoing, considering the complexity of ecosystem interrelationships. Kitowski et al. (2016) argue that "in spite of extensive surveys, many aspects of heavy metal accumulation and transmission in ecosystems remain unclear" and cite further studies by Helander et al. (2009), Stankovic et al. (2013) and Kitowski et al. (2014). Kitowski et al. (2016) also support the choice of raptors as good models for the study of chemical pollution of bird physiology, as they are apex predators, frequently sedentary, with wide distributions, and are consumers of other animals that in turn ingest contaminants from plants, water and other, yet smaller animals (also cited are Martin et al., 2008; Castro et al., 2011; Rajamani & Subramanian, 2015; Golden et al., 2016). These qualifications also apply to waterbirds, which may play the same role as raptors in riverine and coastal areas and have similar ecological interactions (Campbell, 2015). Basically, "metal concentrations in birds are good indicators of the level of heavy metal concentration in their diet, which is strongly associated with the degree of contamination in the exploited ecosystem" (Kitowski et al. (2016, 146, see also Pain et al., 1997; Castro et al., 2011).

This chapter examines the key ecological issues of chemical pollution applied to raptors and waterbirds in selected, disparate ecosystems. All these species may be considered predators near the apex of their various food chains, as they kill other animals (from species as large as antelopes to those as small as insects, fish and worms) which in their turn may also feed on smaller species or on vegetable materials. Raptors and waterbirds at the top and lower levels of the food chain are susceptible to the chemicals ingested by prey species due to the processes of concentration and accumulation, both indirectly (e.g., lead from bullets fired at prey species, pollution of prey species foods from human industrial actions) and deliberately (e.g., human poisoning of prey species carcasses to eliminate predators). This chapter's focus is not exhaustive, but necessarily selective, picking some species studied under toxicology and environmental chemistry topics. This covers specific elements such as lead, copper and zinc, and also chemical compounds such as pesticides and veterinary chemicals, all of which play major roles in raptor and waterbird survival. The research methods used for studying the impacts of these chemicals on raptors and waterbirds are also discussed. The research methods play the most important role in the assessment and prevention of the negative impacts on the selected birds and their related ecologies.

CHEMICAL ELEMENTS

Chemical elements to be considered are listed in Table 2 below. Individual elements may have the same impact as compounds, as they may combine with other elements and create compounds. These are not exhaustive but represent some of the main chemical agents that compromise the heath and existence of raptors and waterbirds. Some studies have indicated that raptors may have several chemical elements in their systems, hence intensive study may be required to determine which element or compound has the most impact on the bird's health (Campbell, 2015). Therefore, while some studies seek to discover the impacts of one element on the selected species, others seek to discover all the chemical elements in the studied birds. For example, Kitowski et al. (2017) looked at chemical impacts on White-tailed sea eagles, using livers from 22 dead eagles in Eastern Poland. The birds were classified by sex, age (adults over 2 years old, immature less than 2 years old), based on iris color, plumage and gonad development (Kitowski et al., 2017; Helander et al., 1989, 2009; Forsman, 1999). After mineralization using laboratory techniques, and a solution of elements was derived, inductively coupled plasma optical emission spectrometry (ICP-OES) with a Thermo Scientific iCAP Series 6500 analyzer with a charge injection device (CID) was used for the identification of elements. The liver samples were tested with a control sample and divided into two for duplicate analysis and the various, potentially toxic chemicals were identified.

The impacts of the discovered chemicals could be judged through referral to published findings regarding the chemical levels that might adversely affect birds. For example, lead is one of the more dangerous elements for birds. Lead hepatic (liver) concentrations above 6 mg/kg^{-1} dw (6 ppm) have resulted in subclinical toxicity (below the level for visible symptoms) (Pain et al., 1995; Helander et al., 2009). Lead levels equal to or above 10 - 15 mg/kg^{-1} dw (dry weight) (10 – 15 ppm) generally represent poisonous levels (Franson, 1996; Martin et al. 2008). These are based on the findings of research that links hepatic lead concentrations in raptors to the lead poisoning and subclinical effects (Pain et al., 1995; Franson, 1996; Martin et al., 2008; Kitowski et al., 2017). For cadmium, levels at ≥3 mg/kg^{-1} dw have been found to result from increased environmental exposure (Burgat, 1990; Battaglia et al., 2005). For mercury, Shore et al. (2011) argued that hepatic mercury concentrations greater than 6.7 mg/kg^{-1} dw (6.7 ppm) might affect avian reproduction, and mercury values above 67.0 mg/kg^{-1} dw might even kill non-marine birds. For selenium, Kitowski et al. (2017) argue that "selenium at the level of 4–10 mg/kg^{-1} dw is considered the background concentration in avian liver tissue" and cite a study by Ohlendorf (1989), where "a hepatic Se concentration above 10 mg/kg^{-1} dw was taken as exceeding the background level." For zinc, Kitowski et al. (2017) point to the study by Honda et al. (1990) in which 200 mg/kg^{-1} dw was defined as the background concentration in liver tissue (Kitowski et al., 2017).

Table 2. The main chemical elements and their impacts

Element	Symbol	Threshold	Impact on raptors, waterbirds and prey species (e.g., *Passerines*)
Lead	Pb	2-6 ppm dw[1 5]	"Gradual weight loss, polyuria/polydipsia, depression, weakness, anorexia, and/or rapid anemia"[3]
Zinc	Z	2 ppm[3]	"Passive regurgitation, lethargy, weakness, weight loss, anemia, cyanosis, polyuria/polydipsia, diarrhea, regurgitation, neurologic signs, feather picking, anorexia, and/or hemoglobinuria."[3]
Mercury	Hg	1+ ppm ww[2 5]	Reduced reproductive output, weak immune function, avoidance of high-energy behaviors, death with higher doses[4]
Cadmium	Cd	3 ppm dw[1]	"Affects the endocrine system, kidneys, reproduction, moulting, haemoglobin formation and growth"[5]
Arsenic	As	0.5 ppm ww[6]	Liver, kidney problems[6] "Muscular incoordination, debility, slowness, jerkiness, falling hyperactivity, fluffed feathers, drooped eyelid, huddled position, unkempt appearance, loss of righting reflex, immobility, seizures"[7]
Chromium	Cr	4 ppm dw[7]	Gastrointestinal and immune system, respiratory tract, male reproductive system damage, miscarriage, low birth weight, and development problems of the skeleton, reproductive system[8]
Copper	Cu	4 – 6 ppm[8]	Gastrointestinal tract damage, hyperplasia of the forestomach mucosa, liver and kidney damage, immune system damage, decreased embryo and fetal growth.[8]
Selenium	Se	3 ppm ww[11]	Emaciation, breast muscle atrophy, absence of fat, lung fluids, excess body gelatinous or serous fluid, swollen livers, flaccid heart, congested and enlarged spleens, head feather loss.[9]

The threshold represents the level at which the chemical begins to cause toxic reactions in the animals. Kitowski et al. (2016) suggest that the conversion from wet (ww) to dry weight (dw) 4.0.

Sources: 1. Pérez-López et al. (2008), Scheuhammer (1987), Nighat et al. (2013); 2. Weir et al. (2018); 3. Huang & Mayer (2020); 4. Whitney & Cristol (2017); 5. Gruz et al. (2019); 6. National Research Council 1977;7. Eisler (1988); 8. Agency for Toxic Substances and Disease Registry, 2019, 2020a; 9. Albers et al. (1996), O'Toole & Raisbeck, (1997).

Note that, for example 10 mg/kg = 10 μg/g = 10 ppm.

In another study, Kitowski et al. (2016) examined concentrations of lead, cadium, nickel, chromium and mercury in the livers of six species of raptors, represented by 58 specimens captured in Eastern Poland. The six species, in order of size where the White-tailed eagle *Haliaeetus albicilla*, Linnaeus, 1758 (2 specimens); common buzzard *Buteo buteo*, Linnaeus, 1758 (31 specimens); Goshawk *Accipiter gentilis* Linnaeus, 1758 (9 specimens); Marsh harrier *Circus aeruginosus*, Linnaeus, 1758 (3 specimens), Common kestrel *Falco tinnunculus*, Linnaeus, 1758 (3 specimens) and Sparrowhawk *Accipiter nisus*, Linnaeus, 1758 (10 specimens). The findings indicated that higher mercury and chromium levels were correlated with passerine predation, while higher hepatic concentrations of lead and cadmium indicated links to small mammal predation. This illustrated the sources of these chemicals, and the links with interspecies and intraspecies variations in feeding habits.

There were significant differences in the lead concentrations in the livers of Sparrowhawks and Common buzzards, but no differences in the cadmium levels.

Sparrowhawks had higher levels of mercury than goshawks (possibly due to the use of mercury in antifungal products for seed dressing, which eventually entered the bodies of small birds) despite both raptors being bird hunters (Kitowski et al., 2016). Water-sourced prey was not a key mercury source, consumption of larger avian prey reduced mercury accumulations (e.g., in goshawks), larger birds accumulated less mercury (e.g., the common buzzard and white-tailed eagle), cadmium and lead accumulations correlated with small and medium sized mammal consumption, and the study concluded that "prey preference has an effect on the heavy metal load in raptor species" (Kitowski et al., 2016, 5). Two studies, Helander et al. (2009) and Nadjafzadeh et al. (2013), are cited that found that White-tailed eagles respond to declining fish availability by switching to game carrion, increasing their ingestion of lead, due to the lead deposits from bullets in hunted game. This study showed that raptors may change their food preferences, and this may alter their vulnerability to metal poisoning and that human activities may alter these results.

In another study, Van Wyke et al. (2001, 57) recorded concentrations of eight elements (calcium, cobalt, chromium, copper, iron, manganese, nickel and zinc) and three toxic metals (aluminum, lead and strontium) in tissue samples from African white-backed (*Gyps africanus*, Salvadori, 1865), Cape griffon (*Gyps coprotheres*, Forster, 1798) and lappet-faced (*Torgos tracheliotos*, Forster, 1791) vultures in South Africa. Using blood samples from live nestling African white-backed vultures in two breeding populations, adults from a wildlife reserve and captive immature birds, the findings indicated concentrations of metals (except for Sr and Mn), varied between sampling localities and between captive and wild vultures. Metal accumulation (especially strontium) was highest in fatty tissues and bones, in agreement with findings from previous research. Levels of chromium, nickel and lead levels in the dead vultures were mostly higher than the average found in other research, but similar levels were found for some other substance concentrations for other species. The study concluded these results for each of two wild populations of *G. africanus* and the other species contribute as base-line data for later comparative assessments (Van Wyke et al., 2001).

Lead Toxicity and Research Methods

Lead toxicity is cited as possibly the most important dangerous heavy metal impact for birds (Cruz-Martinez, et al. 2012). Lead is recognized as a neurotoxin in excess amounts for both animals and people (Campbell, 2015). This is because as a neurotoxin means it can accumulate in bones, flesh, and hence the nervous system, impairing neurological functions and facilitating brain damage and behavioral change. Lead in the digestive tract of raptors is absorbed into the blood (Redig & Arent, 2008) and evidence indicates that in some cases it can take on the role of calcium in chemical processes (Haig et al., 2014). This can disrupt the functions of calcium, with serious consequences for calcium homeostasis, signal transmissions, cholinergic nerve cells and the cerebellum, and creating anemia through blood cell impacts (Haig et al., 2014).

Behavioral changes may occur, impacts including impaired flight, walking, foraging, digestion and strength (De Francisco et al., 2003).

Lead toxicity has been recorded as an important neurotoxin for some species of raptors in North America, especially Bald eagles. This may be because there are high levels of human-induced pollution from heavy industry, and the use of lead bullets in hunting (Campbell, 2015). Numerous studies based in the United States have documented these issues. For example, Cruz-Martinez, et al. (2012) using the case study of Bald eagles admitted to a veterinary clinic over 40 years (The Raptor Center (TRC) of the University of Minnesota from January 1996 to December 2009), noted "90% of the bald eagles received each year (120-130) for all types of problems have elevated lead residues in their blood...20–25% percent of these eagles have sufficiently high levels to cause clinical lead poisoning. Most of these birds die or are euthanized. In the last 24 years, over 500 eagles received or admitted to our clinic have either died or had to be euthanized due to lead poisoning." Spent lead ammunition from shotguns and rifles in waterfowl and other hunted birds was blamed for the high levels of poisoning, revealed by the correlation between poisoning and the hunting seasons (Cruz-Martinez, et al., 2012).

The role of lead-poisoned waterfowl in eagle population decimation led to a 1986 United States Federal Law banning lead ammunition for waterfowl hunting. However, Kramer and Redig (1997) found the law ineffective, as the number of poisoned eagles did not decline, despite hunter compliance, hinting at another source found to be eagle-scavenged shot deer and discarded deer gut piles left by hunters. Another study covering the period 1996-2009, evaluating the hypothesis of deer carcasses, used four epidemiological parameters: (1) links between poisoned eagles and the deer hunting season in Minnesota, Wisconsin, and Iowa; (2) correlations between animal recovery locations within deer-hunting zones; (3) lead isotope ratio analysis of metal fragments in eagle gastrointestinal tracts and blood lead levels; and (4) comparison of eagle kidney copper concentrations in lead-exposed and non-exposed birds (Cruz-Martinez, et al. (2012). This study used extracted livers, to explain why lead levels in bald eagles remained high after the ban on lead shot for waterfowl hunting, and the possibility that lead from bullets from the hunting of white-tailed deer (*Odocoileus virginianus*, Zimmermann, 1780) was responsible. The data was based on studies of 1,277 Bald eagles admitted for rehabilitation from January 1996 through December 2009 (334 were found to have elevated lead levels) and a predictive model was used to ascertain high lead levels using the predictors of deer hunting season, eagle age and recovery location linked to hunting zones. Copper concentrations in eagles with high lead levels and from eagles with background lead levels were also assessed, because most high velocity rifle bullets used for deer hunting were jacketed in copper.

Lead levels and poisoning were also found to be spatially and temporarily correlated with the hunting seasons, deer hunting rifle zones, and adult eagle age, with high levels of matching isotopic signatures in the lead samples (hence sourced from ingestion of lead) and blood and fragment samples within the isotope ratio of ammunition (Cruz-Martinez, et al., 2012), see also Church et al., 2006). Copper levels correlated with lead levels (Cruz-Martinez, et al., 2012). Copper-jacketed bullets were detected by the high kidney copper concentrations in lead-exposed eagles (see the section on copper). Findings indicated that poisoning was due to spent ammunition in deer and shotgun pellets in smaller pheasants, squirrels, and rabbits, hence ammunition sourced lead "is an important, if not the primary, source of lead exposure for Bald eagles" (Cruz-Martinez, et al., 2012). These findings were significant because they contributed to the larger public debate on lead poisoning and possible regulation of deer hunting activities, like those in reaction to waterfowl hunting (Campbell, 2015).

In another study, Manning et al. (2018) looked at lead poisoning in Bald eagles, using 93 bald eagles recorded with severe lead intoxication to compare internal lesions with blood, liver, kidney, and/or bone lead concentrations. Findings indicated that gross lesions linked to lead toxicity were commonest within the heart (55%), detectable by "multifocal myocardial pallor and rounding of the apex" (Manning et al. 2018, 289). Gross lesions in the brain "included petechiae or hemorrhagic necrosis" (14%) birds), while histologic lesions were in the heart (82%), brain (63%), and eyes (26%) (Manning et al. 2018, 289). The authors further argued that "lead toxicity in Bald eagles is characterized by fibrinoid necrosis of small- to medium-caliber arteries, most commonly affecting the heart, brain, and eyes"; furthermore, the gross and histologic lesions are "consistent with ischemia caused by a primary vascular injury." The findings also indicated that blood lead concentrations above 4 ppm, and higher liver lead concentrations were linked to more frequent heart lesions. They concluded that "the presence of fibrinoid arterial necrosis and parenchymal degeneration, necrosis, and/or hemorrhage within the heart, brain, and/or eyes is suggestive of lead toxicity in Bald eagles and warrants evaluation of liver or bone lead concentrations" (Manning et al. 2018, 289).

In a Californian study, Kelly et al. (2014) looked at 48 carcases mostly of raptor species (21 Golden eagles, 23 Turkey vultures and 4 Common ravens), which were either found alive (45) or dead (3, two Golden eagles and one Turkey vulture); of the 45 living birds 14 died at the centers and 31 were killed due to their condition. Blood samples and carcases were examined, using the parameters of date of accession, extraction location, admission factor, age class, clinical presentation, treatments and outcome for each bird. Mortality factors were determined using the location and severity of lesions, radiography screening (for trauma and metallic opacities in the gastrointestinal tract, and heavy metal analysis), and "microscopic examination of tissues, including brain, spinal cord, peripheral nerves, bone, muscle, tongue, heart, trachea, lung, liver, kidney, gonads, spleen, adrenals, thyroid/ parathyroid glands, oesophagus, ventriculus, proventriculus, pancreas, intestines and bursa of fabricius" (Kelly et al. 2014, 2). Rodenticides, organophosphate pesticides, strychnine and zinc among other substances were also checked. Mortality was found to be primarily due to lead poisoning (17% of the Turkey vultures and 14% of the eagles) with some of the latter recoding pathological changes, elevated hepatic lead concentrations and anticoagulant rodenticide intoxication. Four Turkey vultures (8% of the total sample) died of anticoagulant rodenticide intoxication (Kelly et al., 2014).

Canadian studies of eagles recorded similar issues to those found in the United States, possibly due to the similarity of human environmental interactions and ecological systems. Owen (2019) notes that Golden eagles (*Aquila chrysaetos*, Linnaeus, 1758) are treated for lead poisoning at the Orphaned Wildlife Rehabilitation Society in Delta, B.C. and that some experts argue for the cessation of lead ammunition for hunting. Owen (2019) argues that "the Canadian government banned lead for hunting waterfowl two decades ago, and non-toxic shot is now required to hunt most migratory game birds across the country. Lead shot is also prohibited for all types of hunting in 55 national wildlife areas… but pellets and bullets containing lead are still widely used to hunt large game and upland game birds, such as grouse or pheasant." Furthermore, the "Canadian government estimates that 40 to 80 tonnes of lead are used every year for hunting activities in Canada" (Owen, 2019). A cited study in western Canada (1986 to 1998) found that 10 per cent of 546 eagles were lead-poisoned, and that Golden eagles were more affected than bald eagles (Owen, 2019).

Lead poisoning, resulting from the swallowing of ammunition cartridges, has also been crucial to the survival of the California condor (Cade, 2007; Campbell, 2015). As argued by Cade (2007, 2125) "the scientific evidence that California condors (*Gymnogyps californianus,* Shaw, 1797) are frequently sickened and killed by lead poisoning from spent ammunition supports the conclusion that current levels of lead exposure are too high to allow reintroduced condors to develop self-sustaining populations in the wild in Arizona and, by inference, in California." This lead-related mortality had influenced the decline of the California condor population during the 1980s, and during the post- reintroduction period (Snyder & Snyder, 2000; Kelly & Johnson 2011). Research findings have indicated lead accumulation in the blood of condors during and outside hunting seasons (Hall et al., 2007; Hunt et al., 2007; Sorensen et al., 2007). Such findings influenced the ban on lead ammunition in the condor's California range, and other, supportive findings that detected high lead levels in the bodies of other species in the condor's range, such as Golden eagles (Pattee et al. 1990) and in ravens in the Greater Yellowstone Area (Craighead et al., 2009).

The New World Turkey vulture (*Cathartes aura*, Linnaeus, 1758), another common victim of the lead ammunition in animal carcasses, may suffer after ingestion from weakness and uncoordinated movements, including flight or death (Carpenter et al., 2003; Campbell, 2015). Turkey vultures in California have recently been recorded with higher lead poisoning levels, mostly correlated with higher blood lead concentrations during the deer hunting season and the intensity of wild pig hunting, this despite the possibly violated ban on lead ammunition in Californian hunting (Kelly & Johnson, 2011). Canada records similar issues (Martin et al., 2008); Clark and Scheuhammer (2003) reported that 184 carcasses of 16 species (including Turkey vultures, Red-tailed hawks, Great horned owls, and Golden eagles) were found with evidence of lead ammunition poisoning from game bird and mammal consumption. Martin et al. (2008) add that the dissection of 225 carcasses from 19 species of raptors in southern Ontario, found that after bone, kidney and liver examination, Turkey vultures had the 'highest mean concentrations of lead in bone and kidney compared to other raptor species' (Martin et al. 2008: 96).

Eurasian studies have also recorded similar result. For example, some studies have found important results for the Golden eagles (*Aquila chrysaetos*, Linnaeus,

1758) in the European Alps (Madry et al. 2015). These authors note that the main issue concerns Golden eagles scavenging on either non-retrieved carcasses or offal, both of which are discarded by hunters. This study addressed an important question: did the three Golden eagles discovered with severe lead poisoning represent a one-off situation or a generalized trend (perhaps at sublethal levels) in the Alpine region? The methodology involved the measurement of body burdens (body burden is the term for the concentration (or amount) of chemical in the body at any given time, ScienceDirect, 2020) in Golden eagles (36 specimens) in relation to their prey and a comparison of the results with those of nearby similar Eagle owls (19 specimens) in relation to their prey (marmots *Marmota marmota*, Linnaeus, 1758 and *Capra ibex*, Linnaeus, 1758) for eagles, and birds, rodents and Snow hares *Lepus timidus*, Linnaeus, 1758) for eagle owls. The study acknowledged that while eagles might feed on carcasses, eagle owls generally fed on live-caught prey (Madry et al. 2015).

Findings indicated that soft tissue lead levels were higher in golden eagles than in eagle owls. Bone lead levels were 10 higher in eagle bones than in owl bones. Prey species for both eagles and owls both had low lead concentrations. The next question concerned the source of the sublethal lead in Golden eagles, from the general environment or from ammunition in the carcasses of prey species. Lead isotope ratios were examined, the findings indicating that the lead isotope signatures of the golden eagle bones were like those of the ammunition but were different from the signatures of prey animals' bones, and those of eagle owls and soil. Additionally, the isotope signatures were unchanged with increasing bone lead concentration in any of the species examined. The conclusions were that eagles in the Alps generally pick up lead from spent ammunition in both carcasses and offal in sublethal and sometimes lethal quantities over several years (Madry et al. 2015).

Lead poisoning from bullets in accessed carcasses is dangerous to African vultures as well (García-Fernández et al., 2008; Cade, 2007; Dobrowolska & Melosik, 2008; Gangoso et al. 2009). Kruger et al. (2018) report on the lead poisoning of the Bearded vulture (*Gypaetus barbatus*, Linnaeus, 1758) in southern Africa, also based on examination of lead levels in bone and tissue samples, taken from carcasses with gun ammunition. The research question concerned the extent to which such poisoning contributed to the critically endangered status of this species. The methodology used blood lead levels in from six live birds (2017) five live birds (2009) and eight dead birds (with evidence that possibly lead exposure and accumulation was the factor for the deaths). The findings supported the conclusion that "the levels of lead may be detrimental to the survival and fecundity of this small and declining population" of Bearded vultures with recommendations being the banning of "hunting and culling with lead ammunition, which is the most likely source of this pollutant. Such actions may reduce the population's susceptibility to other threats, which may be compounded by high lead levels, and help ensure the success of planned reintroduction programs" (Kruger et al., 2018, 491).

In Europe, Spanish vultures have also been strongly affected (Rodriguez-Ramos et al., 2009: 235). Rodriguez-Ramos et al. (2009: 235; see also Iñigo & Atienza, 2007) argue that "this exposure may have increased after the ban on abandoning carcasses of domestic ruminants in the field due to the bovine spongiform encephalitis (BSE) crisis, both because the vultures consume hunting bag residues more frequently and because malnutrition may lead to mobilisation of lead stores." Lead toxicosis contributed to severe hypochromic anemia, disorientation, ataxia, impaired landing and posterior paresis. Donázar et al. (2002: 89) also found high frequencies of lead poisoning in Egyptian vultures in the Canary Islands, which included sub-clinical and clinical intoxication levels, which likely resulted from the ingestion of lead shot. Around 16% of the surveyed birds had over 0.2 ppm of lead, which at higher levels could be sufficient for serious or fatal consequences. Sympathetic arguments have called for lead pellets to be replaced by steel or molybdenum/tungsten alloy bullets (Donázar et al., 2002).

Anglers using lead tackles pose another problem in North America (Bedrosian et al., 2012). Grade (2017, 155) note that "ingested lead fishing tackle has been documented in 28 species of North American birds, as a risk factor for 75 species and as a leading cause of death for Common loons" (*Gavia immer*, Brunnich, 1764) (see also Pokras & Chafel, 1992; Blus, 1994; United States Environmental Protection Agency, 1994; Scheuhammer & Norris, 1995; Anderson et al., 2000; Stone & Okoniewski, 2001; Scheuhammer et al., 2002; Franson et al., 2003; Sidor et al., 2003; Strom et al., 2009; Grade, 2011). The study by Grade (2017, 157) used dead specimens of common loons, and assessments were made using the methodology of Sidor et al. (2003) and Franson et al. (2003). This study used measurements of lead tackle parts present in the gizzard and gastrointestinal tract, based on the parameters listed below. It also acknowledged a limitation of this type of methodology, as "ingested tackle erodes in the gizzard because of grinding action and stomach acids; and, in the case of jigs, hooks or other attachments break off and often dissolve or are passed prior to death" (Grade, 2017, 157; see also Pokras et al. 2009).

The cause of death was classified as lead tackle toxicosis, based on five conditions: (1) "tissue, blood, or body fluids were tested for lead and exceeded thresholds at which clinical signs of lead poisoning have been observed" (Grade, 2017, 157; see also Sidor et al., 2003, Franson & Pain 2011); (2) the necropsy form showed the death as due to lead toxicosis, justified by the discovery of lead tackle in the gastrointestinal tract, or evidence of clinical pathology identifiable as lead toxicosis; (3) a tackle item from the loon's gastrointestinal tract, revealed to be lead; (4) a radiograph revealing a tackle part, such as a lead sinker (a weight used with a fishing hook to increase its sinking rate, rate of sink, anchoring ability or casting distance) or jig (a hooked, lead sinker, covered to attract fish, jerked in vertical motion) in the loon's gastrointestinal tract; and (5) a report including evidence of lead toxicosis (Grade, 2017, 157). The methodology recorded the findings in the gizzard, using LeadCheck SwabsTM (3M, St. Paul, MN, USA), which can detect lead in metals at levels >0.2% (2,000 ppm; 3M 2011) (Grade, 2017).

Mercury and Research Methods

Mercury is generally described as a very dangerous chemical element, this assessment based mostly on the impact on human health (Clarkson et al., 2003; Campbell, 2015). Henny and Elliott (2018, 337) note that mercury "toxicity depends on whether mercury is in the organic or inorganic form. Only a small percentage of inorganic mercury is absorbed, but almost all organic mercury is absorbed by the intestine... inorganic mercury produces methylmercury (MeHg), which fish accumulate from water and their diet; nearly all mercury in fish flesh is MeHg. MeHg can adversely affect developing neural tissue in birds, with fish-eating birds being especially vulnerable." In the United States human exposure to mercury may occur through eating fish that contain methylmercury, a compound reacting from bacterial reactions with mercury in water, soil, or plants. Both elemental mercury and methylmercury affect the nervous system, and even inhalation of mercury vapor results in severe symptoms and in some cases death. These results may occur in both people and animals (Brean, 2019).

Mercury has been considered a dangerous chemical for raptors for several decades and has increasingly been traced in raptors and waterbirds, these being tested through blood and feathers (Campbell, 2015; Henny & Elliott, 2018). An important problem in the 1960s and 1970s, was the use of alkyl mercury as a fungicide for seed dressing for agricultural use, which killed some raptors (Berg et al. 1966, Jenson et al., 1972). For example, Kitowski et al. (2016) argue that small birds, including passerines (about 20% of the prey of sparrowhawks) are a source of mercury for raptors, as they consume grains with high concentrations of mercury-based fungicides. Alkyl mercury was used during the 1940s and was later banned as a seed dressing in Sweden in 1966 (Johnels et al. 1979). Henny and Elliott (2018) described the main contemporary sources of mercury: these include the atmospheric deposition from coal-fired power plants, mostly from the Arctic, northeastern U.S. and southeastern Canada, which affect fish and fish-eaters; mining operations in the Amazon basin (see also Nriagu et al. 1992); and local mining around the world, either for mercury or gold and silver extraction.

In one study, Barnes (2019, cited in Brean, 2019) tested 700 peregrines for mercury in Washington, Maryland, Southern Nevada, and the Gulf Coast of Texas, and found that all the birds tested positive for mercury. Peregrines are "one of the most impacted species when it comes to contamination," Barnes noted, because they feed on other birds that have been exposed to mercury. An identified contributory factor for the peregrines in the Lake Mead, Colorado River (Arizona and Nevada), was the Eared grebe (*Podiceps nigricollis*, Brehm, 1831), a common prey species that might be exposed to mercury in Utah's Great Salt Lake. Barnes et al. (2019, 1) note that "peregrines have the potential to be a particularly useful biomonitor because they are globally distributed, nest in proximity to a diverse assortment of aquatic and terrestrial habitat types (e.g., freshwater rivers and lakes, marine systems, estuaries, wetlands, arctic tundra, temperate forests, and open desert), are long-lived with high breeding-site fidelity, and have an extremely diverse avian diet" (see also White et al., 2013).

Kitowski et al. (2016) cite mercury poisoning for sparrowhawks in Poland, buzzards in Spain and the Czech Republic and Spanish and German goshawks (Houserova et al., 2005; Castro et al., 2011). In the study by Kitowski et al. (2016) a difference between goshawks and sparrowhawks was discerned: "the strong correlation between mercury and the consumption of small birds…implies that consumption of bigger avian prey by Goshawks allows them to avoid the accumulation of large amounts of mercury in the livers, in contrast to the Sparrowhawks specialized in catching small birds (Kitowski et al., 2016; see also Newton, 1986). Waterbodies and waterborne prey are also cited as sources of mercury poisoning in raptors (see also Kalisinska et al., 2014). White tailed eagles are cited as studied in this regard, through liver examination (Falandysz et al., 2001), and mercury levels been found to have been reduced in some areas due to lowered heavy metal emissions (Kalisinska et al., 2014).

A study by Weir et al. (2018) focussed on the rodenticide brodifacoum (formula $C_{31}H_{23}BrO_3$) and mercury and the effects of these on certain raptor species, including Red-tailed hawks (*Buteo jamaicensis*, Gmelin, 1788), Red-shouldered hawks (*Buteo lineatus*, Gmelin, 1788), Barred owls (*Strix varia*, Barton, 1799), Great horned owls (*Bubo virginianus*, Gmelin, 1788), and ospreys *Pandion haliaetus*, Linnaeus, 1758). This study used liver samples from the raptor carcasses from a non-profit rehabilitation center in Charlotte, North Carolina, USA. The livers were analyzed for the total mercury residue (mg/kg, dry weight) and five anti-coagulant rodenticides (µg/kg wet weight). Mercury was analyzed using a direct mercury analyzer approach and rodenticides were quantified by liquid chromatography–mass spectrometry (LC-MS). The findings indicated that mercury residues were high in piscivorous (fish eating) birds (ospreys 15.09 mg/kg and great blue herons 6.93 mg/kg, dry weight) and "relatively high in red-shouldered hawks and one eastern screech owl tested" (Weir et al., 2018). Six samples (three ospreys, one each of great blue heron, red-shouldered hawk, eastern screech owl) were above a health threshold of 1 mg/kg (wet weight). Brodifacoum, the only rodenticide consistently detected, exceeded 75% in Barred owls, Great horned owls, and Red-shouldered hawks. Crucially, 69% of the owl samples were either within or above a threshold of brodifacoum residue indicating a 10-20% risk of acute toxicity. The study concluded that brodifacoum was a more serious risk than mercury for most raptor species (possibly except for the osprey and to a lesser extent the red-shouldered hawk), due to the lower levels of mercury ingestion.

Chromium and Research Methods

Chromium toxicity may result when an organism is exposed to certain types of chromium, for example hexavalent chromium (chromium (VI), which is generally toxic when inhaled or ingested, and is hemotoxic or haemotoxic (destructive of red blood cells, Merriam-Webster, 2020), genotoxic ("the process by which an agent produces a deleterious effect on DNA and other cellular targets that control the integrity of genetic material," Lopez et al., 2012)., and carcinogenic (producing or tending to produce cancer, Merriam-Webster, 2020). Ingested hexavelent chromium damages blood cells by causing oxidation reactions, leading to hemolysis and sometimes kidney and liver failure (Xie et al., 2017). Speer and Wise (2018) argue that "hexavalent chromium [Cr(VI)] is a valuable industrial resource but ubiquitous environmental contaminant. Humans can be exposed to Cr(VI) occupationally and through environmental sources. It is known Cr(VI) causes lung cancer and other health issues including reproductive and immune dysfunction."

In raptors, chromium has been detected as a factor for several physiological problems. However, few studies record serious impacts or extremely high levels of chromium. For example, Kitowski et al. (2016) note that in other studies, Marsh harriers, kestrels, buzzards and sparrowhawks in southern Italy (from preying on passerines) had low hepatic concentrations of chromium, always less than 0.40 mg/kg dw (Zaccaroni et al., 2008). Tawny owls (*Strix aluco*, Linnaeus, 1758), had mean hepatic concentrations slightly over 0.30 mg/kg dw. In north Italy, the Little owls (*Athene noctua*, Scopoli, 1769) had a concentration of 0.29 mg/kg ww (about 1.16 mg/kg dw.) (Zaccaroni et al. 2003). Both owl species preyed in passerines, apparently an important source of chromium. This was like that of Sparrowhawks in their study (Kitowski et al., 2016). Falandysz et al. (2001) also recorded measured mean Cr concentrations of 0.087 mg/kg dw in three White-tailed eagles from the south Baltic, which was less than the findings of Kitowski et al. (2016). A study by Manning et al. (2000) found a low chromium hepatic value for one Australian White-bellied sea-eagle (*Haliaeetus leucogaster*, Gmelin, 1788) of 0.04 mg/kg ww (about 0.16 mg/kg dw). In contrast, for White-backed vultures (*Gyps africanus*, Salvadori, 1865) from the Republic of South Africa, Van Wyk et al. (2001) estimated a higher average concentration of 19.57 mg/kg dw of hepatic Cr.

Nickel Toxicity and Research Methods

Nickel is a heavy metal that occurs throughout the environment in small amounts (Das et al. 2018). Sources most soil types and volcanic emissions, generally bound with oxygen or sulfur, through oxides or sulfides. A principal source of nickel is industrial production and processing, recycling and disposal, which may be discharged into the atmosphere through mining, power plant operation and/or incinerators, industrial production of rubber, plastic, nickel-cadmium battery and electroplating industries (Das et al. 2018). Mudd et al. (2010, 9) notes that nickel's uses, including "stainless steel, alloys, electroplating and rechargeable batteries" are derived from either mostly sulfide or to a less often laterite-type ores, but recently increased extraction has been from the laterite ores which create energy and greenhouse gas emission costs, and serious environmental costs especially in Sudbury in Canada and the Taimyr and Kola Peninsulas in Russia (Mudd et al., 2010, 9). Impacts on people and animals are critically researched: Buxton et al. (2019) note that "notable human health toxicity effects identified from human and/or animal studies include respiratory cancer, non-cancer toxicity effects following inhalation, dermatitis, and reproductive effects. These effects have thresholds, with indirect genotoxic and epigenetic events underlying the threshold mode of action for nickel carcinogenicity."

DeForest et al. (2012, 107) examines the application of the European Union's Existing Substances regulation (EEC 793/93), which "was developed to assess the ecological risks posed by chemical substances such as Ni..." and also looked at "a generic framework for assessing secondary poisoning risks and prescribes the following terrestrial food chain: soil → earthworm → worm-eating bird or mammal." The Nickel Institute (2018) evaluates the impacts of nickel from marine, fresh water and soil. The methodology used a framework based on the ratio of dietary predicted exposure concentrations (PEC) and dietary predicted no effect concentrations (PNEC) (European Chemicals Bureau, 2003). The argument was that "the PEC in this assessment was the estimated Ni concentration in the diet of the representative predatory birds or mammals in each of the food chains evaluated, while the PNEC was the Ni concentration in bird and mammal diets at which no adverse effects are expected" (Nickel Institute, 2018). The ratio of the PEC to PNEC created Risk Characterization Ratio, or RCR, where an RCR <1 indicates low potential risk and an RCR >1 indicates a stronger risk. The formula is: a PEC-to-PNEC ratio, or RCR:

RCR = PEC × RAF/PNEC, where RAF is relative bioavailability.

Bioaccumulation of nickel from soil or water into organisms at the bottom of the food change has been found to be low. The Nickel Institute (2018) argues that "there is no evidence that Ni biomagnifies in aquatic or terrestrial food chains (i.e., Ni concentrations do not increase with increasing trophic level). This contrasts with other chemicals, such as methyl mercury, which tends to biomagnify across multiple trophic levels, thereby resulting in higher trophic level organisms being potentially more susceptible to mercury poisoning." Nickel may even have a opposite reaction, termed "biodilution" when the nickel rather reduces as it moves up the food chain (references include, Campbell et al. 2005; Lapointe & Couture, 2006).

The study uses a tiered approach to estimate possible poisoning risk (Nickel Institute, 2018). In this the first tier represents more conservative assumptions, that may overestimate the potential risk of poisoning. The second tier represents greater refinement which may increase the possibility of poisoning, by assuming for example a wider diet for birds, rather than a diet of organisms that may have high nickel levels such as the common cockle (*Cerastoderma edule*, (Linnaeus, 1758). For marine ecosystems, an example of oystercatchers (*Haematopus* spp, Linnaeus, 1758), in Tier 1, an assumption is that the diet is entirely on the *C. edule* diet, which gives a high risk, but in Tiers 2 and 3, the Risk Characterization Ratio is lower, when the ingestion rate of oystercatcher species and a mixed bivalve diet, are considered. For freshwater ecosystems, the diet of a representative mollusk-eating bird, subsequent Tiers failed to include other secondary poisoning risks, allowing for the food ingestion rate and body weight of mollusk-eating birds, which reduced the RCRs. For soil ecosystems, Tier 1, the analysis was based on a typical worm-eating bird, with subsequent Tiers adjusted for the food ingestion rate of the worm-eating bird and a wider diet including both worms and isopods (Nickel Institute, 2018).

For raptors, there is evidence of nickel ingestion. For example, Kitowski et al. (2016) looked at sparrowhawks and goshawks, which accumulated similar levels of nickel (Ni 0.3 mg/kg dw) through their diet of birds. Also cited are the low nickel levels in one peregrine falcon (0.107 mg/kg dw (Kalisinska et al., 2008) in northwest Poland, and low levels in the livers of White-tailed eagles in the study by Kitowski et al. (2016). However, it is mentioned that White-tailed eagles in a study by Falandysz et al. (2001) in the Baltic sea area had liver- nickel concentrations 56 times higher (13.0 mg/kg dw) than those of the study by Kitowski et al. (2016) in east Poland, leading the latter researchers to speculate that "it seems that White-tailed eagles from east Poland have less opportunities to prey in highly mercury- and nickel-contaminated areas than raptors from the south Baltic Sea coast" (Kitowski et al. (2016, 11).

Cadmium Toxicity and Research Methods

Birds may also suffer from cadmium toxicity. Kitowski et al. (2016) write that cadmium concentrations in birds are usually highest in kidneys, intermediate in muscle and lowest in the livers (see also García-Fernández et al., 1996; Battaglia et al., 2005; Binkowski et al., 2013). However, Kitowski et al. (2016) argue that it is still valid to use the liver to monitor Cd exposure, as this accumulates about 50% of the total cadmium and the body burden is stable in the liver, as this organ is resistant to cadmium toxic effects (Burgat, 1990; Battaglia et al., 2005). In their study, Kitowski et al. (2016) also note that the high concentration of cadmium buzzard livers might be linked to their feeding on voles and other relevant animals, as the findings indicate a higher cadmium levels for small mammal consumption. This possibly indicates voles and other small mammals feeding on fields fertilized with chemicals including cadmium (Beyer, 2000; Louekari et al., 2000).

Coal combustion, including emissions from coal-burning stoves is another important cadmium source, especially in Eastern Europe (Seshadri et al., 2010; Pacyna et al., 2010). Some studies indicated that cadmium levels in buzzards recorded Northern Italy, (Battaglia et al., 2005), Portugal (Carneiro et al., 2014) and Spain (Castro et al. 2011) were lower than in East Poland (Kitowski et al., 2016). Other low levels of liver cadmium concentrations were recorded in goshawks in Germany (Kenntner et al., 2003) and buzzards in Spain (Castro et al., 2011). Kitowski et al. (2016) note that the levels of cadmium for peregrines in Sweden (Ek et al., 2004) were like those of goshawks in their studies and Falandysz et al. (2001) reported lower cadmium levels in the livers of Baltic White-tailed eagles than the goshawks in their study (Kitowski et al., 2016).

Marine environments may also have high levels of cadmium. For example, Parslow and Jefferies (1973) point out that "gannets, in common with other seabirds which feed mainly in offshore waters, contain less cadmium in their livers (mean 4.7 ppm dry weight, seven adults) than do mainly pelagic feeders (e.g., Fulmar Fulmarus glacialis, mean 37 ppm, 11 individuals) or certain littoral and sublittoral mollusc feeders (e.g., Eurasian oystercatcher *Haematopus ostralegus*, mean 29 ppm... Eider *Somateria mollissima*, mean 10 ppm, eight adults). The argument is that the cadmium is derived from the invertebrates upon which the oceanic birds feed, which may be concentrators of cadmium. Rainbow (1997, 195) supports the position of marine invertebrates being possible concentrators of environmental metals: "Trace metals are accumulated by marine invertebrates to body concentrations higher, in many cases orders of magnitude higher, than the concentrations in an equivalent weight of the surrounding sea-water."

Copper Toxicity and Research Methods

Copper toxicity is generally seen as less serious than that of lead (Campbell, 2015). Copper is "an essential cofactor for several oxidative stress-related enzymes including catalase, superoxide dismutase, peroxidase, cytochrome c oxidases, ferroxidases, monoamine oxidase, and dopamine βmonooxygenase" (Tchounwou et al., 2012, 134, see also Harvey & McArdle, 2008; Stern, 2010; Agency for Toxic Substances and Disease Registry, 2020a). Manto (2014, 327) notes that "copper is a cofactor of proteins and enzymes (called cuproenzymes) involved in fundamental mechanisms, such as energy generation, oxygen transportation, hematopoiesis, cellular metabolism and signal transduction." Therefore, copper is not primarily a toxic element, being a nutrient within metalloenzymes in "hemoglobin formation, carbohydrate metabolism, catecholamine biosynthesis, and cross-linking of collagen, elastin, and hair keratin" (Tchounwou et al., 2012, 134). Toxicities may however be generated when copper transitions between the oxidized Cu(II) and a reduced states Cu(I), and possibly human cellular damage and Wilson disease (Chang et al., 1996; Harvey & McArdle, 2008; Tchounwou et al., 2008; Stern, 2010).

However, the use of copper jacketed bullets in hunting of potential prey animals and resultant raptor-accessed carcasses has fueled studies on possible negative copper impacts on raptors (Cruz-Martinez, et al., 2012). Franson et al. (2012) examined copper toxicity in American kestrels (*Falco sparverius*, Linnaeus, 1758). The study methodology used orally administered copper into birds, rather than naturally abstracted copper sources. The 16 kestrels (there were also 10m controls, which were sham gavaged – supplied with a feeding tube with no copper administered) were orally administered 5 mg Cu/g body mass copper pellets nine times during a 38-day period. Most birds retained the pellets for more than an hour, but 69% regurgitated the pellets in the 12-h monitoring period. The findings indicated that hepatic copper concentrations were higher in Cu-administered kestrels than in the controls, but no difference in treated and control bird's blood Cu concentrations.

Another important finding was that hepatic copper and the protein metallothionein concentrations were significantly correlated, but that metallothionein was higher in male birds that received copper than in controls and for female birds those that received copper were similar to controls. Hepatic copper and metallothionein concentrations in kestrels were significantly correlated. Crucially, there were no clinical signs, "no treatment effect on body mass; concentrations of Cu, hemoglobin, or methemoglobin in the blood; or Cu concentrations in kidney, plasma biochemistries, or hematocrit" (the percentage of the blood that consists of red blood cells) (Franson et al., 2012, 145). The conclusions were that ingested Cu pellets "pose little threat to American kestrels (and presumably phylogenetically related species), although the retention time of pellets in the stomach was of relatively short duration" (Franson et al., 2012, 145). Furthermore, "birds expected to regurgitate Cu fragments with a frequency similar to kestrels are not likely to be adversely affected by Cu ingestion, but the results of our study do not completely rule out the potential for toxicity in species that might retain Cu fragments for a longer time" (Franson et al., 2012, 145).

Copper jacketed lead bullets have been compared with normal lead bullets in terms of their impact on birds, with some variable results (Kallok, 2013). Kallok's (2013) study examined injured, sick and dead bald eagles delivered to the University of Minnesota's Raptor Center, some of which die, and others recover and are released. A critical history comments on the events leading to the Minnesota State and Federal bans on lead shot for hunting in 1987 and 1991 respectively. In Canada, waterfowl hunters were required to use nontoxic shot on and after 1999. Lead shot in waterfowl was initially blamed, but later lead in deer gut piles was identified as a source of lead ingestion for birds. Copper jackets on soft lead bullets were hypothesized to be insufficient to prevent lead fragmentation into flesh after impact, poisoning eagles that ate the flesh. A Raptor Center study found measurable levels of copper in 63 out of 89 examined eagle carcasses. Suggested alternatives were based on copper bullets, which studies found did not fragment into flesh on impact (although lighter, copper bullets may penetrate the entire animal), unlike 12-gauge lead shotgun slugs (which left about 28 fragments per shot) and muzzleloader bullets (which left 82 to 141 lead fragments) all several centimeters from the channel of the wound (Kallok, 2013).

Selenium Toxicity and Research Methods

Selenium, a non-metal element, occurring frequently in sulfide ores (where it may replace sulfur, and be produced when these ores are processed) is generally evaluated as only moderately toxic in large amounts; in small amounts it is necessary for animal physiological health (Tchounwou et al., 2012). Selenium combines with other elements to form oxides (selenium dioxide, SeO_2; selenium trioxide, SeO_3, the former dissolving in water or nitric acid to form selenous acid, H_2SeO_3). Selenium toxicity in humans induces gastrointestinal and behavorial disorders, and even neurological damage and even liver cirrhosis, pulmonary edema, or death (Agency for Toxic Substances and Disease Registry, 2020b).

Selenium studies focus on environmental, herbivore and carnivore impacts, in a food chain format. There are also evaluative classifications for the study of such impacts. For example, Lemly (1996) gives a such a procedure, applied to toxic levels of selenium as a threat to fish and wildlife. The justifying argument is that a "toxic threat is expressed as hazard and is based on the potential for food-chain bioaccumulation and reproductive impairment in fish and aquatic birds, which are the most sensitive biological responses for estimating ecosystem-level impacts of selenium contamination" (Lemly, 1996, 19). Five degrees of hazard are defended as possible, dependent on the variables of selenium concentrations in the target environments, possible bird and fish exposure and the possibilities for variable degrees of partial or total reproductive impairment. The five degrees are, on a 1 to 5 scale: no identifiable hazard, and then minimal, low, moderate and high hazard. The five ecosystem components; water, sediments, benthic macroinvertebrates, fish eggs, and aquatic bird eggs, each had a different separate hazard score. Environmental threats are assessed by totalling these scores. Therefore, the least hazardous environment would be calculated by giving each of the five ecosystem components a rating of one, and the highest hazard would be based on a rating of 5 for each of these components. Hence, based on the totals, a score of 5 would be no hazard, score of 6-8 would be minimal hazard, 9-11 would be low hazard, 12-15 would moderate hazard and 16 a high hazard (Lemly, 1996, 19). Additional information would be derived from knowledge of the local ecosystem, including selenium contamination and potential animal exposure. It is concluded that this method "is based on a large body of field and laboratory research data on selenium cycling and toxicity in aquatic ecosystems… a useful tool for assessing the toxic threat of selenium to fish and wildlife" (Lemly, 1996, 35).

Lemly (1993, 83) argues that "it is now possible to formulate diagnostic selenium concentrations in four distinct ecosystem-level components; water, food-chain, predatory fish (consuming fish or invertebrate prey), and aquatic birds." Furthermore, despite the low selenium toxicity, "waterborne selenium concentrations of 2 µg/l or greater (parts per billion; total recoverable basis in 0.45 µ filtered samples) should be considered hazardous to the health and long-term survival of fish and wildlife populations due to the high potential for food-chain bioaccumulation, dietary toxicity, and reproductive effects" (Lemly, 1993, 83). Bioaccumulation occurs, increasing the toxicity of waterborne concentrations below 1 µg/l. At the lower levels of the food chain, some organisms (e.g., zooplankton, benthic invertebrates, and some fishes accumulate selenium with little effect up to 30 µg/g (even to 370 µg/g). Such levels are nevertheless dangerous at the higher levels of the food chain, where for many fish and other wildlife the tolerance level is only 3 µg/g. Lemly (1993, 83) further notes that the thresholds of selenium tolerance for freshwater and anadromous (fish that migrate up rivers from the sea to spawn) fish are: "whole body=4 µg/g; skeletal muscle=8 µg/g; liver=12 µg/g; ovaries and eggs=10 µg/g"… thresholds for aquatic birds are: liver=10 µg/g; eggs=3 µg/g." For impacts on the waterbirds that feed on these fish, measurements of selenium concentrations in gravid ovaries and eggs offer "the most precise way to evaluate potential reproductive impacts to adult fish and aquatic bird populations" (Lemly (1993, 83). Therefore, "resource managers and aquatic biologists should obtain measurements of selenium concentrations present in water, food-chain organisms, and fish and wildlife tissues in order to formulate a comprehensive and conclusive assessment of the overall selenium status and health of aquatic ecosystems" (Lemly (1993, 83).

Spallholz and Hoffman (2002) offer more detail on methodologies for the measurement of selenium in animals. This includes a list of three forms of selenium toxicity. Firstly, there is the formation of CH_3Se- which may create free radicals that affect enzymes and proteins or create superoxide and oxidative stress. Secondly, where there is excess selenium in the form of selenocysteine, selenium methylation metabolism may be inhibited, which may result in the concentration of hydrogen selenide with possible toxic effects on animals. Thirdly, there is a possibility that excess selenium linked to sulfur-containing enzymes and structural proteins can relate to avian teratogenesis (possibly from L-selenomethionine or to a lesser extent L-selenocysteine, the most likely to be ingested by waterbirds). In this respect, a teratogen is a substance that can affect physiological development in young animals and people, especially affecting the fetus and embryo (teratology is the study of abnormalities of physiological development in the young) (Rogers & Kavlock, 1996; Heinz et al., 2011; Bastow & Holmes, 2016). Disputes exist as to the toxicity of L-selenomethionine to embryos, but Spallholz and Hoffman (2002, 27) note that "there is no suggestion from the literature that the levels of L-selenomethionine that would be expected to accumulate in eggs in the absence of environmental concentration of selenium pose harm to the developing embryo." This points to environmental concentration, rather than physiological accumulation as a possible factor for any toxic levels. Nevertheless, "for several species of aquatic birds, levels of Se as selenomethionine in the egg above 3 ppm on a wet weight basis result in reduced hatchability and deformed embryos" (Spallholz & Hoffman, 2002, 27).

The role of selenium-polluted surface waters in waterbird toxicity was explored by Hoffman (2002). Consequences for birds included "mortality, impaired reproduction with teratogenesis, reduced growth, histopathological lesions and alterations in hepatic glutathione metabolism" (Hoffman, 2002, 11). This study focusses on "adverse biological effects of Se in aquatic birds to altered glutathione metabolism and oxidative stress" and refers to "oxidative stress in different stages of the mallard (*Anas platyrhynchos*) life cycle"; consequences of increased Se include increases in plasma and hepatic GSH peroxidase" and "dose-dependent increases in the ratio of hepatic oxidized to reduced glutathione (GSSG:GSH) and ultimately hepatic lipid peroxidation measured as an increase in thiobarbituric acid reactive substances (TBARS)." Findings on mallard ducks (*Anas platyrhynchos*, Linnaeus 1758) indicated links with teratogenesis (4.6 ppm wet weight Se in eggs), reduced duckling growth (15 ppm Se in liver), lowered immune function (5 ppm Se in liver) and histopathological lesions (29 ppm Se in liver) (Hoffman, 2002, 11).

Findings on other species indicate important Se impacts, especially concerning glutathione metabolism: growth reduction, possible immune function decline, increased hatchling liver to body weight and hepatic Glutathione disulfide (GSSG) to Glutathione (GSH) ratios in hatchings, when eggs contained 9 ppm Se (in American avocet, *Recurvirostra americana*, Gmelin, 1789); decrease in hepatic GSH, (Black-necked stilts, *Himantopus mexicanus*, P.L.S.Müller, 1776); emaciation, increased feather losses, histopathological lesions, and increased hepatic GSH peroxidase, loss of hepatic protein-bound thiols and total thiols, and minor increases in GSH (American coot, *Fulica americana*, Gmelin, 1789). A thiol is defined as an organosulfur compound where sulfur replaces oxygen in the chemical structure. Other findings document a positive correlation between hepatic Se concentration and GSH peroxidase activity and a negative correlation between hepatic Se and GSH concentration in diving ducks (*Aythyini* spp), and high blood levels linked to increased plasma GSH peroxidase activity and decreased plasma GSSG reductase activity (Emperor geese, *Anser canagicus*, Sevastianov 1802) (Hoffman, 2002, 11).

Selenium toxicity has been blamed for the decline of the snowy plover (*Charadrius nivosus*, Cassin, 1858) in the south western United States (Ashbaugh et al., 2018). In this study, although vanadium, arsenic, cadmium and lead were also traced in the blood, feathers and potential prey of this species (e.g., of the tiger beetles, *Cicindela circumpicta*, Fitch, 1841, *Cicindela togata*, LaFerté-Sénectère, 1841) selenium was the commonest, and quantities were measured as above expected thresholds. Ashbaugh et al. (2018, 703) note that in "samples greater than instrumentation detection limits, 98% of snowy plover blood and 22% of feather samples were greater than Se toxicity thresholds of 1 ppm ww for blood and 5 ppm dw for feathers (blood quantifiable range: 0.83-15.12 ppm; feathers quantifiable range: 1.90-27.47 ppm)." Variables considered included sex, individual body condition, local tiger beetle Se concentrations, state, and presence of body molt. Tiger beetle Se concentrations were related to individual study sites in Texas. The authors argue that "these results provide some of the first evidence of Se exposure risk for interior snowy plovers nesting in saline lake and alkali flat environments of the SGP" (Ashbaugh et al., 2018, 703).

Another case looked at fish, which may be targeted by waterbirds. Selenium abstraction and poisoning in fish starts with the eggs, which are infected through the diet of the female fish. As the selenium (usually at concentrations above 10 µg/g) is metabolized by the growing fish, through the embryo stages, "biochemical functions may be disrupted, and teratogenic deformity and death may occur" (Lemly, 2002, 39). Lemly (2002) found that 20 fish species which were resident in Belews Lake, North Carolina were affected by selenium in wastewater from a coal-fired power plant during the mid-1970s, to the extent that 19 species were extirpated. Physical examination found "telangiectasia (swelling) of gill lamellae, elevated lymphocytes, reduced hematocrit and hemoglobin (anemia), corneal cataracts, exopthalmus (popeye), pathological alterations in liver, kidney, heart, and ovary (e.g., vacuolization of parenchymal hepatocytes, intracapillary proliferative glomerulonephritis, severe pericarditis and myocarditis, necrotic and ruptured mature egg follicles), reproductive failure (reduced production of viable eggs due to ovarian pathology, and post-hatch mortality due to bioaccumulation of selenium in eggs) and teratogenic deformities of the spine, head, mouth, and fins" (Lemly, 2002, 39). The conclusions highlighted the importance of these findings for the larger food chain: "Bioaccumulation in aquatic food chains causes otherwise harmless concentrations of selenium to reach toxic levels, and the selenium in contaminated sediments can be cycled into food chains for decades" (Lemly, 2002, 39).

Another case study looked at invertebrates, which are even lower on the food chain than fish. Debruyn & Chapman (2007, 1766) argue that most studies of selenium pollution in freshwater ecosystems have been primarily on birds and fish, "with invertebrates most often considered only as dietary sources of Se to higher trophic levels." This research targeted 29 macro-invertebrate species, reviewing links between toxic effects and 156 aqueous, dietary, or internal Se concentrations. The findings documented pronounced differences between the samples according to location. Internal concentrations were the least variable, with about 30-fold variation. Concentrations in the diets varied about 100-fold and those among taxa varied over 1000-fold. There was also much less variation in sublethal effects (about 10-fold lower concentrations) than in lethal effects. Crucially, sublethal effects in the targeted invertebrates were found at 1-30 microg Se/g dry weight in invertebrate tissue. As this range covered some dietary toxicity thresholds for waterbirds and fish, Debruyn & Chapman (2007, 1766) argue that "Se may cause toxic effects in some invertebrate species at concentrations considered to be 'safe' for the organisms consuming them."

Another case study examined the impact of selenium on mallard ducks (O'Toole & Raisbeck, 1997). These authors argue that "selenosis is thought to be a significant problem among waterfowl populations in selenium-contaminated wetlands in the western United States" (O'Toole & Raisbeck, 1997, 330). Two 150-day studies examined the possibility that morphological criteria for selenosis could supplement chemical analysis. The methodology involved the feeding of 48 young male mallard ducks, with proprietary waterfowl rations containing < 1, 10, 25, and 60 ppm selenium (seleno-L-methionine). Another 12 birds were fed twice daily with either a proprietary ration or a selenium-supplemented ration (120 µg/g) for one of the two daily feedings. Selenium in the blood increased from the baseline concentrations, which were < 0.4 µg/ml, to averages of 4.5 µg/ml (10ppm group), 8.9 µg/ml(25ppm group), and 16.0 µg/ml(60ppm group). Ducks in the 60-ppm-group became emaciated and were either killed or died within 22 to 50 days of dietary exposure.

Impacts on this 60ppm group, other than emaciation and general malaise, included "mild to moderate generalized hepatopathy with single-cell necrosis, karyomegaly of hepatocytes, hyperplastic bile duct epithelium, and/or iron accumulation in Kupffer cells" half of the birds. Those given lower dosages acquired lesions (integumentary structures of hard keratin), later developed into gross lesions of bilaterally symmetrical alopecia of the birds' scalp and the dorsal cervical midline, and damaged digital nails, and "necrosis of the tip of the beak (maxillary nail)" (O'Toole & Raisbeck, 1997, 330). These increased with the increased dosages of selenium (for those with 10 ppm selenium, 0%; 25 ppm, 42%; 120 ppm 44%). Histologic lesions also occurred; in the digital and maxillary nails, single-cell to full-thickness necrosis of keratinocytes and multifocal parakeratosis in stratum corneum existed; in "alopecic skin (necrosis of the epidermal collar, inflammation of the feather pulp, and follicular keratosis) were mild" (O'Toole & Raisbeck, 1997, 330). As in some other studies of ingestion, the liver, kidneys and feathers had high selenium concentrations. The study records that "mean hepatic tissue concentrations were 14.5 µg/g (10 ppm group), 29.6 µg/g (25 ppm group), 60.6 µg/g (60 ppm group), 13.0 µg/g (120 ppm split-feed group), and 2.0 µg/g (controls)" (O'Toole & Raisbeck, 1997, 330). This study showed that selenium, perceived by some scholars requiring more studies (Lemly, 1993; Tchounwou et al., 2012) may present dangers at high concentrations and hence would pose risks due to accumulation effects, as recent studies appear to indicate (Ashbaugh et al., 2018).

CHEMICAL COMPOUNDS

Many chemical compounds also affect birds. In this sense, a chemical compound is defined as a substance which is composed of more than one element, with atoms from more than one element connected by chemical bonds. Common chemical compounds affecting birds are illustrated in Table 2. These compounds are usually artificial in that they are made for specific purposes: for example, to kill pests (animals defined as trouble to people, usually insects), termed pesticides; to kill rodents, these termed rodenticides; and veterinary compounds intended to benefit livestock. These compounds may contain chemicals that have different impacts on different species of raptors or waterbirds. These impacts are at least as serious as the impacts from the chemical elements described above. Some of the more dangerous chemical compounds have been banned, but in some cases the impacts prevail (Campbell, 2015).

Table 2. The main chemical compounds and their impacts

Compound	Formula	Impacts on Raptors and Waterbirds
DDT	$C_{14}H_9Cl_5$	Toxic to prey animals, reproductive degradation, shell thinning, endocrine problems.[1] Affects egg laying, pregnancy, lactation/
DDE	$C_{14}H_8Cl_4$	Toxic to prey animals, shell thinning, embryonic and chick risk.[2] abnormal thyroid function, fertility decrease, hatching problems, immune function decline.[2]
Dieldrin/Aldrin	$C_{12}H_8Cl_6O$	Fatal to raptors, including juveniles and breeding adults, levels correlated to levels in prey animals.[3]
PCB	$C_{12}H10_{-x}Cl_x$	Liver, immune, reproductive, system damage, anemia; skin damage (chloracne); liver, stomach, and thyroid gland injuries (hepatocarcinoma thymocyte apoptosis[4]
Diclofenac	$C_{14}H_{11}C_{l2}NO_2$	Renal and visceral gout (crystallization of uric acid in tissues), enteritis[5]
PAH/di-methyl-naphthalene	$C_{12}H_{12}$	Bird/embryo toxicity, with death, developmental abnormalities, egg impacts, clutch/brood abandonment, reduced growth. Trophic level accumulation unlikely.[6]

Sources: 1. Colborn, 1991, NPIC (1999); 2. Newton (1993); 3. Sibly, et al. (2000); 4. Fernie et al. (2001), Crinnion (2011); 5. Campbell (2015); 6. Fingas (2017), Albers (2006).

DDT, DDE and other Pesticides

Pesticides are generally composed of such connected elements, such as the banned pesticide dichlorodiphenyltrichloroethane (DDT) (chemical formula $C_{14}H_9Cl_5$). The creator of this pesticide, Paul Hermann Müller, was awarded the Nobel Prize in

Physiology or Medicine in 1948 "for his discovery of the high efficiency of DDT as a contact poison against several arthropods" (The Nobel Prize in Physiology or Medicine, 1948). DDT was used as a pesticide for insect control in the United States until 1972, when it was cancelled by the United States Environmental Protection Agency (EPA). The National Pesticide Information Center (NPIC) (2011) noted that "bald eagles, ospreys and peregrine falcons were brought to the brink of extinction because their food sources (fish and other birds) were contaminated with DDE, the breakdown product of the insecticide DDT." DDE (dichlorodiphenyldichloroethylene) is formed when hydrogen chloride (dehydrohalogenation) is lost from DDT (chemical formula $C_{14}H_8Cl_4$) (Walker et al., 1954). The change from DDT to DDE is shown in Figure 1 below.

The National Pesticide Information Center (NPIC) (1999) describes DDT as "widely used... because it is effective, relatively inexpensive to manufacture, and lasts a long time in the environment." It is further noted that "DDT was canceled because it persists in the environment, accumulates in fatty tissues, and can cause adverse health effects on wildlife. In addition, resistance occurs in some insects (like the house fly) who develop the ability to quickly metabolize the DDT." Dieldrin, was introduced as an alternative to DDT (Jorgenson, 2001). This is an organochloride developed by 1948 by J. Hyman & Co, Shell Chemical, Denver, then AMVAC, Los Angles as an insecticide, related to aldrin (with a similar formula, $C_{12}H_8Cl_6$) which oxidizes after insect ingestion to form dieldrin. It has strong environment persistence and biomagnifies as it ascends the food chain, a factor for its abolition. PubChem (2020b) defines dieldrin as an

> organochlorine compound resulting from the epoxidation of the double bond of aldrin. It is the active metabolite of the proinsecticde aldrin. It has a role as a xenobiotic and a carcinogenic agent. It is an organochlorine insecticide, an organochlorine compound and an epoxide. It derives from an aldrin... aldrin and dieldrin were widely used pesticides for crops like corn and cotton. Because of concerns about damage to the environment and potentially to human health, EPA banned all uses of aldrin and dieldrin in 1974, except to control termites. In 1987, EPA banned all uses.

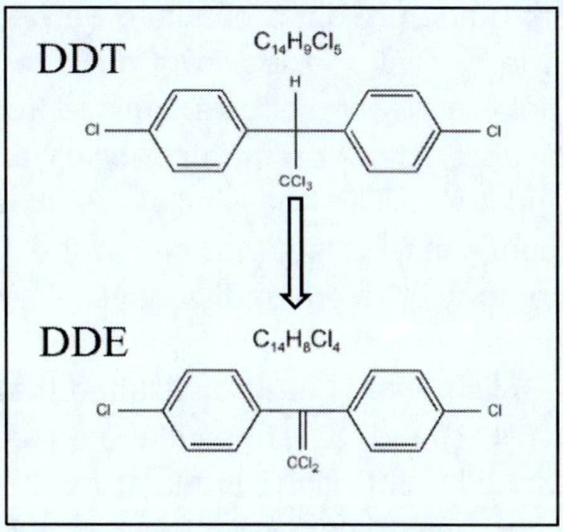

Figure 1. Change from DDT to DDE.

The impacts of DDT on animals included: negative effects on the nervous system, interfering with nerve impulses; hyperexcitability, tremors, incoordination, and convulsions in laboratory-tested animals; fatal doses resulting in liver lesions in the short term and liver changes over the longer term, increased risk of live tumors; long persistence in the environment, the soil half-life being 2 to 15 years and in an aquatic environment about 150 years; and moderately toxic to birds, but highly toxic to aquatic animals including fish and amphibians, affecting the heart, brain and other organs. The impacts on the animals at the bottom of the food chain are particularly important, especially for marine and aquatic animals preyed upon by waterbirds and relevant raptors such as ospreys and bald eagles (Campbell, 2015).

DDT and DDE are sometimes more persistent in the environment than other pollutants, but in some cases such assessments are difficult to establish (Campbell, 2015). For example, García-Fernández et al. (2008, 432) found that in Spain there were many studies, but few overall assessments. Nevertheless, the authors conclude that "excepting DDT and DDE, most organochlorine-based pesticides in raptors from Spain have diminished over the last 2 decades. The concentrations of DDE found in the eggs of various species could in part explain problems in the reproductive success of raptors in Spain." Jürgens et al. (2016, 333) found high levels of DDT in fish in English ecosystems and noted that "some fish contained levels of DDT harmful to them or their predators." They compiled a list, based on relevant literature findings of the threshold levels of DDT for prey animals and associated raptors and waterbirds. These are: "based on egg-shell thinning in ducks" 0.014 ppm (reference, Canadian Council of Ministers of the Environment, 1999); "dietary effects concentration" in ospreys, 0.09 - 0.19 ppm (reference, Hinck et al. (2009); "dietary effects concentration" in Bald eagles, 0.13 ppm (reference, Hinck et al. (2009); and "egg-shell thinning in pelicans," 0.02 ppm (reference, Lazorchak et al., 2003).

British raptors and waterbirds were also affected during the peak years of DDT use in the late 19th century. For example, for raptors, Presst et al. (1970) found that ingestion levels for bird-feeding raptors the levels of PCBs in the liver were 70ppm, which was like like that of DDE. For fresh-water fish-feeding waterbirds in the same region, the figure was 900ppm. The authors point out that in their study "most of the 559 specimens came from predatory birds obtained between April 1966 and August 1968. PCB was found in terrestrial species from most regions of Britain, in all the individual and bulked samples of seabird eggs examined from one west coast and two east coast colonies and in most of the freshwater species collected from the midlands, east and south of England." As DDE had similar levels, it appears both chemical compounds had similar dominance of ecosystems in the 1960s-70s period. Testing used gas-liquid chromatography on livers and kidneys (Presst et al., 1970).

In another example, Parslow and Jefferies (1973) examined the impacts of DDT/DDE on gannets in British waters, and concluded that "measurable residues of polychlorinated biphenyls (PCB) and DDE (the chief breakdown product of DDT found in birds) are present in virtually all British seabirds; many contain residues of other organochlorine materials, such as dieldrin and hexachlorobenzene (HCB), usually in much smaller quantities; PCB and HCB derive chiefly from industrial use, the others mainly from agricultural insecticides." They also noted that as larger fish carry more pollutants, birds such as gannets that eat large fish are vulnerable (examples of such fish are pollack *Pollachius pollachius*, *P. virens* and haddock *Melanogrammus aeglefinus*). Toxicity and mortality are partly dependent on the level ingested and the area of deposition, i.e., the subcutaneous fat or the liver. In health birds there may be low liver deposits, with higher levels in the fat, but stress may mobilize pollutants when fat is utilized, and the result may affect the liver and endocrine system (Parslow & Jefferies, 1973). The study concluded PCBs occurred at higher levels in gannets and other British seabirds than DDE, gannets had higher mean liver levels (e.g., 200 ppm) than other seabirds (e.g., murres or guillemots *Uria aalge*, Pontoppidan, 1763, 116 ppm). In some dead gannets, PCB (also DDE) concentrations in the brain were a fraction of those in the liver. Residues of PCB and DDE also had a significant eggshell thickness (Parslow & Jefferies, 1973).

Other pesticides have been used to deliberately target raptors. Vandenbroucke et al. (2010, 266) write that in Belgium "in the last few years, corpses of birds of prey and other predators have frequently been suspected of malicious poisoning." Supporting examples exist in Flanders, where 162 buzzards (*Buteo buteo*, Linnaeus, 1758) were examined (2003 – 2006) and 91 seemed to have been deliberately killed, by poisoning (n=79) or shooting (n=12) and other species evidently killed by poison (common kestrel, *Falco tinnunculus*, Linnaeus, 1758, sparrowhawk, *Accipiter nisus,* Linnaeus, 1758), red kite (*Milvus milvus*, Linnaeus, 1758), magpie (*Pica pica*, Linnaeus, 1758), carrion crown (*Corvus corone*, Linnaeus, 1758) and pheasant (*Phasianus colchicus*, Linnaeus, 1758). The commonest poison detected was carbofuran, the others being aldicarb, carbaryl, strychnine and chlorophacinone (Vandenbroucke et al. (2010, 266). Hernández and Margalida (2008), record the pesticide poisoning of the Cinereous vulture (*Aegypius monachus*) in Spain from 1990 to 2006., involving 241 incidents and 464 vultures, using Carbofuran, aldicarb, and strychnine in 88% of the recorded incidents. They note about 98% were deliberate poisoning of the birds, and that 83% resulted in death. The chemical formulae of these compounds are displayed in Table 3 below.

Table 3. Pesticides Toxic to Birds

Compound	Formula	Research Assessment for Impacts on Raptors/birds
Carbofuran	$C_{12}H_{15}NO_3$	Carbofuran is highly toxic to birds.[1]
Aldicarb	$C_7H_{14}N_2O_2S$	Indeterminate impacts on birds. [2]
Carbaryl	$C_{12}H_{11}NO_2$	Slight toxicity on birds.[3]
Strychnine	$C_{21}H_{22}N_2O_2$	Highly toxic, but indeterminate results of poisoning.[4]
Chlorophacinone	$C_{23}H_{15}ClO_3$	Serious possibilities for physiology.[5]

Sources: 1. National Pesticide Information Center (1996); 2. Moore et al. (2010); 3. Walker et al. (1991); 4. Wobeser & Blakley (1987). Schmutz et al. (1989). 5. Rattner et al. (2015).

Poisoning, both deliberate and accidental, is an important contributory factor for the decline of Hooded vultures across sub-Saharan Africa, especially in Cameroon, Uganda and Kenya and Southern Africa (Zimmerman et al., 1996; Mundy, 1997;

Thiollay, 2001; Ogada et al., 2010; Ssemmanda & Pomeroy, 2010; Ogada & Buij, 2011; Virani et al., 2011). Vultures are sometimes poisoned in Botswana, as poachers believe that circling vultures will reveal illegal hunting kills (Hancock, 2009, 2010). Vultures may also be poisoned in Namibia, where the poisoning may target problem animals such as lions (*Panthera leo*, Linnaeus, 1758), hyenas (*Crocuta crocuta*, Erxleben, 1777) and jackals (*Canis mesomelas*, Schreber, 1775) (Bridgeford, 2004), or may target vultures (Pain et al. (2003) a situation existing elsewhere (Anderson, 1999; Verdoorn et al., 2004; Roche, 2006). Findings indicate that in places where vultures were not poisoned, their populations did not decline (Mundy et al., 1992). Multiple poisons have been recorded in such events. For example, Van Wyk et al. (2001: 243) found 'the presence of quantifiable residues of 14 persistent chlorinated hydrocarbon pollutants', after collecting 'whole blood, clotted blood, heart, kidney, liver and muscle samples' from white-backed, Cape and lappet-faced vultures. Comparisons between wild adults and nestlings, and captive birds showed lower pesticide levels for captive birds (pesticides included gamma-BHC (lindane), alpha(cis)-chlordane and alphaendosulfan) (Van Wyk et al., 2001, 243).

In the Americas, findings indicate that Turkey vultures are less affected by some chemical compounds, including potent rodenticides such as the compound 1080 (Snyder & Snyder, 1991). Serious threats to Turkey vultures include cyanide, strychnine (used to predators of livestock, such as coyotes) (Campbell, 2015) and pesticides such as dichlorodiphenyldichloroethylene (DDE) which induce eggshell thinning (Wilbur, 1978; Kiff et al., 1983). Burnett et al. (2013) also found eggshell thinning for California condors, when comparing a current cases study in central California with findings from more successful cases from southern California, where pesticide incidence was lower. Examining breeding, captive reared condors from 1997 through 2010, eggs or eggshell fragments were recovered from 12 nests. The mean thickness of shell fragments, without membrane, was found to be 0.46 mm, which was 34% lower than the average thickness of 0.70 mm of fragments recovered from nine successful nests in interior southern California, 2007–2009, and hatching success in the central California case was also lower (20–40%) than in southern California (70–80%). Furthermore, on the eggshells the outer crystalline layer was much thinner or absent, like the southern Californian eggs during the 1960s (when DDE was more prevalent). Weight and water losses during incubation much higher than the normal rate representing successful hatching, and loss rates were correlated with decreasing shell thickness (Burnett et al., 2013).

Diclofenac

The Diclofenac Epidemic, highly published as an example of an avian and ecological catastrophe and human interference with the natural environment, was one of the most important poisoning episodes of the twentieth century, an event termed the "global vulture crises" (Woodford et al. 2008; Olea & Mateo Thomas, 2009; Campbell, 2015). This disaster resulted in the near extinction of six of the 23 recognized vulture species (especially five Indian species), listed as such on the International Union for the Conservation of Nature (IUCN) Red List of Threatened Species. The affected species included the White-rumped vulture (*Gyps bengalensis*, Gmelin, 1788), Slender-billed vulture (*Gyps tenuirostris*, Hodgson, 1844), Indian vulture (*Gyps indicus*, Scopoli, 1786) and Red-headed vulture (*Sarcogyps calvus*, Scopoli, 1786)) (Houston, 1985; Prakash 1999; Prakash & Rahmani, 1999; Virani et al., 2001; Prakash et al., 2003; Clements et al., 2012; Gilbert et al., 2002; Gilbert et al., 2004). Records indicate that 15 years (1992 to 2007) the white-rumped vulture population declined to one thousandth of the previous population, the red-headed vulture's population to below 10% and that of the Slender-billed vulture to about 1,000 birds (Cuthbert et al. 2006; Prakash et al. 2007). The Egyptian vulture (*Neophron percnopterus*) and Red-headed vulture (*Sarcogyps calvus*) were also classified as Endangered and Critically Endangered respectively (Cuthbert et al. 2006; Campbell, 2015).

Symptoms of the Diclofenac poisoning were renal and visceral gout (crystallization of uric acid in the tissues) and enteritis (inflammation of the intestine, especially the small intestine,) (Oaks et al. 2001; Gilbert et al. 2002; Pain et al. 2002). The factor for this was the tendency in the mid-1990s for Indian, Pakistani and Nepalese farmers to treat their cattle and water buffaloes with the non-steroidal anti-inflammatory drug (NSAID) diclofenac. This drug was extremely toxic to vultures that fed on the few carcasses at carcass dumps, which was the traditional method of livestock disposal in South Asia (Green et al. 2004; Oaks et al. 2004; Gross 2006). Vulture deaths followed one to two days after ingestion, after kidney failure—severe renal dysfunction—led to a buildup of urates in the internal organs and increased blood uric acid levels, contributing to anorexia and emaciation and a chalky-white coating over the visceral organs, especially the liver and heart and in in the body tissues.

Bans on diclofenac was passed during 2006 to 2010 in India, Nepal, Pakistan and Bangladesh, resulting a slow vulture population recovery in India and later in Pakistan and Nepal. For example, a breeding colony of Long-billed vultures in Pakistan was regularly monitored before (2003 – 2006) and after (2007 – 2012) and found to have declined (61%) leading up to the ban, and increased (55%) after the ban (Chaudhry et al., 2012). An alternative to diclofenac was meloxicam, which did not damage vulture kidneys (Swan et al., 2006). Studies concluded that "meloxicam is of low toxicity to *Gyps* vultures and that its use in place of diclofenac would reduce vulture mortality substantially in the Indian subcontinent. Meloxicam is already available for veterinary use in India" (Swan et al. 2006: 0395). These measures applied mainly to Asia, and to a lesser extent similar vulture species in Europe and Africa, especially Griffon vultures (Campbell, 2015). There were fewer actions in the Americas, as some species such as the turkey vulture were not affected by diclofenac ingestion at levels over 100 times over the dosage that killed Asian vultures, with no toxicity, visceral gout, renal necrosis, or elevate plasma uric acid (Rattner et al., 2008).

Polychlorinated Biphenyls (PCBs) and Perfluoroalkyl Substances (PFASs)

Polychlorinated biphenyls (PCBs, molecular formula $C_{12}H_5Cl_5$) are another important environmental pollutant. PubChem (2020a) notes:

> Polychlorinated biphenyls are mixtures of up to 209 individual chlorinated compounds (known as congeners). There are no known natural sources of PCBs. PCBs are either oily liquids or solids that are colorless to light yellow. Some PCBs can exist as a vapor in air. PCBs have no known smell or taste. Many commercial PCB mixtures are known in the U. S. by the trade name Aroclor. PCBs have been used as coolants and lubricants in transformers, capacitors, and other electrical equipment because they don't burn easily and are good insulators. The manufacture of PCBs was stopped in the U. S. in 1977 because of evidence they build up in the environment and can cause harmful health effects. Products made before 1977 that may contain PCBs include old fluorescent lighting fixtures and electrical devices containing PCB capacitors, and old microscope and hydraulic oils.

Zhang et al. (2011, 556) describe a study from China, examining the impacts of polychlorinated biphenyls and organochlorinated pesticides on mostly waterbird species (Rallidae, Ardeidae) and doves which could be prey species (Columbidae) from a contaminated region in South China. This study follows a trend examining birds as indicators of the levels of organic pollutants, arguing that "there is exceedingly little data on organochlorinated pesticide (OCPs) and polychlorinated biphenyls (PCBs) residues in bird inhabiting in China." The findings described concentrations of PCBs, dichlorodiphenyltrichloroethane and its metabolites (DDTs) and hexachlorocyclohexanes (HCHs) in the muscle tissues, liver and kidney of birds living near an e-waste recycling site. The bird specimens were examined, using pectoral muscle, liver and kidney tissues, which were Soxhlet-extracted and dissolved in isooctane and the PCBs and OCPs were analyzed using gas chromatography-mass spectrometry. The subsamples for nitrogen stable isotope analysis were then lyophilized (i.e., preserved by removal of water, by freezing and then drying, in a vacuum, at low temperatures, allowing longer term storage) and then analyzed using a flash EA 112 series elemental analyzer (determines the weight percent composition of the studied elements) and an isotope ration mass spectrometer (used to determine the origins of substances, using the relative isotopic abundances of the elements which comprise the material (Zhang et al., 2011, 556). The findings showed that the PCBs had higher concentrations than the other compounds, with possible negative impacts on the target birds. The highest concentrations were in the liver and kidney, compared with the muscle. Crucially, contaminant levels varied according to dietary composition and habitat, and levels increased up the trophic levels for PCBs and DDTs, while HCHs (hexachlorocyclohexane, formally benzene hexachloride, BHC) were level at the interspecies level (Zhang et al., 2011, 556).

Another study on PCBs by Fernie et al. (2009) argued that that only PCBs can affect wild bird reproduction, compared with other environmental contaminants. The methodology was based on a case study of captive American kestrels, which were fed food containing PCBs, with the formula (7 mg/kg body weight/d) before and during the first breeding season. The findings showed that "reproduction changed during, not after, PCB exposure in this two-year study." Kestrels exposed to PCBs produced smaller clutches of chicks, more clutches where the chicks were infertile, later in the season, and in more than half did not produce fledglings; additionally, hatching success was reduced, with half of nestlings dying within 3 days of hatching. The authors concluded that "higher levels of total PCB residues and congeners were associated with later clutch initiation and fewer fertile eggs, hatchlings, and fledglings. We suggest that nonpersistent PCB congeners have a greater influence on reproduction than do persistent congeners." Here as noted by PubChem (2020a) congeners refer to the individual chlorinated compounds that compose the PCBs.

In another study, Barghi et al. (2018, 220) contribute to the perceived gap in knowledge on "on the accumulation and potential risks of persistent chlorinated and fluorinated contaminants in birds." Organohalogen contaminants (OHCs), including PCBs, organochlorine pesticides (OCPs), and perfluoroalkyl substances (PFASs), were measured, using the liver tissues of 69 Korean birds, members of 10 species. Christensen et al. (2019, 147) define perfluoroalkyl substances as "a class of contaminants used in many industrial applications and consumer products. Certain PFAS are regulated or voluntarily limited due to concern about environmental persistence and adverse health effects." Jayaraj et al. (2016, 90) define Organochlorine (OCPs) pesticides as "synthetic pesticides" belonging "to the group of chlorinated hydrocarbon derivatives, which have vast application in the chemical industry and in agriculture...known for their high toxicity, slow degradation and bioaccumulation."

The findings were that PFASs had the highest concentrations, ranging from 0.0054 to 11.3 ppm (median: 0.294) wet weight. Median concentrations of OCPs and PCBs in the samples were 0.147 and 0.82 ppm lipid weight, respectively (Barghi et al., 2018). The concentrations of OHCs were higher in predatory than non-predatory birds, hence diet (on smaller, infected animals) and trophic position (the top of the food chain, allowing accumulation) were key issues. Further findings indicated concentrations of PCBs and DDTs were below toxic threshold levels, "PFOS concentrations in most of the predatory birds exceeded threshold values for adverse health effects..., implying potential health risks to Korean birds" (Barghi et al., 2018, 220). PFOs (perfluorooctane sulfonate) are a form of PFAS (National Institute of Environmental Health Sciences, 2020).

Another study by Luzardo et al. (2014) also looked at a larger sample with slightly different pollutants. These included 57 anthropogenic pollutants (23 organochlorine pesticides (OCs), 18 polychlorinated biphenyls (PCBs) and 16 polycyclic aromatic hydrocarbons (PAHs), using liver samples from 102 specimens of six species raptor species from Wildlife Recovery Centers of the Canary Islands from 2009 and 2012. The objectives were to measure contaminant levels and evaluate the six species as regional bioindicators. The findings indicated that the Eurasian sparrowhawk (*Accipiter nisus*, Linnaeus, 1758), Barbary falcon (*Falco peregrinus pelegrinoides*, Temminck, 1829) and kestrel (*Falco tinnunculus*, Linnaeus, 1758) were more contaminated than the other species, mostly by DDT and dieldrin and that these pollutants were comparatively high.

By contrast, the contamination by PCBs was comparatively low, possibly below threshold levels of toxicity for these contaminants. Polycyclic aromatic hydrocarbons (PAHs) were dependent on the feeding patterns but below threshold levels. The authors argue that their study "represents the first report of contamination by PAHs in all these species and is also the first report of PCB levels in barbary falcons" (Luzardo et al., 2014, 146). Polycyclic aromatic hydrocarbons (PAHs) are defined as chemicals that occur naturally in coal, crude oil, and gasoline, also when these substances and others such as wood, tobacco and garbage are burned, forming small air particles (Environmental Protection Agency, 2013). Examples are petrogenic alkyl polycyclic aromatic hydrocarbons, which are those derived from petroleum and pyrogenic polycyclic aromatic hydrocarbons, which are derived from combustion and anthropogenic sources (Paruk et al. 2014).

Conclusion

This review of environmental toxicity in raptor and waterbird ecology illustrates the complexity and importance for modifications of environmental chemistry and toxicology for birds near the apex of the food web. Six key strands may be identified. Firstly, there are general conclusions on the different elements and compounds in terms of their impacts on waterbirds and raptors. The elements and compounds both follow similar through the food chains up to foraging birds, but and the extent of variation largely depends on human action. Secondly, there is insufficient knowledge of the status of toxicity from both elements and compounds. There are substantial variations at the inter- and intra-species levels in terms of the ingestion of such chemicals on birds, largely due to pronounced differences in the chemical concentrations in prey animals and carcasses. Impacts of toxic chemicals vary and change through time and perhaps space, and this may vary depending on the sources and the species affected. Raptors may also change their target prey, complicating some conclusions.

Thirdly, different species may in some cases have different resistances and reactions to toxic chemicals, while not in others. Therefore, intensive studies must be conduced through geographical space and time. Fourth, some impacts such as those of lead, diclofenac and DDT/DDE were so serious that the chemical compounds were either replaced or seriously restricted. Others, such as mercury may require similar restrictions. Others, such as arsenic or copper require more research. Fifth, some chemicals are important for health, and only have negative impacts at excess levels. Hence, the threshold levels of compounds must be studied, with understanding that these vary by the organs in which they accumulate, such as embryos, livers, kidney and the gastrointestinal. Finally, the study of the impacts of toxic chemicals must be ongoing as knowledge and consequent assessments may change, as more impacts are documented, some chemicals are outlawed, and new chemical compounds are applied to the environment. Future directions may include systems for detoxification of ingested chemicals, continual replacement of toxic chemicals with less dangerous substances and more advanced detection methods. Industrial and agricultural developments may also need more sensitive environmental impact assessments. This chapter made a few selections among a vast number of potential possibilities and hence documents few selected species, ecologies and regions, case studies and technologies for chemical assessment selected. Future research may focus on a wider range of species, ecologies, regions and research tools.

References

Agency for Toxic Substances and Disease Registry (ATSDR). 2020a. *Toxicological Profile for Copper.* Retrieved from https://www.atsdr.cdc.gov/toxprofiles/tp132.pdf.

Agency for Toxic Substances and Disease Registry (ATSDR) 2020b. *Selenium.* Retrieved from https://www.atsdr.cdc.gov/toxprofiles/tp92-c3.pdf.

Agency for Toxic Substances and Disease Registry (ATSDR). 2019. *Toxicological Profile for Chromium.* Retrieved from https://www.atsdr.cdc.gov/ToxProfiles/tp7.pdf.

Albers, P. H. 2006. Birds and polycyclic aromatic hydrocarbons. *Avian and Poultry Biology Reviews* 17(4):125-140.

Albers, P. H., Green, D. E. & Sanderson, C. J. 1996. Diagnostic criteria for selenium toxicosis in aquatic birds: Dietary exposure, tissue concentrations, and macroscopic effects. *Journal of Wildlife Diseases* 32(3): 468-485. https://doi.org/10.7589/0090-3558-32.3.468.

Anderson, W. L., Havera, S. P. & Zercher, B. W. 2000. Ingestion of lead and nontoxic shotgun pellets by ducks in the Mississippi Flyway. *Journal of Wildlife Management* 64(3): 848–857. https://www.doi:10.2307/3802755.

Anderson, M. D. 1999. Africa's hooded vulture: a dichotomy of lifestyle. *Vulture News* 41: 3–5.

Ashbaugh, H. M., Conway, W. C., Haukos, D. A., Collins, D. P., Comer, C. E. & French, A. D. 2018. Evidence for exposure to selenium by breeding interior snowy plovers (*Charadrius nivosus*) in saline systems of the Southern Great Plains. *Ecotoxicology* 27: 703–718. http://www.doi:10.1007/s10646-018-1952-2.

Barghi, M., Jin, X., Lee, S., Jeong, Y., Yu, J. P., Paek, W. K. & Moon, H. B. 2018. Accumulation and exposure assessment of persistent chlorinated and fluorinated contaminants in Korean birds. *Science of the Total Environment,* 645:220-228. https://doi.org/10.1016/ j.scitotenv.2018.07.040.

Barnes, J. G., Doney, G. E., Yates, M. A., Seegar, W. S. & Gerstenberger, S. L. 2019. A broadscale assessment of mercury contamination in peregrine falcons across the northern latitudes of North America. *Journal of Raptor Research* 53(1), 1-13. J*ournal of Raptor Research,* 53(1):1-13. https://doi.org/10.3356/JRR-18-0003.

Bastow, B. D. & Holmes, J. L. 2016. Teratology and drug use during pregnancy. *Medscape*. WebMD. Retrieved from https://emedicine.medscape.com/article/260725-overview.

Battaglia, A., Ghidini, S., Campanini, G. & Spaggiari, R. 2005. Heavy metal contamination in little owl (*Athene noctua*) and common buzzard (Buteo buteo) from northern Italy. *Ecotoxicology and Environmental Safety* 60: 61–66. https://www.doi:10.1016/j.ecoenv. 2003.12.019.

Bedrosian, B., Craighead, D. & Crandall, R. 2012. Lead Exposure in Bald Eagles from Big Game Hunting, the Continental Implications and Successful Mitigation Efforts. *PLoS ONE* 7(12): e51978. https://www.doi:10.1371/journal.pone.0051978.

Berg, W., Johnels, A., Sjöstrand, B., & Westermark, T. 1966. Mercury Content in Feathers of Swedish Birds from the Past 100 Years. *Oikos*, 17(1): 71-83. https://www.doi:10.2307/3564782.

Beyer, W. N. 2000. Hazards to wildlife for soil-borne cadmium reconsidered. *Journal of Environmental Quality* 29: 1380–1384. https://doi.org/10.2134/jeq2000.0047242500 2900050002x.

Binkowski, L. J., Stawarz, R. M. & Zakrzewski. M. 2013. Concentrations of cadmium, copper and zinc in tissues of mallard and coot from southern Poland. *Journal of Environmental Science and Health, Part B,* 48: 410-415.https://www.doi:10.1080/03601234.2013.742725.

Burnett, L. J., Sorenson, K. J., Brandt, J., Sandhaus, E. A., Ciani, D., Clark, M., David, C., Theule, J., Kasielke, S. & Risebrough, R. W. 2013. Eggshell thinning and depressed hatching success of California Condors reintroduced to Central California. *The Condor,* 115(3):477-491. https://doi.org/10.1525/cond.2013.110150.

Blus, L. J. 1994. A review of lead poisoning in swans. *Comparative Biochemistry and Physiology* 108C: 259–267. https://doi.org/10.1016/0742-8413(94)00021-2.

Brean, H. 2019. Peregrine falcons act as mercury bellwether, Nevada study shows. *Las Vegas Review-Journal*. Retrieved from https://www.reviewjournal.com/local/local-nevada/peregrine-falcons-act-as-mercury-bellwether-nevada-study-shows-1640798/.

Bridgeford, P. 2004. Lappet-faced vulture *Torgos tracheliotus*. In Monadjem, A., Anderson, M. D., Piper, S. E. & Boschoff, A. F. (Eds.). *The Vultures of Southern Africa—Quo Vadis?* pp. 28–33. Birds of Prey Working Group, Johannesburg.

Burgat, V. 1990. *Un micropollutant: le cadmium Bulletin Mensuel de l'Office National de la Chasse* 146: 40-42.

Buxton, S., Garman, E., Heim, K. E., Lyons-Darden, T., Schlekat, C. E., Taylor, M. D. & Oller, A. R. 2019. Concise Review of Nickel Human Health Toxicology and Ecotoxicology. *Inorganics* 2019, 7(7), 89; https://doi.org/10.3390/inorganics7070089.

Cade, T. J. 2007. Exposure of California Condors to lead from spent ammunition. *Journal of Wildlife Management* 71(7): 2125–2133. https://doi.org/10.2193/2007-084.

Campbell, M. 2018. *Biological Conservation in the 21st Century: A Conservation Biology of Large Wildlife.* New York, NY: Nova Science Publishers.

Campbell, M. 2015. Vultures: *Their Ecology and Evolution.* Boca Raton: CRC Press.

Campbell, L. M., Norstrom, R. J., Hobson, K. A., Muir, D. C. G., Backus, S. & Fisk, A. T. 2005. Mercury and other trace elements in a pelagic Arctic marine food web (Northwater Polynya, Baffin Bay). *Science of the Total Environment,* 351–352: 247–263. https://doi.org/10.1016/ j.scitotenv.2005.02.043.

Canadian Council of Ministers of the Environment 1999. Canadian tissue Residue Guidelines for the Protection of Wildlife Consumers of Aquatic Biota DDT. *Excerpt from: Canadian environmental quality guidelines*, Winnipeg. Retrieved from http://www.ccme.ca/ publications/ceqg_rcqe.html.

Carneiro. M., Colaco. B., Brandao. R., Ferreira, C, Santos, N., Soeiro, V., Colaco, A., Pires, M. J., Oliveira, P. A. & Lavín, S. 2014. Biomonitoring of heavy metals (Cd, Hg, and Pb) and metalloid (As) with the Portuguese common buzzard (*Buteo buteo*). *Environmental Monitoring and Assessment,* 186: 7011-7021. https://doi.org/10.1007/s10661-014-3906-3.

Carpenter, J. W., Pattee, O. H., Fritts, S. H., Rattner, B. A., Wiemeyer, S. N., Royle, J. A. & Smith, M. N. 2003. Experimental lead poisoning in Turkey Vultures (*Cathartes aura*). *Journal of Wildlife Diseases* 39: 96–104. https://doi.org/10.7589/0090-3558-39.1.96.

Castro, I, Aboal, J. R., Fernandez, J. A. & Carballeira, A. 2011. Use of raptors for biomonitoring of heavy metals: gender, age and tissue selection. *Bulletin of Environmental Contamination and Toxicology,* 86: 347-351. http://www.doi: 10.1007/s00128-011-0220-4.

Chang, L. W., Magos, L., Suzuki, T. (Eds.) 1996. *Toxicology of Metals.* Boca Raton. FL, USA: CRC Press.

Chaudhry, M. Jamshed, I., Ogada, D. L., Malik, R. N., Virani, M. Z, & Giovanni, M. D. 2012. First evidence that populations of the critically endangered Long-billed Vulture *Gyps indicus* in Pakistan have increased following the ban of the toxic veterinary drug diclofenac in south Asia. *Bird conservation international,* 22, 389-397. https://www.doi:10.1017/ S0959270912000445.

Church, M. E., Gwiazda, R., Risebrough R. W., Sorenson, K., Chamberlain, C. P., Farry, S., Heinrich, W., Rideout, B. A., Smith, D. R. 2006. Ammunition is the principal source of lead accumulated by California condors re-introduced to the wild. *Environmental Science & Technology,* 40(19): 6143- 6150. https://www.doi:10.1021/es060765s.

Clarkson, T. W., Magos, L. & Myers, G. J. 2003. Human exposure to mercury: The three modern dilemmas. *The Journal of Trace Elements in Experimental Medicine,* 16(4): 321 – 343. https://doi.org/10.1002/jtra.10050.

Clements, T., Gilbert, M., Rainey, H. J., Cuthbert, R., Eames, J. C., Bunnat, P., Teak, S., Chansocheat, S. & Setha, T. 2012. Vultures in Cambodia: population, threats and conservation. *Bird Conservation International* 1–18. https://www.doi:10.1017/S0959270912000093.

Colborn, T. 1991. Epidemiology of great lakes bald eagles. *Journal of Toxicology and Environmental Health,* 33(4): 395-453. https://doi.org/10.1080/15287399109531537.

Craighead, D., Bedrosian, B., Watson, R. T., Fuller, M., Pokras, M. & Hunt, W. 2009. A Relationship between Blood Lead Levels of Common Ravens and the Hunting Season in the Southern Yellowstone Ecosystem. *Ingestion of Lead from Spent Ammunition: Implications for Wildlife and Humans.* Boise, ID: Peregrine Fund.

Crinnion, W. J. 2011. Polychlorinated Biphenyls: Persistent pollutants with immunological, neurological, and endocrinological consequences. *Alternative Medicine Review,* 16(1): 5 – 13.

Christensen, K. Y., Raymond, M. & Meiman, J. 2019. Perfluoroalkyl substances and metabolic syndrome. *International Journal of Hygiene and Environmental Health,* 222(1): 147-153. https://doi.org/10.1016/j.ijheh.2018.08.014.

Cross, T. L. 2009. *Waterbirds.* New York, NY: W.W. Norton & Company.

Cruz-Martinez, L., Redig, P. T. & Deen, J. 2012. Lead from spent ammunition: a source of exposure and poisoning in bald eagles. *Human–Wildlife Interactions* 6(1):94–104.

Cuthbert, R., Green, R. E., Ranade, S., Saravanan, S., Pain, D. J., Prakash, V. & Cunningham, A. A. 2006. Rapid population declines of Egyptian Vulture (*Neophron percnopterus*) and Redheaded Vulture (*Sarcogyps calvus*) in India. *Animal Conservation* 9(3): 349–354. https://doi.org/10.1111/j.1469-1795.2006.00041.x.

Das, K. K., Reddy, R. C., Bagoji, I. B., Das, S., Bagali, S., Mullur, L., Khodnapur, J. P. & Biradar M. S. 2018. Primary concept of nickel toxicity - an overview. *Journal of Basic and Clinical Physiology and Pharmacology,* 30(2):141-152. http://www.doi:10.1515/jbcpp-2017-0171.

DeBruyn, A. M. H. & Chapman, P. M. 2007. *Environmental Science & Technology* 2007 41 (5), 1766-1770. http://www.doi:10.1021/es062253j.

DeForest, D. K., Schlekat, C. E., Brix, K. V. & Fairbrother, A. 2012. Secondary poisoning risk assessment of terrestrial birds and mammals exposed to nickel. *Integrated Environmental Assessment and Management,* 8(1):107- 119. https://doi.org/10.1002/ieam.248.

De Francisco, N., Troya, J. D. E. & Agüera, E. I. 2003. Lead and lead toxicity in domestic and free-living birds. *Avian Pathology.* 32(1): 3–13. https://www.doi:10.1080/0307945021000 070660.

Dobrowolska, A. & Melosik, M. 2008. Bullet-derived lead in tissues of the wild boar (*Sus scrofa*) and red deer (*Cervus elaphus*). *European Journal of Wildlife Research,* 54(2): 231–235. https://doi.org/10.1007/s10344-007-0134-y.

Donázar, J. A., Palacios, C. J., Gangoso, L., Ceballos, O., Gonzalez, M. J. & Hiraldo, F. 2002. Conservation status and limiting factors in the endangered population of Egyptian vulture (*Neophron percnopterus*) in the Canary Islands. *Biological Conservation* 107(1): 89-97. https://doi.org/10.1016/S0006-3207(02)00049-6.

Eisler, R. 1988. *Arsenic hazards to fish, wildlife, and invertebrates: A synoptic review.* Biological Report 85(1.12) Contaminant Hazard Reviews January 1988 Report No. 12. Laurel, MD: U.S. Fish and Wildlife Service, Patuxent Wildlife Research Center.

Ek, K. H., Morrison, G. M., Lindberg, P. & Rauch, S. 2004. Comparative Tissue Distribution of Metals in Birds in Sweden Using ICP-MS and Laser Ablation ICP-MS. *Bulletin of Environmental Contamination and Toxicology,* 47(2): 259-269. https://doi.org/10.1007/ s00244-004-3138-6.

Eeva, T., Belskii, E. & Kuranov, B. 2006. Environmental pollution affects genetic diversity in wild bird populations. *Mutation research. Genetic toxicology and environmental mutagenesis*, 608(1): 8-15. https://www.doi:10.1016/j.mrgentox. 2006.04.021.

Eeva, T., Lehikoinen, E., & Ronka, M. 1998. Air Pollution Fades the Plumage of the Great Tit. *Functional Ecology,* 12(4): 607-612. https://www.doi:10.1046/j.1365-2435.1998.00221.x.

European Chemicals Bureau, 2003. *Technical guidance document in support of Commission Directive 93/67/EEC on risk assessment for new notified substances, Commission Regulation (EC) 1488/94 on risk assessment for existing substances and Directive 98/8/EC of the European Parliament and of the Council concerning the placing of biocidal products on the market.* Ispra, Italy: European Commission, Joint Research Centre, European Chemicals Bureau.

Evers, D. C., Savoy, L. J., DeSorbo, C. R., Yates, D. E., Hanson, W., Taylor, K. M., Siegel, L. S., Cooley, J. H. Jr, Bank, M. S., Major, A., Munney, K., Mower, B. F., Vogel, H. S., Schoch, N., Pokras, M., Goodale, M. W. & Fair, J. 2008. Adverse effects from environmental mercury loads on breeding common loons. *Ecotoxicology,* 17(2): 69-81. https://www.doi:10.1007/s10646-007-0168-7.

Environmental Protection Agency. 2013. *Polycyclic Aromatic Hydrocarbons (PAHs).* Retrieved from https://www.epa.gov/sites/production/files/2014-03/documents/ pahs_factsheet_cdc _2013.pdf.

Falandysz, J., Ichihashi, H., Szymczak, K., Yamasaki, S. & Mizera, T. 2001. Metallic elements and metal poisoning among White-tailed sea eagles from the Baltic Sea. *Marine Pollution Bulletin,* 42: 1190-1193. https://www.doi:10.1016/s0025-326x(01)00217-x.

Louekari, K., Mäkelä-Kurtto, R., Pasanen, J., Virtanen, V., Sippola, J. & Malm, J. 2000. *Cadmium in Fertilizers: Risks to human health and the environment.* Ministry of Agriculture and Forestry in Finland. Retrieved from http://julkaisut. valtioneuvosto.fi/bitstream/handle/10024/160651/2000_4_Cadmium%20in%20fer tilizers%20-%20Risks%20to%20human%20health%20and%20the%20 environment.pdf.

Fernie, K. J., Smits, J. E., Bortolotti, G. R. & Bird, D. M. 2001. Reproduction success of American Kestrels exposed to dietary polychlorinated biphenyls. *Environmental Toxicology and Chemistry,* 20:776–781. https://doi.org/10.1002/etc.5620200412.

Fingas, M. 2017. The fate of PAHs resulting from in-situ oil burns. *International Oil Spill Conference Proceedings,* 2017(1): 1041-1056. https://doi.org/10.7901/2169-3358-2017.1.1041.

Forsman, D. 1999. *The raptors of Europe and the Middle East: A handbook of field identification.* London: T. & A.D. Poyser.

Frank, R. A. & Lutz, R. S. 1999. Productivity and survival of great horned owls exposed to dieldrin. *The Condor*, 101(2): 331–339. https://doi.org/10.2307/1369996.

Franson, J. C., Lahner, L. L., Meteyer, C. U. & Rattner, B. A. 2012. Copper pellets simulating oral exposure to copper ammunition: absence of toxicity in American kestrels (Falco sparverius). *Archives of Environmental Contamination and Toxicology,* 62(1):145-53. http://www.doi:10.1007/s00244-011-9671-1.

Franson, J. C. & Pain, D. J. 2011. Lead in birds. In Beyer, W. N. & Meador, J. P. (Eds.) *Environmental Contaminants in Biota: Interpreting Tissue Concentrations,* pp. 563–593. Second edition. Boca Raton: CRC Press.

Franson, J. C., Hansen, S. P., Creekmore, T. E., Brand, C. J., Evers, D. C., Duerr, A. E. & DeStefano, S. 2003. Lead fishing weights and other fishing tackle in selected waterbirds. *Waterbirds* 26:345–352. https://www.doi:10.1675/1524-4695(2003)026[0345:LFWAOF] 2.0.CO;2.

Franson, J. C. 1996. Interpretation of tissue lead residues in birds other than waterfowl. In Beyer, W. N., Heinz, G. H. & Redmon-Norwood, A. W. (Eds.) *Environmental contaminants in wildlife: Interpreting tissue concentrations,* 265–279. Boca Raton: CRC Press, Lewis Publishers.

Frantz, A., Pottier, M. A., Karimi, B., Corbel, H., Aubry, E., Haussy, C. Gasparini, J., Castrec-Rouelle, M. 2012. Contrasting levels of heavy metals in the feathers of urban pigeons from close habitats suggest limited movements at a restricted scale. https://doi.org/10.1016/j.envpol.2012.04.003.

Gangoso, L., Alvarez-Lloret, P., Rodríguez-Navarro, A. A., Mateo, R., Hiraldo, F. & Donázar, J. A. 2009. Long-term effects of lead poisoning on bone mineralization in vultures exposed to ammunition sources. *Environmental Pollution,* 157(2): 569-574. https://doi.org/10.1016/j.envpol.2008.09.015.

García-Fernández, A. J., Calvo, J. F., Martínez-López, E., María-Mojica, P. & Martínez, J. E. 2008. Raptor ecotoxicology in Spain: A review on persistent environmental contaminants. *AMBIO: A Journal of the Human Environment* 37(6), 432-439. https://doi.org/10.1579/0044-7447(2008)37[432:REISAR]2.0.CO;2.

García-Fernández, A. J., Sanchez-Garcia, J. A., Gomez-Zapata, M. & Luna, A. 1996. Distribution of cadmium in blood and tissues of wild birds. *Archives of Environmental Contamination and Toxicology,* 30: 252-258. https://www.doi:10.1007/BF00215805.

Gilbert, M., Oaks, J. L., Virani, M. Z., Watson, R. T., Ahmed, S., Chaudhry, J., Arshad, M., Mahmood, S., Ali, A., Khattak R. M. & Khan, A. A. 2004. The status and decline of vultures in the provinces of Punjab and Sind, Pakistan: a 2003 update. pp. 221–234. In: R. C. Chancellor and B. U. Meyburg (eds.). Raptors Worldwide. *Proceedings of the 6th world conference on birds of prey and owls.* Berlin and Budapest: WWGBP and MME/Birdlife Hungary.

Gilbert, M., Virani, M. Z., Watson, R. T., Oaks, J. L., Benson, P. C., Khan, A. A., Ahmed, S., Chaudhry, J., Arshad, M., Mahmood, S. & Shah, Q. A. 2002. Breeding and mortality of Oriental White-backed Vulture *Gyps bengalensis* in Punjab

Province, Pakistan. *Bird Conservation International* 12: 311–326. https://doi.org/10.1017/S0959270902002198.

Golden, N. H., Warner, S. E. & Coffey, M. J. 2016. A review and assessment of spent lead ammunition and its exposure and effects to scavenging birds in the United States. *Reviews of Environmental Contamination and Toxicology,* 237: 123-129. https://www.doi:10. 1007/978-3-319-23573-8_6.

Grade, T. J., Pokras, M. A., Laflamme, E. M. & Vogel, H. S. 2017. Population-level effects of lead fishing tackle on common loons. *The Journal of Wildlife Management* 82(1):155–164. https://www.doi:10.1002/jwmg.21348.

Grade, T. J. 2011. Effects of lead fishing tackle on common loons (Gavia immer) in New Hampshire, 1989–2010. Thesis, University of Wisconsin, Madison, USA.

Green, R. E., Newton, I., Shultz, S., Cunningham, A. A., Gilbert, M., Pain, D. & Prakash, V. 2004. Diclofenac poisoning as a cause of vulture population decline across the Indian subcontinent. *Journal of Applied Ecology* 41: 793–800. https://doi.org/10.1111/j.0021-8901.2004. 00954.x.

Gross, L. 2006. Switching drugs for livestock may help save critically endangered Asian vultures. *PLoS Biol.* 4(3): e61. https://doi.org/10.1371/journal. pbio.0040061.

Grúz, A., Mackle, O., Bartha, A., Szabó, R., Déri, J., Budai, P. & Lehel, J. 2019. Biomonitoring of toxic metals in feathers of predatory birds from eastern regions of Hungary. *Environmental Science and Pollution Research,* 26: 26324–26331. https://doi.org/10. 1007/s11356-019-05723-9.

Hall, M., Grantham, J., Posey, R. & Mee, A. 2007. Lead exposure among reintroduced California condors in southern California. *California condors in the 21st century.* Series in Ornithology No. 2, 163–184. Cambridge, MA: Nuttall Ornithological Club and American Ornithologists' Union.

Hancock, P. 2010. Vulture poisoning at Khutse. Birds and People 27: 7. Hancock, P. 2009. Poisons devastate vultures throughout Africa. *African Raptors Newsletter* 2 (November).

Harvey, L. J. & McArdle, H. J. 2008. Biomarkers of copper status: a brief update. *British Journal of Nutrition,* 99(S3): S10–S13. https://www.doi:10.1017/S000711 4508006806.

Haig, S. M., D'Elia, J., Eagles-Smith, C., Fair, J. M., Gervais, J., Herring, G., Rivers, J. W. & Schulz, J. H. 2014. The persistent problem of lead poisoning in birds from ammunition and fishing tackle. *The Condor.* 116 (3): 408–428. https://www.doi:10.1650/condor-14-36.1.

Haschek, W. M., Rousseaux, C. G. & Wallig, M. A. 2013. *Haschek and Rousseaux's handbook of toxicologic pathology.* San Diego: Academic Press.

Heinz, G. H., Hoffman, D. J., Klimstra, J. D., Stebbins, K. R., Kondrad, S. L., Erwin, C. A. 2011. Teratogenic effects of injected methylmercury on avian embryos. *Environmental Toxicology,* 30(7): 1593-1598. https://doi.org/10.1002/etc.530.

Helander, B., Axelsson, J., Borg, H., Holm, K. & Bignert, A. 2009. Ingestion of lead from ammunition and lead concentrations in white-tailed sea eagles (Ha*liaeetus* *albicilla*) in Sweden. *Sciences of Total Environment* 407: 5555–5563. http://www.doi:10.1016/ j.scitotenv.2009.07.027.

Helander, B., Ekman, B., Hagerroth, J. E., Hagerroth, P. A. & Israelsson, J. 1989. Age-specific field characteristics of the white-tailed sea eagle, *Haliaeetus albicilla.* Var Vagelvarld 48: 319–334. http://www.doi:10.1016/j.scitotenv.2009.07.027.

Henny, C. J. & Elliott, J. E. 2015. Toxicology. In Bird D. M. & Bildstein, K. L. (Eds.) *Raptor Research and Management Techniques Manual,* Second Edition, 329-350. Surrey: Hancock House Publishing, Ltd.

Hernandez, M. & Margalida, A. 2008. Pesticide abuse in Europe: effects on the cinereous vulture (*Aegypius monachus*) population in Spain. *Ecotoxicology* 17: 264–272. https://doi.org/10.1007/s10646-008-0193-1.

Hinck, J. E., Norstrom, R. J., Orazio, C. E., Schmitt, C. J. & Tillitt, D. E. 2009. Persistence of organochlorine chemical residues in fish from the Tombigbee River (Alabama, USA): Continuing risk to wildlife from a former DDT manufacturing facility. *Environmental Pollution,* 157(2): 582-591. https://doi.org/ 10.1016/j.envpol.2008.08.021.

Hoffman, D. J. 2002. Role of selenium toxicity and oxidative stress in aquatic birds. *Aquatic Toxicology* 57(1–2): 11-26. https://doi.org/10.1016/S0166-445X(01) 00263-6.

Honda, K., Marcovecchio, J. E., Kan, S., Tatsukawa, R. & Ogi, H. 1990. Metal concentrations in pelagic seabirds from the North Pacific Ocean. *Archives of Environmental Contamination and Toxicology* 19: 704–711. https://www.doi:10. 1007/BF01183988.

Houserova, P., Hedbavny J., Matejicek, D., Kracmar, S. Sitko, J. & Kuban, V. 2005. Determination of total mercury in muscle, intestines, liver and kidney tissues of cormorant (*Phalacrocorax carbo*), great crested grebe (*Podiceps cristatus*) and Eurasian buzzard (Buteo buteo). *Veterinary Medicine* (Prague), 50: 61-68. https://doi.org/10.17221/5597-VETMED.

Huang, J. & Mayer, J. 2020. *Lead and Zinc Toxicity in Birds*. Retrieved from https://todaysveterinarypractice.com/lead-and-zinc-toxicity-in-birds/.

Houston, D. C. 1985. Indian white-backed vulture (*G. bengalensis*). In Newton, I. & Chancellor, R. D. (Eds.). Conservation Studies on Raptors Technical publication 5, 465–466. Cambridge: International Council for Bird Preservation.

Hunt, W. G., Parish, C. N., Farry, S. C., Lord, T. G. & Sieg, R. 2007. Movements of introduced California Condors in Arizona in relation to lead exposure. In Mee, A., Hall. L. S. & Grantham, J. (Eds.). *California Condors in the 21st Century*, 79–96. Cambridge, MA: Nuttall Ornithological Club, American Ornithologists Union, McLean.

Iñigo, A. & Atienza, J. C. 2007. Efectos del Reglamento 1774/2002 y las decisions adoptadas por la Comisión Europea en 2003 y 2005 sobre las aves necrófagas en la peninsula Ibérica y sus posibles soluciones. *Official report to the European Commission.* SEO/BirdLife 15-06-2007. [Effects of Regulation 1774/2002 and the decisions adopted by the European Commission in 2003 and 2005 on ghouls in the Iberian Peninsula and their possible solutions]

Jayaraj, R., Megha, P. & Sreedev, P. 2016. Organochlorine pesticides, their toxic effects on living organisms and their fate in the environment. *Interdisciplinary Toxicology* 9(3-4): 90–100. https://doi.org/10.1515/intox-2016-0012.

Jenson, S., Johnels, A. G., Olsson, M. & Westermark, T. 1972. The avifauna of Sweden as indicators of environmental contamination with mercury and chlorinated hydrocarbons. *Proceedings of the International Ornithological Congress*. 15:455–465.

Johnels, A., Tyler, G. & Westermark, T. 1979. A history of mercury levels in Swedish fauna. *Ambio,* 8(4): 160-168.

Jorgenson, J. L. 2001. Aldrin and dieldrin: a review of research on their production, environmental deposition and fate, bioaccumulation, toxicology, and epidemiology in the United States. *Environmental Health Perspectives,* 109(Suppl 1): 113–139. https://doi.org/10. 1289/ehp.01109s1113.

Jürgens, M. D., Crosse, J., Hamilton, P. B., Johnson, A. C. & Jones, K. C. 2016. The long shadow of our chemical past – High DDT concentrations in fish near a former agrochemicals factory in England. *Chemosphere,* 162: 333-344. https://doi.org/10.1016/j.chemosphere.2016.07.078.

Kalisinska, E., Gorecki, J., Lanocha, N., Okonska, A., Melgarejo, J. B., Budis, H., Rzad, I. & Goals, J. 2014. Total and methylmercury in soft tissues of white-tailed eagle (*Haliaeetus albicilla*) and osprey (*Pandion haliaetus*) collected in Poland. *Ambio,* 43: 858-870. https://www.doi:10.1007/s13280-014-0533-8.

Kallok, M. A. 2013. *The Case for Copper: Research uncovers problems with lead bullets for deer hunting*. Minnesota Conservation Volunteer. Retrieved from https://www.dnr.state.mn.us/ mcvmagazine/issues/2013/sep-oct/copper.html.

Kelly, T. R., Poppenga, R. H., Woods, L. A., Hernandez, Y. Z., Boyce, W. M., Samaniego, F. J., Torres, S. G. & Johnson, C. K. 2014. *Causes of mortality and unintentional poisoning in predatory and scavenging birds in California*. Veterinary Record Open. http://www.doi:10.1136/vropen-2014-000028.

Kelly, T. R. & Johnson, C. K. 2011. Lead exposure in free-flying turkey vultures is associated with big game hunting in California. *PLoS One* 6(4): e15350. https://doi.org/10.1371/ journal.pone.0015350.

Kenntner, N., Crettenand, Y., Fünfstück, H., Janovsky, M. & Tataruch, F. 2007. Lead poisoning and heavy metal exposure of golden eagles (Aquila chrysaetos) from the European Alps. *Journal für Ornithologie,* 148 (2): 173-177. http://www.doi:10. 1007/s10336-006-0115-z.

Kiff, L. F., Peakall, D. B., Morrison, M. L. & Wilbur, S. R. 1983. Eggshell thickness and DDE residue levels in vulture eggs. pp. 440–458. In Wilbur, A. & Jackson, J. A. (Eds.), *Vulture Biology and Management.* Berkeley: University of California Press.

Kitowski, I., Jakubas, D., Wiącek, D. & Sujak, A. 2017. *Concentrations of lead and other elements in the liver of the white-tailed eagle (Haliaeetus albicilla), a European flagship species, wintering in Eastern Poland Ambio* (2017) 46: 825. https://doi.org/10.1007/s13280-017-0929-3.

Kitowski, I., Sujak, A., Wiącek, D., Strobel, W., Komosa, A., Stobiński, M. 2016. Heavy metals in livers of raptors from Eastern Poland –The importance of diet composition. *Belgian Journal of Zoology.* 146, 3-13. https://www.doi:10.26496/bjz.2016.34.

Kitowski, I., Sujak, A., Wiacek, D., Strobel, W. & Rymarz, M. 2014. Trace element residues in eggshells of Grey Heron (*Ardea cinerea*) from colonies of East Poland. *North-West Journal of Zoology,* 10: 346-354. https://www.doi:10.1016/j.foodchem.2005.06.036.

Kramer, J. & Redig, P. T. 1997. Sixteen years of lead poisoning in eagles, 1980-1995: An epizootiologic view. *Journal of Raptor Research* 31(4):327-332.

Krüger, S. C. & Amar, A. 2018. Lead Exposure in the Critically Endangered Bearded Vulture (*Gypaetus barbatus*) Population in Southern Africa. *Journal of Raptor Research*, 52(4):491-499. https://doi.org/10.3356/JRR-17-86.1.

Kuhn, L. T. 2001. Bone mineralization. In Martin, J. W. (Ed.) *Encyclopedia of Materials: Science and Technology* (Second Edition), 47 - 54. Retrieved from https://epdf.pub/queue/concise-encyclopedia-of-the-structure-of-materials.html.

Lapointe, D. & Couture, P. 2006. Importance of the route of exposure in accumulation and subcellular partitioning of nickel in fathead minnows (*Pimephales promelas*). Presentation P484 at the *27th Annual Meeting of the Society for Environmental Toxicology and Chemistry*, Montreal, Quebec. November 2006.

Lazorchak, J. M., McCormick, F. H., Henry, T. R. & Herlihy, A. T. 2009. Contamination of fish in streams of the Mid-Atlantic Region: An approach to regional indicator selection and wildlife assessment. *Environmental Toxicology* 22(3): 545-553. https://doi.org/10.1002/etc.562 0220312.

Lemly, A. D. 2002. Symptoms and implications of selenium toxicity in fish: the Belews Lake case example. *Aquatic Toxicology,* 57: (1–2): 39-49. https://doi.org/10.1016/S0166-445X(01) 00264-8.

Lemly, A. D. 1996. Assessing the toxic threat of selenium to fish and aquatic birds. *Environmental Monitoring and Assessment.* 43(1):19-35. http://www.doi:10.1007/BF00399568.

Lemly, A. D. 1993. Guidelines for evaluating selenium data from aquatic monitoring and assessment studies. *Environmental Monitoring and Assessment* 28(1):83-100. http://www.doi:10.1007/BF00547213.

López, S. L., Aiassa, D., Benítez-Leite, S., Lajmanovich, R., Mañas, F., Poletta, G., Sánchez, N., Fernanda Simoniello, M., Carrasco, A. E. 2012. Chapter Two - Pesticides Used in South American GMO-Based Agriculture: A Review of Their Effects on Humans and Animal Models. *Advances in Molecular Toxicology* 6: 41 – 75. https://doi.org/10.1016/B978-0-444-59389-4.00002-1.

Luzardo, O. P., Ruiz-Suárez, N., Henríquez-Hernández, L. A., Valerón, P. F., Camacho, M., Zumbado, M., & Boada, L. D. 2014. Assessment of the exposure to organochlorine pesticides, PCBs and PAHs in six species of predatory birds of the Canary Islands, Spain. *Science of the Total Environment,* 15: 472:146-153. https://doi.org/10.1016/ j.scitotenv. 2013.11.021.

Madry, M. M., Kraemer, T., Kupper, J., Naegeli, H., Jenny, H., Jenni, L., Jenny, D. 2015. Excessive lead burden among golden eagles in the Swiss Alps. *Environmental Research Letters,* 10(3):034003.

Manning, L. K., Wünschmann, A., Armién, A. G., Willette, M., MacAulay, K., Bender, J. B., Buchweitz, J. P. & Redig, P. 2018. First lead intoxication in free-ranging bald eagles (*Haliaeetus leucocephalus*). *Veterinary Pathology,* 56(2): 289-299. https://doi.org/10.1177/0300985818813099.

Manning, T., Ross, G. A. & Symons, R. 2008. Environmental contaminants in white-bellied sea-eagles (*Haliaeetus leucogaster*) found in Sydney, Australia. *Australian Journal Ecotoxicology,* 14: 21-30.

Manto, M. 2014. Abnormal copper homeostasis: Mechanisms and roles in neurodegeneration. *Toxics,* 2: 327-345. https://www.doi:10.3390/toxics2020327.

Martin, P. A., Campbell, D., Hughes, K. & McDaniel, T. 2008. Lead in the tissues of terrestrial raptors in southern Ontario, Canada, 1995–2001. *Sciences of Total Environment* 391: 96–103. https://www.doi:10.1016/j.scitotenv.2007.11.012.

Merriam-Webster.com Medical Dictionary, 2020. *Carcinogenic.* Retrieved from https://www.merriam-webster.com/dictionary/carcinogenic.

Merriam-Webster.com Medical Dictionary, 2020. *Hemotoxic.* Retrieved from https://www.merriam-webster.com/medical/hemotoxic.

Moore, D. R., Teed, R. S., Rodney. S. I., Thompson, R. P. & Fischer D. L. 2010. Refined avian risk assessment for aldicarb in the United States. *Integrated Environmental Assessment and Management,* 6(1):83-101. https://doi.org/10. 1897/IEAM_2009-022.1.

Mudd, G. M. 2010. Global trends and environmental issues in nickel mining: Sulfides versus laterites. *Ore Geology Reviews,* 38(1–2): 9-26. https://doi.org/10.1016/ j.oregeorev. 2010.05.003.

Mundy, P. J. 1997. Hooded Vulture Necrosyrtes monarchus. In Harrison, J. A., Allan, D. G., Underhill, L. G., Herremans, M., Tree, A. J., Parker, V. & Brown, C. J. (Eds.). *The Atlas of Southern African Birds,* Vol. 1: Non-passerines, 156–157. Johannesburg: BirdLife South Africa.

Mundy, P., Butchart, D., Ledger, J. & Piper, S. 1992. *The Vultures of Africa.* London: Academic Press.

Nadjafzadeh, M., Hofer, H. & Krone, O. 2013. The link between feeding ecology and lead poisoning in white-tailed eagles. *Journal Wildlife Management,* 77: 48-57. https://www.doi:10.1002/jwmg.440.

Nam, D. H. & Lee, D. P., 2006b. Reproductive effects of heavy metal accumulation on breeding feral pigeons (*Columba livia*). *Science of the Total Environment* (2-3):682-687. https://www.doi:10.1016/j.scitotenv.2006.02.004.

National Institute of Environmental Health Sciences, 2020. *Perfluoroalkyl and Polyfluoroalkyl Substances (PFAS).* Retrieved from https://www.niehs.nih. gov/health/topics/agents/pfc/index.cfm.

National Pesticide Information Center (NPIC).2011. *Ecotoxicology.* Retrieved from http://npic.orst.edu/factsheets/ecotox.html.

National Pesticide Information Center (NPIC).1999. *DDT.* Retrieved from http://npic.orst.edu/factsheets/ddtgen.pdf.

National Pesticide Information Center (NPIC).1996. *Carbofuran.* Retrieved from http://extoxnet. orst.edu/ pips/carbofur.htm.

National Research Council (US) *Committee on Medical and Biological Effects of Environmental Pollutants.* Arsenic: Medical and Biologic Effects of Environmental Pollutants. 5, Biologic Effects of Arsenic on Plants and Animals. Retrieved from https://www.ncbi.nlm.nih.gov/books/NBK231025/.

Newton, I. 1986. *The Sparrowhawk.* Calton: T and AD Poyser Ltd.

Newton, I. 1993. Causes of breeding failure in wild raptors: A review. In Redig, P. Y. (Ed.) *Raptor Biomedicine*, 62 - 70. Minneapolis, MN: University of Minnesota Press.

Nighat, S., Iqbal, S., Nadeem, M. S., Mahood, T. & Shah, S. I. 2013. Estimation of heavy metal residues from the feathers of Falconidae, Accipitridae, and Strigidae in Punjab, Pakistan. *Turkish Journal of Zoology,* 37: 488-500. http://www.doi: 10.3906/zoo-1112-1.

Nobel Prize in Physiology or Medicine 1948. Retrieved from https://www.nobelprize.org/prizes/medicine/1948/summary/.

Nriagu, J. O., Pfeiffer, W. C., Malm, O., Magalhaes De Souza, C. M. & Mierle, G. 1992. Mercury pollution in Brazil. *Nature* 356: 389. https://doi.org/10.1038/356389a0.

Oaks, J. L., Gilbert, M., Virani, M. Z., Watson, R. T., Meteyer, C. U., Rideout, B. A., Shivaprasad, H. L., Shakeel, A., Chaudhry, M. J. I., Arshad, M., Mahmood, S., Ahmad, A. & Khan, A. A. 2004. Diclofenac residues as the cause of vulture decline in Pakistan. *Nature* 10.1038. https://doi.org/10.1038/nature02317.

Ogada, D. L. & Buij, R. 2011. Large declines of the hooded vulture *Necrosyrtes monachus* across its African range. *Ostrich: Journal of African Ornithology* 82(2): 101–113. https://doi.org/doi:%2010.2989/00306525.2011.603464.

Ogada, D. L., Thomsett, S., Virani, M. Z., Kendall, C. & Odino, M. 2010. Raptor road counts in Kenya: with emphasis on vultures. Unpublished report to the Royal Society for the Protection of Birds.

Ohlendorf, H. M. 1989. Bioaccumulation and effects of selenium in wildlife. In Jacobs, L. W. (Ed.) *Selenium in agriculture and the environment,* 133–177. Madison: Soil Science Society of America.

Olea, P. P. & Mateo-Tomás, P. 2009. The role of traditional farming practices in ecosystem conservation: the case of transhumance and vultures. *Biological Conservation* 142(8): 1844–1853. https://doi.org/10.1016/j.biocon.2009.03.024.

O'Toole, D. & Raisbeck, M. F. 1997. Experimentally induced selenosis of adult mallard ducks: Clinical signs, lesions, and toxicology. *Veterinary Pathology,* 34(4): 330-340. https://doi.org/10.1177/030098589703400409.

Owen, B. 2019. Switch ammo to stop lead poisoning in birds, B. C. wildlife experts tell hunters. *The Canadian Press.* Retrieved from https://www.cbc.ca/news/canada/british-columbia/ switch-ammo-lead-poisoning-1.5361718.

Pacyna, E. G., Pacyna, J. M., Sundseth, K., Munthe, J., Kindbom, K., Wilson, S., Steenhuisen, F. & Maxson, P. 2010. Global emission of mercury to the atmosphere from anthropogenic sources in 2005 and projections to 2020. *Atmospheric Environment.* 44. 2487-2499. https://www.doi:10.1016/j.atmosenv.2009.06.009.

Pain, D. J., Bavoux, C. & Burneleau, G. 1997. Seasonal blood lead concentration in marsh harrier *Circus aeruginosus* from Charente- Maritime, France: relationship with the hunting season. *Biological Conservation,* 81: 1-7.

Pain, D. J., Sears, J. & Newton, I. 1995. Lead concentrations in birds of prey in Britain. *Environmental Pollution,* 87: 173–180. https://doi.org/10.1016/0269-7491(94) P2604-8.

Parslow, J. L. F. Jefferies, D. J. & Hanson, H. M. 1973. Gannet mortality incidents in 1972. *Marine Pollution Bulletin*, 4(3): 41-43. https://doi.org/10.1016/0025-326X (73)90303-2.

Paruk, J. D., Long, D., Perkins, C., East, A., Sigel, B. J. & Evers, D. C. 2014. Polycyclic aromatic hydrocarbons detected in common loons (*Gavia immer*) wintering off coastal Louisiana. *Waterbirds,* 37(sp1): 85-93. https://doi.org/10.1675/063. 037.sp111.

Pattee, O. H., Bloom, P. H., Scott, J. M. & Smith, M. R. 1990. Lead hazards within the range of the California Condor. *Condor* 92: 931–937. https://doi.org/10.2307/ 1368729.

Pérez-López, M., Mendoza, M., López-Beceiro, A. & Rodriguez, F. S. 2008. Heavy metal (Cd, Pb, Zn) and metalloid (As) content in raptor species from Galicia (NW Spain). *Ecotoxicology and environmental safety.* 70. 154-62. https://www.doi:10. 1016/j.ecoenv.2007.04.016.

Pokras, M., Kneeland, M., Ludi, A., Golden, E., Major, A., Miconi, R. & Poppenga, R. H. 2009. Lead objects ingested by common loons in New England. *Northeastern Naturalist,* 16(2): 177-182. https://doi.org/10.1656/045.016.0202.

Pokras, M. A. & Chafel, R. 1992. Lead toxicosis from ingested fishing sinkers in adult common loons (*Gavia immer*) in New England. *Journal of Zoo and Wildlife Medicine* 23: 92–97. From www.jstor.org/stable/20460274.

Pollock, C, & Chow, C. 2009. *Heavy Metal Poisoning in Birds.* Retrieved from https://lafeber.com/vet/heavy-metal-poisoning-in-birds/.

Prakash, V., Green, R. E., D. E. Pain, S. P. Ranade, S. Saravanan, N. Prakash, R. Venkitachalam, R. Cuthbert, A. R. Rahmani and A. A. Cunningham. 2007. Recent changes in populations of resident Gyps vultures in India. *Journal of the Bombay Natural History Society* 104: 129–135.

Prakash, V., Pain, D. J., Cunningham A. A. et al. 2003. Catastrophic collapse of Indian white-backed *Gyps bengalensis* and long-billed *Gyps indicus* vulture populations. *Biological Conservation* 109: 381–390.

Prakash, V. 1999. Status of vultures in Keoladeo National Park, Bharatpur, Rajasthan with special reference to population crash in Gyps species. *Journal of the Bombay Natural History Society* 96: 365–378.

Prakash, V. & Rahmani, A. R. 1999. Notes about the decline of Indian Vultures, with particular reference to Keoladeo National Park. *Vulture's News* 41: 6–13.

Prestt, I., Jefferies, D. J. & Moore, N. W. 1970. Polychlorinated biphenyls in wild birds in Britain and their avian toxicity. *Environmental Pollution,* 1(1): 3-26. https://doi.org/10.1016/0013-9327(70)90003-0.

PubChem 2020a. *Aroclor -1254*. Retrieved from https://pubchem.ncbi.nlm.nih.gov/compound/40470.

PubChem 2020b. *Dieldrin.* Retrieved from https://pubchem.ncbi.nlm.nih.gov/compound/ Dieldrin.

Rainbow, P. S. 1997. Trace metal accumulation in marine invertebrates: Marine biology or marine chemistry? *Journal of the Marine Biological Association of the United Kingdom* 77(1): 195-210. https://doi.org/10.1017/S0025315400033877.

Rajamani, J. & Subramanian, M. 2015. Toxicity Assessment on the Levels of Select Metals in the Critically Endangered Indian White-backed Vulture, Gyps bengalensis, in India. *Bulletin of Environmental Contamination and Toxicology*, 94: 722-726. http://www.doi: 10.1007/s00128-015-1548-y.

Rattner, B. A., Horak, K. E., Lazarus, R. S., Schultz, S. L., Knowles, S., Abbo, B. G. & Volker S. F. 2015. Toxicity reference values for chlorophacinone and their application for assessing anticoagulant rodenticide risk to raptors. *Ecotoxicology,* 24(4):720-734. https://www.doi:10.1007/s10646-015-1418-8.

Rattner, B. A., M. A. Whitehead, G. Gasper, C. U. Meteyer, W. A. Link, M. A. Taggart, A. A. Meharg, O. H. Pattee and D. J. Pain. 2008. Apparent tolerance of turkey vultures (*Cathartes aura*) to the non-steroidal anti-inflammatory drug diclofenac. *Environmental Toxicology and Chemistry* 27: 2341–2345. https://doi.org/10.1897/08-123.1.

Redig, P. T. & Arent, L. R. 2008. Raptor toxicology. *Veterinary Clinics of North America: Exotic Animal Practice*. 11 (2): 261–282. https://www.doi:10.1016/j.cvex.2007.12.004.

Roche, C. 2006. Breeding records and nest site preference of Hooded Vultures in the greater Kruger National Park. *Ostrich* 77: 99–101. https://doi.org/10.2989/003065 20609485515.

Rogers, J. M. & Kavlock, R. J. 1996. Developmental toxicology. In Klaassen, C. D. (Ed.), *Casarett & Doull's Toxicology,* (5th ed.). pp. 301-331. New York: McGraw-Hill.

Sanderfoot, O. V. & Holloway, T. (2017). Air pollution impacts on avian species via inhalation exposure and associated outcomes. *Environmental Research Letters,* 12(8): 1 – 16. http://dx.doi.org/10.1088/1748-9326/aa8051.

Scheuhammer, A. M., Money, S. L. & Kirk, D. 2002. Lead fishing sinkers and jigs in Canada: review of their use patterns and toxic impacts on wildlife. *Canadian Wildlife Service Occasional Paper* 108, Ottawa.

Scheuhammer, A. M. & Norris, S. L. 1995. A review of the environmental impacts of lead shotshell ammunition and lead fishing weights in Canada. *Canadian Wildlife Service Occasional Paper* 88, Ottawa.

Scheuhammer A. M. 1987. The chronic toxicity of aluminium, cadmium, mercury, and lead in birds: a review. *Environmental Pollution* 46: 263-295. http://www.doi: 10.1016/0269-7491(87)90173-4.

Schmutz, J. K., Rose, K. A. & Johnson, R. G. 1989. Hazards to raptors from strychnine poisoned ground squirrels. *Journal of Raptor Research,* 23(4):147-151. Retrieved from https://sora.unm.edu/sites/default/files/journals/jrr/v023n04/p00147-p00151.pdf.

ScienceDirect 2020. *Body Burden.* Retrieved from https://www.sciencedirect.com/topics/medicine-and-dentistry/body-burden.

Seshadri, B., Bolan, N. S., Naidu, R. & Brodie, K. J. 2010. The role of coal combustion products in managing the bioavailability of nutrients and heavy metals in soils. *Soil Science and Plant Nutrition,* 10: 378-398. http://dx.doi.org/10.4067/S0718-95162010000100011.

Shore, R. F., Pereira, M. G., Walker, L. A. & Thompson, D. R. 2011. Mercury in nonmarine birds and mammals. In Beyer, W. N. & Meador, J. P. (Eds.) *Environmental contaminants in biota*, 609–642. Boca Raton: CRC Press.

Sidor, I. F., Pokras, M. A. Major, A. R., Poppenga, R. H., Taylor, K. M. &, Miconi, R. M. 2003. Mortality of common loons in New England, 1987 to 2000. *Journal of Wildlife Diseases* 39: 306–315. https://www.doi:10.7589/0090-3558-39.2.306.

Sibly, R. M., Newton, I. & Walker, C. H. 2000. *Effects of dieldrin on population growth rates of sparrowhawks* 1963–1986. https://doi.org/10.1046/j.1365-2664.2000.00516.x.

Snoeijs, T., Dauwe, T., Pinxten, R., Vandesande, F. & Eens, M. 2004. Heavy Metal Exposure Affects the Humoral Immune Response in a Free-Living Small Songbird, the Great Tit (Parus major). *Environmental Contamination and Toxicology,* 46(3): 399–404. https://www.doi:10.1007/s00244-003-2195-6.

Snyder, N. & Snyder, H. 2000. *The California Condor: A Saga of Natural History and Conservation.* San Diego, CA: Academic Press.

Snyder, N. F. R. & Snyder, H. A. 1991. *Birds of Prey: Natural History of North American Raptors.* Stillwater, MN: Voyageur Press.

Sorenson, K., Burnett, J., Mee, A. & Hall, L. 2007. Lead concentrations in the blood of Big Sur California condors. *California condors in the 21st century.* Series in Ornithology No. 2., 185–195. Cambridge, MA: Nuttall Ornithological Club and American Ornithologists' Union.

Spallholz, J. E. & Hoffman, D. J. 2002. Selenium toxicity: cause and effects in aquatic birds. *Aquatic Toxicology,* 57(1-2): 27-37. http://www.doi:10.1016/s0166-445x(01)00268-5.

Speer, R. & Wise, J. P. 2018. Current Status on Chromium Research and Its Implications for Health and Risk Assessment. In *Reference Module in Chemistry, Molecular Sciences and Chemical Engineering.* https://doi.org/10.1016/B978-0-12-409547-2.14283-0.

Ssemmanda, R. & Pomeroy, D. 2010. *Scavenging birds of Kampala: 1973–2009.* Scopus 30: 26–31.

Stankovic, S., Kalaba, P. & Stankovic, A. R. 2013. Biota as toxic metal indicators, *Environmental Chemistry Letters* 12(1): 63-84. https://doi.org/10.1007/s10311-013-0430-6.

Stern, B. R. 2010. Essentiality and toxicity in copper health risk assessment: overview, update and regulatory considerations. *Journal of Toxicology and Environmental*

Health, Part A. 73(2):114–127. https://doi.org/10.1080/15287390903337100.

Stone, W. B. & Okoniewski, J. C. 2001. Necropsy findings and environmental contaminants in common loons from New York. *Journal of Wildlife Diseases* 37:178–184. https://www.doi:10.7589/0090-3558-37.1.178.

Strom, S. M., Langenberg, J. A., Businga, N. K. & Batten, J. K. 2009. Lead exposure in Wisconsin birds. Pages 1–8 in R. T. Watson, M. Fuller, M. Pokras, and W. G. Hunt, editors. *Ingestion of lead from spent ammunition: implications for wildlife and humans.* Boise, Idaho: The Peregrine Fund.

Swan, G., Naidoo, V., Cuthbert, R., Green, R. E., Pain, D. J., Swarup, D., Prakash, V., Taggart, M., Bekker, L., Das, D., Diekmann, J., Diekmann, M., Killian, E., Meharg, A., Chandra Patra, R., Saini, M. & Wolter, K. 2006. Removing the threat of diclofenac to critically endangered Asian vultures. *PLoS Biology* 4(3): 395–402. https://doi.org/10.1371/journal.pbio.0040066.

Tchounwou, P. B., Yedjou, C. G., Patlolla, A. K. & Sutton, D. J. 2012. Heavy metals toxicity and the environment. In Luch A. (Ed.) Molecular, Clinical and Environmental Toxicology. *Experientia Supplementum,* 101. Basel: Springer.

Tchounwou, P., Newsome, C., Williams, J. & Glass, K. 2008. Copper-induced cytotoxicity and transcriptional activation of stress genes in human liver carcinoma cells. *Metal Ions in Biology and Medicine,* 10: 285–290. PMC3058949.

Thiollay, J. M. 2001. Long-term changes of raptor populations in northern Cameroon. *Journal of Raptor Research* 35: 173–186.

U.S. Environmental Protection Agency. 1994. *Lead fishing sinkers: response to citizens' petition and proposed ban.* 40 CFR Part 745. Federal Register 59:11121–11143.

Vandenbroucke, V., Van Pelt, H., De Backer, P. & Croubels, S. 2010. Animal poisonings in Belgium: a review of the past decade. *Vlaams Diergeneeskundig Tijdschrift*, 79: 259-268. http://hdl.handle.net/1854/LU-1039067.

Van Wyk, E., Bouwman, H., van der Bank, H., Verdoorn, G. H. & Hofmann, D. 2001. Persistent organochlorine pesticides detected in blood and tissue samples of vultures from different localities in South Africa. *Comparative Biochemistry and Physiology.* Part C: Comparative Toxicology & Pharmacology, 129(3): 243–264. https://doi.org/10.1016/S1532-0456(01)90201-7.

Van Wyk, E., Van der Bank, F. H. & Verdoorn, G. H. 2001. Selected mineral and heavy metal concentrations in blood and tissues of vultures in different regions of South Africa. *South African Journal of Animal Sciences,* 31: 57-64. http://dx.doi.org/10.4314/sajas.v31i2.3831.

Verdoorn, G. H., van Zijl, N., Snow, T. V., Komen, L. & Marais, E. W. 2004. Vulture poisoning in southern Africa. In Monadjem, A., Anderson, M. D., Piper, S. E. & Boshoff, A. F. (Eds.). *The Vultures of Southern Africa—Quo Vadis?* 195–201. Johannesburg: Birds of Prey Working Group.

Virani, M., Kendall, C., Njoroge P. & Thomsett, S. 2011. Major declines in the abundance of vultures and other scavenging raptors in and around the Masai Mara ecosystem, Kenya. *Biological Conservation* 144: 746–752. https://doi.org/10.1016/j.biocon.2010.10.024.

Virani, M., Gilbert, M., Watson, R., Oaks, L., Benson, P., Kham, A. A. & Baral, H. S. 2001. Asian vulture crisis project: field results from Pakistan and Nepal for the 2000–2001 field season. Reports from the workshop on Indian Gyps vultures. pp. 461–469. In Katzner, T. & Parry-Jones, J. (Eds.). *Proceedings of the 4th Eurasian Congress on Raptors.* Seville: Estación Biológica Donaña, Raptor Research Foundation.

Walker, C. H., Brealey, C. J., Mackness, M. & Johnston, G. 1991. Toxicity of pesticides to birds: the enzymic factor. *Biochemical Society Transactions.* 19 (3): 741. https://www.doi:10.1042/bst0190741.

Walker, K. C., Goette, M. B.; Batchelor, G. S. 1954. Pesticide Residues in Foods, Dichlorodiphenyltrichloroethane and Dichlorodiphenyldichloroethylene Content in Prepared Foods. *Journal of Agricultural and Food Chemistry.* 2 (20): 1034–1037. https://www.doi:10.1021/jf60040a006.

Weir, S. M., Thomas, J. F. & Blauch, D. N. 2018. Investigating spatial patterns of mercury and rodenticide residues in raptors collected near the Charlotte, NC, USA, metropolitan area. *Environmental Science and Pollution Research* (33):33153-33161. https://www.doi: 10.1007/s11356-018-3229-y.

White, C. M., Cade, T. J. & Enderson, J. H. 2013. *Peregrine Falcons of the World.* Barcelona, Spain: Lynx Edicions.

Whitney, M. C. & Cristol, D. A. 2018. Impacts of sublethal mercury exposure on birds: A detailed review. *Reviews of Environmental Contamination and Toxicology.* 244:113-163. https://www.doi:10.1007/398_2017_4.

Wilbur, S. R. 1978. *The California Condor, 1966–1976: A Look at its Past and Future.* Washington DC: Fish and Wildlife Service.

Williams P. L., James R. C., & Roberts S. M. 2014. *Principles of toxicology: environmental and industrial applications.* New York: John Wiley & Sons.

Wobeser, G. & Blakley B. R. 1987. Strychnine Poisoning of Aquatic Birds. *Journal of Wildlife Diseases,* 23(2): 341-343. https://doi.org/10.7589/0090-3558-23.2.341.

Woodford, M. H., C. G. R. Bowden and N. Shah. 2008. *Diclofenac in Asia and Africa—repeating the same mistake? Harmonisation and improvement of registration and quality control of Veterinary Medicinal Products in Africa.* Paris: OIE World Organisation for Animal Health.

Xie, Y., Holmgren, S., Andrews, D. M. K. & Wolfe, M. S. 2017. Evaluating the impact of the U.S. National Toxicology Program: A case study on Hexavalent Chromium. *Environmental Health Perspectives.* 125 (2): 181–188. http://www.doi:10.1289/ehp21.

Zaccaroni, A., Andreani, G., Ferrante, M. C., Carpene, E., Isani, G. & Lusiano, A. 2008. Metal concentrations in the liver and kidney of raptor species from the Calabria region, Italy. *Acta Veterinaria* (Beograd), 58: 315-324. https://www.doi:10.2298/AVB0804315Z.

Zaccaroni, A., Amorena, M., Naso, B., Castellani, G., Lucisano, A. & Stracciari, G. L. 2003. Cadmium, chromium and lead contamination of *Athene noctua*, the little owl, of Bologna and Parma, Italy. *Chemosphere,* 52:1251-1258. https://www.doi: 10.1016/s0045-6535(03)00363-1.

Zhang, X. L., Luo X. J., Liu, J., Luo, Y., Chen, S. J., Mai, B. X. 2011. Polychlorinated biphenyls and organochlorinated pesticides in birds from a contaminated region in South China: association with trophic level, tissue distribution and risk assessment. *Environmental Science and Pollution Research,* 18(4):556-565. https://doi.org/10.1007/s11356-010-0397-9.

Zimmerman, D. A., Turner, D. A. & Pearson, D. J. 1996. *Birds of Kenya and Northern Tanzania.* London: Christopher Helm.

In: Critical Research Techniques in Animal …
Editors: Michael O'Neal Campbell et al.
ISBN: 978-1-53619-846-1

Chapter 5

VULTURE FORAGING AND THE CHEMISTRY OF PUTREFACTION

Michael O'Neal Campbell*
Simon Fraser University, British Columbia, Canada

ABSTRACT

Cathartes (Illiger, 1811) vultures, namely turkey (*Cathartes aura*, Linnaeus, 1758), Greater Yellow-headed (*Cathartes melambrotus*, Wetmore, 1964) and Lesser Yellow-headed (*Cathartes burrovianus*, Cassin, 1845) vultures have superior foraging abilities to other vultures, per their olfactory abilities (sense of smell). However, the actual abilities of these vultures to detect hidden food sources is still under research and few studies have measured and compared the actual differences in foraging success between the *Cathartes* vultures and other vulture species. This article examines current knowledge about vulture foraging, specifically the olfactory abilities of *Cathartes* vultures and the lack thereof for other vulture species. Recent findings from relevant studies are tabulated and analyzed. The evidence indicates that the olfactory abilities of the Turkey vulture may not confer a decisive advantage in foraging success, as later arrivals at food sources may dominate (due to greater numbers in the case of the Black vulture, or superior size and strength in the case of eagles or the King vulture (*Sarcoramphus papa*, Linnaeus, 1758) but it may be a minor factor for the wider range of these species into forests where food sources may be difficult to detect. However, urbanization, deforestation and agricultural expansion have created a more patch dynamic landcover configuration, that may obscure the advantages of strong olfactory abilities. It is concluded that new research methods, such as hidden cameras, elusive drones, strong global positioning systems (GPS) and field techniques may be needed for further investigations. This contributes to avian and landscape ecology research.

* Corresponding Author's E-mail: ecol55@hotmail.com.

INTRODUCTION

Vultures of the Genus *Cathartes* (Turkey vulture, Greater Yellow-headed vulture or Forest vulture, Lesser Yellow-headed vulture or Savanna vulture) have a strong sense of smell, which distinguishes them from all other species of vultures (Houston 1986, 1988; Graves 1992; Campbell, 2015). This enables them to forage in closed canopy forests (Figures 1, 2). In the sensory detection of putrefied food sources, the *Cathartes* vultures must compete with other scavengers with a sense of smell, such as *Canids* and *Felids* (Campbell, 2015). Their strong sense of smell derives from their large olfactory organs that determine the strength of the sense of smell, among the largest of any bird, including rivals such as the Black vulture (Figure 3, see Bang & Cobb, 1968; Snyder & Snyder, 2006). According to Prada and Furton (2018 one *Cathartes* vulture, the Turkey vulture has "one of the largest olfactory bulbs of any bird"; hence "these birds are normally associated with their rapid presence at scenes with decomposing tissue and dead animals." Stager (1964) also found that Turkey vultures could detect both decomposing and fresh animal tissue. Houston (1986) also found that Turkey vultures using scent could distinguish between one day old carcasses and strongly putrefied meat. Detection of unseen food sources is therefore possible, if the foraging vulture detects floating fumes from the food source, through forest canopies, dense shrubbery, leaf piles, roofs, soil layers and even perforated plastic bags and half open dustbin lids (Lowery & Dalquest, 1951). These abilities enable the Cathartes vultures to forage in the dense, canopied neotropical forests (Houston (1986: 318, see also Bang, 1964; Stager, 1964).).

Grigg, et al. (2017) provide one of the "the first measurements of turkey vulture brains, including the size of their olfactory bulbs and numbers of mitral cells, which provide the primary output of the olfactory bulbs"; the findings indicating that "the turkey vulture has olfactory bulbs that are 4× larger and contain twice as many mitral cells as those of the sympatric black vulture (*Coragyps atratus*), despite having brains that are 20% smaller" Grigg, et al. (2017). Figure 2 also shows that Turkey vultures have larger nostrils than Black vultures. Grigg, et al. (2017) further note that "the combination of large olfactory bulbs, high mitral cell counts and a greatly enlarged nasal cavity likely reflects a highly sensitive olfactory system...this sensitive sense of smell allowed the turkey vulture to colonize biomes that are suboptimal for scavenging birds and become the most widespread vulture species in the world" (Grigg et al. 2017).

The study of olfactory abilities in birds is comparatively recent. As noted by Prada and Furton (2018) "as opposed to canines, avian olfactory capabilities have been substantially overlooked by a historical belief that birds are anosmic (i.e., having little or no smell)" (see also Nevitt & Prada, 2015). The study of olfactory abilities in vultures developed after the work of Stager (1964). However, Grigg et al. (2017) note that Few studies in this regard have been conducted on the other *Cathartes* vultures (Campbell, 2015). The Lesser Yellow-headed vulture has a similar olfactory ability to the Turkey vulture (Houston 1986, 1988; Graves 1992). Evidence indicates that when foraging, it flies at lower altitudes than the Turkey vulture (Belton, 1984). Despite its ability to detect hidden food sources, it frequents open landcover, hence its name "Savanna vulture" (Ferguson-Lees and Christie 2001). The larger Greater Yellow-headed vulture forages in moist, dense, lowland forest, avoiding savanna and other open landcover (Campbell, 2015). Schulenberg (2010) argues that the Greater Yellow-headed vulture is "ecologically separated from the other members of the genus, occurring exclusively over large tracts of undisturbed lowland forest in Amazonia and the Guyanas."

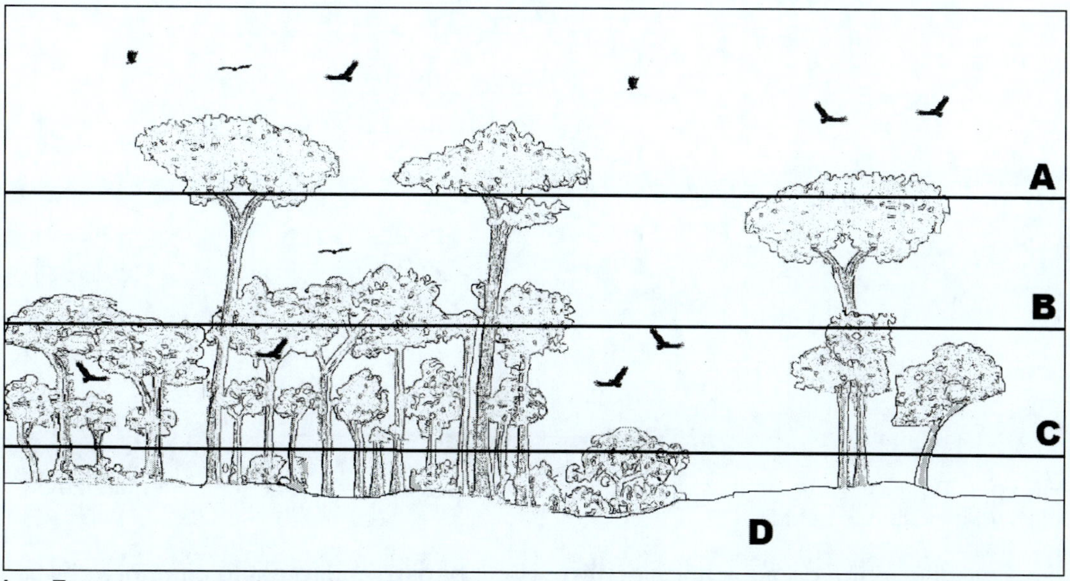

A = Forest emergent layer; B = Forest canopy layer; C = Forest understory, grasses; D = Clearing. Source: Campbell (2015).

Figure 1. Typical soaring patterns for *Cathartes* vultures in the Neotropical forests.

Source: Author's field survey.

Figure 2. Typical forested foraging territory for Turkey vultures in El Salavdor.

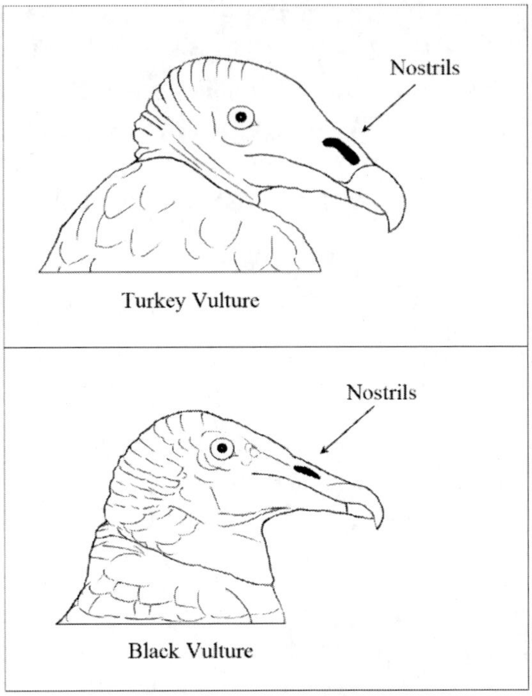

Figure 3. The perforated nostrils of the Turkey and Black vultures. Note the superior size of the Turkey vulture's nostrils.

Foraging in the dense, canopied forests it inhabits require acute olfactory abilities. As argued by Graves (1992: 38), "although the olfactory capacities of the Greater Yellow-headed vulture are unknown, they are thought to be similar to those of its congeners," these being the members of the Genus *Cathartes*. A preference for fresh meat has been noted, which would be even more difficult to detect in closed forests (Snyder & Synder, 2006; Von Dooren, 2011; Wilson & Wolkovich, 2011).Two other common Cathartid vultures, the Black vulture (*Coragyps atratus*, Bechstein 1793) and King vulture (*Sarcoramphus papa*, Linnaeus, 1758) have very weak olfactory powers, hence they almost never forage over dense, canopied forests and in some cases must follow *Cathartes* vultures to food sources (Hilty 1977; Gomez et al. 1994; Schulenberg 2010; Campbell, 2015). The rest of this article looks at the chemical processes of putrefaction, the physiological basis of vulture olfactory abilities and the details of the resultant foraging patterns.

PUTREFACTION

Putrefaction has been defined medically as: "the decomposition of organic matter especially the typically anaerobic splitting of proteins by bacteria and fungi with the formation of foul-smelling incompletely oxidized products" (Merriam-Webster, 2019). Powers (2005, 3) writes that putrefaction is "the consequence of a "population explosion" of xenobiota in the body, as various organisms (both micro- and microbiological endogenous and invasive plants, animals, and fungi) compete for the available energy the decomposing tissues represent." This process occurs parallel to autolysis, another process of tissue breakdown. Autolysis is defined as an "aseptic phenomenon caused by the release and subsequent uncontrolled activity of intracellular enzymes that hydrolytically break down cellular constituents that may serve as catalytic substrates" (Powers, 2005, 3). Powers (2005, 3) further notes that this "autolytic process sets the stage for the subsequent massive transformation of previously (more or less) solid tissue to gas, liquid, and salt products during the septic process of putrefaction." In this process, "catabolically active enzymes are able to act on cellular organelles and molecular components that would not normally serve as substrates"; these enzymes, which are released from the initial subcellular locations begins the process of decomposition (Powers (2005, 3).

Putrefaction occurs during the arguably five stages of decomposition (Powers, 2005) of which the first four may be relevant to vultures, the first when they can detect the scent of fresh meat, or can see the fresh carcass, the last may be irrelevant if no soft tissue survives, except in the case of the bone-eating Bearded vulture (*Gypaetus barbatus*, Linnaeus, 1758) (Campbell, 2015). Decomposition begins with the "fresh stage," which covers the first two days after the death of the organism. During this stage the autolytic processes begin, with minimal or no insect attracted presence and few noticeable changes the carcass. The "bloated" stage occurs two to six days after death and marks the beginning of putrefaction. The term bloated derives from the main characteristic of this phase, that of the production of gas through the start of putrefaction. These gases inflate the tissues, especially the abdomen, due to the activity of anaerobic bacteria and other organisms. Insects may also be attracted to the decay and the excreted fluid production. At this stage, the process of putrefaction may accelerate by the action of microorganisms present within the digestive system during life, which after death may digest the proteins and excrete products, including gases such as hydrogen sulphide, carbon dioxide, and methane, and amines including putrescine and cadaverine, the cause of the odor of decaying flesh (Wisman & Shrira, 2015). Intestinal anaerobic bacteria also change haemoglobin to sulfhemoglobin, gases then transport the sulfhemoglobin through the circulatory and lymphatic systems. The gas dispersal through the body blood vessels and tissues bloating the carcass (Campbell, 2015).

The next stage is the decay phase, which may last from the fifth to the eleventh day. This may be characterized by the weakening and breaking of the abdominal wall, which allows the escape of the previously trapped putrefactive gases, which may attract insects and any animals with olfactory abilities. This is phase of greatest loss of body mass to the body, as insects lay eggs and maggots (larvae) feed on the tissue. The result is the decomposition of the tissue under the skin, leading to skin slippage and hair loss, and openings in the skin that lead to the escape of gases and fluids, and the entry of oxygen leading to aerobic microbe invasion and more insect larvae activity (Campbell, 2015). Maggots may consume much of the soft tissue, and the carcass will become less moist. The last two stages are the post-decay phase, 10 to 25 days after death and involves the reduction to dried remains (depending on the climate); and the dry state, usually more than 25 days after death, when no soft tissue or even odor remains, only hair and bones (Powers, 2005).

The smell of decaying flesh is one of the most important issues for those scavengers with olfactory abilities (Campbell, 2015; Wisman & Shrira, 2015). Hussain et al. (2013, 19579) notes that the smell "is mainly carried by two small aliphatic diamines, putrescine and cadaverine, which are generated by bacterial decarboxylation of the basic amino acids ornithine and lysine." The chemical compound of putrescine is $NH_2(CH_2)_4NH_4$ (molecular formula $C_4H_{12}N_2$) and the preferred IUPAC name is Butane-1,4-diamine. The chemical compound of cadaverine is $NH_2(CH_2)_5NH_2$ (molecular formula $C_5H_{14}N_2$) and the preferred IUPAC name is Pentane-1,5-diamine (Izquierdo et al. 2018). The diamines are organic compounds with two amino groups, and aliphatic refers to a hydrocarbon compound which contains carbon and hydrogen linked in straight or branched trains, or non-aromatic rings (Helmenstine, 2019). Hussain et al. (2013, 19579) further note that "cadaverine and putrescine, two diamines emanating from decaying flesh, are strongly repulsive odors to humans but serve as innate attractive or social cues in other species… It may be expected that animal species feeding on carcasses attribute a more positive valence to diamines, and indeed both putrescine and cadaverine have been reported to be feeding attractants."

Detection of Carrion

Three common methods used to assess *Cathartes* vultures' carrion detection abilities are: (1) the hiding of food sources of varying states of decomposition outside the vultures' sight range; (2) the comparison of their detection abilities with non-*Cathartes* vultures (usually Black and King vultures, which share their habitat); and (3) the soaring altitudes of *Cathartes* vultures compared with non-*Cathartes* vultures. Houston's (1986) study is typical of the first methodology. Carcasses were hidden in dense, Panamanian rainforest, with variations in sky visibility, leaf coverage and state of decomposition (fresh, one and four-day old dead chickens, which compared in size to the normal prey of opossums, sloths and small primates), and measurement of vulture arrival times. The findings were: no correlation between forest-canopy gap size and vultures' carcass detection times; no significant difference in vulture detection time for exposed and leaf-covered carcasses; although there was a significant difference in detection times between fresh and one day old carcasses, there was no significant difference between the one day and four-day old carcasses.

Houston's (1986) study also used the second methodology, in comparing the abilities of the *Carthartes* vultures with non-*Cathartes* vultures. The findings were that Turkey vultures arrived long before Black vultures (Houston, 1986; Campbell, 2015). Houston (1986: 321) concluded firstly that "the results show smell to be the major sense used by Turkey vultures to locate carrion and puts vision in a minor role, for completely hidden food was found as quickly as visible bait"; and "Black vultures do not have a sense of smell, and my observations in other study areas have shown that they are not found in undisturbed forest and cannot locate food in forest conditions unless led there by Turkey vultures' (ibid. 322). Gomez et al. (1994) in comparing the King vulture's olfactory senses with those of the Greater Yellow-headed vulture, noting the superiority of the latter, despite its dominance at carcasses based on its superior size and strength, and consequent abilities in tearing open carcasses for other vultures to feed.

Currently, most studies indicate that King vultures follow the *Cathartes* vultures to carcasses (Houston 1984, 1994; Beason, 2003). Lemon (1991) nevertheless found that that King vultures located carrion in dense forest, where *Cathartes* vultures were absent. However, Lemon (1991: 700) notes that 'Little is known about the foraging behavior or physiology of King Vultures,' although there is a general acknowledgment that it is a forest specialist (see also Schlee, 1995). Mallon et al. (2013, 475) point out that "few accounts exist of Neotropical scavenger succession in conjunction with carcass condition and none exist for Greater Yellow-headed (*Cathartes melambrotus*), Black (*Coragyps atratus*) and King Vultures (*Sarcoramphus papa*) concurrently." Nevertheless, current evidence generally indicates that the Cathartes vultures, are generally the first arrivals at carcasses (Stager 1964; Olrog, 1985; Houston, 1986, 1988; Graves, 1992; Snyder & Snyder, 2006). After arrival, Black and King vultures may be dominant over them, due to the larger flocking behavior in the case of the former and the larger size of the latter (Wallace & Temple 1987; Buckley 1996). The two larger *Cathartes* vultures, the Turkey and Greater Yellow-headed vultures, are generally even in competitive dominance (Hilty, 1977; Schulenberg 2010).

The third methodology, focused on soaring methodology, contrasts the comparatively low flight patterns of most *Cathartes* vultures, which allows them to use the rising air between tall trees, enhancing the detection of putrefied scents. For example, the Greater Yellow-headed vulture, occurs principally in the dense forests of South America, generally flies low over the canopy, capable of detecting small carcasses on the forest floor (Robinson, 1994). Gomez et al. (1994) found that in the forested Amacayacu National Park, Colombia, it was the commonest vulture species with the fastest carcass arrival rate, even in dense forest, finding over 60 percent of provided carcasses. These findings were supported by Graves (1992), who found that the Greater Yellow-headed vulture was the commonest vulture in the dense forest on the east bank of the Rio Xingu, 52 km SSW of Altamira, Patti, Brazil. Other vultures such as the Black vulture stayed outside the closed forest.

Eating and Digestion of Putrid Meat

Vultures' ability to ingest and digest putrid meat is perhaps the most important aspect of their ecology (Campbell, 2015). The issue here is the extent to which they may resist the bacterial toxins in such decaying meat. Roggenbuck et al. (2014)

examining facing and guts of the Turkey and Black vultures found that there was a "remarkably conserved low diversity of gut microbial flora." The samples from the gut and facial skin revealed several operational taxonomic units (OTUs, these are organisms, which may be grouped by similarity of DNA sequence with a specific taxonomic marker gene). For each vulture sampled, the gut samples contained an average of 76 such OTUs, while the facial skin had an average of 528 OTUs (Roggenbuck et al. 2014). Crucially, the dominant bacteria in the gut were Clostridia and Fusobacteria, which are pathogenic to many vertebrates. The authors postulate a "faecal–oral–gut" entry sequence and found that there was a "strong adaption of vultures and their bacteria to their food source, exemplifying a specialized host–microbial alliance." The toxicity of such carcasses emerges after the death of the animal, when the microbes in the body begin participation in the decomposition of the soft tissues. During this process, these microbes begin to excrete toxic metabolites, which make the carcass unsuitable for many carnivorous and omnivorous scavengers.

Vultures could be particularly susceptible to any resultant diseases, as their natural food is carrion, they may prefer putrid carcasses which have softer, more easily breached skin, and they often thrust their heads and necks up to the shoulders into carcasses. Roggenbuck et al. (2014) argue that most bacteria may be killed by chemicals in the acidic gastric passage, but some toxins such as botulinum may reach the hindgut (large intestine). The findings of this paper where that "acidic gastrointestinal tract of vultures is a strong filter of the microbiota ingested from decaying carcasses resulting in a significantly less diverse hindgut flora dominated by Clostridia and Fusobacteria, which are pathogenic to most vertebrates" (Roggenbuck et al. 2014).

These two bacterial groups can contribute to severe diseases in people. Arane & Goldman (2016, 813) note that "fusobacterium is a genus of anaerobic bacteria. Although Fusobacterium infections are rare, they can become severe if not treated promptly... Delayed treatment of an infection caused by Fusobacterium can lead to serious complications." Clostridial bacteria contribute to many diseases in people and cattle, including sudden death syndrome, lockjaw and malignant edema (Mcguirk, 2015; Wright, 2016). These abilities explain why vultures may concentrate on putrid meat that would repel many other scavengers, and may assist smaller species such as the Turkey and Black vultures, which may either require larger species to break the skin of fresh carcasses or have to wait in some cases for advanced putrefaction to soften the carcass skin (Campbell, 2015).

Zepeda Mendoza et al. (2018) support these findings and argue that the holobiont concept suggests that the genetic basis of such adaptation may not only lie within their genomes, but additionally in their associated microbes. Using the same species as Roggenbuck et al. (2014) and the same methodology (examination of the facial skin and large intestine for microbes) and used derived DNA sequencing datasets, "examining the functional potential and taxonomic diversity of their microbiomes, the potential pathogenic challenges confronted by vultures, and the microbial taxa and genes that could play a protective role on the facial skin and in the gut" (Zepeda Mendoza et al. 2018) The findings were of highly dangerous microbes, some known to connect with diseases (for example, dermatitis and pneumonia, which were found mostly on the facial skin, and gas gangrene and food poisoning, which were more prevalent in the gut). An important finding was the discovery of microbes and functions which could assist in the control or even elimination of dangerous microbes. These included genes producing antiparasitics and insecticides on the facial skin, antilisterial bacteria and *Adineta vaga* in the gut, the last being an invertebrate that scavenges on dead bacteria and protozoans. The authors also "suggest a colonization resistance role through biofilm formation played by Fusobacteria and Clostridia in the gut" (Zepeda Mendoza et al. 2018).

Turkey and Black vultures have evidently adapted to the dangers of their risky ecology, their facial and gut microbiome including variable populations of environmental and carcass-derived microbes. However, Zepeda Mendoza et al. (2018) argue that some questions are unanswered: "the presence of a wide variety of microbes reported as pathogenic to non-scavengers (mammals and other birds) without an apparent or a reported pathogenic effect on the vultures calls for deeper study." Clarification is required on the toxicity of the reported microbes to vultures and any possible role as reservoirs. Combined studies of genomic analyses and metagenomics on the host microbes is clearly important for the study of such biological processes. Vultures' ability to resist toxic impacts may thus be due to both the vultures' biology and their microbiomes, an interesting possibility that stands at the cutting edge of current vulture and scavenger ecology (Zepeda Mendoza et al. 2018).

ARTIFICIAL DANGERS IN THE VULTURE DIET

Vultures are often at risk due to their specialised diet (Campbell, 2015). This specialization is noted as a factor for extinction (Sekercioglu et al. 2004). Crucially, "Avian scavengers have the highest percentage of extinction-prone species among avian functional groups" (Ogada et al. 2012: 453). Although vultures are resistant to some pathogens, through their potent stomach acids, these acids are too weak to dispose of artificial products such as plastics and do not protect vultures from poisons such as lead and some pesticides (Campbell, 2015). Therefore, the ingestion artificial products, including plastic products may pose problems. Larger plastic items may contribute to "choking, poisoning, intestinal obstruction, malnutrition and death." (BirdLife International, 2008). As vultures forage for bone fragments for calcium needs and food pellet regurgitation, they may mistake plastic products for bone. Records include china, metal and glass in the nests of White-rumped vultures in Pakistan and metal ingested by Griffon vultures in Israel and Armenia (Ferro, 2000; Houston et al. 2007; Campbell, 2015).

Accidental ingestion commonly affects nestlings, as they lack the ability to regurgitate indigestible objects, resulting in choking, poisoning and even intestinal obstruction and consequent malnutrition (Ferro, 2000). Such accidents among nestlings have severely affected the reintroduced California condors (Campbell, 2015).

Mee et al. (2007) report cases where up to 650 junk items were retrieved from nests, these including glass, metal and plastics. This study further notes that several nestlings were recorded with pieces of plastic piping, cloth, rubber, glass, metal bottle-tops and ammunition cartridges, with health impacts ranging from zinc poisoning to malnutrition, distended and blocked digestive passages and consequent arrested lack feather development and death (Mee et al. 2007).

Lead poisoning is another important problem for vultures (Campbell, 2015). Wani et al. (2015: 55) note that "lead is the most important toxic heavy element in the environment." Lead may enter carcasses through from bullets during hunting seasons (García-Fernández et al. 2005; Dobrowolska & Melosik, 2008; Gangoso et al. 2009; Mateo-Tomás & Olea, 2010). It is a crucial factor for the deaths of some species, such as the California condor during the 1980s and after the reintroduction of this species (Snyder & Snyder, 2000; Cade, 2007; Kelly & Johnson 2011). In Turkey vultures, there occurred the "highest mean concentrations of lead in bone and kidney compared to other raptor species" (Martin et al. 2008: 96). Although in some cases high lead levels were found in the blood of condors outside the hunting season, (Pattee et al. 1990; Hall et al. 2007; Hunt et al. 2007; Sorensen & Burnett, 2007), a ban on the use of lead ammunition in hunting was enacted in 2008 with the condor's California range. This may have affected golden eagles and ravens within the condor's range (Pattee et al. 1990; Craighead et al. 2009). In Old World Accipitrid vultures, there was clinical intoxication in Griffon vultures (*Gyps fulvus*) and Cinereous vultures (Mateo et al. 1997; Mateo et al. 2003; Hernandez & Margalida 2008). Rodriguez-Ramos et al. (2009) describe the clinical symptoms of the lead toxicosis as including disorientation, ataxia and impaired landing, posterior paresis, and severe hypochromic anemia. Donazar et al. (2002: 89) also report poisoning of Egyptian vultures in the Canary Islands, Pesticide poisoning is another danger for vultures.

Pesticides also contain chemicals harmful to vultures (Campbell, 2015). Hernández and Margalida (2008), found incidents of pesticide poisoning in the Cinereous vulture (*Aegypius monachus*, Linnaeus, 1766)) in Spain during 1990–2006. Pesticides composed mostly of three chemical compounds (carbofuran, aldicarb, and strychnine, which accounted for up to 88% of the cases) mortality in this case affected vultures (83%). In South Africa, van Wyk et al. (2001: 243) examined White-backed, Cape and Lappet-faced vultures and found "the presence of quantifiable residues of 14 persistent chlorinated hydrocarbon pollutants." Chemical compounds included gamma-BHC (lindane), alpha(cis)-chlordane and alphaendosulfan. They argued that "although no threat is posed by any of the organochloride pesticides" but monitoring of breeding colonies would be necessary and White-backed vulture nestlings could be used as bioindicators (ibid. 243). Turkey vultures have been found to be resistant to certain natural pathogens and even the strong rodenticide compound 1080 (Snyder & Snyder (1991) but are vulnerable to cyanide and strychnine. Other threats are DDE (dichlorodiphenyldichloroethylene) which contributes to eggshell thinning (Wilbur, 1978; Kiff et al. 1983; Coleman & Fraser, 1989).

The most dangerous dietary threat to vultures was the diclofenac epidemic that affected Asian vultures in the late 20th and early 21st centuries (Campbell, 2015).

Although initially the factor for the vulture deaths was undetermined, eventually the veterinary drug factor was discerned (Woodford et al. 2008; Van Dooren (2010, 2011) notes that six of the 23 recognized vulture species were either threatened or critically endangered on the ICUN Red List of Threatened Species, especially the five species in India, an event part of the "global vulture crises" (Olea & Mateo Thomas, 2009). Affected vultures included those common throughout South and South-East Asia, including the White-rumped vulture (*Gyps bengalensis*, Gmelin, 1788), Slender-billed vulture (*Gyps tenuirostris*, Hodgson, 1844), Indian vulture (Gyps indicus, Scopoli, 1786) and Red-headed vulture (*Sarcogyps calvus*, Scopoli, 1786). The populations of these vultures were severely reduced in India and some neighboring countries. Some studies even claimed they were headed for extinction (Campbell, 2015).

Diclofenac, also termed by the trade name Voltaren and other names, is classified as a nonsteroidal anti-inflammatory drug (NSAID) for treating pain and inflammatory diseases. Administration of the drug to cattle, and vultures ingesting the compound with rotting meat, led to fatal consequences for the vultures. Effects include renal failure, inhibition of uric acid secretion, severe seizures and eventual death (Gilbert et al. 2004; Prakash et al. 2007; Gill, 2009; Clements et al. 2012). The vulture digestive system could not cope with this compound. Nevertheless, later studies found that the Turkey vulture and other *Cathartes* vultures were largely unaffected by this drug (Campbell, 2015).

ADVANTAGES OF OLFACTORY CAPABILITY

The olfactory abilities of the *Cathartes* vultures, especially the Turkey vulture may be a factor for its wide distribution, the widest of all the *Cathartid* vultures. The success of the Turkey vulture may be hypothesized to be at least partly due to the olfactory ability, but other factors may be relevant (Campbell, 2015). The key question is why should the *Cathartes* vultures have olfactory abilities, as other vultures with similar foraging ecologies are able to forage without such abilities? For example, Campbell (2009) examined the urban ecology of the Hooded vulture (*Necrosyrtes monachusis*, Temminck, 1823) an Old World *Accipitrid* species with a similar foraging ecology as the Turkey vulture, in Ghana, West Africa. This species has no olfactory ability, but is very versatile in its foraging habits, entering towns and villages in the savanna and forest regions. The findings were that vultures were strongly attracted to meat and waste production and human presence (butchers' shops in dense urban substrates), with some attraction to other urban and rural areas; vultures frequently engaged in observational perching near food sources or dense urban substrate, hovering or very low soaring and rapid diving to the ground to take food, usually human refuse or small mammal carcasses. They also invaded windows and doors, took food from butchers' tables and talked under stalls and small buildings. Such close and intense foraging, usually successful, showed the possibilities for medium-sized vultures in critical foraging areas, without olfactory abilities (Campbell, 2009).

However, in a comparative study, Holland et al. (2017) predicted that "turkey vultures will range across larger areas and spend a greater proportion of time in flight than black vultures, considering the lighter wing loading capacity of turkey vultures, their enhanced sense of smell, and higher likelihood of avoiding competitive interactions at carcasses." Hence, olfactory ability was considered one of three factors that might give the Turkey vulture an ecological advantage over the Black vulture. The findings were that Turkey vultures foraged and ranged further than the Black vulture; "turkey vultures spend significantly more time in flight than black vultures across all temporal scales measured (monthly, seasonally, and annually)" this expected "considering their lighter wing-loading as the energetic demands of flight are lower for turkey vultures and they are likely to succeed in finding carcasses with random flight searches due to enhanced olfaction as compared to black vultures."

Campbell (2014) tested the hypotheses that Turkey vultures would be commoner than Black vultures in the forested areas, due the former bird's sense of smell and that Black vultures would avoid canopied forests in favor of urbanized and open landcover, due to their lack of olfactory ability. These hypotheses were based on a wide supporting literature (Stewart 1978; Garrett & Dunn 1981; Unitt, 1984; Olrog, 1985; Hill & Neto, 1991; Buckley, 1996, 1999; Kiff, 2000, Stolen, 2000, Feguson Lees & Christie 2001, Airola 2011; Novaes & Cintra, 2013). The findings were that Turkey vultures were more common than Black vultures in canopied forest, although both were much commoner in dense urban areas, including markets, abattoirs and fish unloading zones (Campbell, 2014). Therefore, indications are that olfactory abilities confer an advantage for *Cathartes* vultures only in forested areas, increasing foraging areas, but this may be offset with landcover changes such as deforestation, urbanization and agricultural development (Campbell, 2015).

Competition with mammal scavengers with a sense of smell is an important issue for the success of vultures, which may also influence their foraging ranges (Campbell, 2015). Canids, including wolves, coyotes, foxes and feral domestic dogs are the main species in this regard, due to their extremely sharp sense of smell (Campbell, 2015). In North America, the commonest mammalian scavengers are coyotes *(Canis latrans*, Say, 1823), raccoons (*Procyon lotor*, Linnaeus 1758) and opossums (*Didelphis virginiana,* Gray 1821) (Wilmers et al. 2003a, b; DeVault et al. 2011). Coyotes are known for using their sense of smell in hunting, which is strong enough to track even living prey (Lawson et al. 2019) and would obviously be stronger than *Cathartes* vultures, as they are dominant over the much larger California condors, even when the condors were in large flocks (Koford, 1953).

Coyotes are known to increase their scavenging activities when wolves (*Canis lupus*, Linnaeus, 1758) are present (Switalski, 2003). Taking Yellowstone National Park as a case study, Switalski (2003, 985) found that with wolf reintroduction, "coyotes began feeding on carcasses throughout the year," but in these activities the trade offs for coyotes were "increased vigilance, decreased rest, and a higher risk of being killed"; the conclusion was that "changes in the behavior of coyotes in response to the reintroduction of this large carnivore may ultimately have wide-ranging cascading effects throughout the ecosystem" (Switalski, 2003, 985). Vultures may experience increased competition from such developments (Campbell, 2015). Another study by Prior & Weatherhead (1991, 1550) in southern Ontario found that "despite a diversity of avian and mammalian species directly or indirectly competing with turkey vultures for carrion, only coyotes (*Canis latrans*) were found to occasionally deplete experimental food patches sufficiently to limit use by turkey vultures." American crows (*Corvus brachyrhynchos*, Brehm, 1822) were the commonest competitors for vultures, but they had lower food requirements and were subordinate to vultures.

Conclusion

This article has examined some of the current issues concerning the olfactory abilities of *Cathartes* vultures and the chemical bases of their physiology. Current research indicates that olfactory abilities in *Cathartes* vultures confer significant advantages over the non-*Cathartes* vultures, especially as both groups may be equally resistant to possible diseases, share equal flight abilities and the non-*Cathartes* vultures may detect and shadow *Carthartes* vultures in food source access. Current issues concern the impacts of landcover changes, such as urbanization and improved sanitation in such areas, deforestation and agricultural expansion, which may complicate the advantages enjoyed by *Cathartes* vultures. These vultures are evidently strongly adapted to forested landcover but may adapt to food sources hidden by new structures such as gutters, gardens, dense urban parks, road kills in treed avenues and household refuse in insecure containers. Questions emerge whether *Cathartes* vultures may maintain any advantages in foraging abilities over non-*Cathartes* vultures in such environments. Future research may therefore be more focussed on the impacts of human modified landcover on vulture ecology.

References

Airola, D. A. 2011. Dynamics of an Urban Turkey Vulture Roost in Sacramento, California. *CVBC Bulletin*, 1 - 8.

Arane, K., & Goldman, R. D. (2016). Fusobacterium infections in children. *Canadian Family Physician Medecin de Famille Canadien*, 62(10): 813–814.

Avery, M. L. 2004. Trends in North American vulture populations. In Timm, R. M. & Gorenzel, W. P. (Eds.) *Proceedings of the 2lst Vertebrate Pest Conference*, 116–121. University of California, Davis.

Bang, B.G . 1964. The nasal organs of the black and turkey vultures: A comparative study of the *Cathartid* species *Coragyps atratus atratus* and *Cathartes aura septentrionalis* (with notes on *Cathartes aura falklandica*, *Pseudogyps engalensisa* and *Neophronp ercnopterus*). *Journal Morphology*, 115(2):153-83. https://doi.org/10.1002/jmor.1051150203.

Bang, B. G. & Cobb, S. 1968. The size of the olfactory bulb in 108 species of birds. *Auk*, 85: 55 - 61. https://doi.org/10.2307/4083624.

Beason, R. C. 2003. *Through a Birds Eye: Exploring Avian Sensory Perception.* Washington, D. C.: USDA Wildlife Services.

Belton, W. 1984. Birds of Rio Grande do Sul, Brazil. Part 1. Rheidae through Furnariidae. *Bulletin of the American Museum of Natural History*, 178: 369–636.

BirdLife International. 2008. *Anthropogenic Junk Ingestion Leads to High Nestling Mortality in Vultures and Condors.* Retrieved from http://birdlaa8.miniserver.com/dz2/sowb/casestudy/anthropogenic-junk-ingestion-leads-to-high-nestling-mortality-in-vultures-and-condors.

Buckley, N. J. 1996. Food finding and the influence of information, local enhancement, and communal roosting on foraging success of North American vultures. *Auk*, 113: 473–488.

Buckley, N. J. 1999. Black vulture (Coragyps atratus). In Poole, A. & Gill, F. (Eds.) *The Birds of North America*, No. 411. Philadelphia: The Birds of North America, Inc.

Cade, T. J. 2007. Exposure of California Condors to lead from spent ammunition. *Journal of Wildlife Management*, 71(7): 2125–2133. http://www.doi:10. 2193/2007-084.

Campbell, M. 2015. *Vultures: Their Evolution, Ecology and Conservation.* Boca Raton: CRC Press.

Campbell, M. 2014. The impact of urbanization and agricultural development on vultures in El Salvador. *Vulture News*, 66: 16 – 28.

Campbell, M. 2009. Factors for the presence of avian scavengers in Accra and Kumasi, Ghana. *Area*, 41(9): 341–349. https://doi.org/10.1111/j.1475-4762.2008.00870.x.

Clements, T., Gilbert, M., Rainey, H. J., Cuthbert, R., Eames, J. C., Bunnat, P., Teak, S., Chansocheat, S. & Setha, T. 2012. Vultures in Cambodia: population, threats and conservation. *Bird Conservation International*, 23(1): 7 – 24. https://doi.org/10.1017/S0959270912000093.

Coleman, J. S. & Fraser, J. D. 1989. Habitat use and home ranges of Black and Turkey Vultures. *Journal of Wildlife Management* 53:782–792. https://www.doi:10.2307/3809213.

Craighead, D., Bedrosian, B., Watson, R. T., Fuller, M., Pokras, M. & Hunt, W. 2009. *A Relationship between Blood Lead Levels of Common Ravens and The Hunting Season in the Southern Yellowstone ecosystem. Ingestion of Lead from Spent Ammunition: Implications for Wildlife and Humans.* Boise, Idaho: Peregrine Fund.

Donazar, J. A., Palacios, C. J., Gangoso, L., Ceballos, O., Gonzalez, M. J. & Hiraldo, F. 2002. Conservation status and limiting factors in the endangered population of Egyptian vulture (Neophron percnopterus) in the Canary Islands. *Biological Conservation*, 107(1): 89–97. https://doi.org/10.1016/S0006-3207(02)00049-6.

DeVault, T. L., Olson, Z. H., Beasley, J. C. & Rhodes, Jr., O. E. 2011. Mesopredators dominate competition for carrion in an agricultural landscape. *Basic and Applied Ecology*, 12: 268–274. https://www.doi.org/10.1016/j.baae.2011.02.008.

Dobrowolska, A. & Melosik, M. 2008. Bullet-derived lead in tissues of the wild boar (Sus scrofa) and red deer (Cervus elaphus). *Eurasian Journal of Wildland Resources*, 54: 231–235. https://doi.org/10.1007/s10344-007-0134-y.

Ferguson-Lees, J. & Christie, D. A. 2001. *Raptors of the World*. London: Christopher Helm.

Ferro, M. 2000. Consumption of metal artefacts by Eurasian Griffons at Gamla Nature Reserve, Israel. *Vulture News* 43: 46–48.

Gangoso, L., Álvarez-Lloret, P., Rodríguez-Navarro, A., Mateo, R., Hiraldo F. & Donázar, J. A. 2009. Long-term effects of lead poisoning on bone mineralization in vultures exposed to ammunition sources. *Environmental Pollution*, 157(2): 569–574. https://www.doi:10.1016/j.envpol.2008.09.015.

García-Fernández, A. J., Martínez-López, E., Romero, D., María-Mojica, P., Godino, A. & Jiménez, P. 2005. High levels of blood lead in griffon vultures (Gyps fulvus) from Cazorla natural park (Southern Spain). *Environmental Toxicology*, 20: 459–463. https://doi.org/10.1002/tox.20132.

Garrett, K. & Dunn, J. 1981. *Birds of Southern California: Status and Distribution*. Los Angeles: Los Angeles Audubon Society.

Gilbert, M., Oaks, J. L., Virani, M. Z., Watson, R. T., Ahmed, S., Chaudhry, M. J. I., Arshad, M., Mahmood, S., Ali, A., Khattak, R. M. & Khan, A. A. 2004. The Status and Decline of Vultures in the provinces of Punjab and Sind, Pakistan; a 2003 update. In Chancellor, R. D. & Meyburg, B. V. (Eds.). *Raptors Worldwide*, 221–234. Berlin: Raptors Worldwide, WWGBP/MME.

Gill, V. 2009. *New drug threat to Asian vultures*, BBC News December 9, 2009.

Gomez, L. G., Houston, D. C., Cotton, P., Tye, A., Luis, G., Houston, D. C., Cotton, P. & Tye, A. 1994. The role of greater yellow-headed vultures *Cathartes melambrotus* as scavengers in neotropical forest. *Ibis*, 136(2): 193–196. https://doi.org/10.1111/j.1474-919X.1994. tb01084.x.

Graves, G. R. 1992. Greater yellow-headed vulture locates food by olfaction. *Journal of Raptor Research*, 26(1): 38–39. Retrieved from https://sora.unm.edu/sites/default/files/journals/jrr/v026n01/p00038-p00039.pdf.

Grigg, N. P., Krilow, J. M., Gutierres-Ibanez, C., Wylie, D. R., Graves, G. R. & Iwaniuk, A. N. 2017. Anatomical evidence for scent guided foraging in the turkey vulture. *Scientific Reports*, 7:17408. https://doi.org/10.1038/s41598-017-17794-0.

Hall, M., Grantham, J., Posey, R., Mee, A. & Hall, L. 2007. Lead exposure among reintroduced California condors in southern California. California condors in the 21st century. *Series in Ornithology* No. 2, 163–184. Cambridge, MA: Nuttall Ornithological Club and American Ornithologists' Union.

Helmenstine, A. M. 2019. *Aliphatic Hydrocarbon Definition*. Retrieved from https://www.thoughtco.com/definition-of-aliphatic-hydro carbon-604763.

Hernandez, M. & Margalida, A. 2008. Pesticide abuse in Europe: effects on the cinereous vulture (*Aegypius monachus*) population in Spain. Ecotoxicology, 17: 264–272. http://www.doi: 10.1007/s10646-008-0193-1.

Hill, J. R. & Neto, P. S. 1991. Black vultures nesting on skyscrapers in southern Brazil. *Journal of Field Ornithology*, 62(2):173-176. https://www.jstor.org/ stable/4513621.

Hilty, S. L. 1977. *A Guide to the Birds of Colombia*. Princeton: Princeton University Press.

Holland, A. E., Byrne, M. E., Bryan, A. L., DeVault, T. L., Rhodes, O. E., Beasley, J. C. 2017. Fine-scale assessment of home ranges and activity patterns for resident black vultures (Coragyps atratus) and turkey vultures (Cathartes aura). *PLoS ONE*, 12(7): e0179819. https://doi.org/10.1371/journal.pone.0179819.

Houston, D. C. 1984. Does the King Vulture Sarcoramphus papa use a sense of smell to locate food? *Ibis*, 126: 67–69. https://doi.org/10.1111/j.1474-919X.1984.tb03665.x.

Houston, D. C. 1986. Scavenging efficiency of turkey vultures in tropical forest. *Condor*, 88: 318–323. https://doi.org/10.2307/1368878.

Houston, D. C. 1988. Competition for food between Neotropical vultures in forest. *Ibis*, 130: 402–417. https://doi.org/10.1111/j.1474-919X.1988.tb08815.x.

Houston, D. C. 1994. Cathartidae (New World Vultures). In del Hoyo, J., Elliott, A. & Sargatal, J. (Eds.). *Handbook of the Birds of the World*, 24–41. Barcelona, Spain: Lynx Edicions.

Houston, D. C., Mee, A. & McGrady, M. (2007) Why do condors and vultures eat junk? The implications for conservation. *Journal of Raptor Research*, 41: 235–238. https://www.doi:10.3356/0892-016(2007)41[235:WDCAVE]2.0.CO;2.

Hunt, W., Parish, C., Farry, S., Lord, T., Sieg, R., Mee, A. & Hall, L. 2007. Movements of introduced California condors in Arizona in relation to lead exposure. California condors in the 21st century. *Series in Ornithology*, 2: 79–96. Cambridge, MA: Nuttall Ornithological Club and American Ornithologists' Union.

Hussain, A., Saraiva, L. R., Ferrero, D. M., Ahuja, G., Krishna, V. S., Liberles, S. D., & Korsching, S. I. 2013. High-affinity olfactory receptor for the death-associated odor cadaverine. *Proceedings of the National Academy of Sciences of the United States of America*, 110(48), 19579–19584. http://www.doi:10.1073/pnas. 1318596110.

Izquierdo, C., Gomez-Tamayo, J. C., Nebel, J-C., Pardo, L. & Gonzalez, A. 2018. Identifying human diamine sensors for death related putrescine and cadaverine molecules." *PLOS Computational Biology*, 14 (1): e1005945. http://www.doi: 10.1371/journal.pcbi. 1005945.

Kelly, T. R. & Johnson, C. K. 2011. Lead exposure in free-flying Turkey Vultures is associated with big game hunting in California. PLoS One, 6(4): e15350. http://www.doi:10.1371/journal.pone. 0015350.

Kiff, L. F. 2000. The current status of North American vultures. In Chancellor, R. D. & Meyburg, B.-U. (Eds.) *Raptors at Risk*, 175–189. Berlin and Surrey, Canada: WWGBP/Hancock House.

Kirk, D. A. & Mossman, M. J. 1998. Turkey Vulture (Cathartes aura). In Pool, A. F. & Gill, F. B. (eds.) *The Birds of North America*. Ithaca, NY: Cornell Lab of Ornithology. https://doi.org/ 10.2173/bna.339.

Koford, C. B. 1953. *The California Condor*. National Audubon Society Research Report 4: 1–154.

Lawson, R. R., Fogarty, D. T. & Loss, S. R. 2019. Use of visual and olfactory sensory cues by an apex predator in deciduous forests. *Canadian Journal of Zoology*, 97: 488 - 494, https://doi.org/10.1139/cjz-2018-0134.

Lemon, W. C. 1991. Foraging behavior of a guild of Neotropical vultures. *The Wilson Bulletin*, 103(4): 698–702. Retrieved from https://sora.unm.edu/sites/default/files/journals/wilson/v103n04/p0698-p0702.pdf.

Lowery, G. H. & Dalquest, W. W. 1951. *Birds from the State of Veracruz, Mexico*. University of Kansas Publications, Museum of Natural History 3: 531–649. Retrieved from https://mczbase.mcz.harvard.edu/publication/25453.

Mallon, J. M., Swing, K. & Mosquera, D. 2013. Neotropical Vulture Scavenging Succession at a Capybara Carcass in Eastern Ecuador. *Ornitologia Neotropical*, 24: 475–480. Retrieved from https://sora.unm.edu/node/133394.

Martin, P., Campbell, D., Hughes, K. & McDaniel, T. 2008. Lead in the tissues of terrestrial raptors in southern Ontario, Canada, 1995–2001. *Science of the Total Environment*, 391: 96–103. https://doi.org/10.1016/j.scitotenv.2007.11.012.

Mateo, R., Molina, R., Grifols, J. & Guitart, R. 1997. Lead poisoning in a free ranging Griffon Vulture (*Gyps fulvus*). *Veterinary Record*, 140: 47–48. http://dx.doi.org/10.1136/vr.140.2.47.

Mateo, R., Taggart, M. & Meharg, A. A. 2003. Lead and arsenic in bones of birds of prey from Spain. *Environmental Pollution*, 126: 107–114. http://www.doi:/10.1016/S0269-7491(03)00055-1.

Mateo-Tomás, P. & Olea, P. P. 2010. When hunting benefits raptors: a case study of game species and vultures. *European Journal of Wildland Resources*, 56: 519–528. http://www.doi:10.1007/s10344-009-0341-9.

Mcguirk, S. M. 2015. *Managing Clostridial Diseases in Cattle*. University of Wisconsin: School of Veterinary Medicine.

Mee, A., Rideout, B. A., Hamber, J. A., Todd, N., Austin, G., Clark, M. and Wallace, M. P. (2007) Junk ingestion and nestling mortality in a reintroduced population of California Condors *Gymnogyps californianus*. *Bird Conservation International*, 17: 119–130. https://doi.org/10.1017/S095927090700069X.

Merriam-Webster 2019. *Putrefaction*. Retrieved from https://www.merriam-webster.com/dictionary/putrefaction.

Nevitt, G. A. & Prada, P. A. 2015. The chemistry of avian odors: An introduction to best practices. In Doty, R. L. (Ed.). *Handbook of Olfaction and Gustation*, 565–577. New York, NY: Wiley-Liss.

Novaes, W. G. & Cintra, R. 2013. Factors influencing the selection of communal roost sites by the Black Vulture *Coragyps atratus* (Aves: *Cathartidae*) in an urban area in Central Amazon. *Zoologia* 30(6): 607–614. Retrieved from https://www.scielo.br/pdf/zool/v30n6/ aop1413.pdf.

Ogada, D. L., Torchin, M. E., Kinnaird M. F. & Ezenwa, V. O. 2012. Effects of vulture declines on facultative scavengers and potential implications for mammalian disease transmission. *Conservation Biology*, 26: 453–460. http://www.doi:10.1111/j.1523-1739.2012.01827.x.

Olea, P. P. & Mateo-Tomás. P. 2009. The role of traditional farming practices in ecosystem conservation: the case of transhumance and vultures. *Biological Conservation*, 142(8):1844–1853. https://doi.org/10.1016/j.biocon.2009.03.024.

Olrog, C. C. 1985. *Status of Wet Forest Raptors in Northern Argentina.* In ICBP Technical Publication, 191–197.

Pattee, O., Bloom, P., Scott, J. & Smith, M. 1990. Lead hazards within the range of the California Condor. *Condor*, 92: 931–937. http://www.doi:10.2307/1368729.

Powers R. H. 2005. The decomposition of human remains. In: Rich J., Dean D. E., Powers R. H. (eds) *Forensic Medicine of the Lower Extremity*, 3 – 15. Forensic Science and Medicine. Humana Press.

Prada, P. A. & Furton, K. G. 2018. Birds and dogs: Toward a comparative perspective on odor use and detection. *Frontiers in Veterinary Science*, 5:188. https://doi.org/10.3389/fvets. 2018.00188.

Prakash, V., Green, R. E., Pain, D. J., Ranade, S. P., Saravanan, S. et al. 2007. Recent changes in populations of resident Gyps vultures in India. *Journal of the Bombay Natural History Society*, 104: 129–135. https://doi.org/10.1017/S09592709 17000545.

Prior, K. A. & Weatherhead, P. J. 1991. Competition at the carcass: opportunities for social foraging by turkey vultures in southern Ontario. *Canadian Journal of Zoology*, 69(6): 1550-1556. https://doi.org/10.1139/z91-218.

Robinson, S. K. 1994. Habitat selection and foraging ecology of raptors in Amazonian Peru. *Biotropica*, 26: 443–458. https://doi.org/10.2307/2389239.

Rodriguez-Ramos, J., V. Gutierrez, U. Höfl e, R. Mateo, L. Monsalve, E. Crespo and J. M. Blanco. 2009. Lead in Griffon and Cinereous Vultures in Central Spain: Correlations between clinical signs and blood lead levels. Extended abstract in Watson, R. T., Fuller, M., Pokras, M. & Hunt, W. G. (Eds.) *Ingestion of Lead from Spent Ammunition: Implications for Wildlife and Humans*, 236. The Peregrine Fund, Boise. http://www.doi10.4080/ilsa.2009.0213.

Roggenbuck, M., Schnell, I. B., Blom, N., Baelum, J., Bertelsen, M. F., Sicheriz-Ponten, T., Sørensen, S. J., Gilbert, M. T. P., Graves, G. R. & Hansen, L. H. 2014. The microbiome of New World vultures. *Nature Communications*, 5. https://doi.org/10.1038/ncomms6498.

Schlee, M. 1995. Nest records for the king vulture (*Sarcoramphus papa*) in Venezuela. *Journal of Raptor Research*, 29(4): 269–272.

Schulenberg, T. S. (Ed.) 2010. *Greater Yellow-headed Vulture* (*Cathartes melambrotus),* Neotropical Birds Online). Cornell Lab of Ornithology, Ithaca. Retrieved from Neotropical Birds Online: http://neotropical.birds.cornell.edu/ portal/species/overview?p_p_spp=118 556.

Sekercioglu, C. H., Daily, G. C. & Ehrlich, P. R. 2004. Ecosystem consequences of bird declines. *Proceedings of the National Academy of Sciences*, USA 101: 18042–18047. http://www.doi:10.1073/pnas.0408049101.

Snyder, N. F. R. & Snyder, H. 2000. *The California Condor: A Saga of Natural History and Conservation*. San Diego: Academic Press.

Snyder, N. F. R. & Snyder, H. 2006. *Raptors of North America: Natural History and Conservation*. Beverly, MA: Voyageur Press.

Sorenson, K. J. & Burnett, J. 2007. Lead concentrations in the blood of Big Sur California condors. In Mee, A., Hall, L. S. & Grantham, J. *California Condors in the 21st century. Special publication of the American Ornithologists' Union and Nuttall Ornithological Club.*

Stager, K. E. 1964. *The role of olfaction in food location by the turkey vulture (Cathartes aura). Contributions in Science*, 81:1–63. https://www.biodiversity library.org/page/52104707#page/475/mode/1up.

Stewart, P. A. 1978. Behavioral interactions and niche separation in Black and Turkey vultures. *Living Bird*, 17:79-84.

Stolen, E. D. 2000. Foraging behavior of vultures in central Florida. *Florida Field Naturalist*, 28(4): 173 - 181. Retrieved from https://sora.unm.edu/sites/default/files/FFN_28-4p1173-181Stolen%5B1%5D.pdf.

Switalski, T. A. 2003. Coyote foraging ecology and vigilance in response to gray wolf reintroduction in Yellowstone National Park. *Canadian Journal of Zoology*, 81(6): 985-993. https://doi.org/10.1139/z03-080.

Unitt, P. 1984. The Birds of San Diego County. *San Diego Society of Natural History Memoir* 13.

Van Dooren, T. 2010. Vultures and their people in India: equity and entanglement in a time of extinctions. *Manoa*, 22: 130–146. https://www.muse.jhu.edu/article/407433.

Van Dooren, T. 2011. *Vulture*. London, UK: Reaktion Books Ltd.

Van Wyk, E., Bouwman, H., van der Bank, H., Verdoorn, G. H. & Hofmann, D. 2001. Persistent organochlorine pesticides detected in blood and tissue samples of vultures from different localities in South Africa. *Comparative Biochemistry and Physiology*, 129(3): 243–264. http://www.doi:10.1016/S1532-0456(01)90201-7.

Wallace, M. P. & Temple, S. A. 1987. Competitive interactions within and between species in a guild of avian scavengers. *Auk*, 104: 290–295. Retrieved from https://sora.unm.edu/sites/ default/files/journals/auk/v104n02/p0290-p0295.pdf.

Wani, A. L., Ara, A. & Usmani, J. A. 2015. Lead toxicity: a review. *Interdisciplinary Toxicology*, 8(2), 55–64. http://www.doi:10.1515/intox-2015-0009.

Wilbur, S. R. 1978. *The California Condor, 1966–1976: A Look at its Past and Future*. Washington DC: Fish and Wildlife Service.

Wilmers, C. C., Crabtree, R. L., Smith, D. & Murphy, K. M. 2003a. Tropic facilitation by introduced top predators: gray wolf subsidies to scavengers in Yellowstone National Park. *Journal of Animal Ecology*, 72: 909–916. https://doi.org/10.1046/j.1365-2656.2003. 00766.x.

Wilmers, C. C., Stahler, D. R., Crabtree, R. L., Smith, S. & Getz, W. M. 2003b. Resource dispersion and consumer dominance: scavenging at wolf and hunter-killed carcasses in Greater Yellowstone, USA. *Ecology Letters*, 6: 996–1003. https://doi.org/10.1046/j.1461-0248.2003.00522.x.

Wilson, E. E. & Wolkovich, E. M. 2011. Scavenging: How carnivores and carrion structure communities. *Trends in Ecology & Evolution*, 26(3):129-135. https://doi.org/10.1016/ j.tree.2010.12.011.

Wisman, A. & Shrira, I. 2015. The smell of death: evidence that putrescine elicits threat management mechanisms. *Frontiers in Psychology*, 6: 1274. http://www.doi:10.3389/ fpsyg.2015.01274.

Woodford, M. H., Bowden C. G. R. & Shah, N. 2008. *Diclofenac in Asia and Africa—Repeating the Same Mistake? Harmonisation and Improvement of Registration and Quality Control of Veterinary Medicinal Products in Africa*. Paris: OIE World Organisation for Animal Health.

Wright, A. 2016. *Clostridial Diseases of Cattle*. Retrieved from https://extension.arizona.edu/sites/extension.arizona.edu/files/pubs/az1712-2016.pdf.

Zepeda Mendoza, M. L., Roggenbuck, M., Manzano Vargas, K., Hansen, L. H., Brunak, S., Gilbert, M. T. P. & Sicheritz-Pontén T. 2018. Protective role of the vulture facial skin and gut microbiomes aid adaptation to scavenging. *Acta Veterinaria Scandinavica*, 60(1):61. http://www.doi:10.1186/s13028-018-0415-3.

In: Critical Research Techniques in Animal … ISBN: 978-1-53619-846-1
Editors: Michael O'Neal Campbell et al.

Chapter 6

SOUTH AND NORTHEAST BRAZIL: DOES A SHARED SPECIES ARISE SHARED HUMAN DIMENSIONS OF CONFLICTS WITH WILDLIFE?

Francine Schulz[1,*] and Cláudia Sofia Guerreiro Martins[2]
[1]Universidade do Vale do Rio dos Sinos,
Mammal Ecology Laboratory, São Leopoldo, RS, Brazil
[2]Institute for the Conservation of Neotropical Carnivores, Atibaia, SP, Brazil

ABSTRACT

Carnivores as big cats are key to maintain the ecosystems equilibrium. They face huge and diverse challenges and have been systematically banished from their natural habitats. The species that managed to survive close to human settlements, as pumas (*Puma concolor*), have been severely threatened mainly because of persecution and poaching in retaliation for depredation episodes on domestic herds.

This Chapter will present human-wildlife-conflicts (HWC), focusing on pumas, in two different Brazilian biomes: the hinterland of Northeast and the Southern Atlantic Forest.

The hinterland of Brazilian Northeast has the largest expanse of the Seasonally Dry Tropical Forest and Woodlands (SDTFW) biome in the New World, known as Caatinga, with nine different ecoregions. As in any other semiarid region, it has an extended dry period in which rainfall is scarce. The southern limit of Atlantic Forest in Brazil has the domain of *Araucaria angustifolia* forest creating a mosaic of landscapes with south Brazilian natural grassland. The regional climate is subtropical with a high annual precipitation, without a marked dry season.

Both study sites are located within or surrounding a polygon of protected areas and as in most Brazilian regions, the implementation of this kind of sites caused troublesome with the local populations. Studies considering the human dimensions

* Corresponding Author's E-mail: francine.schulz@gmail.com.

of nature and conflicts are scarce in both regions and more social sciences approaches are needed to better understand it in South and Northeast Brazil.

Considering these two contrasting biomes and the absence of unified protocols in this field of knowledge, some comparable variables between studies methodologies were selected. The objective was to start to fill the gap in human dimensions of human-puma-conflicts, with South and Northeast Brazil as showcases.

Few similarities were found: farmers agreed that poaching pumas would not be a solution for the problem of livestock depredation; they did not ask for help from any wildlife management institution; and, farmers believed that pumas could be a risk to their livelihood, including husbandry practices.

This information may support conservation and wildlife management strategies. Moreover, unified protocols should be designed and implemented to identify, describe, understand, and predict human perceptions and behaviour in different contexts, especially when dealing with charismatic megafauna.

INTRODUCTION

Cultures across the world have over time developed characteristic ways of interacting with the environment and its flora and fauna (Alves, 2012). Ethnobiology, anthropology, sociology, environmental, social, behavioural, and cognitive psychology, study this kind of interactions and can use knowledge, behaviour, and attitudes for the conservation of natural resources (Alves, 2012; Zimmermann et al., 2010). Tools used for these investigations include structured questionnaires that urge to translate the respondent's thoughts, emotions, feelings, and actions towards the researched subject (Ditt et al., 2009; Holland et al., 2018). The importance of these strategies, focused on human responses, are recognized as crucial to develop trust between managers and local communities and to design conservation plans (Alves, 2012; Holland et al., 2018). Despite the relevance to conservation biology, the conflicts between people and wild animals have been little explored within a human dimensions' framework. For instance, even with a known strong interaction between Brazilian fauna and local populations, HWC studies focusing on peoples' perspective and opinions are scarce, especially in the southern region of Brazil (Alves & Souto, 2011).

Carnivores are essential to maintain the ecological equilibrium as they guarantee the diversity and resiliency of ecosystems where they inhabit. As predators, they control herbivore populations, known as a top-down effect on trophic cascades (Ripple & Beschta, 2006), and once they usually leave some part of their hunted preys, they supply food to maintain the diversity of scavenger and decomposer communities (Elbroch & Wittmer, 2012). Small and mainly large carnivores are hunted and killed worldwide, especially in retaliation to conflicts with rural communities (Cavalcanti, 2009; Holland et al., 2018; Inskip & Zimmermann, 2009; Michalski et al., 2006). Moreover, large animals as big cats have always been subject of human fascination and fear, generating an enormous and strong record of conflicts (Herrmann et al., 2013).

The dynamics of HWC is challenging and solving conflicts between species (humans – any other predator) can be even more complicated (Zimmermann et al., 2010). Lions (*Panthera leo*), tigers (*Panthera tigris*), jaguars (*Panthera onca*) and pumas have lost their habitats and preys, leading them to frequent encounter with humans and domestic herds (Holland et al., 2018). Animals of domestic herds have lost some of their natural instincts, what make them easy target for wild animals as big cats (Crawshaw, 2004). Currently, studies show that the contact with rural populations has resulted in purposeful elimination of large felids that are killed due to depredation of herds or as a trophy hunt (Holland et al., 2018; Mazzoli et al., 2002; Palmeira & Barrella, 2007). This problem is one of the most urgent felids conservation issues and in order to maintain big cats worldwide, the human dimensions and interactions with these animals need to be better understood (Herrmann et al., 2013; Holland et al., 2018; Zimmermann et al., 2010).

Since the cattle breeding was introduced in Latin America, the interactions between people and big cats evolved, changing from fear and amazement for its beauty and strength to hostility and competition (Hoogesteijn & Mondolfi, 1992). In Brazil there are two species of big cats, jaguars and pumas, and both have been hunted especially by farmers in retaliation of losses due to the depredation on domestic herds (Marchini & Crawshaw, 2015). Studies reveal that the increasing of domestic herds' depredation can include: lack of natural preys, animals less capable of hunting as elders or injured that have difficulties to catch natural prey and poor management of herds in the farms (Crawshaw, 2004; Murphy & MacDonald, 2010).

Jaguars and pumas are considered threatened of extinction in Brazil and have the status of 'vulnerable' in the country, according to the Brazilian Environmental Ministry (ICMBIO, 2013a; 2013b). The main factors contributing to the decline of these felids' populations are: loss of habitats due to urban and agricultural expansion; mining; expansion of paved roads and new and modern uses of land, such as renewable energy plants settled in pristine areas, causing deforestation and easier access of poachers to remote areas; and retaliatory and opportunistic poaching (ICMBIO, 2013a; 2013b; de Matos Dias et al., 2020). Some studies of conflicts with felids and people are (or were) being conducted in different regions of Brazil such as Pantanal (e. g. Tortato et al., 2015; Zimmermann et al., 2005), Amazonian Forest (e. g. Michalski et al., 2006; Marchini, 2010), Cerrado (e. g. Palmeira et al., 2008), Caatinga (Martins, 2020) and Atlantic Forest (e. g. Palmeira & Barrella, 2007; Engel et al., 2017). The southern and north-eastern regions of Brazil have the minor number of studies with big cats and HWC. In the south the studies are concentrated in the region of *Foz do Iguaçú* National Park, Atlantic Forest biome (Azevedo, 2008) and in the coastal mountains of southern Brazil between states of Santa Catarina and Rio Grande do Sul (Mazzolli et al., 2002; Sá, 2005; Schulz et al., 2014). In the review done by Lozano et al. (2019) about publications concerning human-carnivore relations between 2000 and 2016, neither the Caatinga as a domain or the semiarid region were mentioned.

The most southern region of Brazil has a strong tradition in the agricultural sector, a stable economy, and a high Human Development Index - HDI, 0.746 (INSTITUTO BRASILEIRO DE GEOGRAFIA, 2020). The state of Rio Grande do Sul has as its main biome the Atlantic Forest in a mosaic with *Campos* (a native grassland) and is one of the most deforested states of Brazil, with just few remaining forested areas (Indrusiak & Eizirik, 2003). In this state, the population of pumas has been reduced to few individuals, inhabiting mainly the edges of the north-eastern plateau and the region of Argentine forests' boundaries in the north-west region, in this last site wandering jaguars are also inhabiting denser forest (Indrusiak & Eizirik, 2003; Oliveira & Cassaro, 2006).

The hinterland of Brazilian Northeast is known as Caatinga Domain, harbouring "the largest and most continuous expanse of the seasonally dry tropical forest and woodland biome in the New World" (Queiroz et al., 2017, p. 23), with experts naming nine different ecoregions (Silva et al., 2017). As in any other semiarid region, it has an extended dry period in which rainfall is scarce. Caatinga is considered a well-defined social region, with its human population deriving of the miscegenation of the native indigenous people with the Portuguese who colonized the area during the 1600s (Théry & Mello, 2005). They have developed strong and resilient cultural traditions based on cattle ranching and subsistence agriculture. It is an exclusively Brazilian biome, with about 912,529 km^2 of territorial range, which represents 10.7% of the national territory (Silva et al., 2017). Jaguars and pumas inhabit the area in small densities, being 'critically endangered' (Morato et al., 2013) and 'endangered' (Azevedo et al., 2013), respectively, due to hunting for illegal fur traffic in the past (Broad, 1988) and conflict with human activities, nowadays. They cooccur with one of the poorest populations of Brazil with low to medium HDI (0.515-0.677, considering two of the municipalities parts of this study, namely Umburanas e Juazeiro, Bahia state) (ATLAS DO DESENVOLVIMENTO HUMANO DO BRASIL, 2013).

In the attempt to solve ecological problems and to protect and restore its biological diversity, Brazil's government created in 2000s the National System of Conservation Units (Federal Law 9985/00, 'SNUC' in Portuguese), the legal framework for Protected Areas. The acronym for Conservation Units is 'UC' in Portuguese, UC being the areas managed by federal, state, or municipal public administration, with no physical borders or fences, but outlined to protect highly relevant ecological areas. The creation of UC in Brazil was usually problematic and caused conflicts with local communities, especially because people who live within the areas must change behaviours and traditional habits in order to protect and maintain the environment (Diegues et al., 2000). This is a national scenario and in practical terms, any conflict between communities within and surrounding protected areas and fauna tends to be harder to solve. In this sense, the social representations of wildlife and wildlife management must be considered before starting any kind of conservation plan. The implementation of UC with land uses restrictions needs to have the participation of local population, trying to make them allies in the process, once just protected areas are not enough to maintain viable ecosystems (Diegues et al., 2000; Mazzolli, 2005).

Considering these two completely different and threatened biomes in Brazil that have in common small subpopulations of big cats, and considering that the studies of the authors were done both in a polygon of Brazilian UC, the main research question that this study propose to address is: Are there any similarities between human dimensions towards pumas among farmers from two geographically contrasting biomes, the Brazilian north-eastern semiarid and the southern Atlantic Forest? The question only cites the pumas because in the south the data is related just to this species, once the jaguars are already regionally extinct. Moreover, pumas in the northeast region may frequent areas closer to human habitation.

Study Areas

North-Eastern Plateau of Rio Grande Do Sul State - Southern Brazil

The study in South Brazil was conducted in a polygon mosaic of UC in the north-eastern state of Rio Grande do Sul, nearly from 50º01'W to 50º42'W and from 28º55'S to 29º33'S. In this region, there are at least 11 UC, and some of these units have restrictions on land uses (Schulz et al., 2014). The UC in the area are under federal, state and municipal government management and belong to the Atlantic Forest Biosphere Reserve (*Reserva da Biosfera da Mata Atlântica* – RBMA, in Portuguese). The RBMA covers 17.0% of Rio Grande do Sul state, it was created due to the big fragmentation of Atlantic Forest in whole Brazilian territory, and it follows a model of international integrated management for the conservation of natural resources and improvement in human's life quality (Duarte & Bencke, 2008).

The area represents the southern limit of Atlantic Forest in Brazil and has the domain of *A. angustifolia* forest (region known as *Planalto das Araucárias*, in Portuguese) creating a mosaic of landscapes with south Brazilian *Campos de cima da Serra* (natural grassland vegetation) (Silva et al., 2009). The regional climate is described as subtropical, with moderate temperatures (the average varies between 14ºC and 20ºC) and high annual precipitation (above 2.000 mm) without a marked dry season (Silva & Anand, 2011). In addition, there are mesoclimate characteristics forged by specific relief, sea proximity and abrupt slopes; in the highest regions fog is a constant phenomenon (Backes et al. 2005).

The primary anthropogenic land uses are cattle grazing, logging practices and agriculture. The 300 km^2 region around São Francisco de Paula city, that had the majority of interviewees, is composed of 30.0% native *Campos* grassland, 25.0% of native *A. angustifolia* forest and 45.0% of other human land use (agriculture and/or commercial tree plantation) (Lang, 2013). The typical landscape of the region is represented in Figure 1.

Figure 1. Typical southern landscape where it can be seen *A. angustifolia* forest (the darkest green forest), native *Campos de cima da Serra* (grassland), *Pinus ssp* forestry (the lightest green forest) and cattle breeding (Duarte & Bencke, 2008).

The four cities where the interviews were applied generate an annual Gross National Product (GNP) of R$ 267,000,000.00 (about U$ 48,500,000.00). The city of São Francisco de Paula has 56.0% of the total value, followed by Cambará do Sul (32.0%), the other cities represent less than 7.0% of the amount each one (Duarte & Bencke, 2008). The economy in the region is mainly based on forestry, traditional livestock, and small agricultural practices. Extensive system is widely used in livestock practices, demanding low economic investments, therefore natural factors are determinant in the processes. In the winter (cold and with episodes of frost in the pastures) some better-quality grass can be given to the cattle to improve its diet. Livestock is considered a traditional activity once it has been taught the same way throughout generations, approximately 200 years, with European settlers and descendants reproducing in the New World the agricultural and pastoral secular systems of the Northern Hemisphere (Duarte & Bencke, 2008). In a minor scale there are pigs, horses, sheep, goats, chickens, and geese raised in the farms usually to the family consumption or to complement livestock activities. The profitable market is focused on cattle, other animals can be sold locally, but not as the main source of income (Duarte & Bencke, 2008).

As common in rural areas the resident population is predominantly male in active age. The demographic density is low with large areas of pastures and plantations with no presence of people. This social isolation sometimes results in a poor quality of life, for instance the education level in the region is also low. However, younger residents are going to school and college, becoming a channel of communication between rural communities and the city (Duarte & Bencke, 2008).

Caatinga - Semiarid of Brazilian Northeast

The study was done in a region called *Boqueirão da Onça* in Portuguese, *Boqueirão* meaning a deep and abrupt valley between two mountains, usually carved by a river, and *onça* being the indigenous word for jaguar. *Boqueirão da Onça* is one of the last caatinga vegetation continuum in the biome Caatinga (caatinga means white bush, related to the dominant color of vegetation during the dry season; capital letter defines the biome where this vegetation is dominant over other physiognomies). The region is in the north of Bahia state and became a polygon of UC in April 2018, one National Park with 347,557 ha of full protection, and one Environmental Protected Area of 505,692 ha allowing sustainable land uses. These UC are managed by the Brazilian federal government throughout *Instituto Chico Mendes de Conservação da Biodiversidade* (acronym ICMBio, in Portuguese). The cities represented in this area are: Sento Sé, Campo Formoso, Umburanas, Sobradinho, Juazeiro and Morro do Chapéu, all in the state of Bahia, with an urbanization rate about 12.2% (MMA, 2007). A typical vegetation of *Boqueirão da Onça* is represented in Figure 2.

As the Caatinga is one of the two most vulnerable Brazilian biomes to the global climate changes especially due to desertification process (Seddon et al., 2016), the region is extremely important to maintain the biological conservation once big predators as jaguars inhabit the area, using the patches between the fragmented caatinga vegetation. On the other hand, the biome Caatinga has one of the lowest number of Brazilian protected areas, just 1.2% (Dryflor et al., 2016; Leal et al., 2005) and vegetational remnants have different sizes and are fragmented, not sufficiently effective to fulfill ecological roles (Castelletti et al., 2003).

Figure 2. Typical Boqueirão da Onça landscape during the rainy season, where it is possible to see caatinga vegetation behind a goat herd grazing in the wild (Martins, 2020).

The human population in the study area is estimated in 32,708 people (MMA, 2007), representing 8.7% from the total people living in the six cities, considering rural and urban settlements (376,170 habitants) (BRASIL, 2010), with HDI between 0.500 and 0.699 (ATLAS DO DESENVOLVIMENTO HUMANO DO BRASIL, 2013).

The main economic activity is extensive goat, sheep, and cattle breading. The herds graze freely in the caatinga vegetation in *fundos de pasto* (communal areas to herds to graze, and some also supporting collection of firewood, native fruits, roots or medicinal plants), property of the same family or of the same traditional community. The state government gives inalienable tenure documents of these areas to families and communities (Alcântara & Germani, 2010). Most of these farmers are people with low income who rely on domestic herds and small agricultural practices, depending totally on the rain to maintain their families. They plant cassava, corn, bean, banana and other few fruit varieties, palms, grass for the cattle, castor beans and sisal to sell in local markets. Technical assistance is rare, almost absent. Most part of households do not have electricity or potable water supply. Schools, when present, are very precarious, it is common to see one teacher supervising different class levels, and if a teenager wants to pursue more studies, needs to move to relative's households in the nearest towns (Martins, 2020).

The historical pattern of human settlements is formed by descendants of enslaved Africans who escaped from their master's and started growing communities in places where they could farm small herds, extract natural products and hunt (Andrade, 1998). Thus, hunting prevails as a tradition to complement the protein in families' weekly meals (Alves et al., 2009; Barbosa et al., 2011).

METHODS

Sampling, Characteristics, and Main Objectives

Study with Pumas in South Brazil – Campos de Cima da Serra

The work in southern Brazil was composed by 45 face-to-face interviews and was approved by an Ethical Committee (protocol number: 051/2011), according to the Brazilian Federal Law (CNS 196/96) that oversees studies involving humans.

As a starting point, three environmental managers from state UC were contacted, they were used as a referee group to find those farmers who had depredation on domestic herds caused by pumas (Bailey, 1982; Schulz et al., 2014). The UC areas were selected according to their location; their administrative building is at the downtown of São Francisco de Paula city, which was easier to communicate. The UC were: *Área de Proteção Ambiental* (APA) *Rota do Sol* (29°23'22.58"S; 50°11'3.90"W), *Estação Ecológica* (ESEC) *de Aratinga* (29°18'S e 29°24'S; 50°11'W e 50°17'W), and *Parque Estadual* (PE) *do Tainhas* (29°05'35.33"S; 50°21'51.55"W) (names are in Portuguese). The first farmers interviewed indicated the next ones, following the snowball method (Bailey, 1982).

All the answers given by respondents were related to the years from 2008 up to 2011. The study started and finished in the year of 2011. The data collection was done in four months, from May to August. The maximum number of interviews per day was four, the most common was two per day, the farms were far from each other, the roads usually were not paved, the main researcher was mostly by herself and people were free to talk about other topics, in a cordial visit and dialogue.

A standard questionnaire with open (the person is free to talk) and closed questions (yes or no/ block of question) was used as protocol for the interviews. An ethnobiological perspective was used to elaborate the questionnaire and conduct the interviews, no personal interferences were done.

The questions were created based in other works with depredation problems in Brazil (Conforti & Azevedo, 2003; Mazzoli et al., 2002; Michalski et al., 2006; Palmeira & Barrella, 2007; Zimmermann et al., 2005) and were organized in six different topics: interviewee's profile; characterization of the farm and surrounding area; characteristics of the herds; depredation by pumas; and perception of interviewees about pumas. At the end of each interview an illustrative board was shown to check if the person could correctly identify a puma among other three felids species (protocol is available upon request). As above stated, jaguar is extinct in the region for a long time. Thus, there is no misunderstanding about the species jaguar and puma amidst the local population.

After the interviews, an informed consent term was given to each interviewee to be signed in two copies: one copy was kept with the researcher and the other with the respondent. This form explicitly showed that the interviewee gave all information freely and that everything that was said would be kept anonymous and used only for academic purposes.

All data collected were quantitatively analyzed. Quantitative approach was used to measure geographic characteristics of farms and sites where pumas attacked; respondents' economic and social profile; financial losses; sort and number of herds; climate characteristics when the attacks occurred; and type of management applied to the herds in the farms. The frequencies of answers from open and closed questions were calculated to each question category, according to the following: Frequency = number of times that one answer was given x 100 / total number of answers. It is important to explain that for open questions, the most repeated ones were analyzed by their frequency. Some hypotheses were also tested using the Person's Correlation, Kruskal-Wallis, and Qui-Square (χ^2) tests, using a statistical trust level of 95.0%.

The main objective of this work was (i) to describe the characteristics of pumas' attacks, (ii) to quantify the financial losses on farms, and, (iii) to have a primary assessment of the conflict between farmers and pumas in the region. For more details you can read Schulz et al. (2014 and 2017).

Study in Northeast Brazil - Boqueirão da Onça, Caatinga

The study and the protocol used were submitted and validated by the University of São Paulo to a Brazilian government platform that oversees ethical aspects in research involving human beings (*Plataforma* Brazil). The clearance was issued on June 28th, 2017, with the number CAAE 68314417.1.0000.5395.

The questionnaire was elaborated following some theoretical ideas of Browne-Nuñez et al. (2013) who worked with human-elephant-conflicts in a similar context (environment, socioeconomic and institutional) in rural communities in Africa. Other strategies to build the protocol were taken from human dimensions of wildlife framework (Dström et al., 2009; Newing et al., 2011; Engel et al., 2016; Nuno & St. John, 2014), and previous personal interactions of the researcher with the communities from *Boqueirão da Onça*.

The questions, apart from the interviewees' profile, were conceived, in a second stage, after a pilot test, to get binary answers ('yes' or 'no'), and organized in three categories, following the theoretical framework of vulnerability to climate changes: 'exposition' (to the felids), 'sensibility', and 'capacity to adaptation'. The study got 134 volunteer participants; all the questionnaires were applied face-to-face. The required pilot test was done in the nearest community for practical reasons. Each interview length about 45-90 minutes. The GeoODK© Android app was used to adapt the questionnaire to a form to be filled with the interviewee, registering the answers quickly, and storing the data on offline mode. This tool facilitated the data transference and organization in spread sheets.

The research was characterized as observational when the interviewer does not interfere in the answers and opinions given (Newing et al., 2011).

Data were analyzed with a statistical trust level of 95.0%. Some descriptive statistical analysis was done to explore the data and to create graphics (average, standard deviation, percentage, and frequency) (Conover, 1999). Other statistical tests were used to assess categoric variables: Qui-Square (χ^2), Fisher exact test, Wilcoxon, Linear Regression and Logistic Regression.

The main objectives of this work were to assess (i) the attitudes of farmers towards jaguars and pumas; (ii) farmer's knowledge about these felids' biology and ecology; (iii) husbandry practices; (iv) farmer's perceptions about the impact of these felids in their livelihoods; and, (v) the existence value of these species. For further information you can read the Doctoral Thesis of Martins (2020).

Social Science's Approach

As before mentioned, social science's approach in ethnobiological and human dimensions of HWC can be especially useful to understand how the communities perceive the environment and the natural resources, biotic and abiotic (Alves, 2012; Manfredo, 2008). However, to accomplish a good result in the collection of information that only people can give, the researchers must find a way to build communities' trust and sometimes adjustments are necessary to improve the process. Thus, pilot studies are highly recommended to test communities' feedback (Ditt et al., 2009).

Standard protocols that guarantee anonymity of people's image and identity are important and can help to approach the interviewees. Sometimes people fear to share sensitive information as presence of hunting where this practice is illegal, for instance, as it is in Brazil. When a legal commitment is established with the person that provides the data during an interview, it is assured the respondents and researchers own legal safety. The study in the south (Schulz et al., 2014) found that this was a good way to reach local farmers' trust, and only one person refused to participate of the research. However, some communities are very suspicious to sign any document because it seems to them that their signature can be used against them. The study in the northeast (Martins, 2020) could not explicitly give a document to the respondents because the pilot test showed they would not participate or would restrain to provide answers in private that they would not bring to public. Also, the literacy level is low in the region, and that seems to contribute to the suspicious people may have to sign any paper.

In Brazil all the research involving people must be checked and approved by an Ethical Committee. An Informed and Volunteer Consent Term (TCLE, acronym in Portuguese) is issued and researchers should use it by giving a copy to the interviewee and keeping other to themselves, signed both by researcher and respondent. As in the example above, sometimes this procedure is not possible because people can refuse to sign it. In this case, as Martins (2020) did, just explain to the interviewee the content of the research and the details of the term, giving the person the choice to answer the questions, to stop anytime without further explanation, and, if requested, the term can be signed and left in the household.

The questions in the protocols must follow an adequate sequence, starting with less sensitive questions, providing time to build trust with the interviewees to eventually get trustful answers to complicated topics (Ditt et al., 2009; Marchini, 2010). The researcher needs to be empathetic with the community, it is important to gather the more information as possible about them. It may help to live near the study area or to have help from someone that has bonds with the local population, this will ease the local acceptance. Otherwise, it may take place what Kumar (1992, p. 3) calls of "rash, clumsy, hit-and-run encounter". In both studies presented here, the researchers were residing in the region and acquainted of many dwellers previously to the data collection.

The language used in the questions must be clear and understandable to respondents' socio-economic-cultural context. In this sense, the common names for animals and local expressions were learnt from locals and used in the interviews. The interviews were conducted as a dialogue to create interaction and trust, preventing doubts about the answers to be given. Clothes with conservationist slogans, institutional logos or with animals' draws were not used by researchers during the approaches and interviews to avoid false or induced answers.

An illustrative board as the one used in the work with pumas in the south can be used as a strategy to check if respondents really know the animal that they were talking about during the interview (Michalski et al., 2006; Schulz et al., 2014).

How the Studies Were Compared and Analyzed

A systematic reading of both questionnaires and primary data was done. The objective was to find similarities in the questions even if a different way to ask was used (these were independent studies applied in different contexts, demanding different approaches and language adjustments). When the meaning of the answer matched in a human dimension variable, the question and the data were selected.

An exploratory analysis of data was done getting frequencies and percentage of the answers to the matching questions in both studies. The comparison between the answers in the two different regions was analysed using a Qui-Square (χ^2) test, considering the significance level of 5.0%.

The answers collected in northeast randomly came from people who had or had not depredation problems with two felids species: jaguars and pumas. The answers collected in the south were intentionally focused on people who had depredation problems and just with pumas.

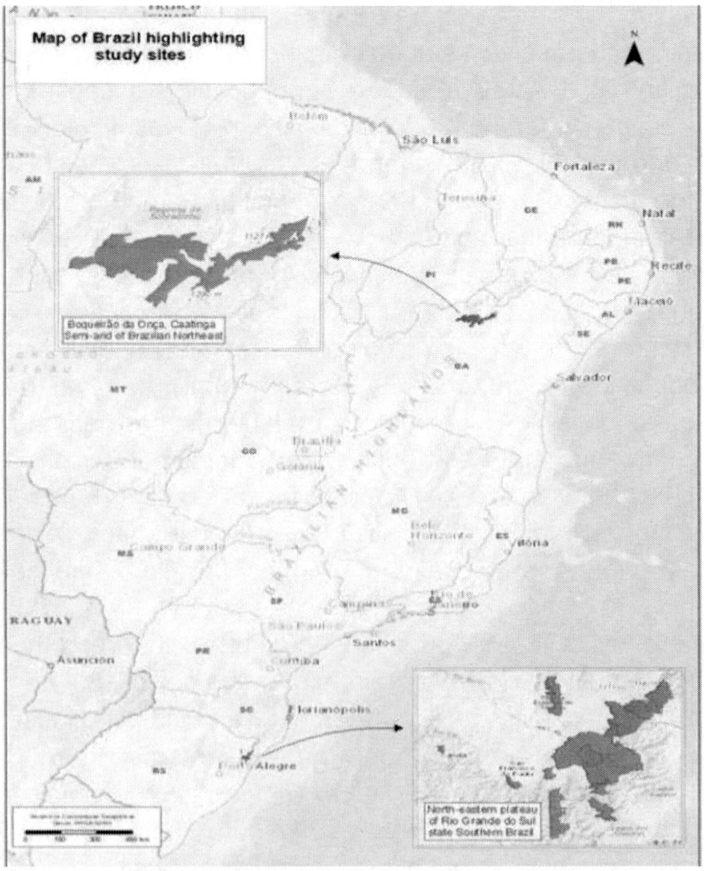

Figure 3. Map of Brazil highlighting the two study sites; in South the north-eastern plateau of Rio Grande do Sul state, Atlantic forest biome with grassland vegetation; in Northeast the hinterland of Bahia state, *Boqueirão da Onça* in the semiarid region, caatinga vegetation.

It is important to highlight that the interviewees in both study sites have a contrasting sociocultural and economic background and reality, as their regions in Brazil are very different historically and geographically (Figure 3).

Considering these scenarios, the main research question emerged: Are there any similarities between human dimensions towards pumas among farmers from two geographically contrasting biomes, the Brazilian North-eastern Semiarid and the Southern Atlantic Forest? Hypothetically, does the in common felid species arise similar human dimensions, in distinct socioecological systems?

RESULTS

Socio-Economic Respondents' Profile

In both studies most respondents were men (South = 86.7%; Northeast = 65.7%). In South respondents' age was between 17 and 83 years old (SD = 15.5), in Northeast between 19 and 90 years old (SD = 16.6). The predominant education level was low for both study sites, lower in Northeast, and the main income based on agricultural or pastoral activities, with 59.6% for South and 36.8% for Northeast. Most interviewees had the land tenure or were close relatives to the owner (South = 75.6%; Northeast = 82.8%). More details about socio-economic information are presented in Table 1.

Table 1. Socio-economic profile of interviewees from South and Northeast

	South (%)	Northeast (%)
Male	86.7	65.7
Female	13.3	34.3
Native from the region	80.0	85.1
Non-native from the region	20.0	14.9
Illiterate	0	31.6
Basic Education	40.0	50.4
High school	37.8	16.5
College	22.2	1.5
Social benefit or retirement as main income	0*	39.8
Agriculture or Livestock as main income	59.6	36.8
Other occupations as main income	40.4	23.4
Animal protein consumption (livestock) at least once a week	NI**	31.6
Animal protein consumption (livestock) more than once a week	NI**	68.4
Water Supply available	NI**	68.7
No water supply	NI**	31.3

* Some were retired but considered their occupation as farmers ** NI: No information collected.

The monthly income in South was considerably low for 33.3% respondents who declared to earn between one (USD 312.50) and two regional minimum wage. Respondents who earned between three and four regional minimum wage represented 17.8%; between five and six represented 26.7%; and just 22.2% of respondents earned more than six regional minimum wage monthly. The medium monthly income per person in the state of Rio Grande do Sul in 2011 was USD 647.43 (about two regional minimum wage) (INSTITUTO BRASILEIRO DE GEOGRAFIA, 2020). In Northeast, the data collected related to income was concerned only with its source: 39.8% declared to depend mainly on governmental social benefits or retirement ("*Bolsa-família*" being the main source, corresponding to a monthly income of approximately USD 36/per child going to school in one household, and rural retirement income, USD 198.15).

Most Important Findings in Both Studies

The study in South showed that there was no correlation between 1) the area of properties; 2) area of native forest within the range of properties; 3) altitude of properties; 4) the proximity to the first UC; and, A) the frequency of pumas' attacks, nor B) the number of domestic animals depredated in the farms. The livestock attacked by pumas was mainly sheep. Most farmers who used to breed sheep reduced or stopped their activity because of the frequency and intensity of pumas' attacks in the years before the study was done in the region (for more details read Schulz et al., 2014).

Most of the interviewees (86.7%) stated that it was necessary to control pumas' attacks but just 64.4% would accept suggestions on husbandry practices to prevent the attacks. The resistance to change husbandry practices was clear, with most of the farmers saying that they would not change their traditional system (55.6%). Results showed that those farmers who would accept recommendations would do it independently of their losses and would do it even if they had to pay for the changes by themselves.

The southern interviewees' attitudes (using a Likert scale of five levels ranging from 'really dislike' to 'really like') towards pumas and pumas' aesthetic appearance (using a Likert scale of five levels ranging from 'really ugly' to 'really pretty') were not related to their financial losses on farms. In other words, financial losses did not influence their opinions of beauty nor what they feel about the species (attitude).

The answers related to pumas' ecological role were influenced by farmers' age and education level, not by their time living in the region, not by their monthly income, nor financial losses on their farms. Thus, an elder and less formal educated southern interviewee had less knowledge about pumas' ecological role.

In Northeast farmers used to have more problems with pumas than with jaguars, however, the perception of risk was higher on those who had problems with jaguars, what might be related to the level of damage that an encounter with a jaguar may cause.

The education level did not influence the answers on knowledge about the species, on the existence value of the species, nor on the attitudes towards wildlife management options.

The age, gender, and place of birth of respondents, did not correlate to their knowledge about the species, perceptions of risk, behavior and beliefs about jaguars and pumas. Only one variable positively influenced farmers' knowledge about the species in Northeast, which was respondents' attendance to meetings beheld in the region about conservation issues.

Another important finding in Northeast was that people who had more felids' sightings were not more or less favorable to their proximity. Sightings and attitudes were not related to each other. On the other hand, people who had problems with pumas, had negative attitudes towards their proximity, seeing them as a threat to their livelihood. Furthermore, this negative attitude towards felids increased about 39.0% with each new problem related to the species in a farm. For more details about the data and data analysis, please read the Doctoral Thesis of Martins (2020).

Data Comparison

Some similarities were found between eleven questions and correspondent answers in the protocols applied in South and in Northeast. Those in common questions are the ones supporting this research, looking for regional human dimensions related to peoples' interactions with the same key species. These questions were related to husbandry, livelihoods, attitudes, species knowledge and perception of risk. The outcoming data of these comparisons is presented in Table 2.

The Qui-Square (χ^2) test showed that:

The perception of an increase in pumas' abundance in the last 5-10 years was higher in South (p-value = 0.002).

In the Northeast, more farmers agreed that they should have a financial compensation for the losses caused by predators (p-value <0.001).

In the South the husbandry practices were comparatively more extensive (herds free to graze and search for water) than in Northeast (p-value = 0.001), and southern farmers were less willing to change husbandry practices to protect their herds (p-value = 0.006).

The northeast interviewees seemed to believe more than southern interviewees that pumas have important roles in nature (p-value = 0.003) and that pumas do not attack without being provoked (p-value = 0.017).

Farmers from Northeast were more willing to learn about pumas than farmers in South (p-value <0.001).

Concerning livestock mortality causes, 'depredation by wild carnivores' and 'accidents/inadequate husbandry' were more present in South and 'other causes' were more pointed in Northeast (p-value = 0.031).

In both study sites farmers similarly agreed that poaching pumas would not be a solution for the problem of livestock depredation (p-value = 0.622). Also, both did not ask for help from any environmental or wildlife management institution (p-value = 0.243), and both believed that pumas can be a risk to their livelihood and traditional husbandry practices (p-value = 0.872). Thus, no association was found between the study site and the variables herewith cited.

Considering these findings, we can assume that the hypotheses elaborated were just partially true. Only three variables among 11 had similar answers in the two sites studied. Thus, there were few similarities between human dimensions towards pumas among farmers from the Brazilian North-eastern Semiarid and farmers from the Southern Atlantic Forest.

Table 2. Comparison between regions (percentage and Qui-Square test)

ANSWERS	SOUTH (N=45)		NORTHEAST (N=134)		p-value
	n	%	n	%	
I asked for help from government environmental institution	8	16.9	3	7.1	0.2425
I did NOT ask for help from any government environmental institution	37	83.1	39	92.9	
I believe that the number of pumas in the region increased in the last 5-10 years	35	77.8	54	48.6	**0.0016**
I do NOT believe that the number of pumas in the region increased in the last 5-10 years	10	22.2	57	51.4	
Pumas have important role in nature	31	68.4	97	89.8	**0.0032**
Pumas do NOT have important role in nature	14	31.6	11	10.2	
Pumas are NOT a threat if not provoked	30	66.4	110	84.6	**0.0174**
Pumas are a threat even if not provoked	15	33.6	20	15.4	
Poaching pumas can solve the problems they cause	8	17.8	16	13.2	0.6217
Poaching pumas will NOT solve the problems they cause	37	82.2	105	86.8	
A financial compensation from the government for depredation events is one of the solutions to problems with pumas	15	33.3	76	65.5	**<0.001**
A financial compensation from the government for depredation events is NOT a solution to problems with pumas	30	66.7	40	34.5	
I would change my husbandry practices to prevent pumas' attacks (in an ideal scenario)	25	55,6	69	80.2	**0.0055**
I would NOT change my husbandry practices to prevent pumas' attacks (in an ideal scenario)	20	44,4	17	19.8	
My farm has extensive management system for livestock	35	78.6	46	47.4	**0.0013**
My farm has semi-intensive management system for livestock	10	21.4	51	52.6	
It would be good to learn more about pumas	26	57.8	119	94.4	**<0.001**
It is irrelevant to learn more about pumas	19	42.2	7	5.6	
I had to stop farming because of pumas/ pumas are a risk to my livelihood	29	65.3	72	61.5	0.8722
I had NOT to stop farming because of pumas/ pumas are NOT a risk to my livelihood	16	34.7	45	38.5	
Main causes of herd's mortality in the farms cited:					
Diseases	19	42.9	28	37.8	**0.0310**
Depredation caused by wild predators	10	21.4	7	9.5	
Accidents or inadequate husbandry	6	13.4	5	6.8	
Other causes	10	22.3	34	45.9	

DISCUSSION

More would have been possible if questionnaires had been intentionally thought and designed to converge in more human dimension aspects. That is a constraint pointed out for many researchers working with HWC in Brazil, the lack of unified protocols.

However, considering what could be compared, farmers from South showed more negative answers towards more topics. One of the possible explanations may be the difference in the connection that people establish with the environment and its elements, based on their personal values and the existence value given to wildlife. In Geography the concept of topophilia gathers 'perceptions', 'attitudes', 'values' and 'world view' and it comprehends the affective connection between the person and her place or physical environment (Tuan, 1974, p. 5). In South most of the interviews were applied to landowners and many of them used to be urban dwellers, and even when they were living on farms, they had employees to manage the herds. In Northeast, most of respondents owned small properties or were part of a traditional group breeding livestock in shared land, they were used to incursions in caatinga, either for hunting bushmeat, or to collect native fruits or water from streams, what seemed to connect them more or differently to the environment and its biodiversity, intentionaly or casually. Topophilia as a feeling depends on time and personal interaction with nature to be developed. It may explain these results, with southern farmers showing more attachment to land and properties that provide them financial profit than to natural areas. In a pre-scientific age, "people adjust themselves to nature, nowadays they reign over it" (Tuan, 1974, p. 287). Also, to acknowledge that "collection-gather, hunting and pastoralism were, in fact, complementary techniques to obtain food that evolved simultaneously" (Ingold, 1996) helps to understand the stronger and sometimes opposite relations established between farmers from Northeast with their livestock and with pumas: they complained and reacted to the losses, but they acknowledged the ecological importance and the hunting natural behaviour of pumas. Both relations result of the historic occupancy of the Northeast hinterland, before and after colonial domain (Andrade, 1998).

Another variable that can explain the negative scenario in South, comparing to Northeast, is that the interviews were applied to farmers who had faced depredation, while in Northeast interviewees were randomly selected, no matter this interaction had taken place or not. Usually those who had problems with wildlife show more aggressive and/or defensive feelings and behaviors. This fact could be seen in Northeast study area where negative attitudes toward jaguars and pumas increased 39.0% among those farmers who had problems with the species (Martins, 2020). Kansky and Knight (2014) and Heberlein (2012) explain that when someone is exposed to a species and that results in a negative interaction, exposition becomes experience and can be a strong predictor for attitude. If the experience is powerful it will be kept in memory, easily accessible once questioned, stressing again attitudes ('prominence').

The lowest income in Northeast can explain their claim for financial compensation for the losses caused by pumas, higher when compared to the southern farmers. In Northeast, in average, the herds are small, and often are the saving account of the family for an emergency. Thus, the attack of a puma may compromise the families' welfare and arise their perception of vulnerability related to predators' proximity. This scenario happens in some countries in Africa and Asia where communities with low incomes demand for compensation schemes to solve predators' problems even not believing that the government will pay fair prices or with a good effectivity (Dickman, 2011). Compensation schemes probably would not be effective in Brazil, a huge country in territory, with difficult access to remote but populated regions and with structural problems such as lack of equivalent professionals to North-American wildlife biologists who are prepared and entitled to check depredation episodes (Crawshaw, 2004). Moreover, at least in Northeast, due to extensive management of livestock and long distances covered by domestic animals, it is difficult to find the carcass, and if they are found it is difficult to clearly identify the *causa mortis*, because of the high temperatures and presence of scavengers (Cavalcanti et al., 2015). Some countries in Africa that have compensation schemes for depredations episodes often have complains from the communities and some are being rethought (Nyhus, 2016).

Despite of people asking for compensation schemes from government, in both studied regions in Brazil, farmers did not believe in the governmental authorities responsible for environmental issues including wildlife management. This becomes evident in their answers when they had depredation problems in their farms, they did not call for government or any other help. It is important to remember that the creation of some protected areas in Brazil deals with the challenge of communities living within and surrounding these areas (Diegues et al., 2000), which causes uncertainty, lack of trust and conflicts between stakeholders.

Farmers from both studied sites pointed out pumas as a threat to their livelihood, in South meaning a threat to the traditional husbandry practices. In South, half of interviewees had stopped to breed sheep because of pumas' depredation episodes by the year the study was done. In Northeast, one of the poorest regions of the country with many socio-economic problems, predators as pumas may add threat to rural communities once they can cause livestock losses to already vulnerable smallholders. In regions were husbandry practices are extensive (herds free to find water and graze), for economic constraints or secular traditions, carnivores are seen as a threat and thus, are being persecuted and killed in high rates as a preventive measure (Holland et al., 2018). Losses due to depredation, even with different economic impacts and in different socio-cultural contexts, have to be pondered over a multi-criteria analysis considering the effort that people have to make to coexist with wildlife (Dickman, 2011; Holland et al., 2018; Nyhus, 2016; Zimmermann et al., 2010). What are the cultural and economic factors involved? Why and how should communities change their livelihoods? They must get information, knowledge, training, assistance, and support, but first they need to be heard and their values, emotions, feelings, and norms must be considered by researchers, managers, and policymakers. Likewise hunting, dealing with cultural capital linked to practices that cannot be changed without effort and suffering, would not make sense to change only because of economic motif (Fernandéz-Armesto, 2010, p. 109).

Farmers from South showed more resistance to change their husbandry practices, even in an ideal scenario, when compared to Northeast farmers. The historical occupation in the region of *Campos de Cima da Serra* in the south occurred by a miscegenation of indigenous population with Portuguese settlers, followed by people from Brazilian southeast (called *Bandeirantes* in Portuguese) who used to pursue red Indians in the lack of slaves from Africa to slave labor on farms or to fight their battles against Imperial Brazil (Ambrosini et al., 2012). Since the 1700s, livestock has been a traditional activity in the region, managed extensively (Ambrosini et al., 2012). Once local population is historically established and the indigenous were mostly extinct by the new settlers, it is possible to infer that people hold a personal sense of herds' management, territory, and nature domination (Tuan, 1974). These facts can make them more suspicious of any novelty or change that may compromise the "progress" hardly conquered by their ancestors. Martins (2020) brought these psychological variables of neuro-system I and neuro-system II to the discussion in her study area in Northeast: "the probability fallacy versus plausibility" and "affections heuristic" (Kahneman, 2011). The ideal scenario to change husbandry is desired, possible, and detailed (puma-proof pens, water, and fodder supply the whole year) ('plausibility'), however, with low probability to be achieved by small farmer's own means ('probability fallacy'). Additionally, the comfort with the known, the attachment to a traditional husbandry system versus a "new technology" makes the first being highly valued and pointed as low risk by respondents, in opposition to the second one. Considering the same reasoning, once southern farmers started to have more problems with depredation attacks in the last years, they believed that the number of pumas increased. For them, the husbandry system has been successful for centuries, thus the problem is not livestock management but pumas' population abundance.

In Northeast predators and diseases were less mentioned as main causes of herds' mortality that in South. However, in Northeast long dry seasons and severe droughts are regular problems to farmers, contrasting with South where it rains regularly during the whole year and farmers do not face problems to supply water to their herds. In this context, drought threatens more Northeast livestock than carnivores. Even in South, predators were mentioned after diseases as main causes for livestock losses. Several studies reported that losses due to carnivores are about 3.0% of the whole threats faced by domestic herds in a year, being other causes related to herds' management, natural factors, and diseases more representative (Palmeira et al., 2008; Zimmermann et al., 2005).

The lowest income and education level in Northeast did not influence more negatively the answers related to pumas' ecological role. On the other hand, comparing interviewees of the same geographic region in South, the answers showed that as higher the education level of farmers was, more positives were the responses about pumas' ecological role. Zimmermann et al. (2005) in Pantanal and Conforti and Azevedo (2003) in the Atlantic Forest, also found that the education level affected respondents' opinion about the ecological importance of big cats in the same way as demonstrated in South study area. However, considering just the Northeast context, education level was not significantly correlated to knowledge about species. Martins (2020) suggested that this may be explained by the poor content about Brazilian wildlife in school curricula and/or the unbalance between other field areas compared to biology, ecology, natural history, etc. Thus, the author suggested that the above stated variables are independent: neither knowledge on species depends on education level nor on the existence of educational facilities as schools. Instead of being a constraint, following the guidelines of Brazilian Environmental Education Policy, researchers and conservationists working within and surrounding protected areas may take advantage of them as "education for conservation spaces" (MMA, 2005), what is even more desirable in regions with difficult access to knowledge and education offered by regular institutions.

Sillero-Zubiri et al. (2006) discussed that sometimes values associated with the use of wildlife can overcome education and economic factors. Considering this information and the data gathered in the results in this Chapter, we can infer that cultural influences can be stronger than education level and income level between contrasting regions, as Northeast and South Brazil are. Nevertheless, these educational influences need to be deeper investigated in similar cultural contexts to weight their strength, for instance in a vulnerability to HWC index.

Conclusion and Recommendations

Human-wildlife interactions, especially conflicts, are extremely important to biodiversity conservation programs. Conservation will not happen without peoples' engagement - their opinions, knowledge, attitudes, behavior, norms, and perceptions must be exhaustively investigated (Holland et al., 2018). However, unified protocols are required to better identify, describe, understand, and predict conflicts with carnivores, independently of geographic locations and socio-economic and cultural aspects.

Understanding the roots of conflicts between communities and felids as pumas is determinant to the conservation of these species in South and Northeast Brazil, as well as in other places where they still inhabit. Human dimensions may vary between geographic and socio-cultural and economic contexts or even between similar communities. It is urgent to find patterns and propose an index of vulnerability to HWC at different scales and levels of complexity (Nyhus, 2016). What can be done in communities more affected by the proximity of big cats? What causes the highest frequency and severity of conflicts in some communities? How can the conscious and manageable side of the conflict change habits, attitudes, and behaviors to influence positively the ecological functions and ecosystem services provided by wildlife? These questions need to be addressed and answered. It is necessary to deepen the studies comparing different sites and regions that have interactions with the same wildlife species, mainly charismatic carnivores such as pumas, to better understand psychological, socio-economic and cultural aspects that can lead to certain human behaviors.

This Chapter brought some shared variables, even though the protocols used for data collection were different. Just three similarities among 11 variables were found, what may explain the importance of socio-economic and cultural contexts in the human dimensions of wildlife. The same species did not arise the same responses from diverse and geographically far Brazilian regional communities. More suitable and comparable variables will be found once the aim of many researchers of having unified experimental design and protocols is achieved. Meanwhile, the relevance of the species, its prominence in peoples' minds, proximity of human settlements and conservation status worth the exercise of displaying differences and similarities of human dimensions of human-puma-conflicts in different Brazilian sites. This may be considered to support management plans for protected areas in both regions studied here, South and Northeast, and in other parts of Brazilian territory as well.

Acknowledgments

The authors are grateful to Rogério Dell'António, geographer, who kindly prepared Figure 3, the map of the study sites; Juliana Luz Passos Argenton, M.Sc. in Statistics who helped with the statistical analysis; and all the respondents from both Brazilian regions who shared their time and information to make this research possible. We also thank Chester Zoo for the sponsorship in Northeast work and SEMA-RS (*Secretaria Estadual do Meio Ambiente*) for the logistic support in South.

References

Alcântara, D. M. & Germani, G. I. (2010). As comunidades de fundo e fecho de pasto na Bahia: luta na terra e suas espacializações. *Revista de Geografia*, 27(1), 40–56. [The background communities and pasture closing in Bahia: struggle on land and its spatializations. *Geography Magazine*]

Alves, J. J. A., Araujo, M. A., & Nascimento, S. S. (2009). Degradação da Caatinga: uma investigação ecogeográfica. *Revista Caatinga*, 22 (3), 126–135. [Caatinga degradation: an ecogeographic investigation. *Caatinga Magazine*]

Alves, R. R. N., & Souto, W. M. S. (2011). Ethnozoology in Brazil: current status and perspectives. *Journal of Ethnobiology and Ethnomedicine*, 7, 22. doi:10.1186/1746-4269-7-22.

Alves, R. R. N. (2012). Relationships between fauna and people and the role of ethnozoology in animal conservation. *Ethnobiology and Conservation*, 1, 1–69. doi: 10.15451/ec2012-8-1.2-1-69.

Ambrosini, L. B., Miguel, L. de A., & Filippi. E. E. (2012). Evolução e diferenciação dos sistemas agrários nos Campos de Cima da Serra: Origem dos pecuaristas familiares produtores do queijo serrano. *Desenvolvimento e Meio Ambiente*. 26, 171-187. doi: 10.5380/dma.v26i0.26460. [Evolution and differentiation of agrarian systems in Campos de Cima da Serra: Origin of family farmers producing Serrano cheese. *Development and Environment*]

Andrade, M.C. (1998). *A terra e o homem no Nordeste*, 6.ª ed., Editora Universitária da UFPE: Recife, 334 p. [*The land and man in the Northeast*]

Atlas do Desenvolvimento Humano do Brasil. (2013). (2019, July 15) Retrieved from: http://www.atlasbrasil.org.br/2013. [*Atlas of Human Development in Brazil.*]

Azevedo, F. C. C. (2008). Food habits and livestock depredation of sympatric jaguars and pumas in the Iguaçu National Park area, South Brazil. *Biotropica*, 40, 494–500. doi: 10.1111/j.1744-7429.2008.00404.x.

Backes, A., Prates, F. L., & Viola, M. G. (2005). Produção de serapilheira em Floresta Ombrófila Mista, em São Francisco de Paula, Rio Grande do Sul, Brasil. *Acta Botanica Brasilica, 19*(1), 155-160. doi:10.1590/S0102-33062005000100015. [Litterfall production in a Mixed Ombrophilous Forest, in São Francisco de Paula, Rio Grande do Sul, Brazil. *Acta Botanica Brasilica*]

Bailey, K. D. (1982). *Methods of Social Research.* The Free Press, New York, NY.

Barbosa, J. A. A., Nobrega, V. A., & Alves, R. R. N. (2011). Hunting practices in the semiarid region of Brazil. *Indian Journal of Traditional Knowledge*, 10(3):486–490.

BRASIL. Bahia 2010. Tabela do Estado da Bahia do Censo demográfico de 2010. *Diário Oficial da União*, 04 nov. 2010. (2019, July 15). Retrieved from: https://www.ibge.gov.br/estatisticas/sociais/populacao/9662-censo-demografico-2010.html?edicao=9665&t=resultados. [Table of the State of Bahia from the 2010 Demographic Census. *Official Gazette of the Union*]

Broad, S. (1988). Species account. In: S. Broad, R. Luxmoore, M. J. (Eds.). *Significant trade in wildlife: a review of selected species in CITES appendix II* (pp. 80-109). Cambridge, UK: IUCN/ CITES.

Browne-Nunez, C., Jacobson, S. K., & Vaske, J. J. (2013). Beliefs, attitudes, and intentions for allowing elephants in group ranches around Amboseli National Park, Kenya. *Wildlife Society Bulletin*, 37(3), 639–648. doi: 10.1002/wsb.296.

Castelletti, C. H. M., Da Silva, J. M. C., Tabarelli, M., & Santos, A. M M. (2003). Quanto ainda resta da Caatinga? Uma estimativa preliminar. *Ecologia e Conservação da Caatinga* (pp. 777–796). Recife, Brasil: Editora da Universidade Federal de Pernambuco. [How much is left of the Caatinga? A preliminary estimate. *Caatinga Ecology and Conservation*]

Cavalcanti, S. M. C. (2009). Manejo e controle de danos causados por espécies da fauna. In L. J. Cullen, R. Rudran, and C. V. Padua (Eds.), *Métodos de Estudos em Biologia da Conservação & Manejo da Vida Silvestre* (pp. 617-632). Curitiba, PR: UFPR. [Management and control of damage caused by fauna species. In L.J. Cullen, R. Rudran, and C.V. Padua (Eds.), *Methods of Studies in Conservation Biology & Wildlife Management*]

Cavalcanti, S. M. C., Oliveira T. G. de, & Paula, R. C. de. (2015). Identificação do agente causador do problema. In: S. M. C. Cavalcanti, R. C. de Paula, and R. L. Gasparini-Morato (Eds.), *Conflitos com mamíferos carnívoros - uma referência para o manejo e a convivência* (pp. 19-28). Brasília: Instituto Chico Mendes de Conservação da Biodiversidade, ICMBio. [Identification of the agent causing the problem. In: S.M.C. Cavalcanti, R.C. de Paula, and R.L. Gasparini-Morato (Eds.), *Conflicts with carnivorous mammals - a reference for management and coexistence*]

Conforti, V. A., & Azevedo, F. C. C. (2003). Local perceptions of jaguars (*Panthera onca*) and pumas (*Puma concolor*) in the Iguaçu National Park area, south Brazil. *Biological Conservation*, 111, 215-221. doi: 10.1016/S0006-3207(02)00277-X.

Conover, W. J. (1999). *Practical Nonparametric Statistics*. 3ª ed. John Wiley & Sons Inc. Nova Iorque.

Crawshaw, P. G. (2004). Depredation of domestic animals by large cats in Brazil. *Human Dimensions of Wildlife*, 9, 329–330. doi: 10.1080/10871200490505774

de Matos Dias, D., Ferreguetti, Á. C., & Rodrigues, F. (2020). Using an occupancy approach to identify poaching hotspots in protected areas in a seasonally dry tropical forest. *Biological conservation*, 251, 108796. doi:10.1016/j.biocon. 2020.108796

Dickman, A. J., Macdonald, E. A., & Macdonald, D. W. (2011). A review of financial instruments to pay for predator conservation and encourage human–carnivore coexistence. *Proceedings of National Academy of Science of the United States of America*, 108, 13937–13944. doi: 10.1073/pnas.1012972108.

Diegues, A. C., Arruda, R. S. V., Silva, V. C. F., Figols F. A. B., & Andrade, D. (2000). *Biodiversidade e Comunidades Tradicionais no Brasil*. São Paulo, Brazil: NUPANB-USP, PROBIO-MMA & CNPq. [*Biodiversity and Traditional Communities in Brazil*]

Ditt, E. H., Mantovani, W., Padua, C. V., & Bassi, C. (2009). Entrevistas e aplicações de questionários em trabalhos de conservação. In L. J. Cullen, R. Rudran, and C. V. Padua (Eds.), *Métodos de Estudos em Biologia da Conservação & Manejo da Vida Silvestre* (pp. 617-632). Curitiba, PR: UFPR. [Interviews and application of questionnaires in conservation work. In L.J. Cullen, R. Rudran, and C.V. Padua (Eds.), *Methods of Studies in Conservation Biology & Wildlife Management*]

Dryflor, B.-R K., Delgado-Salinas, A., et al., (2016). Plant diversity patterns in neotropical dry forests and their conservation implications. *Science*, 353(6306), 1383-1387. doi:10.1126/science.aaf5080

Dström, H. E., et al., (2009). *The EU in the world: future research agendas*. v. 19.

Duarte, M. M., & Bencke, G. A. (2008). *Plano de Manejo da Estação Ecológica Estadual de Aratinga. Final Report.* Fundação Zoobotânica do Rio Grande do Sul, Porto Alegre, Brazil. [*Aratinga State Ecological Station Management Plan*. Final Report.]

Elbroch, L. M., & Wittmer, H. U. (2012). Table scraps: inter-trophic food provisioning by pumas. *Biology Letters*, 8, 776–779. doi: 10.1098/rsbl.2012.0423.

Engel, M. T., Vaske, J. J., Bath, A. J., & Marchini, S. (2016). Predicting Acceptability of Jaguars and Pumas in the Atlantic Forest, Brazil. *Human Dimensions of Wildlife*, 21(5), 427–444. doi: 10.1080/10871209.2016.1183731.

Engel, M. T., Vaske, J. J., Marchini, S., & Bath, A. J. (2017). Knowledge about big cats matters: insights for conservationists and managers. *Wildlife Society Bulletin*, 41(3), 398-404. doi: 10.1002/wsb.798.

Fernandez-Armesto, F. *Comida – uma história*. 2ª ed. (2010). Rio de Janeiro: Record. [*Food – a story*]

Heberlein, T. A. (2012). *Navigating Environmental Attitudes*. New York, NY: Oxford University Press.

Herrmann, T. M., Schüttler, E., Benavides, P., Gálvez, N., Söhn L., & Palomo N. (2013). Values, animal symbolism, and human-animal relationships associated to two threatened felids in Mapuche and Chilean local narratives. *Journal of Ethnobiology and Ethnomedicine*, 9, 41. doi: 10.1186/1746-4269-9-41.

Hoogesteijn, R., & Mondolfi, E. (1992). *El jaguar, tigre Americano*. Editores, Armitano. [*The jaguar, American tiger*]

Holland, K. K., Larson, L. R., & Powell, R. B. (2018). Characterizing conflict between humans and big cats *Panthera* spp: a systematic review of research trends and management opportunities. *Plos One*, 13(9), e0203877. doi:10.1371/journal.pone.0203877.

ICMBIO. (2013a). *Sumário Executivo Do Plano De Ação Nacional Para a Conservação Da Onça-Parda*. p. 1–7. [*Executive Summary of the National Action Plan for the Conservation of the Puma.*]

ICMBIO. (2013b). *Plano de ação nacional para a conservaçao da onça-pintada*. [*National action plan for the conservation of the jaguar.*]

Indrusiak, C., & Eizirik, E. (2003). Carnívoros. In C. S. Fontana, G. A. Bencke, & R. E. Reis. *(Eds.), Livro Vermelho das Espécies da Fauna Ameaçada de Extinção no Rio Grande do Sul* (pp. 523-525). Porto Alegre, RS: EDIPUCRS. [*Red Book of Endangered Fauna Species in Rio Grande do Sul*]

Ingold, T. (1996). Growing plants and raising animals: an anthropological perspective on domestication, In: D. R. Harris (Ed.), *The origins and spread of agriculture and pastoralism in Eurasia* (pp. 12-24). London, England: UCL Press.

Inskip, C., & Zimmermann, A. (2009). Human-felid conflict: a review of patterns and priorities worldwide. *Oryx*, 43, 18–34. doi: 10.1017/S003060530899030X.

INSTITUTO BRASILEIRO DE GEOGRAFIA - IBGE. *Panorama.* (2020, June 20). Retrieved from: https://cidades.ibge.gov.br/brasil/rs/panorama.

INSTITUTO BRASILEIRO DE GEOGRAFIA - IBGE. *Panorama. Censo Demográfico: Renda – Rio Grande do Sul*; 2010. (2020, June 20). Retrieved from: http://www.ibge.gov.br/home/estatistica/populacao/censo2010/indicadores_sociais_municipais/indicadores_sociais_municipais_tab_uf_zip.shtm. [*Panorama. Demographic Census: Income – Rio Grande do Sul*]

Kahneman, D. (2011). *Pensar depressa e devagar*. Bloco Gráfico Lda.: Maia. Translated by: Pedro Vidal.

Kansky, R., & Knight, A. T. (2014). Key factors driving attitudes towards large mammals in conflict with humans. *Biological Conservation*, 179, 93–105. doi: 10.1016/j.biocon.2014.09.008.

Kumar, N. (1992). *Friends, Brothers and Informants: fieldwork memories of Banaras.* Berkeley, CA: University of California Press.

Lang, M. (2013). *Change of land use and land cover by silvicultural expansion on the Campos de Cima da Serra, Rio Grande do Sul, Brazil*. (Master Thesis) Available from: Technische Universita¨t Mu¨nchen, Germany (2002/03-2008/09).

Leal, I. R., da Silva, J. M. C., Tabareli, M., & Lacher Jr, T. E. (2005). Changing the course of biodiversity conservation in the caatinga of northeastern Brazil. *Conservation Biology*, 19 (3), 701–706. doi: 10.1111/j.1523-1739.2005.00703.x.

Lozano, J., Olszańska, A., Morales-Reyes, Z., Castro, A. A., Malo, A. F., Moleon, M., & Kansky, R. (2019). Human-carnivore relations: A systematic review. *Biological Conservation, 237*, 480-492. doi:10.1016/j.biocon.2019.07.002.

Manfredo, M. J. (2008). *Who Cares About Wildlife? Social Science concepts for Exploring Human-Wildlife Relationships and Conservation Issues.* New York, NY: Springer. doi: 10.1007/978-0-387-77040-6.

Marchini, S. (2010). *Human dimensions of the conflicts between people and jaguars (Panthera onca) in Brazil.* (Doctoral Dissertation). Available from: Department of Zoology, University of Oxford: http://ethos.bl.uk/OrderDetails.do?uin=uk.bl .ethos.543030.

Marchini, S. & Crawshaw, P. G. Jr. (2015). Human–Wildlife Conflicts in Brazil: a fast-growing issue. *Human Dimensions of Wildlife*, 20, 323-328. doi:10.1080/ 10871209.2015.1004145.

Martins, C. S. G. (2020). *As dimensões humanas das relações entre pessoas e onças-pintadas (Panthera onca) e onças-pardas (Puma concolor) na Caatinga.* (Doctoral Dissertation). Available from: Department of Ecology, University of São Paulo: https://teses.usp.br/teses/disponiveis/91/91131/tde-04052020-161452/pt-br.php. [*The human dimensions of relationships between people and jaguars (Panthera onca) and pumas (Puma concolor) in the Caatinga*]

Mazzolli, M., Graipel, M. E., & Dunstone, N. (2002). Mountain lion depredation in the Southern Brazil. *Biological Conservation*, 105, 43–51. doi: 10.1016/S0006-3207(01)00178-1.

Michalski, F., Boulhosa, R. L. P., Faria, A., & Peres, C. A. (2006). Human–wildlife conflicts in a fragmented Amazonian forest landscape: determinants of large felid depredation on livestock. *Animal Conservation*, 9, 179–188. doi: 10.1111/j.1469-1795.2006.00025.x.

Morato, R. G., Mello Beisiegell, B., Ramalho, E. E., Campos, C. B., & Boulhosa, R. L. P. (2013). Avaliação do risco de extinção da Onça-pintada *Panthera onca* (Linnaeus, 1758) no Brasil. *Biodiversidade Brasileira*, 3(1), 122–132. [Extinction risk assessment of the jaguar Panthera onca (Linnaeus, 1758) in Brazil. *Brazilian Biodiversity*]

MMA – MINISTÉRIO DO MEIO AMBIENTE. (2005). *Encontros e caminhos: formação de educadores ambientais e coletivos educadores*. Brasília, DF: MMA. [*Meetings and paths: training environmental educators and educator collectives*]

MMA – MINISTÉRIO DO MEIO AMBIENTE, SBF – SECRETARIA DE BIODIVERSIDADE E FLORESTAS. (2007). *Áreas prioritárias para conservação, uso sustentável e repartição de benefícios da biodiversidade brasileira: Atualização – Portaria MMA nº 9, de 23 de janeiro de 2007.* (Série Biodiversidade, 31). Brasília: MMA. [*Priority Areas for Conservation, Sustainable Use and Benefit Sharing of Brazilian Biodiversity: Update – MMA Ordinance No. 9, of January 23, 2007*]

Murphy, T., & MacDonald, D. W. (2010). Pumas and people: lessons in the landscape of tolerance from a widely distributed felid. In D. M. MacDonald and A. J. Loveridge (eds.), *Biology and conservation of wild felids* (pp. 431-452). Oxford, England: University of Oxford.

Newing, H., Eagle, C. M., Puri, R. K., & Watson, C. W. (2011). *Conducting Research in Conservation: social science methods and practice*. Abingdon, Oxon: Routledge, Taylor and Francis Group.

Nuno, A., & St. John, F. A. V. (2014). How to ask sensitive questions in conservation: A review of specialized questioning techniques. *Biological Conservation*, 189, 5–15. doi: 10.1016/j.biocon.2014.09.047

Nyhus, P. J. (2016). Human-wildlife conflict and coexistence. *Annual Review of Environment and Resources*, 41,143-171. doi:10.1146/annurev-environ-110615-085634.

Oliveira, T. G., & Cassaro, K. (2006). *Guia de Campo dos felinos do Brasil*. São Paulo, Brasil: Instituto Pró-Carnívoros; Fundação Parque Zoológico de São Paulo; Sociedade de Zoológicos do Brasil; Pró-Vida Brasil. [*Brazil Cat Field Guide*]

Palmeira, F. B. L., & Barrella, W. (2007). Conflitos causados pela predação de rebanhos domésticos por grandes felinos em comunidades quilombolas na Mata Atlântica. *Biota Neotropical*, 07, 119–128. doi: 10.1590/S1676-06032007 000100017. [Conflicts caused by predation of domestic livestock by large cats in quilombola communities in the Atlantic Forest. *Neotropical Biota*]

Palmeira, F. B. L., Crawshaw, P. G. J., Haddad, C. M., Ferraz, K. M. P. M. B., & Verdade, L. M. (2008). Cattle depredation by puma (*Puma concolor*) and jaguar (*Panthera onca*) in central-western Brazil. *Biological Conservation*, 141, 118–125. doi: 10.1016/j.biocon.2007.09.015.

Queiroz, L. P. de, Cardoso, Fernandes, D. M. F., & Moro, M. F. (2017). Diversity and evolution of flowering plants of the Caatinga Domain. In: J. M. C. da Silva et al. (Eds.), *Caatinga Dry Forest*. (pp. 23). New York, NY: Springer International Publishing.

Ripple, W. J., & Beschta, R. L. (2006). Linking a cougar decline, trophic cascade, and catastrophic regime shift in Zion National Park. *Biological Conservation*, 33, 397–408. doi: 10.1016/j.biocon.2006.07.002.

Schulz, F., Oliveira, L. R., & Printes, R. C. (2014). Depredation of domestic herds by pumas based on farmer's information in Southern Brazil. *Journal of Ethnobiology and Ethnomedicine*, 10, 73. doi: 10.1186/1746-4269-10-73.

Schulz, F., Engel, M. T., Bath, A., & Oliveira, L. R. (2017). Human-Wildlife Interactions: The Case of Big Cats in Brazil. In O. N. Campbell (Ed.), *Biological Conservation in the 21th Century: a conservation biology of large wildlife* (pp. 31-57). New York, NY: Nova Science Publishers.

Seddon, A. W., Macias-Fauria, M., Long, P. R., Benz, D., & Willis, K. J. (2016). Sensitivity of global terrestrial ecosystems to climate variability. *Nature*, 531(7593), 229–232. doi:10.1038/nature16986.

Sillero-Zubiri, C., Sakumar, R., & Treves, A. (2006). In D. Macdonald, and K. Service (Eds.), *Key Topics in Conservation Biology* (pp. 255-272). Oxford, England: Wiley Blackwell.

Silva, J. M. C., Barbosa, L. C. F., Leal, I. R., & Tabarelli, M. (2017). The Caatinga: understanding the challenges. In: J. M. C. da Silva et al. (Eds.), *Caatinga Dry Forest.* (pp. 3-19). New York, NY: Springer International Publishing.

Silva, L. C. R., & Anand, M. (2011). Mechanisms of Araucaria (Atlantic) forest expansion into southern Brazilian grasslands. *Ecosystems*, 14, 1354–1371. doi: 10.1007/s10021-011-9486-y.

Théry, H., & Mello, N. A. (2005). *Atlas do Brasil: disparidades e dinâmicas do território.* São Paulo, Brasil: Editora Universidade de São Paulo. [*Atlas of Brazil: disparities and territorial dynamics*]

Tortato, F. R., Layme, V. M. G., Crawshaw, P. G., & Izzo, T. J. (2015). The impact of herd composition and foraging area on livestock predation by big cats in the Pantanal of Brazil. *Animal Conservation*, 18, 539-547. doi: 10.1111/acv.12207.

Tuan, Y. F. (1974). *Topofilia – um estudo de percepção, atitudes e valores do meio ambiente.* New Jersey, NJ: Prentice-Hall Inc.: Englewood Cliffs. [*Topophilia - a study of perception, attitudes and values of the environment*]

Zimmermann, A., Walpole, M. J., & Leader-Williams, N. (2005). Cattle ranchers' attitudes to conflicts with jaguars in the Pantanal of Brazil. *Oryx*, 39, 406–412. doi: 10.1017/S0030605305000099.

Zimmermann, A., Baker, N., Inskip, C., Linnell, J. D. C., Marchini, S., Odden, J., Rasmussen, G., & Treves, A. (2010). Contemporary views of human wildlife conflict. In: J. T. du Toit, R. Kock, & J. C. Deutsch (Eds.), *Wild Rangelands: Conserving Wildlife While Maintaining Livestock in Semi-Arid Ecosystems* (pp. 129-151). London, England: Wiley-Blackwell. doi: 10.1002/9781444317091.ch6.

In: Critical Research Techniques in Animal ...
Editors: Michael O'Neal Campbell et al.
ISBN: 978-1-53619-846-1

Chapter 7

CONFLICTS WITH PUMAS IN SOUTHERN BRAZIL: THE FARMERS' VIEW

Francine Schulz[1], Rodrigo Cambará Printes[2] and Larissa Rosa de Oliveira[1]

[1]Universidade do Vale do Rio dos Sinos,
Mammal Ecology Laboratory, São Leopoldo, RS, Brazil
[2]Instituto Chico Mendes de Conservação da Biodiversidade (ICMBio) -
Núcleo de Gestão Integrada Aparados da Serra Geral,
Cambará do Sul, RS, Brazil

ABSTRACT

Large carnivores such as pumas are often killed in conflicts with humans because they prey on domestic livestock. Habitat loss, partly driven by the increasing use of traditional pasture systems, makes livestock vulnerable to puma attacks. The aim of this study was to describe the conflict between local farmers and pumas in a mosaic of Protected Areas in southern Brazil. We hypothesized that the farmers' perception, knowledge and attitudes towards the conflict with pumas is affected by socioeconomic variables, such as age, education, monthly income and farming experience. Forty-five face-to-face interviews with local farmers were performed in 2011, using a structured questionnaire with 16 open and 26 closed questions focusing on the perceptions of farmers. Our results show that the majority of the local population considered the conflict with pumas a serious problem and thought that attacks by pumas on domestic herds should be controlled with the involvement of government authorities. Financial losses caused by puma attacks on farms did not influence the attitudes of farmers, and knowledge about pumas was more influenced by social variables such as age and educational level. Meetings with the local Rural Consulting Council revealed that conflict with pumas still remains in the region. In this context, a long- term educational program with local farmers is highly recommended, focused on engaging the community in the discussion about possible mitigations tools. Conservation of wildlife depends on the ability to provide decision makers with both academic and traditional knowledge which could build bridges between the community and Environmental Agencies.

INTRODUCTION

Human–wildlife conflicts occur across the world and are a fast-growing issue in Brazil (Marchini & Crawshaw, 2015). Negative interactions between humans and wildlife are increasing in both frequency and severity, mostly due to uncontrolled human population growth and expansion in areas under cultivation and livestock husbandry, mainly around Protected Areas (Madden & McQuinn, 2014; Nyhus, 2016). The frequency of encounters and other interactions between people and wildlife usually results in conservation problems that vary from negative perceptions and attitudes, to the persecution and killing of the wildlife involved (Nyhus, 2016). The perception of damage, however, varies between stakeholders and, therefore, cannot be considered independently from psychological and socio-cultural variables (Manfredo et al., 2014).

Understanding the views of different people is both urgent and fundamental for any conservation strategy to succeed (Schultz, 2011) and here we explore some concepts: 1) Human perceptions are analyzed as beliefs, whether rational or irrational, held by an individual about the chance of occurrence of something (Swim et al., 2009); 2) Beliefs are based upon specific value orientations (Stern, 2000); 3) Attitude is considered as the unfavorable or favorable judgment that an individual shows about a situation, action or object (Vaske & Manfredo, 2012); 4) Knowledge can be empirical when it comes from stories or myths or factual when something is assumed to be correct or wrong (Perry et al., 2014). Together all these elements, attitude, perception and knowledge determine how a person will act, defining the behavior (Stern, 2000).

Ethnobiology is one of the sciences that studies these interactions and can use popular knowledge and attitudes for the conservation of natural resources (Souza & Alves, 2014). One of the ethnobiological tools is the structured questionnaire that is used to interpret the respondent's thoughts and actions about the research subject (Ditt et al., 2009; Holland et al., 2018). Ethnozoology, as a branch of this science, studies the roots of the first relationships between humans and other animals from the past and into the present, analyzing the different interactions that human cultures maintain with animals (Alves & Souto, 2011). The importance of ethnobiological studies and social science research strategies focused on human responses have been recognized as crucial to developing trust with local communities, and for developing conservation plans (Alves, 2012; Holland et al., 2018).

The last two centuries have witnessed the decline in numbers of various mammal species which interact with people. This is probably driven by a growing human population, which has led to increasing habitat loss (Manfredo et al., 2014). Due to these issues, large carnivores such as pumas (*Puma concolor*) have lost their habitats and prey, leading them to coexist with human populations and domestic livestock

(Holland et al., 2018). During domestication domestic animals have lost most of their natural instincts, which makes them easy prey for wild animals such as big cats (Marchini & Crawshaw, 2015). Preying on domestic livestock has led to negative interactions between species such as pumas and humans. This has become a serious conservation problem for these species which are often severely threatened.

Large carnivores are repeatedly killed worldwide due to conflicts with rural communities (Cavalcanti, 2009; Holland et al., 2018; Inskip & Zimmermann, 2009; Michalski et al., 2006). Studies have shown that contact between rural human populations and large felids has resulted in purposeful elimination of these animals which are killed preventively to protect domestic herds, or are hunted for sport (Holland et al., 2018; Mazzoli et al., 2002; Palmeira & Barrella, 2007). This kind of conflict is one of the most urgent wildcat conservation issues, thus, human perceptions and interactions with these animals need to be more studied to conserve felid populations worldwide (Herrmann et al., 2013; Holland et al., 2018; Zimmermann et al., 2010).

In Brazil, pumas coexist with jaguars (*Panthera onca*) and both are being poached by farmers in retaliation for preying on domestic livestock (Engel et al., 2017). Some studies on this type of conflict have been conducted in different regions of Brazil such as in the *Pantanal* (e. g. Tortato et al., 2015; Zimmermann et al., 2005), the Amazonian Forest (e. g. Michalski et al., 2006; Marchini, 2010), *Cerrado* (e. g. Palmeira et al., 2008) and the Atlantic Forest (e. g. Palmeira & Barrella, 2007; Engel et al., 2017). In south Brazil, studies have been concentrated in the region of *Foz do Iguaçú* National Park (Azevedo, 2008) in the Atlantic Forest and in the coastal mountains of *Campos*, grassland and native *Araucaria angustifolia* forest (Mazzolli et al., 2002; Schulz et al., 2014).

The southern region of Brazil has a strong agricultural tradition. The state of Rio Grande do Sul, where this study was done, is one of the most deforested states in Brazil, with just a few remaining forest areas (Indrusiak & Eizirik, 2003). The population of pumas there has been reduced to a small number of individuals, inhabiting mainly the edges of the northeastern plateau and the region in the border with the Argentine forests, where there are also wandering jaguars inhabiting mainly the denser side of the Argentine forest (Indrusiak & Eizirik, 2003; Oliveira & Cassaro, 2006).

Despite their relevance to conservation biology, the conflicts between people and wild animals have been little explored from an ethnobiological perspective. Even with well-known and frequent interactions between Brazilian fauna and local populations, ethnobiological studies are scarce, especially in southern Brazil (Alves & Souto, 2011). In this context, this study reports on the conflict between local people and the depredation of livestock by pumas in the southern region of Brazil, focusing on the views of farmers about the losses on their farms, the conflict in the region, and towards the pumas themselves.

We hypothesized that 1) the farmers who gave up breeding any kind of livestock would be the same farmers that suffered a higher frequency of puma attacks as well as those who suffered higher economic losses; 2) the younger farmers with higher education levels would have more knowledge about pumas; 3) the farmers living for a longer time in the region and with a higher monthly income would have more knowledge about pumas; 4) the respondents who had higher economic losses caused by pumas would hold the most negative attitudes towards pumas; 5) the farmers who thought that it was necessary to control puma attacks on farms would accept suggestions to avoid these attacks; and 6) the farmers who would accept suggestions to avoid puma attacks would be the same farmers that had suffered higher economic losses.

METHODS

Study Area

This study was conducted in a mosaic of Protected Areas in the South Brazilian plateau, in the northeastern state of Rio Grande do Sul, Brazil, from approximately 50º01'W to 50º 42'W and from 28º55'S to 29º33'S. This region is known in Portuguese as the *Planalto das Araucárias* and contains 11 Protected Areas (Figure 1). The regional climate has been described as subtropical, with moderate temperatures (annual average between 14ºC and 20ºC) and high annual precipitation (more than 2.000 mm) without a marked dry season (Duarte & Bencke, 2008; Silva & Anand, 2011). The northeastern region of this state represents the southern limit of the Atlantic Forest biome in Brazil, and is characterized by forest–grassland mosaics composed of native *Campos* grassland and native *A. angustifolia*, forming some slopes covered with forest and plateaus surrounded by steep cliffs (Duarte & Bencke, 2008; Silva & Anand, 2011). Fire and grazing are the main disturbances that prevent forest expansion (Müller et al., 2012). Afforestation of grasslands with pine plantations, logging of *A. angustifolia* forests, and large-scale conversion of native vegetation into agricultural land have also altered the landscape (Behling & Pillar, 2007).

The economy in the region is mainly based on forestry, traditional livestock rearing and small-scale agricultural practices. An extensive system is widely used for livestock husbandry, which demands low economic investment and relies on natural sources of food and water to provide for the livestock. Raising livestock in the region has been practiced for several generations and is considered a traditional way of life (Duarte & Bencke, 2008). The human population density is low (16,8 hab/km^2), with large areas of pastures and plantations (Duarte & Bencke, 2008).

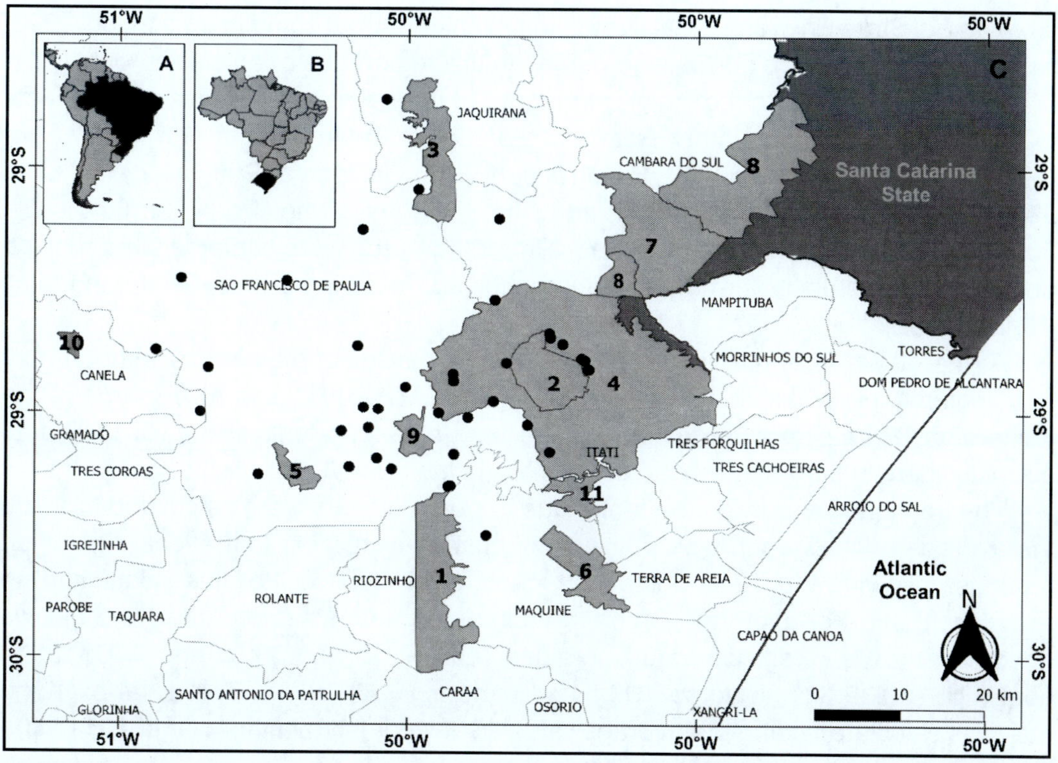

Figure 1. Study area in southern Brazil: where the black dots represent the interview sites, and each number refers to a different Protected Area: 1 – *Riozinho* State Environmental Protection Area; 2 – *Aratinga* State Ecological Station; 3 – *Tainhas* State Park; 4 – *Rota do Sol* State Environmental Protection Area; 5 – *Ronda* Municipal Park; 6 – *Serra Geral* State Biological Reserve; 7 – *Aparados da Serra* National Park; 8 – *Serra Geral* National Park; 9 – *São Francisco de Paula* National Forest; 10 – *Canela* National Forest; 11 – *Pró-Mata* Centre for Research and Nature Conservation.

Data Collection and Analyses

We conducted 45 face-to-face interviews with local farmers in 2011 guided by a standard questionnaire, with 26 closed and 16 open questions (Supplementary Material). These questions were distributed in five different categories: (1) the socio-economic profile of the interviewees; (2) characteristics of puma attacks on farms; (3) farmer's perceptions; (4) farmer's knowledge, including an illustrative board; and (5) farmer's attitudes.

To conduct the interviews and efficiently collect reliable data we followed the same methodology and strategies of Schulz et al., (2014). All the answers given by respondents related to the years of 2008 until 2011, and notes were taken as faithfully as possible of the respondent's answers. Despite the time-lapse since the interviews were conducted, this is still a timely study since, according the local Rural Consulting Council the conflict continues to the present day (E. E. Fleck, Substitute manager of

Rota do Sol State Environmental Protection Area, personal communication, July 2020) and no similar research addressing pumas in the region has been carried out.

As a starting point to the study, managers of three Protected Areas were selected to be the first contacts: the *Rota do Sol* State Environmental Protection Area (29°23'22.58"S; 50°11'3.90"W), the *Aratinga* State Ecological Station (29° 18'S and 29°24'S; 50°11'W and 50°17'W) and the *Tainhas* State Park (29°05'35.33"S; 50°21'51.55'W). The managers of these Protected Areas were contacted and provided details of the first local farmers to be interviewed, following a snowball methodology (Bailey, 1982).

At the end of the interviews, an illustrative board (Supplementary Material) with four pictures of Brazilian felids – jaguar, puma, ocelot (*Leopardus pardalis*) and jaguarondi (*Puma yagouaroundi*), was shown to respondents to test if they could correctly identify a puma and had some knowledge about local wildlife.

The economic losses on the farms due to puma attacks were calculated based on the average price of each type of livestock animal, according to the local agricultural markets, following the methodology of Schulz et al., (2014). For this calculation, the respondents provided information on the number of animals depredated and this was multiplied by the reference value obtained (calf = U$ 255.68; sheep = U$132.39; goat = U$132.39; young horse= U$ 549.43 and pig = U$48.30). The dollar exchange rate was calculated using the average rate (U$ 1 = R$ 1.76) in the year of 2011, when the interviews were conducted in Brazil.

The educational level of respondents was divided into two categories to run the analysis: low, for those who were educated until the end of primary school and high, for those who had high school and college degrees.

Monthly income was also separated into two categories to run the analysis: low, for those earning between one (U$ 312.50) and four (U$ 1,250.00) times the minimum Brazilian wage, and high for those earning more than five times the minimum wage monthly (U$ 1,562.00).

At the end of each interview every respondent signed a consent form to declare that all information was given freely and the data would be used anonymously and only for research purposes. One copy was kept with the interviewer and the other with the respondent.

The study was approved by an Ethical Committee (protocol number: 051/2011), according to a Brazilian Federal Law (CNS 196/96) concerning studies that involve human participants.

All data collected from different questionnaire sections were quantitatively analyzed, using the frequency of the answers. For open questions, the most repeated answers were grouped and analyzed by their frequency.

All hypotheses mentioned in the introduction were tested using the software BioStat 5.0 for chi-Square and Person's Correlation tests.

RESULTS

Interviewee's Profile

The age of the respondents varied between 17 and 83 years old (SD = 15.5). The majority were men (86.7%), the owners of the farms (64.4%) and had lived in the region for 30 years or more (66.7%). The predominant educational level of interviewees was until completion of high school (37.8%) and the main occupation and source of income was based on agricultural practices (59.6%). The monthly income was notably low for 33.3% of the interviewees who declared that they earned between one and two times the regional minimum wage (U$ 312.50 and U$ 625.00), while just 22.2% of respondents earned more than U$ 1,875.00 dollars monthly. The median monthly income per person in the state where the study was carried out was U$ 647.43 in 2011 (IBGE, 2010). More details of the socio-economic profile of respondents are given in Table 1.

Table 1. Interviewee's socio-economic profile. N = number of respondents. % = percentage of answers

Categories		N	%
Place of the interview	On the farm	26	57.8
	In the city	19	42.2
Status on farm	Owner	29	64.4
	Wife or son of the owner	13	28.9
	Herd keeper	3	6.7
Gender	Male	39	86.7
	Female	6	13.3
Age	≤20	3	6.7
	21-40	3	6.7
	41-60	21	46.7
	61-80	16	35.6
	>80	2	4.4
Time living in the region (years)	≤5	2	4.4
	6-10	1	2.2
	11-20	6	13.3
	21-30	6	13.3
	>30	30	66.7
Educational level	Incomplete basic education	10	22.2
	Basic education	8	17.8
	High school	17	37.8
	College	10	22.2
Main occupation	Small-scale farmer	8	17.0
	Cattle breeder	20	42.6
	Merchant	4	8.5
	Other	15	32.0
Monthly income (minimum wage)	1-2	15	33.3
	3-4	8	17.8
	5-6	12	26.7
	>6	10	22.2

Characteristics of Puma's Attacks on Farms

Fifteen respondents (33.3%) suffered economic losses in more than one year of the study and twenty-eight (62.2%) had animals killed by pumas only in one specific year. Taking into account all years with attacks reported by respondents, we found that attacks were most frequent in 2008 (Table 2), and sheep farmers were the most affected by puma attacks.

All respondents were also asked about how they knew which animal was responsible for the attacks on their animals and they gave correct answers about the puma's pattern of carcass consumption, footprints and ecological behavior such as covering carcasses with leaves and sticks (Cavalcanti et al., 2015).

Table 2. Depredation scenario and economic losses caused by pumas according to local farmers in the study area and in the respective year

Year	Number of farms	Number of domestic animals killed	Losses (U$)
2008	26	155	23,515.34
2009	18	102	15,077.27
2010	16	150	22,915.91
2011	12	54	8,646.59

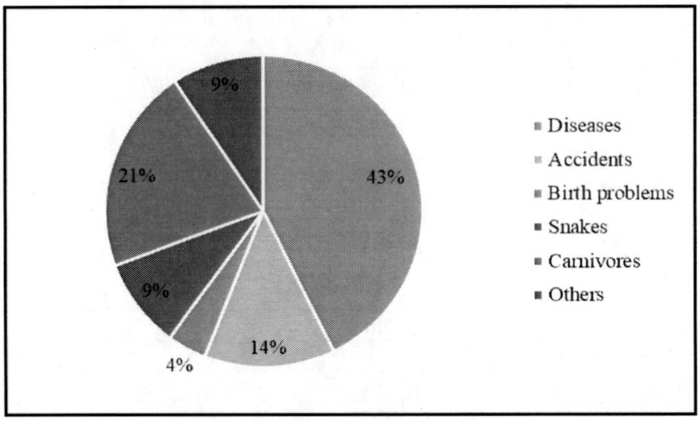

Figure 2. Main cause of livestock mortality in a year according to the local farmers interviewed.

Fifty eight percent of those farmers who suffered some kind of economic losses related to the depredation of domestic animals decided to stop breeding sheep and goats, the most attacked livestock by pumas. There was a positive correlation between the number of farmers who stopped breeding some sort of domestic animal and the frequency of puma attacks on the property ($r = 0.9755$; g.l. = 2; $p < 0.05$). In the same way, there was an association between the financial losses caused by pumas and the attitude to stop breeding sheep on farms ($X2= 7.190$; g.l. = 2; $p < 0.05$). Thus, our results corroborated the first hypothesis, which suggests that farmers who gave up breeding any kind of animal were the same farmers that suffered a higher frequency of puma attacks and those with higher economic losses. However, diseases were the most cited problem (43.0%) when farmers were questioned about the main cause of mortality among their livestock each year (Figure 2), with carnivores (including pumas) only being mentioned by 21.0% of the respondents.

Farmers' Perceptions

Most respondents (80.0%) said they did not know the number of pumas living in the region. When asked about the number of pumas living in the area during the past 10 years (from 2000 to 2001), 78.8% answered that the puma population had increased. The reasons for this perception of abundance were related to 46 answers, the most frequent being that "the pumas were released by the government inside the Protected Areas in the region" (Figure 3). Only three interviewees said that puma population had decreased, because they did not see pumas killing animals next to, or inside, their farms.

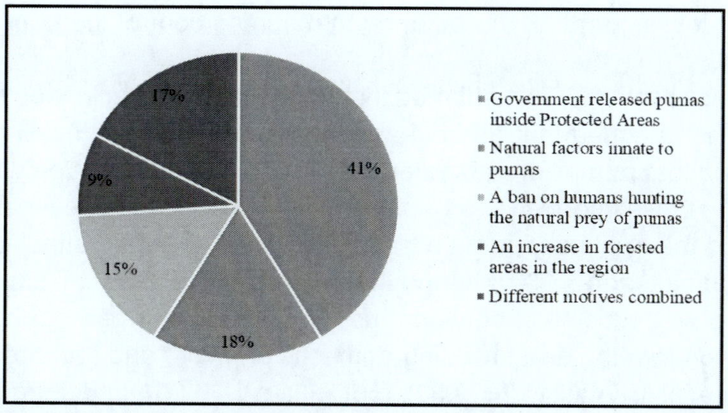

Figure 3. The perception of respondents regarding the increase in the puma population in the region during the last 10 years.

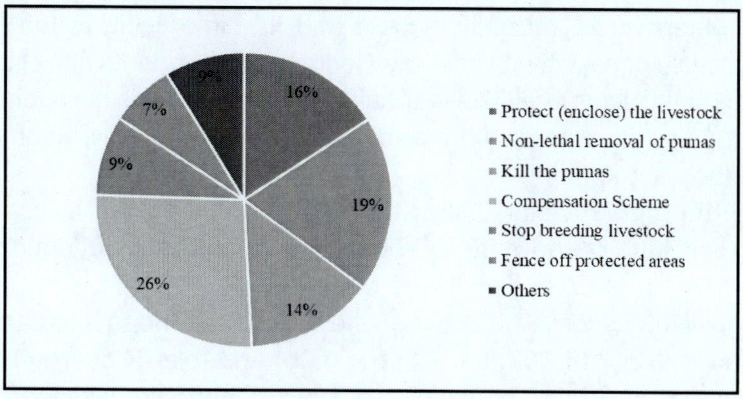

Figure 4. Opinions given by the respondents on how to solve the problem of depredation on domestic livestock by pumas on local farms.

The opinions about what could be done to solve the problem of pumas preying on domestic animals included the compensation scheme (26.0%) and non-lethal removal of pumas (19.0%), which were the most cited options (Figure 4).

When questioned about who should pay for the losses caused by pumas, the majority (73.3%) believed that the government should provide a compensation scheme, while other respondents said that nobody should be responsible or that the farmers should bear the losses (26.7%).

Farmers' Knowledge

Responses to the question about the importance of pumas in nature were: 26.7% stated they have no importance at all; 44.4% thought they have some importance simply because they exist; 15.6% did not know; and 13.3% believed they are very important.

When asked about the role of pumas in nature, 37.8% of people answered that pumas kill other animals to feed themselves; 28.9% said pumas just cause damage and economic losses; and 28.9% affirmed that pumas control the populations of their natural prey.

Thirty respondents (66.7%) affirmed that pumas do not attack people without being provoked, 31.1% claimed that as long as a puma is not teased it will not attack you, and 31.1% said that pumas are afraid of people and they always flee when they see a human. Among those interviewees who said a puma could attack even if not provoked, 41.2% claimed that a hungry puma always attacks, 17.6% said attacking is an innate behavior for pumas, and 5.9% affirmed that a female with cubs will always attack.

There was no association between the monthly income of respondents that was based mostly on farming and breeding domestic animals, and their knowledge about the importance of pumas ($X2= 3.32$; g.l. = 3; $p > 0.05$) and their role in nature ($X2 = 4.244$; g.l. = 3; $p > 0.05$). There was also no association between the time that respondents had lived in the region and their knowledge about the importance of pumas ($X2= 3.492$; g.l. = 3; $p > 0.05$) or their role in nature ($X2 = 5.737$; g.l. = 3; $p > 0.05$). In other words, monthly income and the time living in the region did not influence, positively or negatively, the knowledge of respondents about the importance and the role of pumas in nature. These results refute hypothesis number 2, that farmers living in the region for a longer time and with a higher monthly income would have more knowledge about pumas.

However, there was a significant association between the educational level of respondents and their knowledge about the importance of pumas in nature ($X2 = 9.719$; g.l. = 3; $p < 0.05$) and their role ($X2 = 12.449$; g.l. = 3; $p < 0.05$). There was also a significant association between the age of respondents and their knowledge of the importance ($X2 = 14.607$; g.l. = 6; $p < 0.05$) and role ($X2 = 16.482$; g.l. = 6; $p < 0.05$) of pumas in the wild. The lower the educational level was and the older the respondents were, the lower was their knowledge about the importance and role of pumas in nature. Thus, hypothesis number 3 was corroborated as younger farmers with higher educational levels demonstrated greater knowledge about pumas.

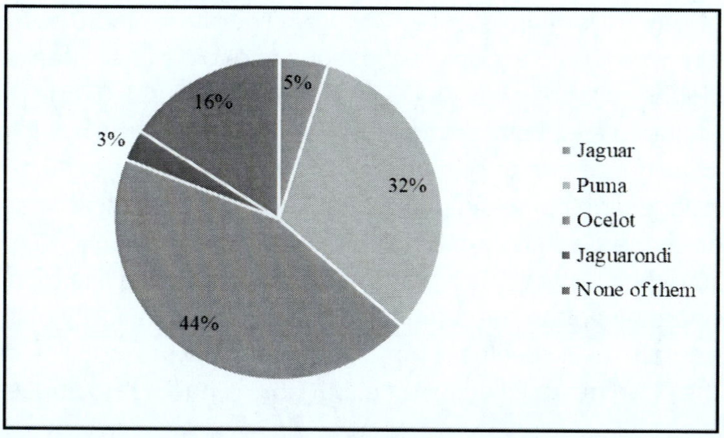

Figure 5. Felids sighted by interviewees at least once in the region.

The action of killing a puma was mentioned as illegal by 73.3% of respondents, as illegal or legal depending on the situation by 22.2%, and 4.4% said they did not know.

The questions related to the illustrative board showed that most respondents could correctly recognize a puma as 42 (93.3%) indicated Figure B (a puma) as the correct option. One interviewee (2.2%) pointed to Figure A, that was a jaguar, and two (4.4%) respondents indicated Figure D, a jaguarondi. Regarding the question about which felid of the four options they had already seen at least once in the region, pumas (62.2%) and ocelots (44.4%) were the most frequently cited (Figure 5).

Farmers' Attitudes

The dominant attitude (presented as the feeling of the respondents) towards pumas was indifference (53.3%), followed by a negative attitude (26.7%) and a positive attitude (20.0%). The economic losses caused by pumas were not associated with the attitude of respondents towards them (X2 = 0.528; g.l. = 4; $p > 0.05$). In other words, the negative attitudes towards pumas were not directly related to the financial losses caused by them on farms. Thus, hypothesis number 4, that respondents who suffered higher economic losses caused by pumas would hold the most negative attitudes towards them, was not confirmed. The majority of respondents (86.7%) said that puma attacks on domestic livestock must be controlled or stopped somehow.

There was no association between those farmers who found it necessary to control the attacks and those who would accept suggestions to avoid the attacks (X2 = 0.865; g.l. = 1; $p > 0.05$). In other words, even those who would not listen to any advice thought it was necessary to control the situation. Thus, hypothesis number 5, that farmers who thought it was necessary to control puma attacks on farms would accept suggestions to avoid these attacks, was refuted.

Although 64.4% of respondents said they would accept suggestions on how to avoid puma attacks if they received technical visits from experts in animal control, 44.4% said that they would not change husbandry practices, and 68.9% declared that they would not pay for any suggested change. Therefore, there was no association between farmers who would accept suggestions to avoid puma attacks and those who had suffered higher economic losses ($X2 = 3.746$; g.l. = 2; $p > 0.05$). Those who would accept suggestions to avoid attacks would do so regardless the magnitude of financial losses they had suffered due to pumas. Thus, hypothesis number 6, that farmers who would accept suggestions to avoid puma attacks would be the same as those that had suffered higher economic losses, was also refuted.

Twenty-six respondents (57.8%) affirmed that they would like to learn more about pumas if they were given the opportunity, indicating the possibility of implementing an educational program with the local population despite the conflict. All respondents declared that they had never killed a puma, but 55.6% had heard about puma poaching in the area.

Discussion

Felids have always been subject of human fascination and fear, and there is a long history of conflicts (Herrmann et al., 2013). Understanding the dynamics of a human–wildlife conflict is challenging, and resolving conflicts can be even more difficult, especially when carnivores such as big cats are involved (Zimmermann et al., 2010). The existence of the conflict between pumas and farmers in the northeastern plateau in south Brazilian was clear, since many people in the area were able to provide information and suggest farmers who would be willing to be interviewed. Furthermore, the losses on farms and the frequency of puma attacks were determinants to the decision to stop sheep breading on some farms. On the other hand, the economic losses caused by puma attacks on farms did not influence in the acceptance of suggestions on how to avoid the puma attacks on farms, and the small number who would accept suggestions were willing to pay for it by themselves. Knowledge about pumas was most strongly influenced by social variables such as age and educational level, but not by monthly income or by time spent living in the region.

Most respondents did not suggest how many pumas were living in the region (perception of abundance), but at the same time most of them (78.8%) said that the number of pumas had increased in the region over the last ten years (from the year the study was done: 2000-2001). The releasing of pumas by Environmental Agencies was cited as the main reason (41.0%) for this growth, followed by natural factors, prohibition of hunting activities and the increase in forested areas. Similar results were found next to the *Foz do Iguaçu* National Park, where the respondents also suggested that Environmental Agencies had released jaguars and pumas in the area (Conforti & Azevedo, 2003). As a potential explanation for this kind of perception the authors stated that a recent television documentary had shown some jaguars being released in this park after being fitted with GPS collars. These animals were originally captured inside the park and were not brought in from elsewhere, but probably due to the low level of education, people did not understand the context (Conforti & Azevedo, 2003). Similar confusion could be inferred in our study area, since people believed that the government had released pumas in the area. Nevertheless, since there are no studies about puma abundance in the area, we cannot say whether there are more or fewer individuals in the area that in previous years. The perception of farmers that the puma population was increasing during the study period could be related to the frequency of puma attacks on livestock in local farms, especially because most respondents said that pumas started to attack about ten years before the interviews were carried out. Mazzoli (2005) described a similar situation in the state of Santa Catarina, in a region with an analogous landscape and cultural background.

A great number of respondents (66.7%) had a good knowledge of the risk of pumas, declaring that they do not attack humans in normal situations. They said that a puma will attack only if it feels provoked or scared. Similarly, the respondents of Conforti & Azevedo (2003), living in the area next to *Foz do Iguaçu* National Park, did not consider pumas as a potentially dangerous animal to humans. However, some studies in North America have shown that, although rare, puma individuals which are less capable of hunting, injured or old can attack humans if they have the opportunity (Murphy & MacDonald, 2010), but this was not mentioned by any of the interviewees.

When questioned about who should pay for the losses caused by pumas, the federal government was the most cited agency in the answers. At the time the study was carried out, there was a strong conflict in the region between the local community, which was formed mainly of farmers, and Federal Environmental Agencies. This conflict was caused by the perceived negative impact of environmental laws on local rural industries such as livestock rearing, potato cultivation or harvesting timber from pine plantations. Despite this, there were a large number of people living within some Protected Areas that were required to move out according to the law enforcement and they were struggling against this (Duarte & Bencke, 2008).

The role of pumas in nature was classified as controlling natural prey species by only 21.3% of the respondents, and this was the answer with the strongest conservation focus among the alternatives. Conforti & Azevedo (2003) found similar results (24.3%) for the same question about jaguars and pumas in the Atlantic forest. The monthly income of respondents and their time living in the region were not associated with the knowledge respondents had about the role and importance of pumas in nature. On the other hand, the educational level of respondents and their age were associated with their knowledge about pumas. Zimmermann et al., (2005) and Conforti & Azevedo (2003) also found that the educational level and age of respondents had the largest effect on the opinions of respondents about the ecological importance of large felids in the *Pantanal* and Atlantic forest. According to Conforti (2006) older people tend to be more negative towards carnivores because they grew up in a time when people expressed fewer positive attitudes towards nature and animals in general. A low educational level was also associated with a more negative and utilitarian view of other carnivores, such as wolves in the studies of Bjerke et al.,

(1999). Herzon & Mikk (2007) found that farmers in the hinterland of some countries in Europe who had a low educational level, did not have knowledge about the role or importance of animals in nature in general.

Usually, people closely connected to agricultural cultures tend to evaluate nature based on utilitarian principles so that, if something does not bring direct benefits and also causes financial losses, its presence will not be accepted (Diegues et al., 2000). This could be exactly the cause of the responses to pumas in the study area, because they are the biggest carnivores in the region, and have the potential to cause financial losses to farms. Moreover, pumas hunting prey species was also spontaneously cited in a negative way by respondents, and this corroborates the utilitarian value that local people have about nature and fauna. Pumas were also cited as hunters of some prey species that are taken from inside the farms and this could indicate that pumas are seen as potential competitors, even if this view is unconsciously held (Foster et al., 2014).

We can assume that even though the respondents even did not agree with the environmental laws, they knew that killing wild animals was illegal in Brazil. Most farmers said that killing a puma is not allowed by law but this is incorrect. The Brazilian Federal Law 9.605/1998 about Environmental Crimes states that is not a crime to kill an animal to protect a human life or livestock. However, all the respondents said they had never killed a puma.

Sixty two percent of respondents said they had already seen a puma. This, together with the correct identification of pumas by recognition of illustrations, and correctly describing the feeding pattern of pumas when questioned, indicates that respondents gave correct and trustful interviews. However, some of the attacks described by respondents as being carried out by pumas could have been confused with wild boar (*Sus scrofa*) or even domestic dogs attacks, in cases when the carcasses were discovered in an advanced state of decomposition.

The predominant attitude towards pumas was indifference or generally negative and attitudes were not associated with financial losses on the farms of respondents. Llhanos et al., (2020) also found generally more neutral or negative attitudes towards pumas in their research in Patagonia, Argentina. It is important to consider that perceptions and attitudes towards a species are not always related to economic factors, but can be related to sociocultural beliefs, based on personal experiences or come from information given by past generations (Engel et al., 2017; Hill, 2004).

The suggestions to control puma attacks highlighted perspectives towards Environmental Agencies in the area. In almost all their answers the government was involved in some way: either funding a compensation scheme, removing pumas, using lethal control or even creating fenced Protected Areas.

Considering the suggestions given by the respondents, the creation of a compensation scheme could be seen as complicated, because it requires a qualified and trained team to go promptly to the farms and check the evidence of attacks and investigate the carcasses before they decompose (Crawshaw, 2004). There is only one case in Brazil in which financial compensation was offered to local farmers due to jaguar and puma attacks on cattle herds in the *Pantanal* biome. However, this was not paid by a national environmental agency, but instead by a local non-governmental organization, the Jaguar Conservation Fund (Silveira, 2004). The compensation covered a period of one year and the recipients from eleven farms reported 51 cases of depredation on their livestock. Among these cases only 14.0% were not confirmed as attacks. However, in total the losses caused by jaguars and pumas represented just 0.19% from the total number of livestock animals on the farms. Silveira (2004) suggests that the compensation scheme was a good measure. However, the Brazilian government does not have adequate structures in place to conduct such a program, and it is likely that not all farmers would be satisfied with the money given because the amount would have to follow a payment structure according to the type of livestock, without considering other livestock characteristics that affect prices in the rural market such as age, sex or breed (Crawshaw, 2004). Analyses of schemes which essentially pay for tolerance of wild predators have suggested that paying for local communities to conserve wildlife may be more effective, especially in poor communities (Woodroffe et al., 2007; Dickman et al., 2011).

The removal or translocation of problematic animals was also suggested by local farmers, but it has been shown to be inefficient in a great number of cases, because felids that had learned how to attack and feed on easy prey such as domestic animals in one area, will possibly do it again in their release site. For this reason, release sites should be chosen carefully, with consideration given to the natural food resources available, the proximity to human settlements, and the distance from the original site as felids are territorial and can return. Usually, a translocation will just transfer the problematic animal to another site, and the lethal removal of a felid can encourage another animal to move into the now empty territory (Cavalcanti, 2009; Conforti & Azevedo, 2003; Holland et al., 2018).

The most frequent solution to conflicts with carnivores suggested in the literature is to invest in mitigation strategies. This can minimize the losses and encourage farmers to modify their husbandry practices through the involvement of local people in the process (Cavalcanti, 2009; Holland et al., 2018; Quigley et al., 2015). It has been demonstrated that small changes in livestock management can be effective against attacks by pumas, such as fencing the most vulnerable animals close to houses during the night (Holland et al., 2018; Mazzolli et al., 2002). However, some respondents to the present study either stopped keeping sheep or reduced the number of sheep in the herds because they thought was not profitable to change husbandry practices. Reducing herd size could allow farmers to keep their sheep close to their properties or in a closed site at night, thereby decreasing the frequency of puma attacks on farms (Schulz et al., 2014).

Although 64.4% of respondents said they would accept suggestions in order to avoid puma attacks, 44.4% said they would not change any husbandry practices, and 68.9% declared they would not pay for any possible solution suggested by specialists. The reluctance of farmers to change husbandry practices and pay for husbandry improvements, were the same problems found in a study in an area next to our study area, in Santa Catarina state (Mazzolli et al., 2002), and in the Brazilian *Pantanal* (Palmeira & Barela, 2007; Zimmermann et al., 2005). According to Manfredo (2008) it does not matter which species is the focus of the conflict, the perceptions and attitudes of people will determine the magnitude and resolution of the problem. This problem is not just about people and fauna, but also about people and people, because farmers probably will resist changing habits and husbandry practices, and ultimately hold negative attitudes towards conservationists and their goals (Diegues et al., 2000). For this reason, if the mitigation work is mainly carried out by people perceived as being local elites, it may inadvertently increase local jealousies and create more antagonism towards conservation groups and wildlife (Dickman, 2010). Therefore, studies of human wildlife conflict need to consider human-human conflicts as well as human-animal conflicts in their proposed solutions (Hill, 2004).

The last point that deserves attention is the fact that the conflict in the region today, ten years after the data for this study was collected, still remains. In a personal communication one of the managers from a State Environmental Agency in the area reported that there are current stories of puma attacks on farms in the region, but attacks are at a relatively low level because a great number of farmers gave up sheep breeding (E. E. Fleck, Substitute manager of *Rota do Sol* State Environmental Protection Area, personal communication, July 2020). Even minor losses need to be monitored and minimized to avoid the poaching of pumas in the region. As Sillero-Zubiri et al., (2006) suggested, education and information in general can improve the tolerance of people if their perceptions of the threat can be brought down to more realistic levels and their vulnerability to carnivore attacks reduced, by improving husbandry practices, for example (Sillero-Zubiri et al., 2006). Despite some negative assessments towards pumas in the region, more than half of the interviewees desired information about pumas, indicating that pumas at least attract the curiosity of farmers. Consequently, local educational programs are highly recommended, because educational level and age were significant determinants in the knowledge of respondents about pumas.

Acknowledgments

The authors are grateful to all the respondents who shared their time and information to make this research possible. We also thank SEMA-RS (*Secretaria Estadual do Meio Ambiente*) for the interview logistic support; family and friends for the company in field work; and Marlon Ferraz da Rosa who kindly did Figure 1.

References

Alves, R. R. N., & Souto, W. M. S. (2011). Ethnozoology in Brazil: current status and perspectives. *Journal of Ethnobiology and Ethnomedicine*, 7, 22. doi:10.1186/1746-4269-7-22.

Alves, R. R. N. (2012). Relationships between fauna and people and the role of ethnozoology in animal conservation. *Ethnobiology and Conservation*, 1, 1–69. doi: 10.15451/ec2012-8-1.2-1-69.

Azevedo, F. C. C. (2008). Food habits and livestock depredation of sympatric jaguars and pumas in the Iguaçu National Park area, South Brazil. *Biotropica*, 40, 494–500. doi: 10.1111/j.1744-7429.2008.00404.x.

Bailey, K. D. (1982). *Methods of Social Research*. The Free Press, New York, NY.

Behling, H., & Pillar, V. D. (2007). Late quaternary vegetation, biodiversity and fire dynamics on the southern Brazilian highland and their implication for conservation and management of modern Araucaria forest and grassland ecosystems. *Philosophical Transactions of the Royal Society Biological Sciences,* 362, 243–251. doi:10.1098/rstb.2006.1984.

Bjerke, T., & Kalten, orn B. (1999). The relationship of ecocentric and anthropocentric motives to attitudes towards large carnivores. *Journal of Environmental Psychology,* 19, 415-421.

Cavalcanti, S. M. C. (2009). Manejo e controle de danos causados por espécies da fauna. In L. J. Cullen, R. Rudran, and C. V. Padua (Eds.), *Métodos de Estudos em Biologia da Conservação & Manejo da Vida Silvestre* (pp. 617-632). Curitiba, Brazil: UFPR. [Management and control of damage caused by fauna species. In L.J. Cullen, R. Rudran, and C.V. Padua (Eds.), *Methods of Studies in Conservation Biology & Wildlife Management*]

Cavalcanti, S. M. C., Oliveira T. G. de, & Paula, R. C. de. (2015). Identificação do agente causador do problema. In: S. M. C. Cavalcanti, R. C. de Paula, and R. L. Gasparini-Morato (Eds.), *Conflitos com mamíferos carnívoros - uma referência para o manejo e a convivência* (pp. 19-28). Brasília, Brazil: Instituto Chico Mendes de Conservação da Biodiversidade, ICMBio. [Identification of the agent causing the problem. In: S.M.C. Cavalcanti, R.C. de Paula, and R.L. Gasparini-Morato (Eds.), *Conflicts with carnivorous mammals - a reference for management and coexistence*]

Conforti, V. A., & Azevedo, F. C. C. (2003). Local perceptions of jaguars (*Panthera onca*) and pumas (*Puma concolor*) in the Iguaçu National Park area, south Brazil. *Biological Conservation*, 111, 215-221. doi: 10.1016/S0006-3207(02)00277-X.

Conforti, V. A. (2006). Local perceptions toward large carnivores in livestock raising areas. In R. G. Morato (Ed.), *Manejo e conservação de carnívoros neotropicais* (pp. 182-191). São Paulo, Brazil: IBAMA Press. [*Management and conservation of neotropical carnivores*]

Crawshaw, P. G. (2004). Depredation of domestic animals by large cats in Brazil. *Human Dimensions of Wildlife*, 9, 329–330. doi: 10.1080/10871200490505774.

Dickman, A. J. (2010). Complexities of conflict: the importance of considering social factors for effectively resolving human–wildlife conflict. *Animal Conservation,* 13(5), 458-466. doi: 10.1111/j.1469-1795.2010.00368.x.

Dickman, A. J., Macdonald, E. A., & Macdonald, D. W. (2011). A review of financial instruments to pay for predator conservation and encourage human–carnivore coexistence. *Proceedings of National Academy of Science of the United States of America*, 108, 13937–13944. doi: 10.1073/pnas.1012972108.

Diegues, A. C., Arruda, R. S. V., Silva, V. C. F., Figols F. A. B., & Andrade, D. (2000). *Biodiversidade e Comunidades Tradicionais no Brasil.* São Paulo, Brazil: NUPANB-USP, PROBIO-MMA & CNPq. [*Biodiversity and Traditional Communities in Brazil*]

Ditt, E. H., Mantovani, W., Padua, C. V., & Bassi, C. (2009). Entrevistas e aplicações de questionários em trabalhos de conservação. In L. J. Cullen, R. Rudran, and C. V. Padua (Eds.), *Métodos de Estudos em Biologia da Conservação & Manejo da Vida Silvestre* (pp. 617-632). Curitiba, Brazil: UFPR.[Interviews and application of questionnaires in conservation work. In L.J. Cullen, R. Rudran, and C.V. Padua (Eds.), *Methods of Studies in Conservation Biology & Wildlife Management*]

Duarte, M. M., & Bencke, G. A. (2008). *Plano de Manejo da Estação Ecológica Estadual de Aratinga. Final Report.* Fundação Zoobotânica do Rio Grande do Sul, Porto Alegre, Brazil. [*Aratinga State Ecological Station Management Plan. Final Report.*]

Engel, M. T., Vaske, J. J., Marchini, S., & Bath, A. J. (2017). Knowledge about big cats matters: insights for conservationists and managers. *Wildlife Society Bulletin*, 41(3), 398-404. doi: 10.1002/wsb.798.

Foster, R. J., Harmsen, B. J., Macdonald, D. W., Collins, J., Urbina, Y., Garcia, R., & Doncaster, C. P. (2014). Wild meat: a shared resource amongst people and predators. *Oryx, 50*, 63-75. doi:10.1017/S003060531400060X.

Herrmann, T. M., Schüttler, E., Benavides, P., Gálvez, N., Söhn L., & Palomo N. (2013). Values, animal symbolism, and human-animal relationships associated to two threatened felids in Mapuche and Chilean local narratives. *Journal of Ethnobiology and Ethnomedicine*, 9, 41. doi: 10.1186/1746-4269-9-41.

Hill, C. M. (2004). Farmers' Perspectives of Conflict at the Wildlife–Agriculture Boundary: Some Lessons Learned from African Subsistence Farmers. *Human Dimensions of Wildlife: An International Journal,* 9, 279-286. doi: 10.1080/10871200490505710.

Holland, K. K., Larson, L. R., & Powell, R. B. (2018). Characterizing conflict between humans and big cats *Panthera* spp: A systematic review of research trends and management opportunities. *Plos One,* 13(9), e0203877. doi:10.1371/journal.pone.0203877.

Indrusiak, C., & Eizirik, E. (2003). Carnívoros. In C. S. Fontana, G. A. Bencke, & R. E. Reis. *(Eds.), Livro Vermelho das Espécies da Fauna Ameaçada de Extinção no Rio Grande do Sul* (pp. 523-525). Porto Alegre, Brazil: EDIPUCRS. [*Red Book of Endangered Fauna Species in Rio Grande do Sul*]

Inskip, C., & Zimmermann, A. (2009). Human-felid conflict: a review of patterns and priorities worldwide. *Oryx,* 43, 18–34. doi: 10.1017/S003060530899030X.

Instituto Brasileiro De Geografia - IBGE. *Panorama. Censo Demográfico: Renda – Rio Grande do Sul*; 2010. (2020, June 20). Retrieved from: http://www.ibge.gov.br/home/estatistica/populacao/censo2010/indicadores_sociais_municipais/in dicadores_sociais_municipais_tab_uf_zip.shtm. [*Panorama. Demographic Census: Income – Rio Grande do Sul*]

Madden, F., & McQuinn, B. (2014). Conservation's blind spot: The case for conflict transformation in wildlife conservation. *Biological Conservation,* 178, 97-106. doi:10.1016/j.biocon.2014.07.015.

Manfredo, M. J. (2008). *Who Cares About Wildlife? Social Science concepts for Exploring Human-Wildlife Relationships and Conservation Issues.* New York, NY: Springer. doi: 10.1007/978-0-387-77040-6.

Manfredo, M. J., Vaske, J. J., Rechkemmer, A., & Duke, E. A. *Understanding Society and Natural Resources: Forging New Strands of Integration across the Social Sciences.* (2014). New York, NY: Springer. doi: 10.1007/978-94-017-8959-2.

Marchini, S. (2010). *Human dimensions of the conflicts between people and jaguars (Panthera onca) in Brazil.* (Doctoral Dissertation). Available from: Department of Zoology, University of Oxford: http://ethos.bl.uk/OrderDetails.do?uin=uk.bl.ethos.543030.

Marchini, S. & Crawshaw, P. G. Jr. (2015). Human–Wildlife Conflicts in Brazil: a fast-growing issue. *Human Dimensions of Wildlife*, 20, 323-328. doi:10.1080/10871209.2015.1004145.

Mazzolli, M., Graipel, M. E., & Dunstone, N. (2002). Mountain lion depredation in the Southern Brazil. *Biological Conservation*, 105, 43–51. doi: 10.1016/S0006-3207(01)00178-1.

Mazzoli, M. (2005). *Efeito de gradientes de floresta nativa em sistemas agropecuários. sobre a diversidade de mamíferos vulneráveis.* Report: Relatório técnico WWF-Brasil, Brasília, Brazil. [*Effect of native forest gradients on agricultural systems. on the diversity of vulnerable mammals.*]

Michalski, F., Boulhosa, R. L. P., Faria, A., & Peres, C. A. (2006). Human–wildlife conflicts in a fragmented Amazonian forest landscape: determinants of large felid depredation on livestock. *Animal Conservation*, 9, 179–188. doi: 10.1111/j.1469-1795.2006.00025.x.

Müller, S. C., Overbeck, G. E., Blanco, C. C., Oliveira, J. M. & Pillar, V. D. (2012). South Brazilian forest-grassland ecotones: dynamics affected by climate, disturbance and woody species traits. In: R. W. Myster (Ed.), *Ecotones between forest and grassland.* New York, NY: Springer Press. doi:10.1007/978-1-4614-3797-0_7.

Murphy, T., & MacDonald, D. W. (2010). Pumas and people: lessons in the landscape of tolerance from a widely distributed felid. In: D. M. MacDonald and A. J. Loveridge (Eds.), *Biology and conservation of wild felids* (pp. 431-452). Oxford, England: University of Oxford.

Nyhus, P. J. (2016). Human-wildlife conflict and coexistence. *Annual Review of Environment and Resources*, 41, 143-171. doi:10.1146/annurev-environ-110615-085634.

Oliveira, T. G., & Cassaro, K. (2006). *Guia de Campo dos felinos do Brasil.* São Paulo, Brasil: Instituto Pró-Carnívoros; Fundação Parque Zoológico de São Paulo;

Sociedade de Zoológicos do Brasil; Pró-Vida Brasil. [*Field Guide of the Brazilian Cats.*]

Palmeira, F. B. L., & Barrella, W. (2007). Conflitos causados pela predação de rebanhos domésticos por grandes felinos em comunidades quilombolas na Mata Atlântica. *Biota Neotropical*, 07, 119–128. doi: 10.1590/S1676-0603200700 0100017. [Conflicts caused by predation of domestic livestock by large cats in quilombola communities in the Atlantic Forest. *Neotropical Biota*]

Palmeira, F. B. L., Crawshaw, P. G. J., Haddad, C. M., Ferraz, K. M. P. M. B., & Verdade, L. M. (2008). Cattle depredation by puma (*Puma concolor*) and jaguar (*Panthera onca*) in central-western Brazil. *Biological Conservation*, 141, 118–125. doi: 10.1016/j.biocon.2007.09.015.

Perry, E. E., Needham, M. D., Cramer, L. A., & Rosenberger, R. S. (2014). Coastal resident knowledge of new marine reserves in Oregon: the impact of proximity and attachment. *Ocean & Coastal Management,* 95, 107–116. doi: 10.1002/ wsb.798.

Quigley, H., Hoogesteijn R., Hoogesteijn, A., Foster, R., Payan, E., Corrales, D., Salom-Perez, R., & Urbina, Y. (2015). Observations and preliminary testing of jaguar depredation reduction techniques in and between core jaguar populations. *Parks,* 21, 63-72. doi: 10.2305/IUCN.CH.2014.PARKS-21-1HQ.en.

Schultz, P. W. (2011). Conservation means behavior. *Conservation Biology,* 25, 1080-1083. doi: 10.1111/j.1523-1739.2011.01766.x.

Schulz, F., Oliveira, L. R., & Printes, R. C. (2014). Depredation of domestic herds by pumas based on farmer's information in Southern Brazil. *Journal of Ethnobiology and Ethnomedicine,* 10, 73. doi: 10.1186/1746-4269-10-73.

Sillero-Zubiri, C., Sakumar, R., & Treves, A. (2006). In: D. Macdonald, & K. Service (Eds.), *Key Topics in Conservation Biology* (pp. 255-272). Oxford, England: Wiley Blackwell.

Silva, L. C. R., & Anand, M. (2011). Mechanisms of Araucaria (Atlantic) forest expansion into southern Brazilian grasslands. *Ecosystems,* 14, 1354–1371. doi: 10.1007/s10021-011-9486-y.

Silveira, L. (2004). *Ecologia comparada e conservação da onça-pintada (Panthera onca) e onça-parda (Puma concolor), no cerrado e pantanal.* (Doctoral Dissertation). Available from: Department of Animal Biology, University of Brasília, Brasília, Brazil: http://www.gesto.to.gov.br/site_media/upload/gestao/ documentos/Doutorado_Leandro_Silveira_Ecologia_comparada_e_conservacao _da_onca_pintada_e_onca_parda_no_cerrado_e_Pantanal.pdf. [*Comparative ecology and conservation of the jaguar (Panthera onca) and puma (Puma concolor) in the cerrado and wetlands.*]

Souza, J. B., & Alves, R. R. N. (2014). Hunting and wildlife use in an Atlantic Forest remnant of northeastern Brazil. *Tropical Conservation Science,* 7, 145–160. doi: 10.1177/194008291400700105.

Stern, P. C. (2000). Toward a coherent theory of environmentally significant behavior. *The Journal of Social Issues*, 56, 407-424. doi: 10.1111/0022-4537. 00175.

Swim, J., Clayton, S., Doherty, T., Gifford, R., Howard, G., Reser, J., Stern, P., & Weber, E. (2009). *Psychology and global climate change: addressing a multi-*

faceted phenomenon and set of challenges. Available on internet at http://www.apa.org/science/about/publications/climate-change.aspx.

Tortato, F. R., Layme, V. M. G., Crawshaw, P. G., & Izzo, T. J. (2015). The impact of herd composition and foraging area on livestock predation by big cats in the Pantanal of Brazil. *Animal Conservation*, 18, 539-547. doi: 10.1111/acv.12207.

Vaske, J. J., and Manfredo, M. J. 2012. Social psychological considerations in wildlife management. In: D. J. Decker, S. J. Riley, & W. F. Siemer (Eds.). *Human dimensions of wildlife management* (pp. 43-57). Baltimore, MD: The Johns Hopkins University Press.

Woodroffe, R., Frank, L. G., Lindsey, P. A., Ranah, S. M. K. ole, & Romañach, S. (2007). Livestock husbandry as a tool for carnivore conservation in Africa's community rangelands: a case–control study. *Biodiversity Conservation*, 16, 1245–1260. doi: 10.1007/s10531-006-9124-8.

Zimmermann, A., Walpole, M. J., & Leader-Williams, N. (2005). Cattle ranchers' attitudes to conflicts with jaguars in the Pantanal of Brazil. *Oryx*, 39, 406–412. doi: 10.1017/S003060530500099.

Zimmermann, A., Baker, N., Inskip, C., Linnell, J. D. C., Marchini, S., Odden, J., Rasmussen, G., & Treves, A. (2010). Contemporary views of human wildlife conflict. In: J. T. du Toit, R. Kock, & J. C. Deutsch (Eds.), *Wild Rangelands: Conserving Wildlife While Maintaining Livestock in Semi-Arid Ecosystems* (pp. 129-151). London, England: Wiley-Blackwell. doi: 10.1002/9781444317091.ch6.

Appendix. Standard Questionnaire

1. SOCIO-ECONOMIC PROFILE OF THE INTERVIEWEES:

1.1. First name and nickname: ________________________________

1.2. Gender: ________________________________

1.3. Age: ________________________________

1.4. Lifetime in the region: ________________________________

1.5. Educational level: ________________________________

1.6. Occupation: ________________________________

1.7. Median monthly income: ________________________________

2. CHARACTERISTICS OF PUMA ATTACKS ON FARMS:

2.1. Main causes of herd's mortality in a year: ________________________

2.2. Did you have any loss in your herds because of a wild animal attack from the year of 2008? ( ) no ( ) yes

2.3. Do you know which was the wild animal? ( ) no ( ) yes

2.4. Which was the animal? ( ) puma ( ) ocelot ( ) wild cat ( ) other ________________

2.5. Did you see the wild animal attacking? ( ) no ( ) yes

2.6. How do you know which was the animal? __________________ _____________________

2.7. How many times a puma attacked the property in each year? 2008______2009______2010______2011_______

2.8. How many domestic animals were attacked in each year? 2008______2009______2010______2011_______

2.9. Did you see the carcass of the animals killed? ( ) no ( ) yes

2.10. Were the carcasses covered by leaves or sticks? ( ) no ( ) yes

2.11. What were the carcasses parts that were eaten (most of the times)? __________________

2. 12. What did you do after the attacks occurred? ____________ _________________________

2.13. Did you stop to breed some kind of herd because of puma attacks?

( ) no ( ) yes

3. FARMER'S PERCEPTIONS:

3.1 How many pumas do you think that live in the region? _________ _________________

3.2. How do you think that puma's population changed in the region in the last ten years?

( ) decreased a lot ( ) decreased ( ) it did not change ( ) increased ( ) increased a lot

3.3. Why did it increase or increase a lot? ______________________________

3.4. Why did it decrease or decrease a lot? ______________________________

3.5. What do you think is the right thing to do to solve the problems with pumas attacking domestic herds?

( ) protect/enclose the livestock ( ) non-lethal removal of pumas ( ) kill the pumas ( ) compensation scheme ( ) stop breeding livestock ( ) fence off protected areas ( ) Other:____________

3.6. Who should pay for the losses caused by pumas?

( ) state government ( ) federal government ( ) the town hall ( ) owners ( ) nobody ( ) other suggestions: ____

4. FARMER'S KNOWLEDGE:

4.1. Do you think pumas are important in nature?

( ) they are not important at al ( ) they have some importance ( ) they are very important ( ) don't know

4.2. What do you think is the role of pumas in nature? ____________________

4.3. Do you think a puma can attack a human being without been provoked?

( ) yes ( ) no Why? ____

4.4. What do you think about killing a puma, considering the law?

( ) it is always allowed by the law ( ) can be allowed by the law, depending on the situation ( ) it is never allowed by the law

ILLUSTRATIVE BOARD

4.4. Which of these animals is the puma? A ( ) B ( ) C ( ) D ( ) Didn't recognise ( )

4.5. Have you ever seen one of these animals? ( ) no ( ) yes

4.6. Which one? A ( ) B ( ) C ( ) D ( )

4.7. Do you know their names?

A ____

B ____

C ____

D ____

5. FARMER'S ATTITUDES:

5.1. Identify your feelings towards pumas: ( ) I really dislike ( ) I dislike( ) I am indifferent ( ) I like ( ) I really like

5.2. Do you think that is necessary to control (to stop somehow) puma attacks in the region?

( ) yes ( ) no

5.3. Would you accept technical suggestions to avoid puma attacks?

( ) yes ( ) no

5.4. Would you change your husbandry practices to protect your livestock?

( ) yes ( ) no

5.5. Would you be willing to pay for the necessary changes?

( ) yes ( ) no

5.6. Would you like to learn more about pumas?

( ) yes ( ) no

5.7. Have you ever heard about people illegally killing pumas in the region?

( ) yes ( ) no

5.8. Have you ever tried to kill a puma?

( ) yes ( ) no

Prancha Ilustrativa

Figure A-1. Jaguar Puma.

Figure A-2. Ocelot Jaguarundi.

In: Critical Research Techniques in Animal ...
Editors: Michael O'Neal Campbell et al.
ISBN: 978-1-53619-846-1

Chapter 8

REFORMATION OF AFRICAN ELEPHANT ECOLOGY AND HUMAN QUALITY OF LIFE THROUGH GIS AND REMOTE SENSING TECHNOLOGIES: A CASE OF AFRICAN ELEPHANT ECOLOGY IN BOTSWANA

Ida Dutt*

Simon Fraser University, British Columbia, Canada

ABSTRACT

The African elephant is one of the most precious yet vulnerable mammals of the twenty-first century. The first Great Elephant Census, a pan-African survey conducted in 2016, revealed that African elephant numbers plummeted by at least 144,000 between 2007 and 2014. Botswana has approximately one third of Africa's remaining population of elephants. Between 2014 and 2018, the number of elephant carcasses within the country increased by 593%, of which 385 elephants were confirmed poaching cases between 2017 and 2018. Moreover, between May and June of 2020, the country has seen more than 350 mysterious elephant deaths. This may be the result of habitat loss, increasing human-elephant conflicts, and Botswana's government lifting their hunting ban. Over the past 20 years, the elephant range in Botswana has expanded by 53% causing increasing concerns about the impact of elephants on biodiversity, the viability of other species, and the livelihoods and safety of people living within the elephant range. Unfortunately, in-depth information about the movements of elephants, their social dynamics, habitat use, and impacts on people around these periphery regions are currently limited within the country. Improving our understanding of the African elephant ecology is not only important for maintaining global wildlife diversity, but also for the benefit of human health, quality of life (QOL), and productivity in

---

* Corresponding Author's E-mail: idutt@sfu.ca.

agricultural systems. This chapter focuses on bringing African elephant ecology reformation through GIS and RS technologies within Botswana.

INTRODUCTION

The African elephant (referred as elephant throughout this chapter, see Figures 1 and 2) is one of the most precious yet vulnerable mammalians of the twenty-first century (the International Union for Conservation of Nature's Red List, 2020). The first Great Elephant Census (Allen, 2016), a pan-African survey conducted in 2016, revealed that elephant numbers plummeted by at least 30%, or 144,000 in just seven years between 2007 and 2014 (Figure 3). This decline in elephant population is detrimental for Africa's ecosystems as elephants within savannas and forested areas play a critical role in shaping surrounding landscapes through feeding activities that uproot small vegetation and disperse seeds (Coverdale et al., 2016; Fritz, 2017; Kohi et al., 2011). Given their keystone role, it is critical to implement conservation efforts to maintain healthy populations of elephants throughout their ranges in Africa. Therefore, improving our understanding of elephant ecology is not only important for global wildlife diversity, but for the benefit of human health, quality of life (QOL) and productivity in agricultural systems (Daszak et al., 2000).

This chapter proposes that one of the most efficient and vital tools which can be used to improve our understanding of elephant ecology and human quality of life (hereafter QOL) is the combined use of Geographic Information Systems (GIS) and Remote Sensing (RS) technologies. GIS and RS technologies are a gold mine for answering geospatial questions in the fields of conservation biology and public health research. While GIS serves to capture, store, analyze and visualize geographic data via algorithmic and functional tools, RS technologies serve to capture non-contact information via ultraviolet, visible, and infrared regions of the electromagnetic spectrum through cameras, scanners, satellites, linear arrays, and analyze acquired information by means of visual and digital image processing (Jensen, 2007). The best mode of analyzing information captured by GIS and RS imagery for the animal welfare and public health sectors is via infrared, temperature, visible, and water band images (Table 1); these images help to identify land cover change, analyze wildlife patterns, and perceive human behaviors (ie. migration, urbanization, and agricultural patterns over time) (Campbell, 2018).

These functions of GIS and RS analyses are vital in the amelioration of elephant ecology as they provide areas for wildlife management and conservation research. Analysis of wildlife conducted via RS images in GIS layered maps (Table 1), provide conservationists and public health authorities crucial information regarding the resources required to protect the elephant population from humans and vice versa. It also provides information concerning the physical and psychological traumas experienced by humans due to human-elephant conflicts (HEC), allowing ecologists, policy analysts, and urban planners to allocate resources such as mental and physical health clinics (Ruda et al., 2018). Moreover, identifying land cover change through RS can inform conservationists and health scientists of the impact of urbanization over time and its effect on nearby water bodies (Rodriguez, 1994). This can enable epidemiologists to study the cause of some vector-borne illnesses, leading to disease prevention and betterment of human and wildlife health (Rodriguez, 1994). Furthermore, GIS and RS image analyses (Table 1) can also help understand human behaviors such as migration, urbanization, and agricultural patterns over time and their impact on physical and psychological quality of human health, thereby influencing health care management, policies, strategies, and resource allocations (Mensah et al., 2016).

Table 1a. The use of infrared, temperature, water and visible bands in GIS and RS analyses and their relevance to African elephant ecology reformation and QOL

Type of Electromagnetic Region Band Image	**Function of Electromagnetic Spectrum Region Band Image**	**Relevance to African elephant Ecology and Human QOL**
Near Infrared (bands 432; 421)	• Reflects chlorophyll depicting vegetation in vivid and various shades of red. Coniferous trees are darker shades of red than hardwoods. Deep reds usually indicate healthier vegetation while lighter reds signify grasslands or sparsely vegetated areas. • Urban areas show as cyan blue • Soils range from light to dark shades of brown • Water appears very dark due to the absorption of energy in the near infrared bands	• The redder the shades in urban areas, the better the indicator of improved physical, psychological, emotional, social, and material wellbeing of individuals (Mensah et al., 2016) • Can help with identification of wildlife as it can help detect wildlife based on greenspaces available • Increased vegetation is crucial for food sources for herbivores such as African elephants and for the agricultural purposes for people in nearby villages
Temperature (band 6)	• Darker portions in image represent a cooler temperature • Lighter portions represent hotter weather • Urban concrete areas are hotter than urban green areas. • The downtown areas of large towns are hotter than the suburbs. • Green areas are good for human quality of life. • Urban water cools downtown areas. • Green rural areas are cooler than bare soil	• Temperature is cooler in areas of abundant green spaces which indicates that urban areas that are darker have more green spaces and therefore have better health outcomes in people in those areas • Urban heat islands contribute significantly to the health impact of heat waves (Laaidi et al. 2012) • Reduction of the green cover increases urban temperatures which highlight the potential for improving health and well-being by measures improving the built environment (Gill et al. 2007). • Increased shades of red indicate higher green spaces which indicate more oxygen is being released in those areas which is good for human health • Temperature levels can help identify human (migration) and wildlife behavior (movement for killings and grazing)

Table 1b. The use of infrared, temperature, water and visible bands in GIS and RS analyses and their relevance to African elephant ecology reformation and QOL

Water (band 4)	• Water increasingly absorbs the light making it darker, so black water bodies indicate water purity • Increasing amounts of dissolved inorganic materials in water bodies shift the shade from dark to light • Allows to distinguish between dry and moist soils (barren land and croplands). Darker shades indicate moistness while lighter shades indicate dryness • Croplands and grasslands show higher reflectance (brighter tone) than the forest	• Useful for crop-drought studies indicating levels of famine, financial loss, and stress which are determinants of health. • Conservationists and public health researchers can get an indication of water purity levels from these images to investigate water-borne illnesses as contaminated water and poor sanitation are linked to transmission of diseases within species populations such as those observed in the 350 carcasses found between March and June 2020 (van Aarde et al., 2021)
Visible (bands 321)	• visible bands are used in this combination, so ground features appear in colors as they do in real life to humans. • healthy vegetation is green, • recently cleared fields are very light • unhealthy vegetation is brown and yellow, roads are gray, and shorelines are white.	• This band combination provides the most water penetration and superior sediment and bathymetric information. It is also used for urban studies. • It allows for the entire map of locations to be visualized, allowing the analysis of disease rates, wildlife attacks, and other patterns at a grander scale. This allows for conservationists and policy makers to confer areas of impact and devise prevention strategies and resource allocations

Figure 1. African Elephants. Permission of Roger Brown at Pexels.

Since Botswana has the largest population of elephants (Thouless et al., 2016), this chapter will focus on bringing elephant ecology reformation through GIS and RS technologies within Botswana. According to aerial surveys conducted by Elephants without Borders (EWB) (2019), although the population of elephants across Botswana has been stable from 2010 to 2018, there have been signs of increased elephant deaths in the country (Figure 3). This may be the result of increasing habitat loss, HEC, and Botswana's government lifting their five-year hunting ban (Van Aarde, 2021).

Figure 2. African Elephants. Permission of Taryn Elliott at Pexels.

Due to the rapid increase in human population and infrastructure in Botswana, developments are expanding further into wildlife habitat, forcing increased proximity between human settlements and elephant populations (Gwebu, 2012). These increases in proximity have led to several HECs within Botswana; elephant populations in some areas have plummeted due to retaliatory killings (Western et al., 2009). It is of great significance for long-term solutions to these HEC problems to be found, however, they require in-depth information about the movements of elephants, their social dynamics, and their impacts on people around these periphery regions, all of which are currently limited within the country. Elephant ecology can therefore be summed up in three parts: Habitat, HEC, and Governmental Policies. These aspects of elephant ecology can be studied in-depth via GIS and RS analyses of the elephants that reside in Botswana. Therefore, this chapter conducts an analysis on how GIS and RS research methods can be used to assess elephant habitat patterns, HEC, QOL as well as suggest strategic management and conservation measures.

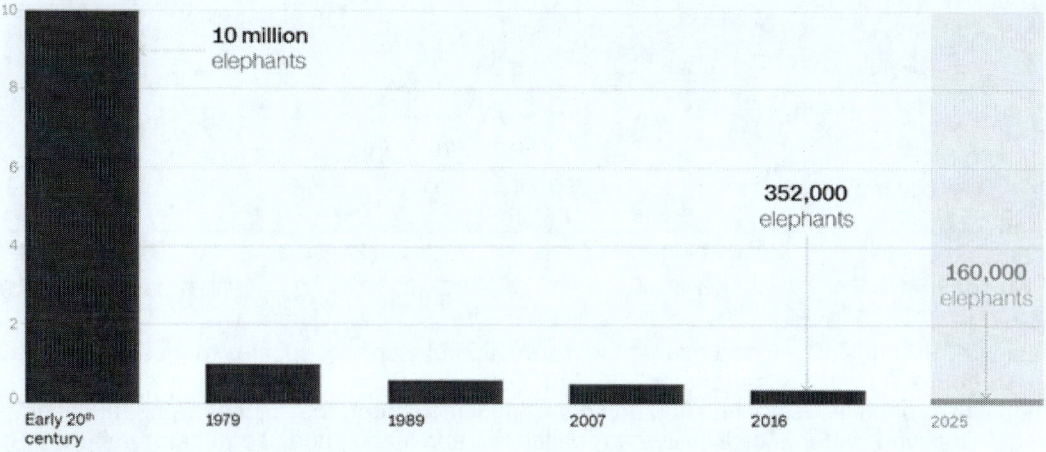

Figure 3. Estimation of the decline of African elephant population from the early 20th century to the year 2025. All values, except for the years 1979 and 2016 are estimates. Graph obtained from McKenzie & Formanek (2016).

ELEPHANT POPULATIONS

The *Elephantidae* family which historically ranged across the American, European, Asian, and African continents, has now been restricted to certain parts in Asia and sub-Saharan Africa (Figure 4) (Calabrese et al., 2017; Sukumar, 2008; Thouless et al., 2016). The population of elephants is estimated to be between 350,000 to 700,000 living in 37 range countries where more than 70% of the geographic range is unprotected (Figure 3 and 4) (Chase et al., 2016). The unprotected ranges within the African continent facilitates activities such as poaching for the illegal ivory trade which continue to decimate Africa's elephant population (Chase et al., 2016; Figure 3-4). This species has been decimated so much so that the International Union for Conservation of Nature (IUCN) lists extant African savanna (*Loxodonta africana*, Blumenbach, 1797) and forest *(L. cyclotis*, Matschie, 1900) elephant species as vulnerable (Blanc, 2016).

Of the 37 African countries this vulnerable species is found in, Botswana is home to approximately one-third of Africa's remaining savanna elephants, making the conservation for the elephant population within Botswana an urgent priority for conservationists (Coverdale et al., 2016; Fritz, 2017; Kohi et al., 2011). Figure 4 is a great representation of how the combined use of GIS and RS technologies can be utilized to monitor, assess, and conserve the elephant population within Botswana; data collected via RS imagery and sensors were added and layered into GIS software to create the cohesive map depicted.

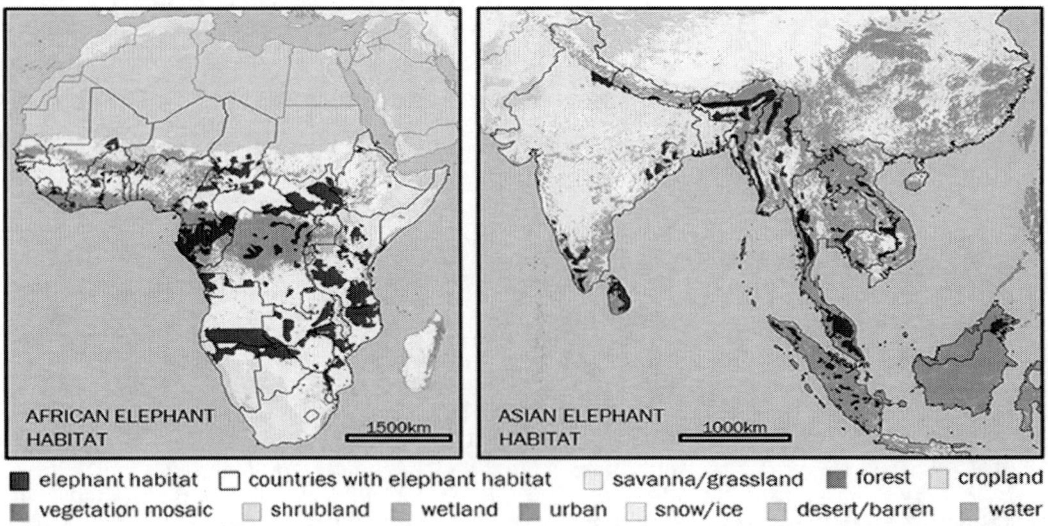

Figure 4. RS and GIS infused map of the distribution of elephant species ranges around the world from 2000-2016 (Shaffer et al., 2019). This map indicates the efficient use of GIS mapping through its use of layering and depicting various features such as elephant habitat, vegetation type, water accessibility and availability, and land space use (urban/rural) from RS gathered information.

ELEPHANT PATTERNS

As the largest of all land mammals, elephant movement patterns within Botswana play an important role in balancing natural ecosystems (Kohi et al., 2011; Coverdale et al., 2016; Fritz, 2017). Based on historically proven data on elephant feeding behavior, this species consumes approximately 150 kg of forage and 190 liters of water daily (Sukumar, 2008). Meeting these basic needs requires a large foraging area to provide a variety of graminoids, shrubs, tree leaves, roots, and fruits in the wet season and woody browse during the dry season (Figure 5-8) (Barnes 1982; Field & Ross, 1976; Vancuylenberg, 1977). Not only is elephant foraging important for elephant health, but also for the keystone role they play in ecosystems by trampling forests and dense grasslands, thus making room for smaller species to coexist (Coverdale et al., 2016; Fritz, 2017; Kohi et al., 2011). During low rainfall seasons, they also dig dry riverbeds which are utilized by the surrounding wildlife (Coverdale et al., 2016). Additionally, elephant herds travel over vast range lands and disperse seeds in their feces which helps generate new green growth (Coverdale et al., 2016; Fritz, 2017; Kohi et al., 2011).

A study conducted by Sekar et al. (2017) found that there was a strong positive correlation between elephant presence in a forest, seed dispersal, and the survival of large-fruiting trees. The depiction of this correlation highlights the impact and influence elephant movement patterns have on their surrounding environments. Therefore, a thorough understanding of elephant migratory patterns is necessary to study their impact on vegetation abundance, vegetation composition, other species biodiversity, and ecosystem productivity. An in-depth understanding of their patterns also enables conservationists and public health researchers to predict the seasons during which elephants are more inclined to range into villages farther than their natural regions. These predictions aid in minimizing HEC and maximizing the management and conservation of the elephant population. Therefore, this section of the chapter depicts how the use of GIS and RS technologies such as mapping, aerial imagery, and Global Positioning System (GPS) tracking collars can play a significant role in the monitoring and detailed analysis of elephant habitat patterns and ecology, thereby providing Botswana with the tools required for the mitigation of HEC and improvement of QOL.

Botswanian Climate

To utilize RS and GIS tools within the context of Botswana's elephant movement patterns, it is important to first understand the country's vegetation and climate. Botswana is a country consisting majorly of savanna grassland of yellow or light brown graminoids, grass cover and woody plants (Figure 5 E and F) (Parsons, 2021). Infrared and visible band RS imageries can be utilized to identify and assess the varying levels of healthy and unhealthy vegetation (Table 1) within Botswana to provide a better understanding of elephant habitat (Figure 5). Through the utilization of RS visible and infrared band images (Table 1), it can be seen that the Botswanian savanna ranges from shrubs and Acacia thorn bush in the southwest regions of Botswana (Figure 5A, E, F) to denser woodland and eventually forests in northern Botswana (indicated by shades of bright red in Figure 5 A) (Parsons, 2021). Mopane woodlands also cover most of northern Botswana, with the exception of the open savanna graminoids and grasslands immediately surrounding the Okavango Delta (depicted by the sparse shades of red in Figure 5 C and D).

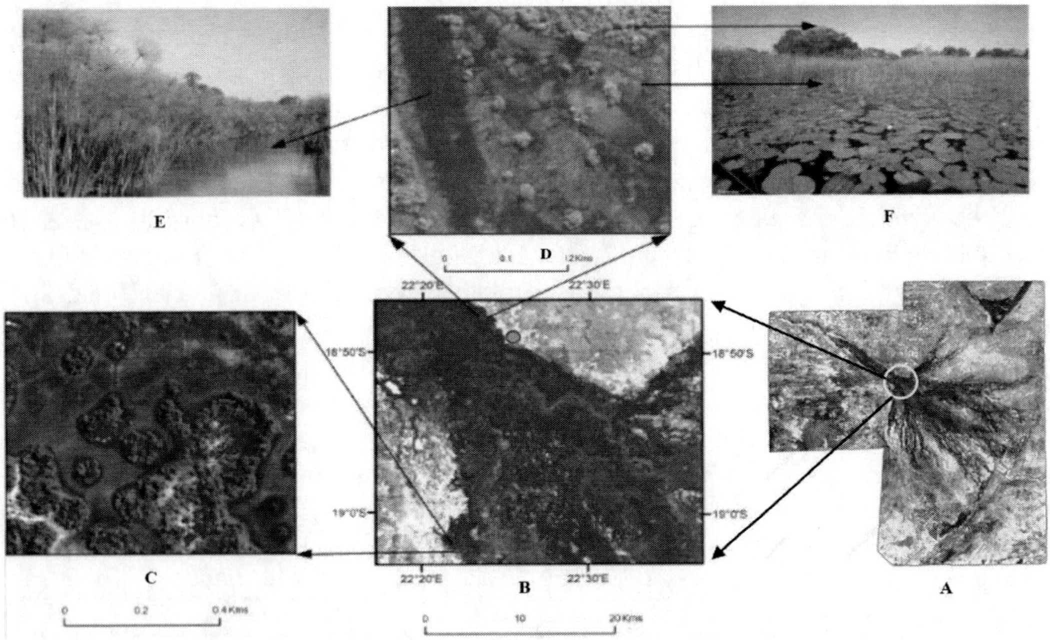

Figure 5. RS Landsat satellite images in the infrared and visual bands (Table 1) as obtained from Mda Information Systems (2018) (A) and Mishra (2014) (B-F). These images depict the vegetation diversity and land cover in the Okavango Delta, in northern Botswana. (A) is an infrared representation of the Okavango Delta from 2018 and depicts the various levels of healthy to unhealthy vegetation (Table 1). (B) is a zoomed-in infrared imagery of the Okavango Delta near the Seronga village (green shaded circle) in Botswana. (C) and (D) are infrared representations of two zoomed in areas of (B). (E) and (F) are visible band RS photographs acquired in the field from parts of (D and include examples of vegetation consumed by elephants.

Furthermore, the annual climate within Botswana ranges from mild dry temperatures during the winter to extremely dry, humid, and hot temperatures during the summer (Parsons, 2021). Winter in Botswana ranges from the months of April to September, during which temperatures fall below 0°C; while summer within the country lasts between the months of October and March with temperatures rising to approximately 34°C (Chipanshi et al., 2003; Parsons, 2021). Within these seasons, Clark (2013) categorized the winter season as three categories of post wet (April), cold dry (May to July) and hot dry (August to September) and the summer season as the wet season (November to March). As opposed to western weather, summer in Botswana is accompanied by gusty winds and rain, which are limited to summer downpours between December and March, marking this the season for plowing and planting for farmers within the country (Chipanshi et al., 2003; Jonsson, 2004; Parsons, 2021). Annual rainfall within Botswana, brought forth by winds from the Indian Ocean, averages to approximately 46 to 64 cm the extreme northern regions to approximately less than 5 inches in the extreme southern regions of Botswana (Jonsson, 2004; Parsons, 2021). Cyclic droughts, lasting approximately 5 to 6 years every two decades, also occur within the country which eliminate harvests, lead to wildlife and livestock starvation, and habitat degradation (Chipanshi et al., 2003; Jonsson, 2004; Parsons, 2021).

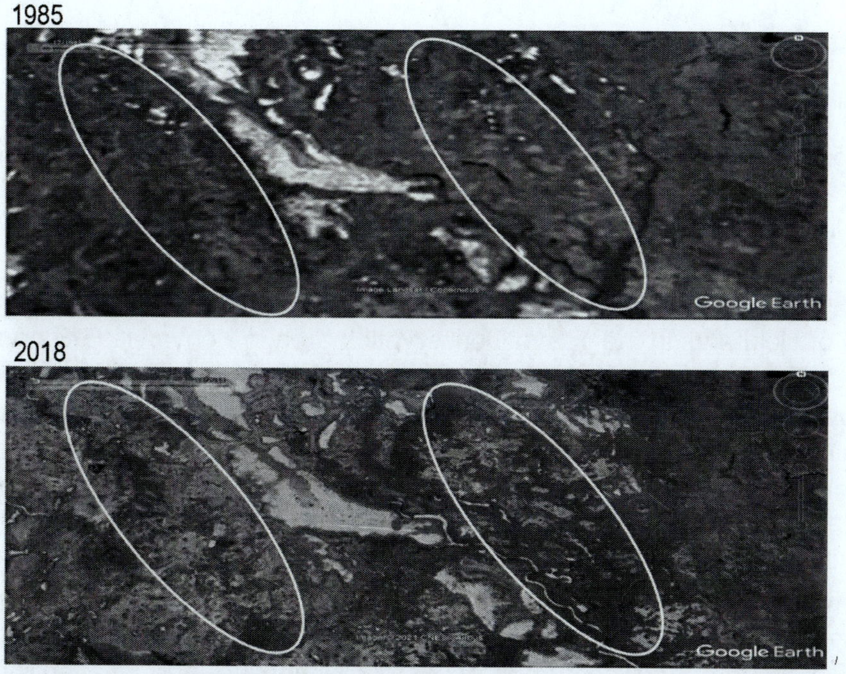

Figure 6. Visible band (table 1) RS images obtained from Google Earth of an area within the Okavango Delta region of Botswana from 1985 and 2018, comparing habitat degradation and water availability. The yellow circles indicate areas of unhealthy and recently cleared vegetation along with waterways exposed to heat over time. The blue shades in the images are a part of the Okavango Delta water body which appears the lightest due to light reflection (table 1). The red dot symbolizes Jedibe Village School.

Visible band RS imagery (Table 1) collected from Landsat images from the years 1985 to 2018 depict how RS can be utilized to analyze the drastic impact climate change and anthropogenic activity have had on the decline of elephant habitat within the Okavango Delta region in Botswana (Figure 6). The lighter shades of green and browns (Table 1) seen in Figure 4 depict the impacts of climate change and anthropogenic activity perfectly; over time, Botswana's vegetation has become increasingly unhealthy due to rising temperatures and has also been cleared for human land space use (Chipanshi et al., 2003; Jonsson, 2004; Parsons, 2021). The urbanization of forest land due to anthropogenic activity is evident within Figure 6 from the year 1985 to 2018, during which Jedibe Village School (indicated by the red dot in Figure 6) was established. Furthermore, the analyses of these RS images depict that the removal of vegetation that formed a canopy of protection over the area led to increased exposure of the waterways to the scorching Sub-Saharan African heat. Increased exposure of sunlight on vegetation within these regions can have detrimental impacts on the ecosystems as they lead to higher evaporation rates of the water sources, increased dehydration within elephants, thereby leading to the elephants travelling farther into human settlements in search for natural resources (Barnes, 1982).

Vegetation and Water Availability

Due to their long-ranging nature, elephants heavily rely upon regular migration over large distances to search for food, water, and social and reproductive partners (Whyte, 2012) (Figures 7-8). These migratory patterns are dependent on the season of the year (Figure 7) and as a result, the availability of vegetation and water (Figure 8). Through the use of radio-telemetry satellite-linked GPS collars (RS technology) on male and female elephants, Wato et al. (2018) analyzed the movement patterns of elephants in the Tsavo Conservation area in Kenya via GIS created maps (Figure 6). The combination of RS captured data in GIS layered maps proved to be an excellent tool of analysis of elephant movement patterns as it enabled the researchers to visually detect that both male and female elephants travel farther for water and remain close to the water sources in dry seasons compared to wet seasons (Wato et al., 2018). Furthermore, these GIS and RS technologies enabled researchers to capture live hourly data to discover that female elephants remain close to the newly located water sources for foraging while male elephants travel farther from the water sources for foraging in dry seasons (Figure 7) (Wato et al., 2018). The results from this study demonstrate and confirm that water distribution and availability is highly correlated to elephant density within an area and that ranging behaviors vary based on the season of the year (Figure 7) (Robson et al., 2017; Wato et al., 2018). These methods could also be implemented within Botswana to identify, monitor, and assess elephant movement patterns.

In addition to water availability impacting elephant movement patterns, Robson et al. (2017) found that vegetation abundance is positively correlated to elephant densities within an area (Figure 8). As seen in figure 6A, elephant density increases as vegetation abundance increases. The regression in elephant density in figure 6A infers that once elephants have met their optimal nutritional needs, they remain in that area until required to move otherwise. However, figure 6B infers that elephant density increases consistently as water availability increases (Robson et al. 2017). This study further confirmed the results from Wato et al. (2018) and propose that while both vegetation abundance and water availability are important in elephant movement patterns, water availability is a stronger indicator of elephant sightings within an area than vegetation abundance (Figure 8) (Robson et al., 2017).

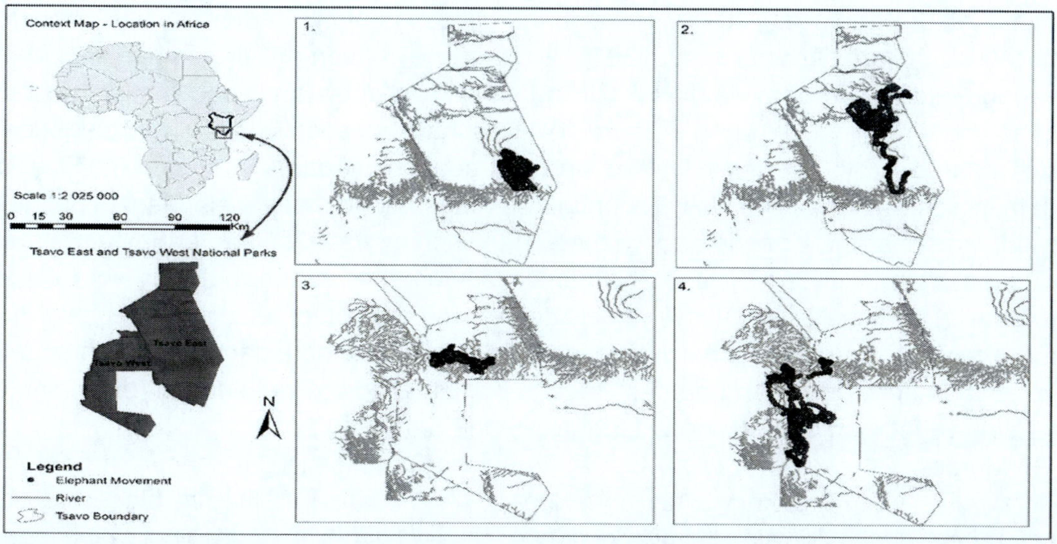

Figure 7. Depiction of male and female African elephant movement patterns in dry and wet seasons. The black dots depict the areas and the hourly GPS data retrieved by the team of researchers for male African elephants in the dry (1) and the wet season (2); and for female elephants in the dry (3) and the wet season (4). In both sexes the individuals' movement paths seem to be tightly anchored on a river in the dry season. GIS maps of African elephant migration patterns in wet and dry season obtained from Wato et al. (2018).

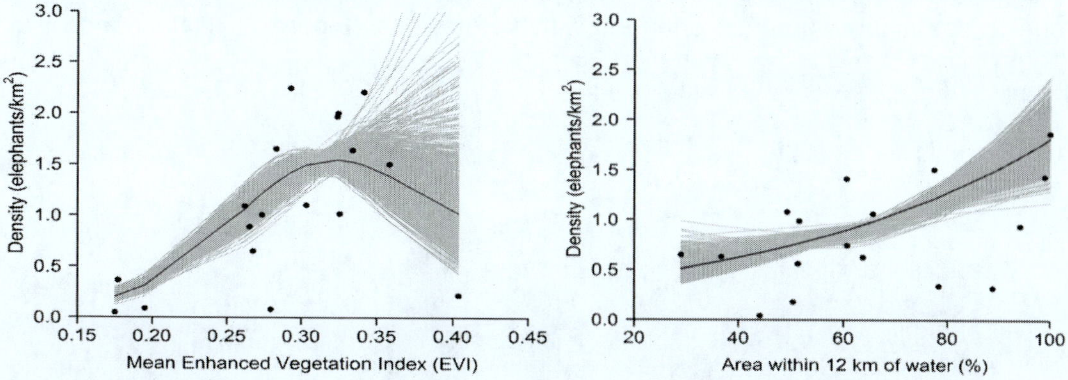

Figure 8. Generalized additive maps (GAM) of the density of African elephant population (number of elephants/km^2) and the mean vegetation index (Graph A) and water availability (Graph B) (Robson et al., 2017).

Seasonal Patterns

While vegetation and water are essential indicators of elephant movement patterns, seasonal migration and foraging patterns analyzed via GIS and RS technologies also provide valuable information regarding elephant ecology. This valuable information enables ecologists, conservationists, and policy analysts to implement preventative and strategic management efforts for the protection of this species (Fullman & Bunting, 2014; Ntumi et al., 2005; Schlossberg et al., 2018; Schmitt et al., 2018; Wato et al., 2018). Through the utilization of RS camera traps, scanners, and GPS collars, Adams et al. (2016) studied the movement of elephants through various wildlife corridors in agricultural, industrial, and open land spaces over the span of 2 years within Botswana (Figure 9). This data was collected and categorized between cold dry, hot dry, post wet, and wet seasons (Adams et al., 2016). The RS data, which was analyzed via GIS configured maps, found that the elephant population within and near the Kasane and Kazungula townships travelled the farthest from their ranging areas and came closest to human settlements during the post wet season (Figure 9) (Adams et al., 2016). It can be inferred that elephants in the region were more inclined to travel farther in this season due to the abundance and richness in vegetation (far from water sources) caused by rainfall in the wet season (examples of healthy vegetation can be seen in Figures 5 and 6) (Clark, 2013).

Moreover, Nellemann et al. (2002) found that during the dry season in northern Botswana, "stem breakage was, on average, almost twice as high in rugged terrain compared with flat terrain at similar distance to water within the same vegetation type." Rugged terrains during the dry season were also found to have very high concentration of the Combretum shrub densities, which are a rich source of elephant forage in the dry seasons (Nellemann et al., 2002; Owen-Smith & Chafota, 2012). These results further suggest that the terrain identifications through GIS and RS technologies (like those seen in figures 3-4) can be useful in elephant movement pattern predictions as they enable identifications of areas which are hotspots for increased elephant densities due to seasonal vegetation changes within Botswana (Figure 7) (Nellemann et al., 2002).

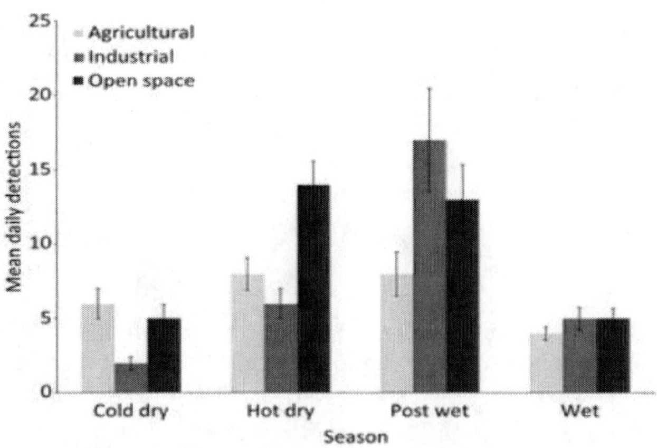

Figure 9. Mean number of African elephants detected per day in wildlife corridors through agricultural, industrial, and open land in the townships of Kasane and Kazungula, in the cold dry, hot dry, post wet and wet seasons (Adams et al., 2016).

Furthermore, several GIS and RS analyses identified that elephant movements are comparatively swifter in unprotected areas such as agricultural, industrial, and open spaces than in protected areas (Adams et al., 2016; Douglas-Hamilton et al., 2005; Graham et al., 2009). This further indicates the resistance elephants have in travelling through highly dense human settlements (Adams et al., 2016; Douglas-

Hamilton et al., 2005; Graham et al., 2009). Elephant resistance towards HEC is further solidified by the fact that elephant usage of corridors is substantially higher during the night than during the day, coinciding with the times humans utilize those land spaces the least and most, respectively (Adams et al., 2016; Douglas-Hamilton et al., 2005; Graham et al., 2009).

GIS and RS Analysis

While traditional methods of monitoring elephant patterns have been helpful in elephant ecology research (Boshe, 1990), the combination of GIS and RS technologies have advanced the analysis of these data immensely. They equip wildlife management with the qualitative and quantitative data required for local authorities to implement conservation strategies. These conservation efforts improve elephant quality of life, their natural environment, and surrounding human settlements (Barnes, 1982; Clark, 2013). Therefore, GIS and RS technologies play a crucial role in understanding elephant patterns within Botswana.

As seen throughout many conservation strategies, RS technologies range from GPS collars to Landsat satellite imagery. The technique of tagging elephants with GPS satellite collars has been used widely and enables ecologists to identify the precise and accurate locations of elephants at hourly, daily, and even yearly time intervals (Figure 7-6) (Sowers et al., 2015). Recent advancements in technology have enabled the use of GPS satellite collars in conjunction with satellite imagery of land spaces to provide high quality spatial and temporal data regarding habitats that have dense canopies and vegetation (Figure 6) (Gadsden, 2019; Sibanda et al., 2016; Sowers et al., 2015). In addition to RS tools, GIS mapping is a powerful tool in elephant ecology research as it enables the presentation of multiple datasets via layers on a common map (Figure 4, Figure 7) (Gadsden, 2019). For instance, the infrared and water band image layers within ArcGIS (Table 1) enable researchers to identify hidden waterways, trail expansions, and wildfire patterns, which collectively impact elephant movements (Gadsden, 2019).

The combination of GIS and RS in the arena of elephant ecology further enables wildlife management teams to model habitat suitability maps, conduct their analyses, and visualize elephant migratory patterns (Gadsden, 2019; Wato et al, 2018). These analyses provide valuable cause and predictor data regarding range metrics, poaching threats, patchiness of resources, vegetation abundance, and water availability, all of which may potentially remain undetected through the sole use of field observations (Figures 3 - 7) (Sekar et al., 2017; Sibanda et al., 2016; Sowers et al., 2015). Moreover, these analyses can highlight causes and impacts of the HEC within Botswana, leading to the implementation of effective strategic management.

HUMAN-ELEPHANT CONFLICT

Shared Spaces

HECs within Botswana have been on the rise within the past several years (Mayberry et al., 2017). With a growth in human population by approximately 2% per year, about two-fifths of the population within Botswana is living in rural areas (BITC, 2021; World Bank, 2019). These rural settlements range from small, scattered hamlets which are sparsely populated to traditional towns which consist of thousands of people (BITC, 2021; World Bank, 2019). This growth in population leads to low-income communities increasingly competing for resources such as food, water, and land space with other humans and elephants (Robson et al., 2017; van Aarde et al., 2021). Moreover, low-income farmers tend to live near forests and protected elephant areas for increased access to irrigation and land spaces that can sustain agricultural practices (Fullman & Bunting, 2014; Pozo et al., 2017). In dry seasons, farmers move even closer to forested areas to have increased access to permanent water sources in order to meet basic household needs, grow crops, obtain bountiful harvests, and sustain their livestock (Fullman & Bunting, 2014; Pozo et al., 2017; Robson et al., 2017).

Through elephant pattern analyses conducted via GIS and RS technologies, it was found that elephants in Botswana most frequently expand their range for vegetation and water availability during the dry and post-wet seasons (Figure 7). This expansion in range and the process of farmers relocating closer to permanent water sources within forests leads to many direct and indirect competitions between humans and elephants within the region. When farmers utilize more land space for agricultural purposes, they conduct activities which are inviting factors for the elephant population. For instance, the removal of forest areas by the cutting and burning of existing vegetation for agricultural expansion and improvement of livestock forage, deprives the elephants from their preferred forage and restricts them to forage in the areas prepared by farmers (examples of cleared vegetation over time can be seen in Figure 6 and Figure 10) (Adams et al., 2016; Barnes, 1982). This leads to increased crop raidings by elephant herds which can be detrimental for the livelihood of low-income farmers since a single elephant can consume a hectare of crops in a relatively short span of time (Adams et al., 2016; Barnes, 1982; WWF, 2008). This leaves financially vulnerable farmers to lose their entire livelihood from a single crop raid overnight (Adams et al., 2016; Barnes, 1982; WWF, 2008). Moreover, fields which are planted adjacent to water sources and holes which are dug to access groundwater also pose as invitational factors for elephants, especially during the dry seasons when water scarcity is high within Botswana (Adams et al., 2016; Barnes, 1982; WWF, 2008).

Due to these increased agricultural activities within forests, the natural habitat of the elephants within Botswana has declined tremendously. The country has experienced significant decline in natural forage, landscape connectivity, as well as in the Elephant population relative to their historical size and range (Figure 3; Figure 6)

(Adams et al., 2016; Barnes, 1982; WWF, 2008). As the elephant natural habitat decreases, their ranging distance within human settlements increases; within Botswana, the elephant range has increased by 53% leading to increased human-elephant proximity and thereby increased conflicts over food and water resources (EWB, 2019). Despite their large size, due to the woody vegetation that surrounds the forested areas, the elephants can go undetected by rural farmers as they blend into the vegetation, rendering easy pathways for crop raidings (Figure 10). As seen through the RS Landsat image of Seronga, a village in Botswana near the Okavango Delta (Figure 10), potential pathways elephants might take to access resources from hamlets, villages, and towns without being detected is depicted.

Figure 10. Visible band (Table 1) image obtained from Google Maps Satellite images indicating a potential African elephant entry pathway from Okavango Delta into Seronga, a village within Botswana. The area below the yellow boundary is a part of Okavango Delta while the area above is in the village of Seronga. The red arrows indicate a potential pathway of elephant entry into a Seronga settlement from inside the Okavango Delta. Blue circles indicate woody browse forage (table 1) to which the African elephant gravitate towards and can move undetected in; the orange circle exemplifies unhealthy vegetation and cleared fields (Table 1).

While the elephant population has been observed to avoid contact with humans, their ventures into human settlements and farms can result in the injury and/or killing of livestock and in some cases even humans (Graham et al., 2009; Pozo et al. 2017; Sitati et al., 2003). "In India, over 100 people are killed by elephants each year, and over 200 people have been killed in Kenya over the last 7 years" (WWF, 2008). Upon these interactions, several retaliatory actions such as the chasing, injury, and killings of elephants increase the HEC (Graham et al., 2009; Sitati et al., 2003). A prime example of HEC over land space within Botswana can be seen in the Okavango Delta region (Figure 5) between the years of 2009 to 2015, during which elephants were responsible for raiding at least 1518 fields, resulting in the retaliatory killing of at least

25 elephants per year within the region (Pozo et al. 2017). Not only are elephant lives at risk due to these conflicts, but the lives of humans living near elephant ranges are at risk (Figure 5). In 2018 alone, elephants were responsible for the deaths of approximately 36 people across Botswana (Chaukura et al. 2020).

In addition to crop raids, HEC within Botswana exist due to the collection of firewood by humans. Firewood is a source of energy for many rural communities as it can be used for cooking, heating, and in some cases as energy used to generate electricity (Sreevani, 2018). Redmore et al. (2020) found that firewood is the sole source of energy for people living near the Okavango Delta. Although these people prefer to remain close to their homes for harvesting firewood, they are required to travel further into elephant habitats for its collection as the increasing settlements have led to proximate trees being cleared for space and building materials (Redmore et al., 2020). Based on these findings, it can be inferred that in order to obtain firewood, people within the region often depend on elephant foraging behavior to uproot trees and branches for firewood harvest purposes. As analyzed via GIS and RS technologies, though elephant foraging patterns unintentionally prepare the firewood harvest for these communities, close proximities between the elephant and humans due to firewood have led to several human casualties. For instance, in 2016 a 65-year-old male ventured into the woodlands behind his home to harvest firewood at night but was killed by an elephant attack near the Okavango Delta region (Redmore et al., 2020). The time of the attack coincides with the findings of Adams et al. (2016) which highlight that elephants travel near human settlements at night to avoid human contact; however, due to the lack of daylight, unforeseen interactions between humans and elephants can startle both parties, leading to attacks.

GIS and RS analyses of the elephant movement patterns such as those studied by Adams et al. (2016), can be of great significance in mitigating the impacts of HEC within Botswana. The data presented from these studies (Adams et al., 2016; Clark, 2013) highlight and visualize the times and seasons during which elephants are most likely to travel near human settlements. These are key pieces of information to equip the Botswanian population with, as they can be used in implementing preventive measures within human patterns of farming and harvesting firewood. Conservationists along with policy makers can bring forth effective changes by implementing policies which utilize the elephant pattern data to enforce preventative measures. Educating the public of information regarding the seasons during which extra precaution is required and the times at which firewood should be collected would not only ensure the protection of the human population, but of the elephant population within the country as well.

In addition to the utilization of GIS and RS technologies to implement policies, veterinary fencing has been widely utilized in efforts to reduce these HEC conflicts within Botswana (Campos-Arceiz et al., 2008; Guldemond & Van Aarde, 2008; Riddle et al., 2010). This mitigatory attempt involves fencing agricultural and protected areas to minimize crop raiding and maximize the protection of vegetation from elephant grazing and trampling (Campos-Arceiz et al., 2008; Guldemond & Van Aarde, 2008; Riddle et al., 2010). However, the majority of habitat and food required for elephants to survive is found beyond the fenced areas (Robson et al., 2017). Therefore, the mitigatory attempt of enclosing elephants within restricted areas can lead to significant issues regarding elephant-to-elephant conflict as well as habitat degradation due to unsustainable proportions of the elephants (Schlossberg et al., 2019). Figure 9, obtained from the combination of a RS visible band image (Table 1) into a GIS layered map, depicts the impact veterinary fencing has had on the elephant population within Seronga, a village near the Okavango Delta (van Aarde et al., 2021). Based on the information provided in this Figure (9), it can be seen that although the measure of fencing has successfully kept the elephant population within the bound region, concentrated elephant forage has resulted in biodiversity losses (Shaffer et al., 2019). Therefore, mitigatory strategies which benefit both, humans and elephants need to be implemented to ameliorate elephant ecology.

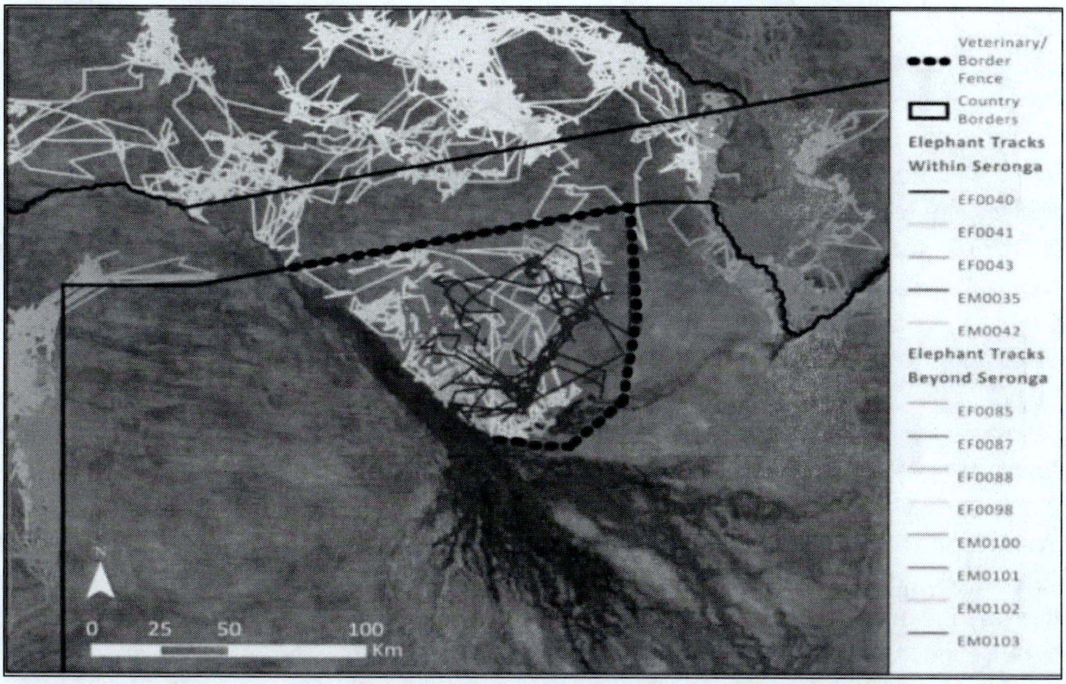

Figure 11. Visible band RS satellite imagery (Table 1) combined with GIS mapping depicting the pathways of GPS tracked elephants over the course of 6 years from 2004 to 2010 (van Aarde et al., 2021).

While current conflict management strategies focus on human prevention through exclusionary veterinary fences, these management approaches merely address the symptoms of shared space HEC rather than the underlying drivers of HEC (Shaffer et al., 2019). While the implementation of veterinary fencing can be helpful in decreasing human-elephant interactions, it is important to ensure that the elephant population has adequate access to water sources. Without proper access given to elephants, the elephants cause damage to the fences in order to retrieve the resources required, thereby nullifying the purpose of the fence (Adams et al., 2016). Therefore, the use of corridors within high conflict areas can be used to tremendously reduce human-elephant interactions within Botswana (Pozo et al., 2017). Through the use of GPS collars and GIS layered maps, Adams et al. (2016) found that wide-ranging corridors in highly dense human settlements such as agricultural, industrial and open-space lands, resulted in reduced HEC. These corridors enabled the elephants to access the resources they require without causing havoc within the human settlements (Adams et al., 2016). Implementation of wildlife corridors, therefore, reduces the burden elephants face due to resource inaccessibility, resulting in reduced HEC and increased QOL.

Human Quality of Life

Although crop depredation and human casualties are the most tangible and publicized costs of the HEC, hidden costs in the form of diminished psychological well-being, disruption of livelihoods, food insecurity, and decreased social interactions are factors of HEC which adversely impact QOL within Botswana (Jadhav & Barua, 2012; Barua et al., 2013). These hidden costs create a barrier for the citizens of Botswana to appreciate and tolerate the elephant population, thus signifying the importance of addressing HEC factors which lower QOL.

Unwanted elephant interactions can cause a significant decrease in the psychological well-being in humans. These unwanted interactions can be analyzed via GIS and RS technologies and have been observed to occur in agricultural land spaces or during the collection of firewood in elephant habitats (Figure 10) (Mayberry et al., 2017). The traumas experienced by these interactions can be the result of the psychological stressors caused by financial burdens brought forth by elephant crop raids, or physical stressors caused by being chased, injured, or nearly killed by elephants (Mayberry et al. 2017). Due to these traumas, perceived risks of elephant interactions also play a significant role in declining the quality of human mental health (Barua, Bhagwat, & Jadhav, 2013; Jadhav & Barua, 2012). Perceived risks and thereby the onset of fear can be triggered due to the abundance in elephant footprints, feces, and signs of uprooted trees and shrubs; all of which serve as continual reminders of the shared human and elephant landscape, thereby increasing stress and decreasing QOL (Barua, Bhagwat, & Jadhav, 2013; Jadhav & Barua, 2012). Furthermore, Redmore et al. (2020) found that perceived risk of elephants within Botswana is more prevalent in females than males; this was solidified through the finding that most females chose to collect firewood in groups as opposed to men who chose to travel alone for firewood harvesting (Figure 12). The perceived risk of elephants from the basis of travelling alone, infers that men may perceive less risk due to the ideals of masculinity which may be prevalent within the country (Redmore et al., 2020). As seen through this study, although perceived risks of HEC may differ by gender within Botswana, the underlying costs and consequences of these perceptions remain the same.

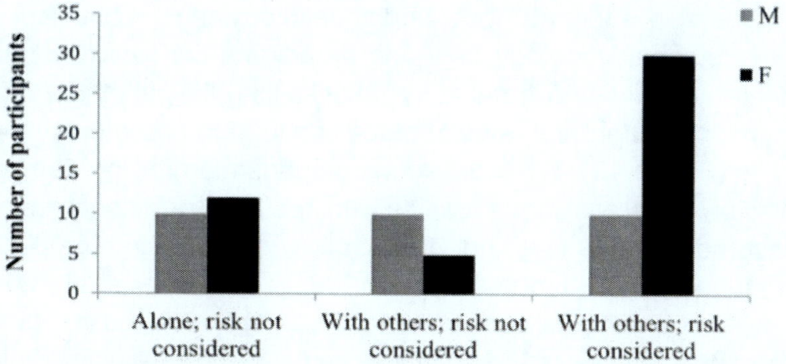

Figure 12. Depiction of gender-based risk perception of elephants within Botswana obtained from Redmore et al. (2020).

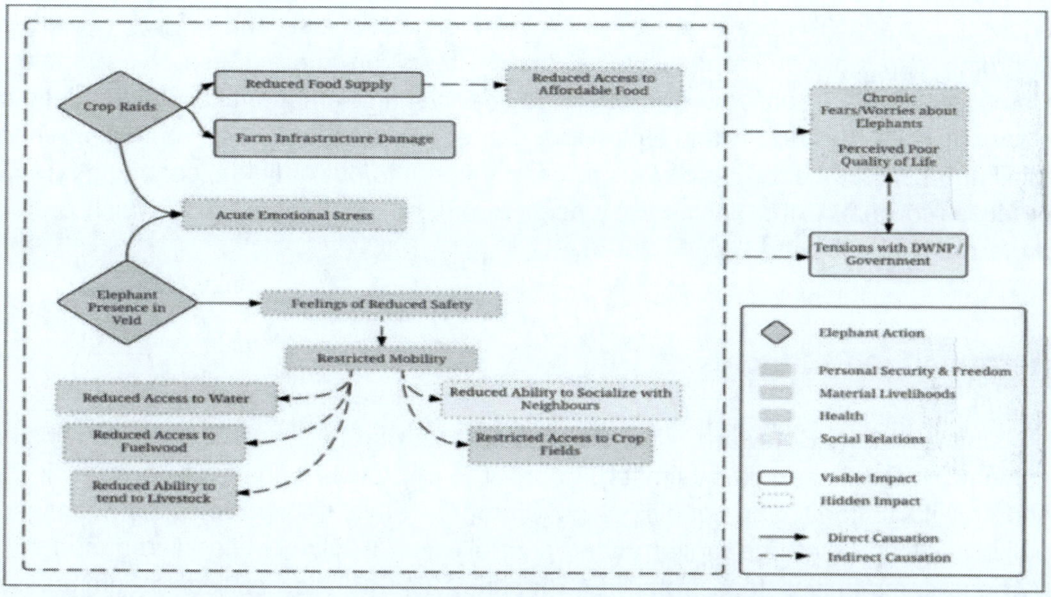

Figure 13. Visual summary of the hidden costs of HEC within Botswana and their impact on QOL as obtained from Mayberry et al. (2017).

Perceived risks from crop degradations and fatalities from elephant attacks result in consequential impacts that amplify family conflicts and untreated mental disorders including alcoholism (Figure 13) (Mayberry et al. 2017). These unresolved issues can then result in the development of intergenerational psychiatric morbidities which adversely impact mental health. Negative health impacts of HEC, therefore, penetrate far deeper than immediate physical threat from elephants, worsen pre-existing mental illness of marginalized people, and lead to newer psychiatric and social pathologies (Figure 13). These conflicts are further perpetuated in institutional spaces of socio-economic and gender-based inequalities (Mayberry et al. 2017). In order to bring forth meaningful and effective changes to elephant ecology, both wildlife conservation and community mental health disciplines need to implement coordinated interventions.

Interdisciplinary paradigms that bridge ecology and QOL are vital in the improvement of elephant ecology as these paradigms can positively impact human wellbeing, thereby mitigating retaliatory attacks on the elephant population. As previously mentioned, elephant raids of crops can lead to detrimental impacts for low-income farmers (Figure 13). When seeking compensation for the damage caused by the elephants, bureaucratic inadequacies can result in additional transaction costs which low-income farmers are not financially equipped to compensate (Barua, Bhagwat, & Jadhav, 2013; Jadhav & Barua, 2012; Mayberry et al., 2017). After the occurrence of elephant raids, considerable costs are incurred for farmers through nocturnal crop and livestock guarding; these costs include reduction in wage-earning activities as well as poor school attendance for children (Barua, Bhagwat, & Jadhav, 2013; Jadhav & Barua, 2012). Nocturnal crop guarding also increases vulnerability to elephant attacks (Adams et al., 2016), as well as exposure to vector-borne diseases such as Malaria (Simon et al., 2013). Moreover, sleep loss and fatigue caused by guarding can be detrimental due to their effects on morbidity (Lock, Bonetti & Campbell, 2018). In order to decrease the financial and emotional burden caused by elephant raids, farmers within Botswana can adopt practiced such as implementing "chili and tobacco-based deterrents to keep elephants out of fields; changing farming practices to ensure that farms are easier to defend; and grow crops which are not preferred by elephants" (WWF, 2008).

Reformation Strategies

Unfortunately, Botswana's governmental response to these visible and invisible costs of HEC has been very limited. In order to improve elephant ecology within the country, its national and municipal governments have implemented compensation policies which reinforce government control over wildlife while disregarding the burdens communities face from HEC (Mosha, 2020). This has led to communities affected by unwanted elephant interactions to increasingly seek more rights to combat HEC in manners they find appropriate (DeMotts & Hoon, 2012). HEC within Botswana can therefore be seen as a point of tension between governing authorities and local communities (DeMotts & Hoon, 2012). While rural communities are aiming for retaliatory methods of management, political strategies are aiming for expanding conservation territories and generating tourism revenues (DeMotts & Hoon, 2012; Mayberry et al., 2017). Thus, effective mitigation of the HEC and improvement of QOL within Botswana requires a shift in the current approach. In order for effective mitigation of the HEC and improvement of QOL to be brought forth, open communication between governmental authorities and the general public is required. When local communities feel that their concerns regarding HEC are heard and valued by decision-makers, there will be an increase in the compliance of elephant conservation efforts (Newton, 2020). This shift in conservation approach will require a unified effort from Botswana's society ranging from individuals, communities, non-governmental organizations, the municipal and national governments, as well international organizations (Mosha, 2020).

In addition to the cooperation of decision makers and local communities, HEC within Botswana can be mitigated through species conservation and urban planning programs. Some examples of programs which can be adopted by the country are the Integrated Planning Approach; Environmental Planning and Management Approach; and the New Urbanism Concept and Sustainable Cities Approach (Mosha, 2020). Equipping Botswana's citizens with information provided from these programs can help reduce poverty, increase participation by women in society, improve governance structures, increase food security, and deliver a sustainable environment for future generations (WWF, 2008). Additionally, GIS and RS analyses of elephant patterns can significantly contribute to the education provided through these programs (EWB, 2019). The modelling and measuring of elephant patterns through GIS and RS analyses (Figure 9, 10. 11) can improve conservation management planning, and thereby improve cultural values, norms, and decision-making regarding the spatial and temporal use of habitat by local communities (Mosha, 2020). This sociocultural component of HEC mitigation via GIS and RS analyses has not been effectively integrated into current elephant management strategies within Botswana; however, it is a vital component in bringing forth effective elephant ecology reformation (Mosha, 2020; Shaffer et al., 2019). These analyses ultimately promote effective planning for all land-use processes and ensure that current and potential HEC are considered in land planning decisions. This can help Botswana's citizens steer clear of unwanted HEC, thereby substantially increasing the livelihoods of communities living with elephants and reducing the financial and psychological burdens of HEC.

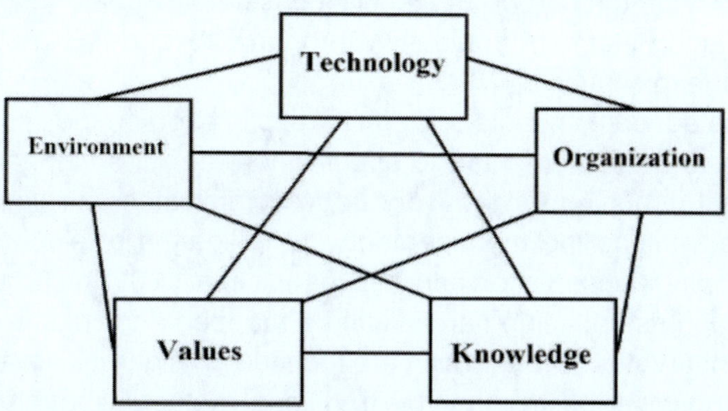

Figure 14. The coevolutionary relationship of technology, geomatics, and the environment within the scope of conservation biology (Norgaard, 1994).

A theoretical model proposed by Norgaard (1994) can be utilized to depict the various interconnected factors involved in effective elephant ecology reformation and improved QOL. Within this framework, technology can be represented by GIS and RS; environment can be represented by the elephant ecology within Botswana; organization can be represented by the management of HEC by municipal and national governments; values can be represented by the ethics of upholding elephant conservation and the prioritization of physical and psychological burden of HEC; and knowledge can be represented by the contribution GIS and RS analyses bring to the understanding of elephant ecology. Each component listed within this framework plays an integral role in reformation strategies.

AFRICAN ELEPHANT EXPLOITATION

While habitat loss and retaliatory killings by farmers pose significant threats to the elephant ecology, elephant exploitation via poaching threatens the species the most (Gadsden, 2019; Schlossberg et al., 2019; Sibanda et al., 2016). Approximately 35,000 elephants are killed every year for their tusks and between the years of 1979 and 2015, poaching for ivory reduced Africa's elephant population from an estimated 1.3 million to around 400,000 (Figure 3) (Schlossberg et al., 2019). In northern Botswana, fresh elephant carcasses increased by 593% from 2014 to 2018 (Figure 3), with approximately 385 elephants poached between 2017 and 2018 and approximately 156 elephants poached in 2018 alone (Schlossberg et al., 2019).

The drive for illegal ivory trade is fueled by demands for ivory jewelry, which are a symbol of wealth and social status (Shaffer & Bishop, 2016). "High demands of ivory have also driven black market values for raw elephant tusks to approximately $1,700 USD per pound, while finished ivory art pieces are being sold for as high as $830,000 USD each" (Shaffer & Bishop, 2016). Mature elephants around the age range of 40 years or older are the most commonly targeted, as tusks grow as with age (Figure 15C). The older the elephant, the larger the tusk, the higher the payout for poachers. This valuable information can be utilized by conservationists to focus their efforts on monitoring the aging elephant population through GIS and RS, as their tusks are in high demand for exploitation.

Due to these adverse impacts of elephant exploitation on the species, appropriate measures need to be implemented to effectively reform the elephant ecology within Botswana. Illegal poaching for the ivory trade complicates calculations for elephant losses in Africa, making the management and mitigation of elephant exploitation a priority in developing conservation efforts (Sibanda et al., 2016). Through the utilization of RS drones, high-risk poaching areas can be mapped. Findings from recent studies indicate that conservation efforts which are focused on high-risk areas through drones are more time and cost-effective than the traditional method of patrolling entire habitats (Shaffer & Bishop, 2016). While traditional methods can cost up to $50,000 USD per week, drone surveillance approximately cost $3000 USD per week (Shaffer & Bishop, 2016). Furthermore, by monitoring high-risk poaching areas with drones, conservationists and researchers were able to identify that more than 85% of poaching incidents occurred in areas of open, low shrub or savanna areas (Shaffer & Bishop, 2016). Due to the vast elephant habitat in Botswana, the utilization of drones to monitor poaching can therefore be very effective in mitigating elephant exploitation and protecting citizens from being attacked by poachers.

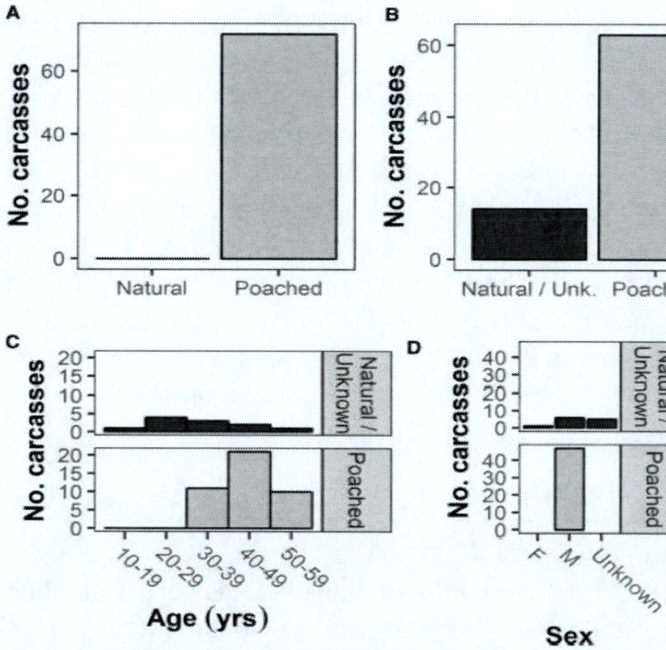

Figure 15. Results from 2018 Carcass Assessments in Northern Botswana (Schlossberg, Chase, & Sutcliffe, 2019). (A) depicts fresh and recent carcasses, (B) depicts old carcasses, (C)depicts the ages of old carcasses, and (D) depicts the sexes of old carcasses.

In addition to RS technologies, GIS can also be impactful in investigating the elephant poaching crisis within Africa. One example of the mitigation of elephant exploitation occurred in Garamba where ecologists used GIS to better understand the movement of poachers (Gadsden, 2019). They did so by tagging the elephants with GPS collars and monitoring the speed by which the elephants moved; the GPS collars provided researchers with live movements, thereby allowing the visualization of any change in natural movement patterns on a digital map (Gadsden, 2019). The digital GIS map displayed the elephant positions and movements as icons on the map and the color of the icons revealed the pace at which the elephant was moving (Gadsden, 2019). Visualization of a hasty convergence of the elephants in a particular part of the park indicated abnormal elephant behavior (Gadsden, 2019). Based on GIS and RS data analysis conducted on elephant patterns beforehand, it was identified that female elephants were inclined to travel in small herds led by matriarchs, while male elephants were inclined to travel solo, until high-risk situation occurred during which males formed small groups with other males (Gadsden, 2019). These patterns of sex-based movements indicated the factors causing exponentially high rates of male elephant poaching compared to their female counterparts (Figure 15 D) and helped identify the rationale behind erratic elephant movements (Gadsden, 2019). The abnormal behavior was confirmed by park rangers as the elephants' seeking safety in number from poachers. The visualization of these patterns in GIS maps further allowed the park ranger to identify locations to intercept the poachers before they could poach the elephants for their tusks (Gadsden, 2019). As a result of GIS and RS analyses of the elephant population, Garamba has witnessed a 97% reduction in elephant exploitation within the last three years (Gadsden, 2019). This example signifies the importance of GIS and RS analyses in mitigating the elephant exploitation occurring within Botswana.

GPS collars have a two-fold benefit in researching elephant ecology. As seen through the previous examples, they can help identify elephant movement patterns and act as deterrents for poachers from killing and transporting animal parts due to the tracking ability of satellites. Additionally, GIS and RS analyses of the elephant habitats can identify potential routes poachers may take to access the elephants. Since elephants remain close to permanent water sources (Barnes, 1982; Vancuylenberg, 1977), monitoring elephants for poaching along waterways can be useful in managing and mitigating the elephant exploitation within Botswana.

HUMAN-ELEPHANT DISEASE ECOLOGY

In addition to HEC and elephant exploitation, Botswana's elephant population is experiencing problems due to increased disease transmission. In just the months of May and June of 2020, the region of Okavango Delta has experienced more than 350 mysterious elephant deaths (van Aarde et al., 2021). Although veterinarians and wildlife experts believe the main cause of death of these elephants to be ingestion of toxic cyanobacteria in water (van Aarde et al., 2021), water toxicity within the region would have depicted more signs of congregated wildlife deaths near the water sources (van Aarde et al., 2021). Therefore, the explanation of sudden deaths such as those seen within Botswana, require exploration into disease ecology.

A noteworthy mention in Figure 9 is that elephants in Seronga, the area in which most of these mysterious deaths occurred, had highly restricted movements. Movements in southwest region were restricted by the Okavango River, while the remaining region was enclosed by veterinary fences (Adams et al., 2016). Given this setting, it is likely that a disease outbreak could spread rapidly through the elephant population. The combined effects of a large and isolated elephant populations, stress due to poaching and other human interactions, and frequent elephant-to-elephant interactions all create favorable conditions for disease to spread. In simple terms, the spread of disease in confined areas such as that depicted in Figure 9 may be deemed as inescapable.

In order to identify an accurate explanation of these deaths, experts suggest sampling carcasses, the soil, and water in their vicinity almost immediately after the elephants die (van Aarde et al., 2021). This can be a daunting challenge in this remote area, where elephant bodies may not be found and analyzed for days or even weeks (Bouché et al., 2011). Due to the intense heat within the region and scavenging animals, key evidence is degraded before the carcasses can be recovered for examination (Schlossberg et al., 2019). Therefore, the utilization of GPS tracking devices and RS surveillance can be greatly useful in remote areas where tracking the location of elephants is crucial in identifying causes of death in timely manners.

Moreover, the analysis of areas using water band imaging can also provide helpful information regarding the water quality within the regions (Table 1). Since scientists are attributing the toxicity within water due to cyanobacteria as the leading cause of death, these water band image analyses can help identify the validity of this claim.

Conclusion and Future Directions

As seen throughout this chapter, the ecological context provided by GIS and RS technologies in their dimensions of elephant movement patterns, identification of changes in land cover, HEC, and poaching, provide detailed information for the strategic management of elephant conservation and human QOL (Robson et al., 2017). They prove to be beneficial to the conservation and public health sectors at large scales such as the national research conducted in Botswana as well as at smaller scales such as villages like Seronga (Figure 11). Some limitations in this paper occurred in the analyses of zoomed-in locations in infrared and visible band imaging as the resolution of the images was inadequate. This may be because they did not have enough pixels to give detailed reflectance of landcover characteristics (Campbell, 2018). This limitation is important to overcome as conservation studies frequently require assessments of land cover to discern changes and possible relations between the vegetation variation and wildlife (Campbell, 2018). In this chapter, this limitation was resolved by using higher resolution satellite images found through Google Maps. Moreover, the use of remote sensing alone as a single instrument to determine elephant impact and habitat alteration can be misleading due "to the influence of ruminants and other ungulates on vegetation structure" (Jordaan, 2012). This limitation can be addressed through the integration of GIS and RS analyses with field observations and assessments. Direct observations of trails and footprints are valuable as they provide information of preferred elephant forage vegetation types and improve the accuracy of large-scale mapping (Jordaan, 2012).

Future research considerations of elephant ecology can explore tagging elephants with satellite microchips instead of GPS collars. The use of microchips in wildlife management studies causes less distractions within wildlife habitats than GPS collars. Moreover, GPS collars can be easily destroyed during poaching activities (Stabach et al., 2020); since microchips are not visible to the naked eye yet still transmit GPS signals, they can deter poachers from exploiting elephants for ivory, as the poachers will be uncertain of which elephants have tracking devices and which do not.

Similar to Botswana, unforeseen interactions of wildlife (ie. Black bear, wolves, and cougars) with humans have resulted in health-related issues in Canada (Frank et al., 2006). These include psychological stressors such as the fear of going into the wild due to attacks; stress due to financial loss because of wildlife; isolation due to being confined solely to safe zones; and anxiety that results from the misinformation of how to deal with the wildlife (Frank et al., 2006). However, the impact of these wildlife attacks has not been thoroughly studied through GIS and RS. This in-depth analysis is crucial to Canadian ecology and human health as it can be an indicator of health determinants, human physical and psychological QOL, and wildlife health.

Acknowledgments

This chapter would not have been possible without the support and guidance of Dr. Michael O'Neal Campbell and Philipa Shauna Dutt.

References

Adams, T., Chase, M., Rogers, T., & Leggett, K. (2016). Taking the elephant out of the room and into the corridor: Can urban corridors work? *Oryx, 51*(2), 347-353. doi:10.1017/S0030605315001246.

Aini, Suhaida & Sood, Alias & Saaban, Salman. (2015). Analysing elephant habitat parameters using GIS, remote sensing and analytic hierarchy process in Peninsular Malaysia. *Pertanika Journal of Science and Technology. 23.* 37-50.

Allen, P. G. (2016). *The Great Elephant Census.* Retrieved from http://www.greatelephantcensus.com/final-report

Barnes, R. F. W. (1982). Elephant feeding behaviour in Ruaha National Park, Tanzania. *African Journal of Ecology*, 20:123-136.

Barua, M., Bhagwat, S. A., & Jadhav, S. (2013). The hidden dimensions of human–wildlife conflict: health impacts, opportunity and transaction costs. *Biological Conservation*, *157*, 309-316.

Blanc, J. (2016). *Loxodonta Africana. The IUCN Red List of Threatened Species* 2016: e.T12392A3339343. https://dx.doi.org/10.2305/IUCN.UK.2008.RLTS.T12392A3339343.en.

BITC. (2021). *What is the minimum wage in Botswana?* Retrieved from https://www.gobotswana.com/what-minimum-wage-botswana.

Bouché, P., Douglas-Hamilton, I., Wittemyer, G., Nianogo, A. J., Doucet, J. L., Lejeune, P., & Vermeulen, C. (2011). Will elephants soon disappear from West African savannahs?. *PLoS One*, *6*(6), e20619.

Campbell, M. O. N. (2018). *Geomatics and conservation biology.* New York: Nova Science Publishers, Inc.

Chaboo, C. S., C. Nyamukondiwa,, T. Bird, R. K. Hitchcock,, M. C. Kelly, & M. Smith. (2019). *Beetle and Plant Arrow Poisons of the San Bushman Nations of Botswana.* National Geographic Society Grant No # NGS-181R-18. Washington, D.C.: National Geographic Society.

Chipanshi, A. C., Chanda, R., & Totolo, O. (2003). Vulnerability Assessment of the Maize and Sorghum Crops to Climate Change in Botswana. *Climatic Change* 61, 339–360. https://doi.org/10.1023/B:CLIM.0000004551.55871.eb.

Clark, H. (2013). *Seasonal movement patterns and home ranges of elephants in northern Botswana* (Doctoral dissertation, MSc thesis. University of Leeds, Leeds, UK).

Dale, R. H. (1996). *Studying Elephants: AWF Technical Handbook Series# 7.*

DeMotts, R., & Hoon, P. (2012). Whose elephants? Conserving, compensating, and competing in northern Botswana. *Society & Natural Resources*, *25*(9), 837-851.

Douglas-Hamilton, I., Krink, T., & Vollrath, F. (2005). Movements and corridors of African elephants in relation to protected areas. *Naturwissenschaften*, *92*(4), 158-163.

Daszak, P., Cunningham, A. A., & Hyatt, A. D. (2000). Emerging infectious diseases of wildlife – threats to biodiversity and human health. *Science*, *287*, 443.

Elephants without Borders (EWB). (2019). *Aerial Surveys*. Retrieved from https://elephantswithoutborders.org/projects/aerial-surveys/.

Field, C. R. & Ross, I. C. (1976). The savanna ecology of Kidepo Valley National park. *East Afrtican Wildlife Journal*, *14*:1-15.

Fullman, T. J., & Bunting, E. L. (2014). Analyzing vegetation change in an elephant-impacted landscape using the moving standard deviation index. *Land*, *3*(1), 74-104.

Gadsden, D. (2019). *African Parks Uses Tracking to Combat Poaching and Protect Animals.* Retrieved from https://www.esri.com/en-us/see/stories/african-parks/new-technologies-battle-poachers-protect-animals-article.

Graham, M. D., Douglas-Hamilton, I., Adams, W. M., & Lee, P. C. (2009). The movement of African elephants in a human-dominated land-use mosaic. *Animal conservation*, *12*(5), 445-455.

Gwebu, T. D. (2012). Botswana's mining path to urbanisation and poverty alleviation. *Journal of Contemporary African Studies*, *30*(4), 611-630.

Jadhav, S., & Barua, M. (2012). The Elephant Vanishes: Impact of human–elephant conflict on people's wellbeing. *Health & place*, *18*(6), 1356-1365.

Jensen, J. R. (2007). *Remote Sensing of the Environment: An Earth Resource Perspective, 2nd Edition.* Upper Saddle River: Pearson Education Inc.

Jonsson, P. (2004). Vegetation as an urban climate control in the subtropical city of Gaborone, Botswana. *International Journal of Climatology: A Journal of the Royal Meteorological Society*, *24*(10), 1307-1322.

Jordaan, M. (2012). The role of remote sensing for sustainable elephant management in South Africa. Four medium sized game reserves as case studies. *ISPRS-International Archives of the Photogrammetry, Remote Sensing and Spatial Information Sciences*, 259-264.

Laaidi K., Zeghnoun A., Dousset B., Bretin P., Vandentorren S., Giraudet E., Beaudeau P. (2012). The impact of heat islands on mortality in Paris during the August 2003 heat wave. *Environ Health Perspect.* *120*(2):254–259. doi: 10.1289/ehp.1103532.

Lock, A. M., Bonetti, D. L., & Campbell, A. D. K. (2018). The psychological and physiological health effects of fatigue. *Occupational medicine*, *68*(8), 502-511.

Mayberry, A., Hovorka, A., & Evans, K. (2017). Well-Being Impacts of Human-Elephant Conflict in Khumaga, Botswana: Exploring Visible and Hidden Dimensions. *Conservation and Society, 15*(3), 280-291. Retrieved March 24, 2021, from http://www.jstor.org/stable/26393296.

Mensah, C. A., Andres, L., Perera, U., & Roji, A. (2016). Enhancing quality of life through the lens of green spaces: A systematic review approach. *International Journal of Wellbeing*, 6(1), 142-163.doi:10.5502/ijw.v6i1.445.

McKenzie, D. & Formanek, I. (2016). *Catastrophic decline of Africa elephant population revealed by most comprehensive survey ever*. Retrieved from https://desdemonadespair.net/2016/08/catastrophic-decline-of-africa-elephant. html.

Mda Information Systems/Science Photo. (2018). "Coloured LANDSAT Image of Okavango Delta, Botswana - Stock Image - E552/0073." *Science Photo Library*, www.sciencephoto.com/media/174750/view/coloured-landsat-image-of-okavango-delta-botswana.

Mishra, N. B. (2014). Wetlands: Remote Sensing. In *Encyclopedia of Natural Resources: Land*. Taylor and Francis: New York, Published online: 21 Oct 2014; 566-574.

Mosha, A. C. (2020). A Reappraisal of Spatial Planning in Botswana. In *Urban and Regional Planning and Development* (pp. 113-128). Springer, Cham.

Nellemann, C., Moe, S. R., & Rutina, L. P. (2002). Links between terrain characteristics and forage patterns of elephants (Loxodonta africana) in northern Botswana. *Journal of Tropical Ecology, 18*(6), 835-844. doi:10.1017/s02664 67402002547.

Newton, K. (2020). Government communications, political trust and compliant social behaviour: The politics of Covid-19 in Britain. *The Political Quarterly*, *91*(3), 502-513.

Norgaard, R. B. (1994). *Development Betrayed: The End of Progress and a Coevolutionary Revisioning of the Future*. Routledge.

Ntumi, C. P., van Aarde, R. J., Fairall, N., & de Boer, W. F. (2005). Use of space and habitat by elephants (Loxodonta africana) in the Maputo Elephant Reserve, Mozambique. *South African Journal of Wildlife Research*, *35*(2), 139.

Owen-Smith, N., & Chafota, J. (2012). Selective feeding by a megaherbivore, the African elephant (Loxodonta africana). *Journal of Mammalogy*, *93*(3), 698-705.

Parsons, N. (2021). *Botswana*. Retrieved from https://www.britannica.com/place/Botswana.

Pozo, R. A., T. Coulson, G. McCulloch, A. L. Stronza, and A. C. Songhurst. 2017. Determining baselines for human-elephant conflict: a matter of time. *PLoS ONE* 12(6):e0178840.

Redmore, L., Stronza, A., Songhurst, A., & McCulloch, G. (2020). Where elephants roam: perceived risk, vulnerability, and adaptation in the Okavango Delta. *Ecology and Society*, *25*(4).

Robson, A. S., Trimble, M. J., Purdon, A., Young-Overton, K. D., Pimm, S. L., van Aarde, R. J. (2017). Savanna elephant numbers are only a quarter of their expected values. *PLoS ONE*, *12*(4): e0175942. https://doi.org/10.1371/journal.pone.0175942.

Rodriguez, M. (1994). Human health data needs and availability in Mexico. In *Appendix 6 of the report of the International Symposium on Core Data Needs for Environmental Assessment and Sustainable Development Strategies*, 2, p. 111.

Ruda, A., Kolejka, J., & Silwal, T. (2018). GIS-Assisted Prediction and Risk Zonation of Wildlife Attacks in the Chitwan National Park in Nepal. *ISPRS International Journal of GeoInformation*, 7(9), 369. doi: 10.3390/ijgi7090369.

Simon, C., Moakofhi, K., Mosweunyane, T., Jibril, H. B., Nkomo, B., Motlaleng, M., ... & Haque, U. (2013). Malaria control in Botswana, 2008–2012: the path towards elimination. *Malaria journal*, *12*(1), 1-9.

Schlossberg, S., Chase, M. J., & Sutcliffe, R. (2019). Evidence of a growing elephant poaching problem in Botswana. *Current Biology*, *29*(13), 2222-2228.

Schmitt, M. H., Shuttleworth, A., Ward, D., & Shrader, A. M. (2018). African elephants use plant odours to make foraging decisions across multiple spatial scales. *Animal Behaviour*, *141*, 17-27.

Sekar, N., Lee, C. L., and Sukumar, R. (2017). Functional Nonredundancy of Elephants in a Disturbed Tropical Forest. *Conserv. Biol.* 31, 1152–1162. doi: 10.1111/cobi.12907

Shaffer, M. J., & Bishop, J. A. (2016). Predicting and preventing elephant poaching incidents through statistical analysis, GIS-based risk analysis, and aerial surveillance flight path modeling. *Tropical Conservation Science*, *9*(1), 525-548.

Shaffer L. J., Khadka, K. K., Van Den Hoek, J., and Naithani, K. J. (2019) Human-Elephant Conflict: A Review of Current Management Strategies and Future Directions. *Front. Ecol. Evol. 6*:235. doi: 10.3389/fevo.2018.00235.

Sibanda, M., Dube, T., Bangamwabo, V. M., Mutanga, O., Shoko, C., & Gumindoga, W. (2016). Understanding the spatial distribution of elephant (Loxodonta africana) poaching incidences in the mid-Zambezi Valley, Zimbabwe using Geographic Information Systems and remote sensing. *Geocarto International*, *31*(9), 1006-1018.

Sitati, N. W., Walpole, M. J., Smith, R. J., & Leader-Williams, N. (2003). Predicting spatial aspects of human–elephant conflict. *Journal of applied ecology*, 40(4), 667-677.

Sowers, M., Fishlock, V. & Manor, T. (2015). *Mapping a future for Kenya's Amboseli elephants.* Available online: http://www.esri.com/esri-news/arcnews/summer15 articles/mapping-a-future-for-kenyas-amboseli-elephants.

Sreevani, P. (2018, August). Wood as a renewable source of energy and future fuel. *In AIP Conference Proceedings, 1992*(1), 040007. AIP Publishing LLC.

Stabach JA, Cunningham SA, Connette G, Mota JL, Reed D, Byron M, et al. (2020) Short-term effects of GPS collars on the activity, behavior, and adrenal response of scimitar-horned oryx (*Oryx dammah*). *PLoS ONE* 15(2): e0221843. https://doi.org/10.1371/journal.pone.0221843.

Sukumar, R. (2008). "Elephants in time and space: evolution and ecology," in *Elephants and Ethics: Toward a Morality of Coexistence* eds C. Wemmer and C. A. Christen (Baltimore, MD: Johns Hopkins University Press), 17–39.

Thouless, C., Dublin, H. T., Blanc, J., Skinner, D., Daniel, T., Taylor, R., ... & Bouché, P. (2016). African elephant status report 2016. *An update from the African Elephant Database.*

van Aarde, R. J., Pimm, S. L., Guldemond, R., Huang, R., Maré, C. (2021). The 2020 elephant die-off in Botswana. *Peer J 9*:e10686 https://doi.org/10.7717/peerj.10686.

Vancuylenberg, B. W. B. (1977). Feeding behaviour of the asiatic elephant in South-East Sri Lanka in relation to conservation. *Biol. Conserv.* 12, 33–54 doi: 10.1016/0006-3207(77)90056-8.

Wato, Y. A., Prins, H. H., Heitkönig, I., Wahungu, G. M., Ngene, S. M., Njumbi, S., & Van Langevelde, F. (2018). Movement patterns of African elephants (Loxodonta africana) in a semi-arid savanna suggest that they have information on the location of dispersed water sources. *Frontiers in Ecology and Evolution*, *6*, 167.

Western, D., Russell, S., and Cuthill, I. (2009). The status of Wildlife in protected areas compared to non-protected areas of Kenya. *PLoS ONE*. *4*:e6140. doi: 10.1371/journal.pone.0006140.

Whyte, I. J. (2012). "The elephant management dilemma," in *Environmental Ethics: What Really Matters, What Really Works 2 Edn*, eds D. Schmidtz and E. Willott (New York, NY: Oxford University Press), 71–84.

Wittemyer, G., Northrup J. M., Blanc, J., Douglas-Hamilton, I., Omondi, P. and Burnham, K. P. (2014). Illegal killing for ivory drives global decline in African elephants. *PNAS*. *111*(36):13117-13121.

World Bank. *Population growth - Botswana*. (2019.). Retrieved from https://data.worldbank.org/indicator/SP.POP.GROW?locations=BW.

WWF. (2008). *Common Ground – Reducing human wildlife conflict.* Retrieved from https://wwfint.awsassets.panda.org/downloads/hwc_final_web.pdf.

In: Critical Research Techniques in Animal ... ISBN: 978-1-53619-846-1
Editors: Michael O'Neal Campbell et al.

Chapter 9

Bio-Environmental Monitoring and Assessment of Species Using eDNA: An International Approach to Research and Future Possibilities

Philipa Shauna Dutt*

Trent University, Peterborough, ON, Canada

Abstract

Genome-based monitoring using environmental DNA (eDNA) has revolutionized the scope of bio-environmental monitoring and assessment globally. While traditional techniques rely heavily on observational and quantitative assessments which render a high degree of error, eDNA samples are obtained via non-invasive techniques and provide detailed and accurate biological information regarding aerial, terrestrial, estuarine, freshwater, and marine populations. The availability of these environmental samples varies based on the degradation rate; high degradation samples may be present in the environment for as little as a day, however, samples with low degradation rates may be accessible in the environment for hundreds of thousands of years. Depending on the type of sample, the use of eDNA in ecological monitoring can provide temporal and spatial information of the species. eDNA is readily used for its time efficient, cost effective, and reliable method of data collection. However, some limitations of this monitoring technique prevent environmental scientists from accurately assessing information concerning the abundance and biomass of populations, especially in marine environments. Therefore, this chapter will focus on the global use and benefits of eDNA in bio-environmental monitoring and assessment while discussing future directions of the integration of both types of ecological monitoring techniques to obtain an accurate image of species population and their environment.

* Corresponding Author's E-mail: philipa.dutt@gmail.com.

INTRODUCTION

Genome-based monitoring has revolutionized the scope of bio-environmental monitoring and assessment (Barnes & Turner, 2016; Thomsen & Willerslev, 2015). Bio-environmental monitoring and assessment plays a vital role in understanding species and their habitats; however, with the increase in anthropogenic activity and its negative impact in ecological habitats, these monitoring and assessment efforts are continually required to mitigate these negative impacts and increase conservation efforts. Anthropogenic activity has altered ecosystems and contributed towards climate change. Climate change affects habitats, causing species to relocate from native to foreign lands, leading to notable changes in their species demographics (Cook et al., 2016; Sinclair et al., 2010). Relocations of species can ultimately lead to extinctions if the species are not well-suited to their new environments (Cook et al., 2016; Sinclair et al., 2010). In order to successfully implement conservation strategies, accuracy of the population distribution data is required (Thomson & Willerslev, 2015). Therefore, it is imperative to implement appropriate monitoring and assessment methods to improve the effectiveness of future conservation efforts.

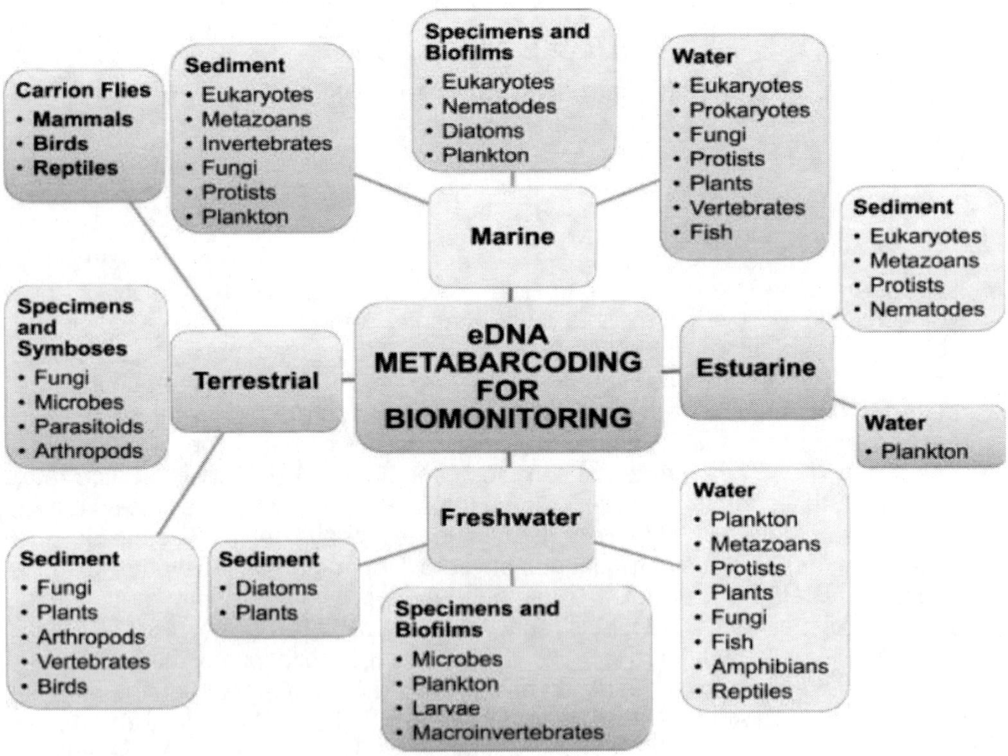

Figure 1. Ecosystem and biodiversity monitoring with eDNA (Ruppert et al., 2019).

Traditional monitoring focuses mainly on observational assessment, counting, and invasive techniques for data collection. However, the continuous relocation of species due to climate change and increased anthropogenic activities have jeopardized the validity of the species distribution patterns determined by traditional techniques (Thomsen & Willerslev, 2015). Alternately, monitoring through genome analysis provides detailed and accurate information by obtaining environmental samples via non-invasive techniques to identify the various species which exist or have existed in specific land spaces (Foote et al, 2012; Thomsen & Willerslev, 2015). Furthermore, the limitations of traditional monitoring are a prevalent hindrance in reliable data collection due to its inability to identify juvenile and cryptic species, leading to an urgent need of implementing alternative methods of data collection (Thomson & Willerslev, 2015). Genome-based monitoring, on the other hand, incorporates the use of environmental DNA (eDNA) which enables the identification of species, including juvenile and cryptic species, in terrestrial, aerial, and aquatic habitats (Figure 1) (Barnes & Turner, 2016; Deiner et al., 2017; Foote et al, 2012; Thomsen & Willerslev, 2015).

eDNA samples from these habitats (Figure 1) are obtained via non-invasive techniques from air, water, hair, urine, feces, ice cores, soil, snow, epidermal cells, gut contents, benthic specimens, and terrestrial and aquatic sediments (Table 1) (Shokralla et al., 2012; Taberlet et al., 2012; Thomsen et al., 2012; Thomsen & Willerslev, 2015). Depending on the type of sample, the use of eDNA in ecological monitoring can provide temporal and spatial information of the species (Figure 2). Figure 2 consists of an accurate visual representation of the versatility within various eDNA samples and the corresponding information that can be obtained upon their analyses.

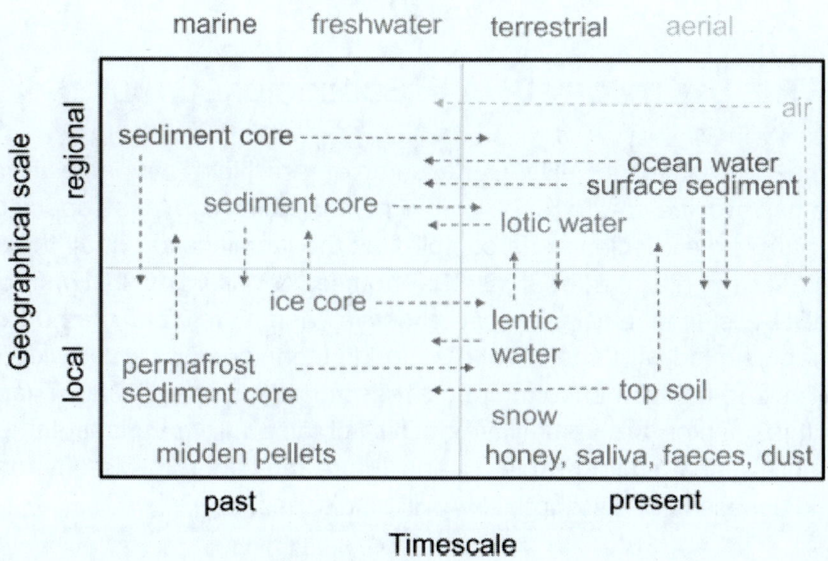

Figure 2. Visual representation of the different types of eDNA samples based on the geographical scale and time scales (Deiner et al., 2017).

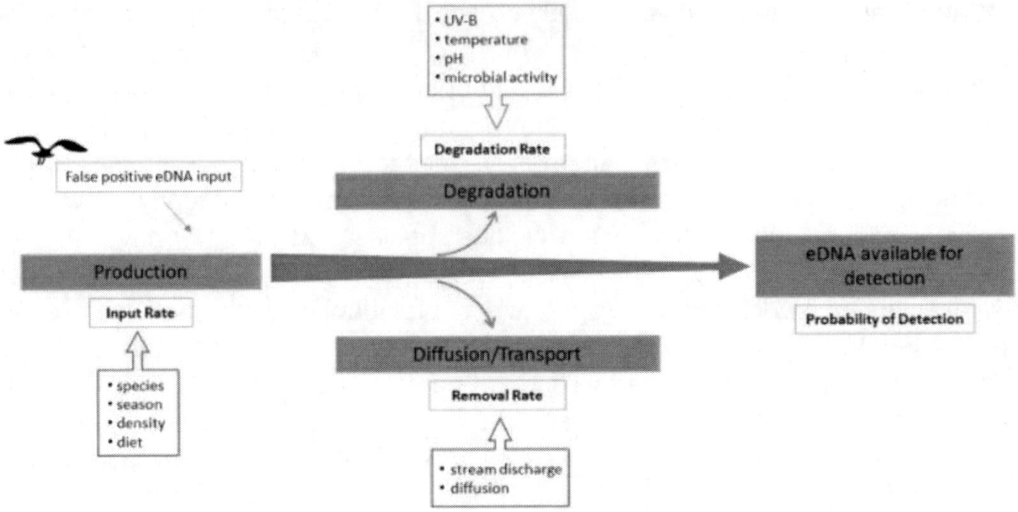

Figure 3. The process of eDNA input, degradation, transport, and availability of analyses (Strickler et al., 2015).

The availability of these eDNA samples (Figure 2) varies based on the degradation rate; high degradation samples may be present in the environment for as little as a day, alternatively, samples with low degradation rates may be accessible in the environment for exponentially longer periods of time ranging from hundreds to thousands of years (Strickler et al., 2015; Thomsen & Willerslev, 2015). Aquatic environments are predominant habitats in which eDNA undergoes high degradation due to several abiotic factors such as high temperature, solar radiation, and low pH; and biotic factors such as microbial activity (Figure 3) (Strickler et al., 2015).

THE UTILITY OF eDNA IN ECOLOGICAL STUDY

Despite the degradation of eDNA, the samples are highly valuable as a small eDNA sample can provide a wide range of information which can be as specific as the identification of a single species or as complex as the identification of all the species in an entire ecosystem (Ruppert et al., 2019; Thomsen & Willerslev, 2015). Moreover, the use of eDNA is time efficient, cost effective, and a reliable method of data collection; when coupled with traditional monitoring techniques it can provide optimal results in bio-environmental monitoring and assessment (Liu et al., 2020; Thomsen et al., 2012). In this chapter, the versatility of eDNA obtained from various samples will be explored on a global scale (Table 1) and future directions in bio-environmental monitoring and assessment utilizing eDNA will be discussed.

Table 1. Summary of the source, country, and timeline of the species identified using eDNA metabarcoding discussed in this paper, along with references of the corresponding studies

Species identified	eDNA sample	Country	Timeline	References
Humans, Cattle, Sheep, Goat, Mouse, and Reindeer	Sand Sediment	Greenland (South-West)	AD 1030 to 1530	Hebsgaard et al., 2009
Humans	Coprolites	USA (Oregon)	12,300 14C years B.P	Gilbert et al., 2008
Conifer Forest	Silty Ice	Greenland	450-800,000 years old	Willerslev et al., 2007
Plants, Mammals, Birds, and pre-human plant taxa	Cave Sediments and Permafrost cores	Siberia and New Zealand	400,000 to 10,000 years old	Hofreiter et al., 2003
Amphibians, Fish, Mammals, Insects and Crustaceans	Freshwater	Denmark, Sweden, Germany, Poland, and Estonia	2009 and 2011	Thomsen et al., 2012
European carp	Freshwater	Australia	2016	Furlan et al., 2019
Nematodes	Marine and estuarine sediments	England	~ 2006	Austen et al., 2006
American Bullfrog	Freshwater	France	2006 and 2008	Dejean et al., 2012
Silver Carp and Bighead Carp	Freshwater	USA (Chicago)	2009-2010	Jerde et al., 2011
Goat, Dog, Cow, Water Buffalo, Sheep, Horse, Kashmir Flying Squirrel, Red Fox, Leopard cat, Monal Pheasant, and Catfish	Feces	Pakistan	2008	Shehzad et al., 2015
Tuna, Tommy Ruff, Whiting, Small-Mouth Hardyhead, Jack Mackerel, Anchovy, Leatherjacket, Barracouta, and Gurnard	Feces	Australia	2007 - 2008	Deagle et al., 2010
Individual grass genera	Air	Netherlands	2012 - 2013	Kraaijeveld et al., 2015
Fungi	Air	Northern and Central Italy	2017	Tordoni et al., 2021
Fungi	Air	Beijing	2014	Tong et al., 2017

To analyze the eDNA samples collected, the DNA is extracted from the samples, amplified, and sequenced using Polymerase Chain Reaction (PCR) and Next Generation Sequencing (NGS) (Behjati & Tarpey, 2013; Garibyan & Avashia, 2013; Thomsen & Willerslev, 2015). PCR is a sensitive method used to amplify a small sample of eDNA to produce millions of copies of isolated specific fragments of DNA for analysis; these provide detailed information regarding the target species (Garibyan & Avashia, 2013). Additionally, quantitative real-time PCR (qPCR) is a predominantly used form of PCR which provides quantitative data about the abundance of specific DNA in a sample (Garibyan & Avashia, 2013). However, PCR can only identify species which have previously been sequenced within the database being used, since the amplified DNA is compared to already existing DNA sequences (Garibyan & Avashia, 2013). While PCR can only detect single species at a time, NGS is versatile and can sequence as small as single species and as large as entire genomes simultaneously (Behjati & Tarpey, 2013; Garibyan & Avashia, 2013). The use of NGS in identifying multiple species instantaneously is also known as eDNA metabarcoding; in most cases DNA is amplified with PCR and analyzed using NGS (Thomsen & Willerslev, 2015).

Species Identification via Aged eDNA

Aged dirt DNA is a form of eDNA composed of unconventional microfossils found in various sources such as fragments of sediments, ice cores, skin flakes, hair, feathers, nails, and fossilised feces which are crucial in identifying historical data regarding species and entire ecosystems (Table 1) (Hebsgaard et al, 2009; Hofreiter et al., 2003; Thomsen & Willerslev, 2015; Willerslev et al., 2003, 2007).

Sand sediment in South-West Greenland

Research shows that less than 2g of dirt DNA either frozen or under unfrozen conditions is adequate to extract information regarding various species and ecosystems (Hebsgaard et al, 2009; Thomsen & Willerslev, 2015). Sand sediment was explored by Hebsgaard et al., (2009) to determine the accuracy of information discovered by dirt DNA samples. The samples were obtained from a site which was unprotected, mostly unfrozen, and open, all of which are unfavorable conditions in terms of DNA preservation (Hebsgaard et al., 2009). Despite these conditions, Accelerator Mass Spectrometry (AMS) dating and eDNA analysis using PCR identified the immigration and emigration of humans, cattle, sheep, goat, mouse, and reindeer from AD 1030 to 1530 (Figure 4). These identifications enabled researchers to conclude that this land space was previously utilized for farming purposes (Hebsgaard et al., 2009). The results of this study show the notable contribution of eDNA in providing archeologists and researchers with the ability to narrate the migration patterns of species. It also brought awareness to the fact that dirt DNA which is unfrozen dirt or near aquatic environments should not be disregarded as unreliable (Hebsgaard et al., 2009). This study shows potential for dirt DNA to be utilized in the study of diet and genetics of ancient species (Hebsgaard et al., 2009).

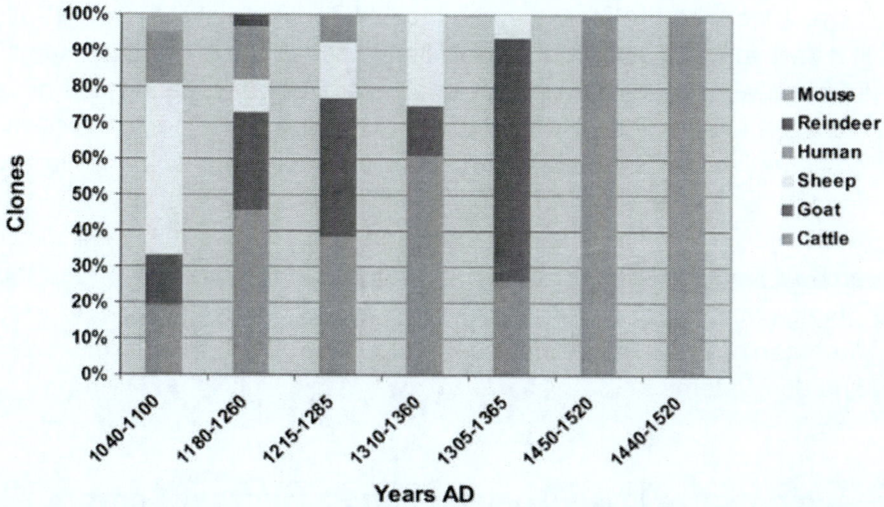

Figure 4. Abundance of species identified via sand sediment eDNA (Hebsgaard et al., 2009).

Coprolites from Oregon, USA

Coprolites are fossilized feces which were utilized as eDNA samples in a study conducted by Gilbert et al., (2008), in which human mitochondrial DNA was extracted for analyses which resulted in the discovery of crucial information regarding the first inhabitation of humans in south-central Oregon. AMS radiocarbon dating, and PCR provided ample information in tracing back the initial inhabitation of the area (Gilbert et al., 2008). Results revealed that humans were present in the area relatively 1000 14C years earlier than the current understanding of the initial inhabitation (Gilbert et al., 2008). To confirm the validity of the data obtained, eDNA samples were sent to laboratories in different regions of the world (Europe and USA) and the conclusive results confirmed that the earliest inhabitation of humans in south-central Oregon was indeed around 12,300 14C yr. B.P (Gilbert et al., 2008). This study proves the importance of using eDNA as a reliable source to discover information regarding ancient history and the initial inhabitation of areas. Limiting factors such as sample contamination and DNA leaching were accounted for and mitigated by conducting additional tests, increasing the credibility of the research (Gilbert et al., 2008). The findings from this study depict the importance in utilizing coprolites as eDNA samples and therefore should be utilized in the future to discover historic evolutionary data for species.

Silty Ice from Greenland

Extensive information regarding historic inhabitation of species and ecosystems is unable to be discovered using traditional techniques in areas such as Greenland due to the lack of access to the fossils embedded under thick layers of ice (Willerslev et al., 2007). However, in a study done by Willerslev et al., (2007), eDNA samples were collected from the cores of silty ice for analyses. PCR and dating techniques were used to analyze the samples which led to the discovery of the earliest evidence of the existence of a native conifer forest, dating back to approximately 2.4 million years ago in Greenland (Willerslev et al., 2007). To yield credible findings, preventative measures were implemented to avoid DNA contamination and species misidentification during the research (Willerslev et al., 2007). In the future, similar and advanced techniques can be implemented on a larger scale to widen the horizon of research in paleoenvironments to obtain a deeper understanding of the evolutionary processes of species and ecosystems.

Cave Sediments from New Zealand and Permafrost Cores from Siberia

Cave sediments are valuable in discovering historic information regarding paleoenvironments as they are composed of a variety of inorganic and organic matter (Hofreiter et al., 2003). The organic substances are generally composed of organismal remains of species that previously inhabited or visited the caves (Hofreiter et al., 2003). The types of remains available at a cave heavily depend on the climate of the cave; cave climates range from humid to dry, depending on the geographical location (Hofreiter et al., 2003).

A study conducted by Willerslev et al., (2003), found paleoenvironmental information via eDNA from samples of Pleistocene and Holocene permafrost in Siberia (1.5-2 million years old) and temperate sediments in New Zealand (0.6-3 thousand years ago). The permafrost samples were obtained by drilling into the cores, whereas the cave samples were obtained via dry silty sediments (Willerslev et al., 2003). The reliability of the data was ensured by preventative measures implemented during sample collection to avoid contamination, and additional tests were conducted to mitigate any contamination which my have occurred (Willerslev et al., 2003). The eDNA samples obtained from each location were analyzed using PCR analysis, resulting in the identification of plant, mammal, bird, and other extinct species including pre-human plant taxa (Willerslev et al., 2003). They also identified fluctuations in species diversity over time (Willerslev et al., 2003). This study further solidified the understanding that microfossils as old as 400,000 years old (Willerslev et al., 2003) can be utilized to obtain accurate information and shifted the paradigm that only macrofossils can provide extensive paleoenvironmental information.

SPECIES IDENTIFICATION AND MONITORING IN AQUATIC ENVIRONMENTS VIA EDNA

Increased anthropogenic activity has had devastating impacts on aquatic environments, leading to aquatic species being the most threatened and endangered species on Earth (Liu et al., 2020; Thomsen et al., 2012). In order to implement effective conservation strategies for aquatic species, a comprehensive understanding of aquatic environments is required. Over the recent years, incorporation of eDNA analysis in biological monitoring and assessment in aquatic environments has increased, resulting in a notable contribution to ecological research globally (Figure 5 & Figure 6) (Liu et al., 2020). Since traditional methods of monitoring mostly rely on visual observations, cryptic or recently emigrated species usually remain undetected; eDNA analysis, on the other hand, provides a greater sensitivity and a wider range of species identification and monitoring, which makes it well suited for aquatic environments (Liu et al., 2020).

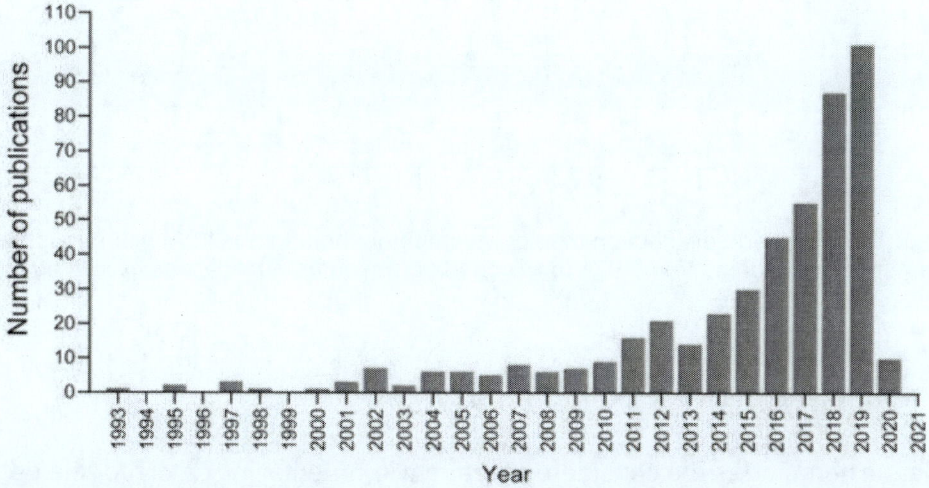

Figure 5. Publications from 1975 – March 2020 on the use of eDNA in eutrophication using keywords, ("environmental DNA" OR "environment DNA" OR "eDNA") AND ("eutrophication" OR "eutrophic" OR "nutrient*" OR "nitrogen" OR "phosphorus" OR "phosphate" OR "bloom" OR "blooms" OR "harmful algae bloom*" OR "HAB*")" (Liu et al., 2020).

Freshwater from Denmark, Sweden, Germany, Poland, and Estonia

In contrast to the study conducted by Willerslev et al., (2003), in which low degradation eDNA detected species up to 400,000 years ago, research by Thomsen et al., (2012), studied eDNA samples which had a high degradation rate where species became undetectable within 2 weeks. This research was conducted to monitor threatened amphibian, fish, mammal, insect, and crustacean species in freshwater habitats to develop and implement effective conservation strategies in a timely manner

(Thomsen et al., 2012). Quantitative data was obtained via traditional techniques which was then compared with the data obtained by qPCR; qPCR detected more species than the traditional approach (Thomsen et al., 2012). The results confirmed that eDNA analysis was a reliable and more sensitive method to monitor the abundance of species in aquatic habitats, especially since traditional methods mainly rely on the visual detection of the species (Thomsen et al., 2012). In freshwater habitats species are immigrating and emigrating at high rates, therefore, the sole use of visual data collection is an unreliable method of initial data collection (Thomsen et al., 2012).

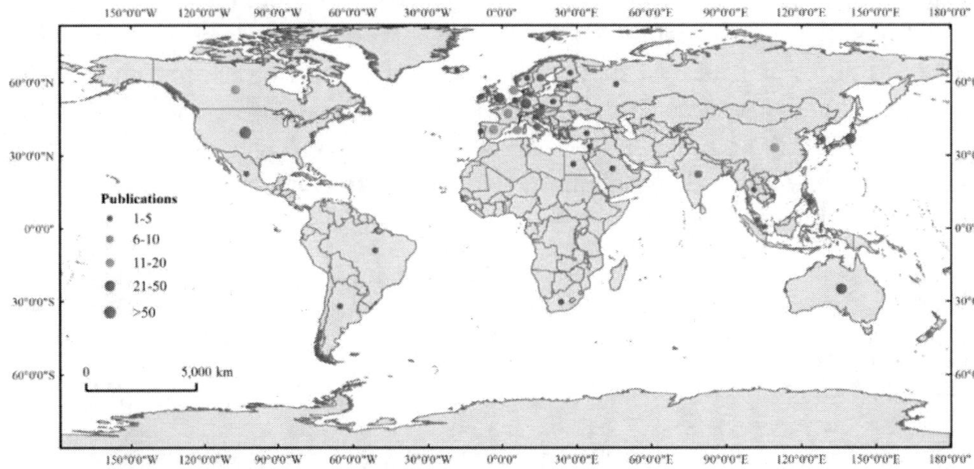

Figure 6. The worldwide distribution map of research on the use of eDNA with eutrophication and harmful algae blooms from 1975 to March 2020 based on publications from Web of Science (Liu et al., 2020).

Marine and Estuarine Sediments from England

Marine nematodes are considered great environmental indicators for the detection of pollution in marine environments; however, they have been neglected in biological research due to the length of time required in the process of identification (Austen et al., 2006). A research conducted in England by Austen et al., (2006), proved the importance of eDNA analysis in aquatic environments, by identifying marine nematodes from marine and estuarine sediments. This research proved that amplification of eDNA samples obtained from marine and estuarine sediments using PCR provide accurate information in a time efficient manner (Austen et al., 2006). Other eukaryotic species were also amplified during the process, however, to mitigate the time required for identification, specific primers were introduced to amplify only the dominant species of nematodes (Austen et al., 2006). This research exhibited the versatility and specificity of eDNA analysis in species identification, which is an asset to develop and implement specific and effective bio-environmental monitoring and assessment strategies.

DETECTION AND MONITORING OF INVASIVE SPECIES VIA EDNA

The increase in invasions of non-native species in various habitats is a prevalent global concern (Vitousek et al., 1997). Invasive species are non-native species that invade habitats, compete with native species for resources, possess the ability to exhibit predatory behaviors, and transmit diseases (Dejean et al., 2012). Once invasive species successfully inhabit an environment, the process of their eradication is extremely arduous and expensive (Dejean et al., 2012). eDNA has revolutionized the monitoring of invasive species as it detects species at low densities (Furlan et al., 2019). Once species are detected at low densities, the efforts to eradicate the species are comparatively inexpensive and highly effective compared to the efforts required to eradicate established communities of species; thereby establishing the importance of eDNA in this area of study.

Freshwater Samples from Australia

European carp are prevalent freshwater invasive species which alter natural ecosystems (Drivere et al., 2005; Rahman, 2015). In order to study and successfully eradicate the species, it is crucial to detect them at low densities (Furlan et al., 2019). Furlan et al., (2019) conducted a study to determine the efficiency of eDNA in detecting European carp at low densities in its non-native habitat within Australia. Water samples were obtained from two lakes; preventative measures were implemented to avoid the contamination of eDNA samples and to avoid sampling from areas where the eDNA was highly scattered (Furlan et al., 2019). Furthermore, PCR analysis was conducted, and the results indicated that European carp was in fact detected in extremely low densities which had remained undetected via traditional methods (Furlan et al., 2019). This research suggests that eDNA analysis plays a crucial role in detecting invasive aquatic species at low densities, which provides reliable information regarding the initiation and termination processes of eradication (Furlan et al., 2019). This information is crucial for ecologists as terminating the eradication efforts beyond the required time can lead to unnecessary costs, alternatively, terminating the efforts prior to complete eradication can lead to repop

Freshwater Samples from France

American bullfrogs are invasive species which have been categorized as one of the worst species in the world to invade habitats (Dejean et al., 2012). In 2012, Dejean et al., conducted a study in France to compare the efficiency of traditional methods and eDNA in monitoring American bullfrogs (Dejean et al., 2012). To study the species, water samples were obtained and analyzed using PCR; precautionary measures to avoid sample contamination or false positives were implemented in the research

(Dejean et al., 2012). Results found that traditional methods identified the species in 14% of the total sampling sites, however eDNA identified the species in 75% of the sites (Dejean et al., 2012). For adequate management and control of invasive species, they need to be detected at low densities to prevent the detrimental damage they cause to native habitats (Bohmann et al., 2014; Dejean et al., 2012). Results of this study support that eDNA is highly sensitive to species detection at low densities, therefore, the study confirmed that eDNA is a reliable method in early detection of invasive species (Dejean et al., 2012).

Freshwater Samples from USA (Chicago)

Silver carp and Bighead carp are invasive species which harm fish communities while hindering the use of waterways by humans (Jerde et al., 2011). These species were previously reported to nearly invade the Laurentian Great Lakes (Jerde et al., 2011). Great Lakes are the key source of available freshwater in the west; therefore, it is imperative to maintain appropriate monitoring to ensure the quality of water for the health of the aquatic species and humans (Grannemann et al., 2000). Jerde et al., (2011), conducted research in the Mississippi River Basin to determine the effectiveness of eDNA in detecting and monitoring Silver carp and Bighead carp. Water samples were collected in a consistent manner to prevent sample contamination and preventative measures were implemented to avoid false positives (Jerde et al., 2011). eDNA samples were analyzed using PCR and invasive species were detected at low densities; around 8 months prior to the first capture of the species by fishermen (Jerde et al., 2011). The fact that the species were identified via eDNA several months prior to the first physical evidence of presence of the species, indicates that eDNA plays an essential role in the detection of invasive species. The study proved the significance of eDNA in bio-environmental monitoring and suggested that further studies are required to improve the sensitivity of species detection using eDNA to lower monitoring costs in the future.

Understanding Diet and Predation Patterns in Species via eDNA

To comprehensively understand a species and its habitat, it is beneficial to discover diet and predation trends (Deagle et al., 2010). Information gathered through diet analysis provides sufficient data to confidently conclude predation trends (Shehzad et al., 2015). Diet of herbivores and carnivores can be explored via analysis of eDNA samples obtained from gastrointestinal contents and feces (Ruppert et al., 2019). Furthermore, metabarcoding of eDNA samples provide a cost-effective and in-depth analysis of the diet and predation patterns as opposed to the labor and financially exhaustive traditional techniques (Ruppert et al., 2019).

Fecal Samples from Pakistan

Fecal samples are great indicators of species feeding patterns, which provide insight regarding their predation trends (Shehzad et al., 2015). A research done by Shehzad et al., (2015) investigated eDNA from fecal samples to determine the diet and predation patterns of common leopards (*Panthera pardus*) in Pakistan to mitigate human-leopard conflict and maximize conservation efforts of the threatened species. Fecal samples were collected and the eDNA was amplified using PCR followed by metabarcoding using NGS (Shehzad et al., 2015). Metabarcoding of the fecal samples suggested that the leopard mainly preyed on goat, dog, cow, and water buffalo, and occasionally preyed on the remaining species seen in Figure 7 (Shehzad et al., 2015).

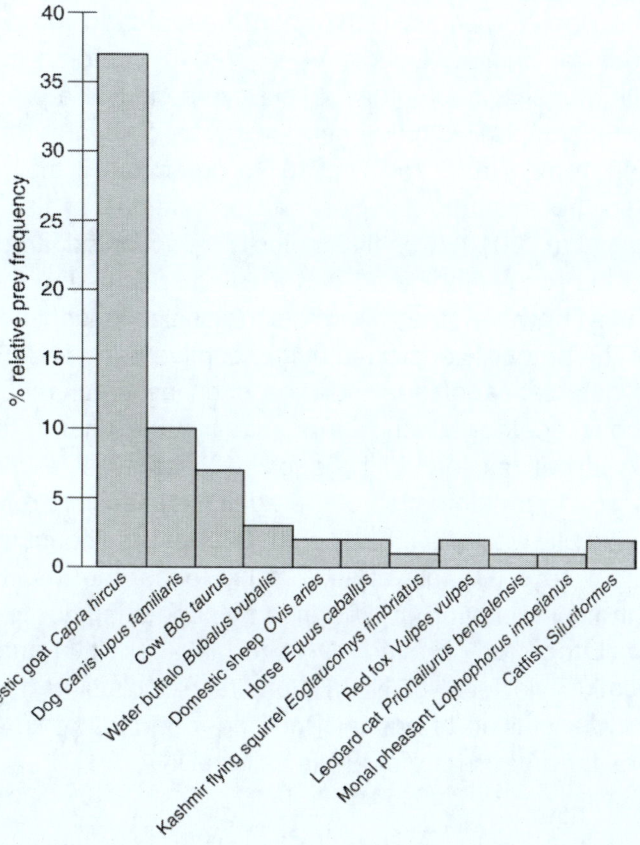

Figure 7. Relative frequency of prey items in the diet of the common leopard In Pakistan obtained via metabarcoding of fecal eDNA samples (Shehzad et al., 2015).

The data obtained from this research provided a detailed framework of predation trends of leopards in Pakistan and assisted in developing appropriate techniques to increase targeted conservation efforts of the threatened species (Shehzad et al., 2015). Moreover, the importance and reliability of the use of eDNA metabarcoding in studying predation trends in species was highlighted through this research. Although the study provided ample information regarding the general diet of the leopard, it was later found that these species attack more sheep in one attack than discovered from the research study; such limitations can have implications on the development of conservation strategies. This research (Shehzad et al., 2015) involved data collection for roughly over a week and a half and it is a prime example of the importance of using eDNA in conjunction with traditional methods; eDNA provided the initial framework of data regarding leopard diet, nevertheless traditional methods can be implemented to further obtain a detailed analysis of predation trends to improve future conservation strategies.

Fecal Samples from Australia

In contrast to Shehzad et al., (2015), who researched the diet of species using feces in their natural habitat, Deagle et al., (2010), studied the validity of eDNA metabarcoding in diet analysis by feeding captive penguins a predetermined diet and analyzing the fecal samples. Captive penguins were fed predetermined meals which consisted of tuna, tommy ruff, and whiting, to determine if the results of the eDNA analysis reflected the accurate species and proportions of fish species fed to the penguins (Deagle et al., 2010). Fecal samples were collected, amplified, and analyzed using PCR, and the results indicated that the fish species fed to the penguins were detected accurately, however, there were discrepancies in the amount of food that was detected versus the amount consumed by the captive penguins (Deagle et al., 2010). eDNA analysis detected additional species such as small-mouth hardyhead, jack mackerel, anchovy, leatherjacket, barracouta, and gurnard; these species were present in the wild but not fed during captivity (Deagle et al., 2010). Upon further investigation regarding the detection of the additional species, researchers found that based on the metabolic rate of each penguin, DNA of prey consumed can be detected in feces for up to 4 days (Deagle et al., 2010). Hence, the results of the study were skewed due to the consumption of other fish by penguins prior to captivity (Deagle et al., 2010). Overall, the study proved eDNA to be a sensitive and reliable source for species identification via fecal eDNA. Future research using this technique can be enhanced with the seclusion of animals for longer periods of time prior to assessing the fecal samples for an accurate analysis using eDNA.

Air Analysis Using eDNA

Breathing is necessary to sustain human life; without air, humans can only last a matter of minutes before collapsing (Irving, 1934). Air hosts a multitude of unseen particles such as fungal spores, microbes, and pollen, many of which can cause allergies and illnesses, therefore, it is imperative to conduct studies on air quality to ensure the safety of human health, as mankind relies on it for survival (Hurley et al., 2019). Through its ability to identify invisible particles, eDNA can enhance the study of airborne allergens, pathogens, and pollution, all of which directly impact humans (Deiner et al., 2017; Ruppert et al., 2019). Once the quality of the air is assessed in detail, future ecological and human health implications can be prevented or mitigated in a timely manner.

Air in the Netherlands

Pollen allergy is prevalent in humans globally. Allergens can cause health implications such as allergic rhinitis, asthma symptoms, and respiratory diseases (Biedermann et al., 2019; Ortega Rosas et al., 2020). The seriousness of the allergic reactions in humans vary by the genera of pollen, therefore, it is imperative to determine the genus of the pollen to implement accurate preventative measures during flowering season (Kraaijeveld et al., 2015). Allergen research using traditional methods involves the use of microscopes, and since allergens are morphologically similar under a microscope, the exact genera of the grass species is unidentifiable using traditional methods, additionally, the cost of labor involved in traditional monitoring is expensive (Kraaijeveld et al., 2015). To detect the allergens responsible for causing allergies, the allergens ought to be identified at the genus level which is impossible for some species of grass using traditional methods; therefore, in this study, Kraaijeveld et al., (2015) researched a mixed sample of airborne grass pollen using eDNA analysis. Pollen samples were collected using Burkard traps in which air was sucked into a vacuum from where the air particles stuck to a tape (Kraaijeveld et al., 2015). The collected pollen was then analyzed using PCR and NGS methods (Kraaijeveld et al., 2015). Results of the research found that four out of the six grass species were identified accurately, suggesting that eDNA is more reliable than traditional methods in analyzing air samples, however, further research is required to increase the sensitivity of the methods implemented to analyze eDNA samples (Kraaijeveld et al., 2015).

Air in Northern and Central Italy

Fungal spores in the air are small enough for humans to inhale and cause health implications such as allergic reactions, tremors, lung inflammation, respiratory diseases, infectious diseases, kidney failure, and cancer (Andersen et al., 2009; Sorenson, 1999). Therefore, it is imperative to conduct studies on fungal spores to obtain data which can be utilized in preventing or mitigating human health implications in a timely manner. Tordoni et al., (2021), conducted a study in Northern and Central Italy to compare traditional techniques with eDNA metabarcoding in detecting fungal spores and their abundance. Samples were collected using an aspiration pump and sticky tape and analyzed using a microscope for morphological analysis (traditional method) and metabarcoding (eDNA analysis) (Tordoni et al., 2021). Results indicated that eDNA was more sensitive and detected considerably more genera than traditional methods (Figure 8), however, traditional methods were more successful in detecting variance in species (Tordoni et al., 2021). The researchers further found that operator bias in traditional methods, and PCR primer and amplification biases in eDNA possibly contribute to some limitations in this study (Tordoni et al., 2021). This research is a great illustration of the importance of using traditional methods in conjunction with eDNA metabarcoding for successful bio-environmental monitoring and assessment.

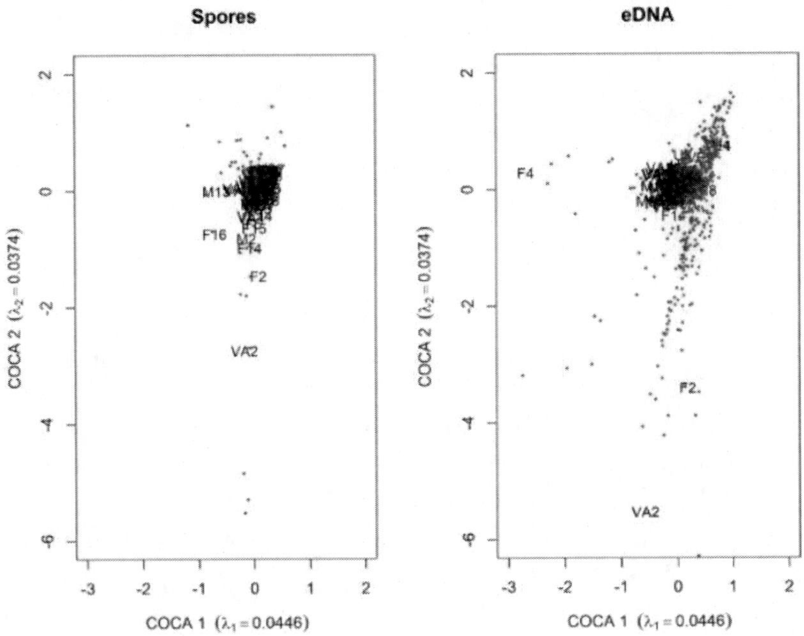

Figure 8. A symmetric Co-CA (co-correspondence analysis) showing the patterns between weighted average scores of two assemblages derived from morphological analyses of spores as seen on the left and DNA metabarcoding on the right. The black writing signifies the location of sampling and the red dots signify the taxa identified (Tordoni et al., 2021).

Air Samples in Beijing

In contrast to the research by Tordoni et al., (2021), where fungal spores were detected in an open environment, Tong et al., (2017) conducted a study on aerial fungi in a closed hospital environment. Hospital air consists of numerous aerial pathogens and microorganisms which cause health implications (Tong et al., 2017). Traditional monitoring methods, which mainly consist of culture analyses, have been unsuccessful in detecting a full range of airborne microorganisms in hospital settings (Tong et al., 2017). Therefore, in this research, traditional methods are compared to eDNA in the detection of airborne fungi (Tong et al., 2017). Air samples were obtained using air sampling pumps from the outpatient department, emergency room, surgical intensive care unit, and respiratory intensive care unit, and analyzed using cultivation (traditional method), and PCR and metabarcoding (eDNA) (Tong et al., 2017). Only bacteria were able to be cultured using traditional methods, whereas eDNA detected fungi, archaea, bacteria, and viruses (Figure 9) (Tong et al., 2017). The results show the significance of using eDNA for bio-environmental monitoring and assessment, especially in a hospital setting, as it identifies many harmful microorganisms that cause and spread diseases in an environment where most of the people are already immunosuppressed. The use of eDNA in a hospital setting can help prevent and mitigate health implications and ultimately save more lives.

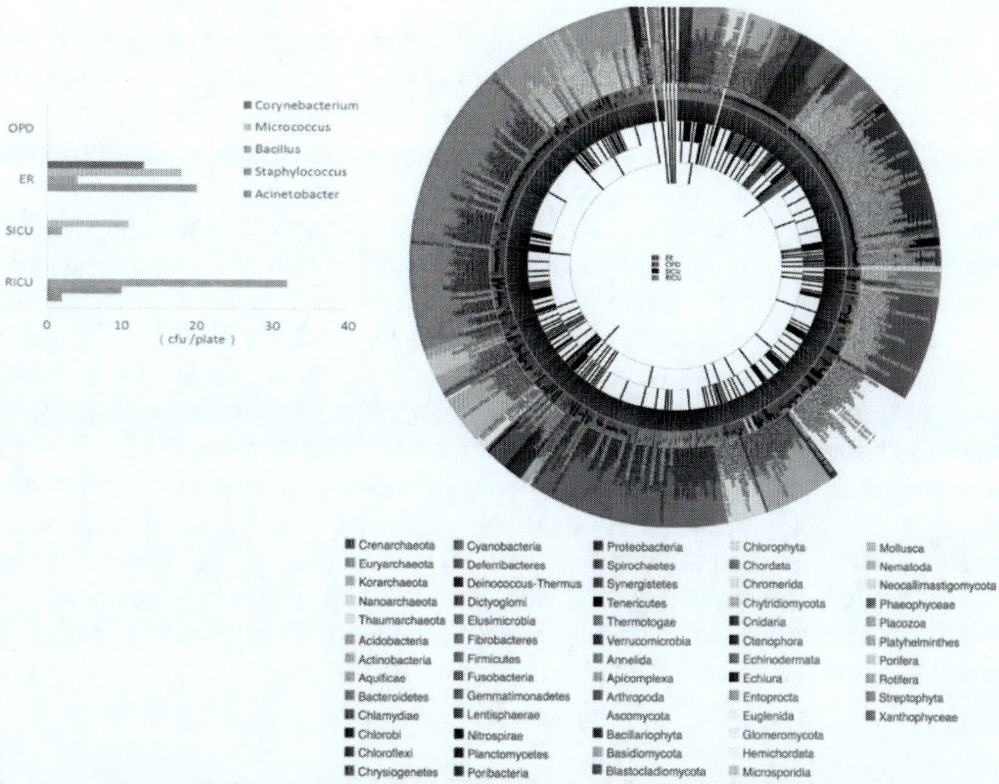

Figure 9. Bacteria identified by culture (traditional method) on the left, and the range of fungi, species of Archaea, bacteria, and viruses identified using eDNA (right side) (Tong et al., 2017).

LIMITATIONS PF THE USE OF EDNA

Some of the main limitations in the use of eDNA in species abundance include false positive/negative results, degradation of samples, inaccurate taxonomic resolution, and outdated databases (Bohmann et al., 2014; Cristescu & Herbert, 2018; Pilloid et al., 2014; Ruppert et al., 2019). False positive results and biases can be caused by the contamination of samples during sample collection or handling (Bohmann et al., 2014). To prevent contamination, samples should be sealed at the site of collection and ideally, multiple sampling sites should not be visited consecutively to ensure eDNA is not being transferred from the previous site (Thomsen & Willerslev, 2015). Furthermore, PCR amplifies billions of copies of DNA which can cause laboratorial contamination of samples; therefore, it is imperative to ensure the working area is always clean, and if possible, amplify the samples in different laboratories to minimize chances of laboratorial contamination (Thomsen & Willerslev, 2015). In some cases, false positives can not be prevented, for instance, when bird species release their DNA through feces and feathers as they fly over the aquatic environments (Figure 3) (Strickler et al., 2015).

In order to mitigate such contaminations, additional testing and sequencing should be conducted to test and sequence for potential known contaminants in the sampling site or laboratory (Gilbert et al., 2008). Moreover, false positives can also be caused by inhibitions, sequencing errors, and misinterpretation of results in eDNA (Bohmann et al., 2014). Inhibitions are caused by the presence of inhibitors which are mostly common in humic substances and other organic substances such as sediment, contaminated water, and soil samples; there are multiple methods used to mitigate this limitation, however, the mitigation techniques depend on the type of inhibitor (Hunter et al., 2019; Thomsen & Willerslev, 2015). False positives can impact the raw sequence data which can lead to misinterpretation of data; therefore, it is crucial to ensure data is filtered prior to the development of the raw sequence data for PCR analysis (Thomsen & Willerslev, 2015). Moreover, eDNA degradation is another source of inaccurate data and is exponentially increased by UV radiation, temperature, and light conditions (Pilloid et al., 2014). Water samples are the most prone to DNA degradation, therefore, to prevent degradation, samples should be filtered as soon as possible; the best approach is through the conduction of on-site filtration, additionally, the samples should be assessed in a timely manner, and over-centrifuging should be avoided (Rogers & Bendich, 1989; Yamanaka et al., 2016).

Conclusion and Future Directions

eDNA has revolutionized the scope of bio-environmental monitoring and assessment. Through its non-invasive techniques, it aids in the early detection of species which improves the implementation of conservation efforts in a timely manner. It is a non-invasive technique which provides extensive information regarding aquatic and terrestrial species. Depending on the type of sample, the use of eDNA in ecological monitoring has the potential to provide temporal and spatial information of the species. In contrast to traditional monitoring methods, eDNA samples possess the ability to provide information regarding higher trophic levels and invasive species in aquatic environments due to the richness of information available from water samples (Dejean et al., 2012; Goodwin et al., 2017; Thomsen et al., 2012). Moreover, the cost of eDNA in ecological monitoring is substantially lower than traditional methods, however, the overall cost of data collection and analysis depends on the type of monitoring project (Thomsen & Willerslev, 2015). The benefits of eDNA in environmental monitoring surpass those of traditional methods, however, due to the infancy stages of development of eDNA in environmental monitoring and assessment, there are many areas that require more research. Future research in eDNA should focus on improving species abundance detection, as well as reducing primer and amplification bias. Further studies need to be conducted to ensure that most limitations are considered and mitigated, until then, eDNA methods should be implemented in conjunction with traditional methods to ensure accurate collection and interpretation of data and results (Deiner & Altermatt, 2014).

References

Andersen, G. L., Frisch, A. S., Kellogg, C. A., Levetin, E., Lighthart, B., & Paterno, D. (2009). Aeromicrobiology/air quality. *Encyclopedia of Microbiology*, 11 – 26. https://doi.org/10. 1016/B978-012373944-5.00166-8.

Austen, P. B. M. C., Bilton, D. T., Lambshead, P. J. D., Rogers, A. D., & Smerdon, G. R. (2006). Molecular detection of marine nematodes from environmental samples: overcoming eukaryotic interference. *Aquatic Microbial Ecology*, 44(1), 97-103.

Barnes, M. A., & Turner, C. R. (2016). The ecology of environmental DNA and implications for conservation genetics. *Conservation genetics*, 17(1), 1-17.

Behjati, S., & Tarpey, P. S. (2013). What is next generation sequencing? *Archives of Disease in Childhood-Education and Practice*, 98(6), 236-238.

Biedermann, T., Winther, L., Till, S. J., Panzner, P., Knulst, A., & Valovirta, E. (2019). Birch pollen allergy in Europe. *Allergy,* 74(7), 1237-1248.

Bohmann, K., Evans, A., Gilbert, M. T., Carvalho, G. R., Creer, S., Knapp, M., Yu, D. W., & de Bruyn, M. (2014). Environmental DNA for wildlife biology and biodiversity monitoring. *Trends in ecology & evolution*, 29(6), 358–367. https://doi.org/ 10.1016/j.tree.2014.04.003.

Cook, J., Oreskes, N., Doran, P. T., Anderegg, W. R., Verheggen, B., Maibach, E. W., Carlton, J. S., Lewandowsky, S., Skuce, A. G., Green, S. A., Nuccitelli, D., Jacobs, P., Richardson, M., Winkler, B., Painting, R., & Rice, K. (2016). Consensus on consensus: a synthesis of consensus estimates on human-caused global warming. *Environmental Research Letters*, 11(4), 048002. http://dx.doi.org/ 10.1088/1748-9326/11/4/048002.

Cristescu, M. E., & Hebert, P. D. (2018). Uses and misuses of environmental DNA in biodiversity science and conservation. *Annual Review of Ecology, Evolution, and Systematics,* 49, 209-230.

Deagle, B. E., Chiaradia, A., McInnes, J., & Jarman, S. N. (2010). Pyrosequencing faecal DNA to determine diet of little penguins: is what goes in what comes out?. *Conservation Genetics*, 11(5), 2039-2048.

Deiner, K., Bik, H. M., Mächler, E., Seymour, M., Lacoursière-Roussel, A., Altermatt, F., Creer, S., Bista, I., Lodge, D. M., de Vere, N., Pfrender, M. E. & Bernatchez, L.

(2017). Environmental DNA metabarcoding: Transforming how we survey animal and plant communities. *Molecular ecology*, 26(21), 5872-5895.

Dejean, T., Valentini, A., Miquel, C., Taberlet, P., Bellemain, E., & Miaud, C. (2012). Improved detection of an alien invasive species through environmental DNA barcoding: the example of the American bullfrog *Lithobates catesbeianus*. *Journal of applied ecology*, 49(4), 953-959.

Driver, P. D., Closs, G. P. & Koen T. 2005. The effects of size and density of carp (*Cyprinus carpio* L.) on water quality in an experimental pond. *Archiv fur Hydrobiologie (Arch Hydrobiol)* 163:117–131. doi: 10.1127/0003-9136/2005/0163-0117.

Foote, A. D., Thomsen, P. F., Sveegaard, S., Wahlberg, M., Kielgast, J., Kyhn, L. A., Salling, A. B., Galatius, A., Orlando, L. & Gilbert, M. T. P. (2012). Investigating the potential use of environmental DNA (eDNA) for genetic monitoring of marine mammals. *PloS one*, 7(8), e41781.

Furlan, E. M., Gleeson, D., Wisniewski, C., Yick, J., & Duncan, R. P. (2019). eDNA surveys to detect species at very low densities: A case study of European carp eradication in Tasmania, Australia. *Journal of Applied Ecology*, 56(11), 2505-2517. https://doi.org/10.1111/1365-2664.13485.

Garibyan, L., & Avashia, N. (2013). Polymerase chain reaction. *The Journal of investigative dermatology*, 133(3), 1–4.

Gilbert, M. T., Jenkins, D. L., Götherstrom, A., Naveran, N., Sanchez, J. J., Hofreiter, M., Thomsen, P. F., Binladen, J., Higham, T. F., Yohe, R. M., 2nd, Parr, R., Cummings, L. S., & Willerslev, E. (2008). DNA from pre-Clovis human coprolites in Oregon, North America. *Science* (New York, N.Y.), 320(5877), 786–789. https://doi.org/10.1126/science.1154116.

Goodwin KD, Thompson LR, Duarte B, Kahlke T, Thompson AR, Marques JC and Caçador I (2017) DNA Sequencing as a Tool to Monitor Marine Ecological Status. *Front. Mar. Sci.* 4:107. doi: 10.3389/fmars.2017.00107.

Grannemann, N. G., Hunt, R. J., Nicholas, J. R., Reilly, T. E., & Winter, T. C. (2000). The importance of ground water in the Great Lakes. *USGS Water Resources* Investigations.

Hayer, C., Klymus, K., Thompson, N., Paukert, C., Chapman, D. & Richter, C. (2016). *Environmental DNA (eDNA) and invasive silver carp (Hypophthalmichthys molitrix): detection of spawning in the Missouri River.* Conference: Midwest Fish and Wildlife ConferenceAt: Grand Rapids, Michigan.

Hebsgaard, M. B., Gilbert, M. T. P., Arneborg, J., Heyn, P., Allentoft, M. E., Bunce, M., Munch, K., Schweger. C. & Willerslev, E. (2009). 'The Farm Beneath the Sand'– an archaeological case study on ancient 'dirt'DNA. *Antiquity*, 83(320), 430-444.

Hofreiter, M., Mead, J. I., Martin, P., & Poinar, H. N. (2003). *Molecular caving. Current Biology,* 13(18), R693-R695.

Hunter, M. E., Ferrante, J. A., Meigs-Friend, G., & Ulmer, A. (2019). Improving eDNA yield and inhibitor reduction through increased water volumes and multi-filter isolation techniques. *Scientific Reports*, 9(1), 1-9.

Hurley, K. V., Wharton, L., Wheeler, M. J., Skjøth, C. A., Niles, C., & Hanson, M. C. (2019). Car cabin filters as sampling devices to study bioaerosols using eDNA and microbiological methods. *Aerobiologia,* 35(2), 215-225.

Irving, L. (1934). On the ability of warm-blooded animals to survive without breathing. *The Scientific Monthly*, 38(5), 422-428.

Jerde, C. L., Mahon, A. R., Chadderton, W. L., & Lodge, D. M. (2011). "Sight-unseen" detection of rare aquatic species using environmental DNA. *Conservation Letters,* 4(2), 150-157.

Kraaijeveld, K., De Weger, L. A., Ventayol García, M., Buermans, H., Frank, J., Hiemstra, P. S., & Den Dunnen, J. T. (2015). Efficient and sensitive identification and quantification of airborne pollen using next-generation DNA sequencing. *Molecular ecology resources*, 15(1), 8-16.

Liu, Q., Zhang, Y., Wu, H., Liu, F., Peng, W., Zhang, X., Chang, F., Xie, P., & Zhang, H. (2020). A Review and Perspective of eDNA Application to Eutrophication and HAB Control in Freshwater and Marine Ecosystems. *Microorganisms*, 8(3), 417. https://doi.org/10.3390/ microorganisms8030417.

Ortega Rosas, C. I., Calderón-Ezquerro, M. D. C., & Gutiérrez-Ruacho, O. G. (2020). Fungal spores and pollen are correlated with meteorological variables: effects in human health at Hermosillo, Sonora, Mexico. *International journal of environmental health research*, 30(6), 677-695.

Pilliod, D. S., Goldberg, C. S., Arkle, R. S., & Waits, L. P. (2014). Factors influencing detection of eDNA from a stream-dwelling amphibian. *Molecular Ecology Resources,* 14(1), 109-116.

Rahman, M. M. (2015) Role of common carp (Cyprinus carpio) in aquaculture production systems, *Frontiers in Life Science*, 8:4, 399-410, doi: 10.1080/21553769.2015.1045629.

Rogers, S. O., & Bendich, A. J. (1989). Extraction of DNA from plant tissues. In *Plant molecular biology manual* (pp. 73-83). Springer, Dordrecht.

Ruppert, K. M., Kline, R. J., & Rahman, M. S. (2019). Past, present, and future perspectives of environmental DNA (eDNA) metabarcoding: A systematic review in methods, monitoring, and applications of global eDNA. *Global Ecology and Conservation*, 17, e00547.

Schenk, J., Kleinbölting, N., & Traunspurger, W. (2020). Comparison of morphological, DNA barcoding, and metabarcoding characterizations of freshwater nematode communities. *Ecology and evolution*, 10(6), 2885-2899.

Shehzad, W., Nawaz, M. A., Pompanon, F., Coissac, E., Riaz, T., Shah, S. A., & Taberlet, P. (2015). Forest without prey: livestock sustain a leopard *Panthera pardus* population in Pakistan. *Oryx*, 49(2), 248-253.

Shokralla, S., Spall, J. L., Gibson, J. F., & Hajibabaei, M. (2012). Next-generation sequencing technologies for environmental DNA research. *Molecular ecology*, 21(8), 1794-1805.

Sinclair, S. J., White, M. D., & Newell, G. R. (2010). How useful are species distribution models for managing biodiversity under future climates? *Ecology and Society*, 15(1).

Sorenson, W. G. (1999). Fungal spores: hazardous to health? *Environmental Health Perspectives*, 107(suppl 3), 469-472.

Strickler, K. M., Fremier, A. K., & Goldberg, C. S. (2015). Quantifying effects of UV-B, temperature, and pH on eDNA degradation in aquatic microcosms. *Biological Conservation*, *183*, 85-92.

Taberlet, P., Coissac, E., Hajibabaei, M., & Rieseberg, L. H. (2012). Environmental DNA. *Molecular ecology*,21(8), 1789-1793.

Thomsen, P. F., & Willerslev, E. (2015). Environmental DNA–An emerging tool in conservation for monitoring past and present biodiversity. *Biological conservation*, 183, 4-18.

Thomsen, P. F., Kielgast, J., Iversen, L. L., Wiuf, C., Rasmussen, M., Gilbert, M. T., Orlando, L., & Willerslev, E. (2012). Monitoring endangered freshwater biodiversity using environmental DNA. *Molecular ecology*, 21(11), 2565–2573. https://doi.org/ 10.1111/ j.1365-294X.2011.05418.x.

Tong, X., Xu, H., Zou, L., Cai, M., Xu, X., Zhao, Z., Xiao, F., & Li, Y. (2017). High diversity of airborne fungi in the hospital environment as revealed by meta-sequencing-based microbiome analysis. *Scientific reports*, 7(1): 1 - 8. https://doi.org/10.1038/srep39606.

Tordoni, E., Ametrano, C. G., Banchi, E., Ongaro, S., Pallavicini, A., Bacaro, G., & Muggia, L. (2021). Integrated eDNA metabarcoding and morphological analyses assess spatio-temporal patterns of airborne fungal spores. *Ecological Indicators*, 121, 107032.

Vitousek, P. M., Mooney, H. A., Lubchenco, J., & Melillo, J. M. (1997). Human domination of Earth's ecosystems. *Science*, 277(5325), 494-499.

Willerslev, E., Cappellini, E., Boomsma, W., Nielsen, R., Hebsgaard, M. B., Brand, T. B., & Collins, M. J. (2007). Ancient biomolecules from deep ice cores reveal a forested southern Greenland. *Science*, 317(5834), 111-114.

Willerslev, E., Hansen, A. J., Binladen, J., Brand, T. B., Gilbert, M. T., Shapiro, B., Bunce, M., Wiuf, C., Gilichinsky, D. A., & Cooper, A. (2003). Diverse plant and animal genetic records from Holocene and Pleistocene sediments. *Science* (New York, N.Y.), 300(5620), 791–795.

Yamanaka, H., Motozawa, H., Tsuji, S., Miyazawa, R. C., Takahara, T., & Minamoto, T. (2016). On-site filtration of water samples for environmental DNA analysis to avoid DNA degradation during transportation. *Ecological Research*, 31(6), 963-967.

In: Critical Research Techniques in Animal … ISBN: 978-1-53619-846-1
Editors: Michael O'Neal Campbell et al.

Chapter 10

WILDING OR WORRYING? PLACE MEANING IN A REWILDING LANDSCAPE

Rachel Orchard and Carijn Beumer [*]
Maastricht University, Department of Health, Ethics and Society, Maastricht, Netherlands

ABSTRACT

Public acceptance of 'place change' is an issue of increasing significance in the UK as post-Brexit policy changes push more landowners to provide ecosystem services. This research looks into the existing conservation project of Knepp Wildland to contribute to understanding the effects of place change on 'place meaning' for the local community. Place meaning looks beyond specific attitudes, to engage in the emotive subjective experience of places for people. The results were collected by means of a qualitative questionnaire distributed through purposive sampling online community groups and finding other contacts through snowballing. Theoretical thematic analysis was conducted using Jacobs and Buijs' (2011) place meaning categories: beauty, functionality, attachment, biodiversity and risk, with two additional categories: emotions-in-place and temporality. The main finding is that in the case of transitioning from an agricultural to a rewilding landscape, place change affects place meaning by triggering renegotiation of place meaning for people themselves and their relationship to the landscape. This can also be understood as causing a change in cultured naturalness, which reflects a community's local identity and way of life, over time. This renegotiation of place meaning included an increasing sense of ecological loss, changing view of visible stewardship, and increasing appreciation of wildlife. Influences on how place change affected place meaning were perceptions on access, wilderness, and belonging. The results also indicate that over time the place meaning derived from Knepp Estate's rewilding landscape is predominantly positive, particularly evident from the emotions-in-place expressed.

[*] Corresponding Author's E-mail: carijn.beumer@maastrichtuniversity.nl.

Keywords: rewilding, place meaning, landscape, agriculture, conservation, biodiversity

INTRODUCTION

The Problem of Public Acceptance

Breaking with the romanticised agricultural landscape of rural England is bound to upset local residents given this 'idyllic' image has been considered the norm since the Victorian era (Park & Selman, 2011). However, contrary to this static stereotype, landscapes are always evolving (Drenthen, 2018). Furthermore, due to pressing climate change, conservation concerns and agricultural subsidy restructuring in post-Brexit UK, land use and landscapes are inevitably facing more radical imminent change (Downing & Coe, 2018). Given the increasing emphasis on public participation in planning processes for land use, the public's opinions are of paramount importance in determining the countryside's future landscapes (Dockerty et al., 2006). Therefore, Park and Selman (2011) call for better understanding of the public's attitudes concerning openness to change in rural landscapes.

Recently, 'rewilding' has grown in popularity as a conservation method, although it remains a controversial concept (Drenthen, 2018). It has become an integral part of discussions for the post-Brexit '*agricultural transition plan'*, which aims to reward the development of 'public goods', such as thriving wildlife, good soil health, better air and water quality and adaptive capacity to climate change (GOV.UK, 2020a).

Rewilding stems from a non-anthropocentric grounding and advocates for allowing natural processes to repair ecosystems (Drenthen, 2018), although its precise definition is variable (Jørgensen, 2015). Oostvaardersplassen practises Frans Vera's interpretation of rewilding, whereby grazing animals create a mosaic landscape of woodland and open space that generates biodiversity (Tree, 2018). Inspired by this project, the owners of Knepp Estate, a formerly industrial farm on unproductive clay soil in West Sussex, decided to pioneer a similar project in the UK two decades ago (Tree, 2018). Nowadays, it is widely perceived as a conservation success, due to its abundance of biodiversity, including a myriad of endangered species (de Bellaigue, 2020). As such, the landscape has transitioned from an industrial agricultural landscape to a rewilding landscape, although the owners prefer the term 'wilding' (Tree, 2018). However, the owners admit that public acceptance is a key problem as rural residents are often reluctant to embrace changes to what they perceive as the 'natural' landscape (Burrell & Greenaway, 2011). Better understanding of whether public acceptance has grown as the rewilding landscape matures will indicate whether this problem persists or if, in this case, time is a healer.

People's attitudes towards intervention in their 'normal' landscape have been shown to derive predominantly from their place meaning (Jacobs & Buijs, 2011). Place meaning is a socio-psychological concept that adds emotion to the concept of place attachment, within the wider framework of sense of place (Davenport & Anderson, 2005). It refers literally to the "meanings that one attributes to places or the beliefs about a spatial setting" (Stedman, 2002, p.566). A place-based approach enables the subjective experience to be better understood by going deeper than relatively superficial 'attitudes' because of its emotional element (Davenport & Anderson, 2005). Therefore, using the concept of place meaning provides a better understanding of the sentiments behind local residents' openness or otherwise to the rewilding landscape change.

In light of the problem of public acceptance, this research aims to better understand whether and how landscape changes due to rewilding -- using Knepp Estate as a case study -- affect place meanings for the local community. The term 'affect' is sufficiently open to accommodate all indications of changes to peoples' place meaning, be it negatively, positively or not at all. As such, the results will give a voice to the community of local residents so they can share their impressions of the wilding project and will fill the gap in knowledge about how place meaning evolves and adapts to landscape changes caused by rewilding. Therefore, the main question of this study is: *how has the change in landscape over the past two decades of the Knepp Wilding project affected place meaning for the local community?*

Relevance: Rewilding Rife in Post-Brexit Restructuring?

This research is relevant for several reasons. Firstly, changes in land-use policies are starting to be implemented in the Post-Brexit UK. The Agriculture Act 2020 will likely increase the number of rewilding projects, following Knepp's example, as landowners and farmers embrace new subsidies for ecosystem services (GOV.UK, 2020b). In anticipation of this landscape change, this research contributes to understanding the impact of rewilding as an alternative land use on locals' place meaning. The case study of Knepp is particularly relevant given its age and pioneering role in transitioning from agriculture to rewilding in a densely populated part of the country (West Sussex). Therefore, it is an example to others seeking to diversify their land use.

Secondly, even the most adamant rewilder, Monbiot (2015), asserts that rewilding requires "active consent" from the local community. Place-based approaches are increasingly acknowledged as a strong basis for participatory planning of land management (Kil et. al, 2014; Jacobs & Buijs, 2011). Although Knepp Wildland did not conduct participatory planning, the place meanings of the local community will be informative for participatory planning in other similar projects. Thirdly, whilst there has been a range of case study investigations on community reactions to conservation projects, most are situated in the Global South in relation to governments and organisations creating national parks (Rodríguez-Izquierdo et al., 2010; Holladay & Ormsby, 2011; Allendorf et al., 2006; Ngonidzashe Mutanga et al., 2015). Therefore, this research is also relevant in extending knowledge on community sentiments to conservation projects in the Global North, which are distinct from the Global South due to the difference in environmental discourses and power dynamics (Guha, 2000).

Lastly, Cheng et al., (2003) argue that, once formed, place meanings are confirmed, challenged, and negotiated through experiences. As such, there is the postulation that "place changes affect place meanings… but few have investigated this phenomenon" (Davenport & Anderson, 2005, p. 630). For example, Greider and Garkovich (1994, p.14) claim that changes in environment "require a renegotiation of the meaning of both themselves as people and their relationships to the environment." By looking at place changes due to rewilding, this research contributes to investigating the phenomenon of whether and how place change affects place meanings.

CONCEPTUAL GROUNDING

Before focusing on the Knepp Estate case study, this chapter will first explore the key concepts of sense of place, place meaning, landscape change, and rewilding. As candidly stated by Davenport and Anderson (2005), "the human-environment relationship is complex" (p. 627). Given this research seeks to contribute to the field that explores this complex and varied relationship, it is worthwhile setting out some underlying principles. Therefore, this research is grounded conceptually on four basic tenets relating to the human-environment relationship, formed by Davenport and Anderson (2005, p. 627):

1. Places manifest the physical characteristics of a setting, activities and experiences in a setting, social phenomena and processes, and individual interpretations.
2. People assign meanings to places and derive meaning in their lives from places.
3. Some place meanings translate into strong emotional bonds that influence attitudes and behaviours within the context of those places.
4. Place meanings are maintained, challenged, and negotiated in natural resource management and planning.

This section will map out the existing relevant literature on the concepts of place meaning and landscape change. First, however, the broader concepts of sense of place and place attachment will be explored to contextualise the concept of place meaning.

Sense of Place and Place Attachment

Sense of place is interpreted in a variety of ways. Some theorists perceive sense of place as synonymous to place attachment (Patterson & Williams, 2005; Jorgensen & Stedman 2001, etc.), but a more common approach is viewing place attachment as a narrower concept, which consists of place identity and place dependence (e.g.,

Davenport & Anderson, 2005, Williams & Vaske 2003). In line with the latter interpretation, sense of place is understood as an umbrella construct for a host of concepts that reflect the subjective processes and experiences through which people identify, depend on, attach to and assign emotional meaning to places (Chapin & Knapp, 2015; Kil et al., 2014; Jacobs & Buijs, 2011). Sense of place is not fixed as it reflects the fluid nature of people's relationship to their environment, depending on a person's socio-demographic context and change over time (Davenport & Anderson, 2005; Proshansky et al., 1984).

As understood in the more common approach, place attachment is broadly defined as the emotional bond between people and places, with place identity focusing on the more symbolic elements and place dependence looking at the more tangible aspects (Davenport & Anderson, 2005). Place attachment has gained prominence as a mechanism for community participation and planning (Manzo & Perkins, 2006) and better understanding of environmental behaviour and concern (Vorkinn and Riese, 2001; Vaske & Kobrin, 2001). However, Stedman (2002) warned that measuring the strength of place attachment alone is limited and that the meanings assigned to place must also be researched. Hence the concept of place meaning has evolved to complement place attachment and go beyond it by emotionally engaging in the subjective experience of sense of place (Davenport & Anderson, 2005). To clarify, place meaning and place attachment may influence each other, but they are treated as conceptually distinct elements of the people-place relationship (Kil et al., 2014).

Place Meaning

Place meaning seeks to capture the subjective, emotional and symbolic aspects of people-place relationships (Davenport & Anderson, 2005). Manzo (2005) researched the processes by which people develop place meanings and the findings emphasised the impact of both positive and negative emotions and experiences-in-place. This results in a frequent occurrence of ambivalent feelings to places considered significantly meaningful (Manzo, 2005). Bleam's (2018) research gives light to the temporal role in place meanings. By investigating place meanings for conservation volunteers at a national park in Arizona, Bleam (2018) found that time of year and time of day are additional temporal lenses, alongside course of life, for understanding place meaning.

Williams (2014) asserts that there is no one approach that can capture all facets of place meaning. As such, there exists a variety of approaches, categories or dimensions that constitute place meaning (Bleam, 2018). Davenport et al., (2010) labelled six dimensions, which overlap significantly with Smith et al.,'s (2011) identification of seven factors: ecological integrity, economic dependency, place dependence, family identity, community identity, place identity, self-expression, which were also used in Kil et al.,'s (2014) research. Jacobs and Buijs (2011) found five categories of place meanings: beauty (aesthetic judgments), functionality (ways of use), attachment (feelings of belonging), biodiversity (meanings pertaining to nature), and risk (worries about current or future events). Whilst these clearly still overlap, some of the dimensions are more concrete, such as 'aesthetic judgments,' and others are more interpretive, for example 'ecological integrity.'

Place meanings are also socially acquired through the historic relationship between place and family or community (Kil et al., 2014). This can be seen through the communal place meaning concept of 'cultured naturalness,' coined by Hull et al., (2001) to reflect a community's local identity and way of life. Interviewing members of a community living by a forest, Hull et al., (2001) found that their landscape was considered natural but also steeped in symbols of their local culture. A key component of 'cultured naturalness' is how landscapes tell the story of local life and a key indicator is access to the landscape, in this case the local forest (Hull et al., 2001). This research is poignant as respondents assumed outsiders valued the wild qualities of nature more than the cultured naturalness. As such, the respondents resented outsiders' input in forest management, indicating a strong communal sense of belonging and ownership (Hull et al., 2001). Kil et al., (2014) affirms the importance of community identity, stating that it "represents the degree to which natural landscapes symbolise the local identities, character, and culture that lead to community association and pride" (p. 478).

Place meanings are a set of more general cognitions for framing more specific attitudes on issues (Jacobs & Buijs, 2011). Kil et al.,'s (2014) research on the influence of place meanings on intentions to participate in planning of natural resource management showed that people's valuation of landscapes is formed by both tangible and intangible elements of their place meanings. Jacobs and Buijs (2011) investigated stakeholders' place meanings and attitudes towards proposed water management intervention and their results affirm that stakeholder attitudes are, to a large extent, derived from place meanings. This demonstrates the importance of place meanings in understanding and responding to place interventions and changes. It should be noted that a large proportion of research on place meaning and place change is looking at proposed change or intervention, through the use of scenarios (Jacobs & Buijs, 2011; Van den Berg & Koole, 2006; Dockerty et. al, 2006). Therefore, there is a need for further research specifically on place meaning in retrospective and ongoing place change, where there is concrete visual change in landscape.

Landscape Change

Landscape change has been investigated both as a more general phenomenon (Park & Selman, 2011) and as a more place and/or method-specific phenomenon (Westling et al., 2014; Lewis, 2008; Marshall et al., 2019). In the former case, Park and Selman (2011) surveyed residents in Sheffield on their attitudes and openness to landscape change and received a predominantly negative response, although younger people appeared more open than older. On the other hand, Westling et al., (2014) looked at the influences on perceptions regarding a river restoration project retrospectively (14 years later) and found a generally more positive response. In this case, the familiarity of the landscape along with the notion of control of the landscape were notable influences on perceptions. In addition, the residents' attachment to the environment was strengthened by the opportunity to view wildlife (Westling et al., 2014).

Research has also been conducted on cases of landscape change due to forest management method change (Lewis, 2008), renewable energy infrastructure (Devine-Wright & Howes, 2010) and reef loss (Marshall et al., 2019). Lewis' (2008) research highlighted the community concern for the cultural appropriateness of landscape conditions and the desire for visible care and stewardship. Devine-Wright and Howes (2010) found the level of trust in key actors to be a moderating variable between negative attitudes to the proposed wind farm and place attachment. Marshall et al., (2019) discovered that the experience of ecological grief is dependent on place meanings and moderated by the other concepts of sense of place along with place-based pride and derived well-being. These case studies all feed into better understanding of the relationship between people and landscape change. However, there is yet to be research that investigates landscape change from farming to rewilding and how it affects place meaning for the local community.

Rewilding

The terminology of 'rewilding' implies returning to a landscape unmanaged by humans, and therefore 'wild,' which not only repairs ecosystems but also reconnects humans to nature (Lorimer, 2015). However, the requirement for often human-less 'wilderness' for such a reconnection has been criticised by Cronon (1995), as this tends to be elitist and detracts from appreciating living organisms in more human settings. Furthermore, contrary to reconnecting, opponents of rewilding perceive this notion as creating spaces that ignore and destroy the cultural heritage of the landscape (Drenthen, 2018). Hence the academic debate on rewilding is often dichotomised along the boundaries of the nature/ culture binary. However, Drenthen (2018) noted that both sides draw their reasoning and motivation from both nature and culture. In addition, Drenthen (2018) argues that accepting the landscape as ever-evolving enables both rewilders and heritage-holders to mesh their visions in shaping future landscapes.

Another criticism of rewilding is the potential for human-wildlife conflicts, which are prone to escalate when locals feel that wildlife's needs and values are prioritised over their own (Madden, 2004). Given the non-anthropocentric grounding of rewilding and the practicalities of carrying out rewilding resulting in the removal of man-made elements in the landscape, as a concept it can be misinterpreted as striving to exclude humans (Drenthen, 2018). In addition, the (re-)introduction of species that local people are no longer or never have been familiar with and which are more loosely managed than the norm of agricultural management, can be perceived as threatening people's safety and property (Madden, 2004). However, human-wildlife conflicts are frequent manifestations of underlying conflicts between humans (Dickman, 2010). Furthermore, the problem with labelling tensions between priorities as 'human-wildlife conflicts' is that it positions wildlife as antagonists to humans (Peterson et al., 2010).

The relevance of this ongoing academic debate is paramount to the decisions being made regarding conservation projects and the attitudes of those directly affected by these projects. Given the growing popularity of rewilding as a conservation method, there is worry that it will become merely a cliché form of rebranding or replace the more species-specific conservation methods (Barkham, 2017). In addition, the negative attitudes of some local communities towards existing proposals or projects have been strongly vocalised, such as the animal rights resistance at Oostvaardersplassen (Barkham, 2018) and the sheep farmers resistance in the Lake District (Meanwell, 2015). On the other hand, there is evidence of rewilding receiving much support (Schepers & Jepson, 2016), although more so by urban dwellers than rural residents, as found in Van den Berg and Koole's (2006) study in the Netherlands. As such, there is no consensus on local communities' attitudes towards rewilding projects given each project has its own interpretation and strategy, hence why this paper looks exclusively at the project of Knepp Wildland.

Knepp Wildland

Much of the existing research conducted at the Knepp Wildland project has had an ecological focus to monitor the site's biodiversity, for example bird and butterfly surveys (Greenaway, 2013). However, there has also been attention paid to the reaction of the local community, as mentioned in the owner's book *Wilding* (Tree, 2018) and the farmer-focused research of Reboah (2019). Looking at the findings of these sources gives an impression of the views already expressed towards the project. However, it is important to keep in mind that the owners are in a unique position to receive and respond to the reaction of locals and Reboah's (2019) research only looks at one, albeit very important, group of perspectives.

Tree (2018) speaks of a very negative reaction from the local community. In citing the disapproval of nearby residents (p. 130), she refers to the findings of a student's survey of local villagers, although it is not clear when this survey was conducted. The consensus presented was that the locals were not against conservation and had strong affiliations with wildlife but felt the rewilding project did not belong to the historic agricultural landscape of Sussex (Tree, 2018). Sentiments of irresponsibility and laziness were strongly conveyed, particularly regarding the southern block, which at first did not have animals grazing so the vegetation was left unmanaged. This indicates that the survey referred to was most likely conducted before 2009, when cattle were introduced to the southern block. In addition, Tree (2018) refers to multiple letters written to the Parish Council, newspapers, and her and her husband as owners, expressing dismay at the "wasteland" (p. 131, see also p. 130, p. 138, p. 144). In this way, Tree (2018) reinforces the impression of a negative reaction to the Wildland project by local people. As such, there is a need to see whether the reactions and feelings of the local residents have changed as the project has developed over the last two decades, and also to learn what could be expected Post-Brexit, when many more rewilding projects are likely to be launched.

Reboah (2019) conducted interviews of local farmers and used the sense of place framework to identify the socio-psychological barriers to rewilding for farmers. By using the place identity, place attachment and place dependence concepts, the research found three barriers to rewilding for farmers. Namely, the sacrifice of agricultural practices, the loss of the 'good farmer' status, and conservationists' reinforcement of an 'us' vs. 'them' dynamic (Reboah, 2019). The quotations used in the research give the impression that not only did the farmers dislike the change in landscape and diminishment of their role as 'farmers,' but that they also believed that the public would be "horrified" (p. 21) by an unkept landscape. Researching the current impressions of the local public will inform farmers as to whether the public are becoming more accepting of the rewilding landscape with time or whether they are indeed still "horrified" by the lack of human care.

The novelty of the Knepp Wildland project has also attracted media attention and several articles about the project comment on the reaction of local residents. From those who "despise the untamed... landscape" (Thring, 2018) and reports of harming the animals as sabotage (Thring, 2018), to those who appreciate the public benefits of flood mitigation and footpaths (de Bellaigue, 2020), there appears to be a range of views on the project. Although the initial reaction by locals is consistently portrayed as being negative (de Bellaigue, 2020; Thring, 2018), the vast majority of recent articles portray the project in a positive light, praising its ecological restoration, climate change mitigation and reconnection to nature (Barkham, 2017; Marren, 2007; Monbiot, 2015). As such, it is important to bear in mind that the media's approval of the project may be reflective of, or have influenced a change in, the local residents' attitudes.

METHODOLOGY

Design

To address the research question: *how has the change in landscape over the past two decades of the Knepp Wilding project affected place meaning for the local community?* a qualitative approach was used. The qualitative worldview looks to expose the complexity of subjective views (Miles and Huberman, 1994) and is thus the most relevant paradigm for investigating place meaning, which is evidently a subjective concept (Jacobs & Buijs, 2011). More specifically, the research falls within what Schroeder (1996) describes as the "ecology of the heart" line of study. This is qualitative research on place meaning that extends understanding about human-environment relationships to look beyond the physical and into the symbolic and emotional (Davenport & Anderson, 2005).

Study Site

The study focuses on the local community, namely residents, workers and farmers, surrounding Knepp Estate in West Sussex, in South-East England. The local area is demarcated to the three villages bordering Knepp Estate, namely Shipley, Dial Post, and West Grinstead (Appendix 1.1).

It is assumed that due to its proximity, those within this area would be aware of the estate, its rewilding project and, consequently, its changed landscape. Views of Knepp Estate are shown in Figures 1, 2 and 3 below.

Figure 1. Longhorn calves at Knepp Estate by Rachel Orchard.

Figure 2. Scrub landscape at Knepp Estate by Rachel Orchard.

Figure 3. Public footpaths through Knepp Estate by Rachel Orchard.

Qualitative Questionnaire

The main method for data collection was a qualitative questionnaire that required written responses via Google Forms. The use of this method was in light of the Covid-19 lockdown regulations, as it was not possible to conduct semi-structured interviews in person. However, given online surveys were used in Park and Selman's (2011) research on landscape change attitudes in England, the method was a suitable alternative to interviews. Park and Selman (2011) structured their online survey in three parts: (a) socio-demographic characteristics; (b) attitudinal statements; and (c) eco-centric and anthropocentric environmental values. However, their questions were formatted quantitatively. Therefore, for a qualitative question format, we looked to Davenport and Anderson's (2005) semi-structured interviews on place meaning, where they used open-ended questions to direct respondents. Taking inspiration from these precedents and adapting their methods to suit the theory of place meaning and the case of Knepp Estate's rewilding landscape change, we structured the qualitative questionnaire (Appendix 1.2) in the following way:

1. *Socio-demographic characteristics.* The questionnaire started with quantitative socio-demographic questions to engage respondents and establish background information. The four multiple-choice questions asked about the participant's proximity to Knepp Estate, length of residence in the local area, role in the community (resident, farmer or part of local business), and regularity of visiting Knepp Estate. The responses from this section were not closely analysed, as the research does not aim to demarcate between different groups

in the community. However, it was available to be used as context when relevant to a respondent's answers.

2. *Emotive and attitudinal statements*. Six open-ended questions requiring written responses enquired about the respondent's emotions and attitudes towards the rewilding project. The first question asked about their feelings towards the project and whether these have changed with time, to address the emotive aspect of place meaning. The next three questions asked about their initial reaction, their views on the current landscape, and whether these views align with those of their neighbours. Among other things, the responses to these questions gave an indication as to the community level variations on place meaning. The last two questions asked about benefits and concerns, to provoke the respondents to think about the rewilding project's implications on the local community.
3. *Context-relevant eco-centric and anthropocentric values*. The remaining three open-ended questions asked about relevant environmental views that provide an indication on the respondent's eco-centric and anthropocentric values. These questions relate to Knepp's influence on environmental interest and notion of 'natural' landscape, as well as the respondent's perception on the role of conservation in farming. They provide additional information on whether the project is viewed as a source of knowledge and whether respondents felt the project challenged their place meaning. Lastly, there was the option to offer 'additional thoughts' at the end of the questionnaire, to give respondents the opportunity to share any other views in an unstructured manner.

Data Collection and Limitations

With initial selected contacts and primary purposive sampling, we relied on snowballing for more respondents. The main limitation of the research has been the online character of the qualitative survey, which did not allow for deeper discussions or follow-up questions. As the responses have been provided anonymously, there is some uncertainty concerning whether all participants really belonged to the local community, and whether some may have completed the questionnaire more than once. However, as this was an independent research (not related to the Knepp Wildland project or any other project) there has been no incentive for participants to distort the results. In addition, the demographic of participants was limited to members of the community who have access to the Internet. However, nowadays it can be assumed that most of the local community are present on online platforms; and as the research aims to gain an insight on subjective views rather than map the whole community's place meaning, it is not a problem that the survey did not reach everyone living locally. A possible bias in the results of the research is that responding to the form was more likely to appeal to those with stronger opinions on the project.

Theoretical Thematic Analysis

The data collected was analysed using theoretical thematic analysis, which enables the process to focus within the data on what is relevant to the research question rather than give a rich description of the whole data set (Braun & Clarke, 2006). Much of the previous place meaning-related research has used inductive thematic analysis (Davenport & Anderson, 2005; Bleam, 2018). However, this research was an opportunity to advance the field and build on previous research by using an existing set of place meaning categories as guidance for our analysis, whilst remaining open to new themes. In this research, the analysis focuses primarily on the semantic or explicit level, with minor use of latent level, in order to observe the significance and implications of the data's patterns in relation to the existing theory of place meaning as opposed to forming a new framework (Braun & Clarke, 2006).

The theoretical thematic analysis was structured using Jacobs and Buijs' (2011) five categories of place meaning. As mentioned above in the section on Place Meaning, the five categories of place meanings that emerged from their research and which were used to guide the thematic analysis of this data set are: beauty (aesthetic judgments), functionality (ways of use), attachment (feelings of belonging), biodiversity (meanings pertaining to nature), and risk (worries about current or future events) (Jacobs & Buijs, 2011). Although other research also offered suitable categories, such as Smith et al., (2011), Jacobs and Buijs' (2011) categories lend themselves more to semantic level analysis than the other more interpretive indicators. In addition, the categories allowed room for change over time, through the risk and functionality categories.

The questionnaire responses were coded using a different colour for each category and going through the responses separately for each category. In this way, the responses were coded five times over with a focused category in mind. At the end, there was a final round of coding with an additional colour to highlight any additionally relevant information that fell outside the five categories and could form new themes. The place meaning categories were a useful structure for the thematic analysis, as it maintained the relevance of the analysis to the theory of place meaning and its relation to place change.[1]

RESULTS

The qualitative survey received 29 responses, with a wide diversity of views. Most respondents identified themselves as local residents and there was a fairly even spread of respondents' length of residence in the area, with a large majority living within 5 miles of Knepp Estate and regularly visiting it. In this section the responses

[1] More details on the socio-demographic characteristics can be found in Appendix 1.4. Furthermore, a table of results, which consists only of direct quotes, is presented in Appendix 1.5 to accurately reflect the full array of responses and convey the subjective experience of respondents.

will be presented within the place meaning framework to provide a better understanding of the respondents' views. Extensive use of quotes is used, as the respondents' views are best expressed in their own words. The respondent is indicated using their randomly assigned number order; for example, as below 2, 3, 4 etc. In the additional round of coding for finding other themes relevant to the research question, two more categories became evident, namely respondents' emotions in place and temporal comments.

Beauty

All respondents provided aesthetic judgments on the Knepp Wilding project with a mixture of negative and positive meanings becoming evident from their views on the landscape. Respondents 1, 3, 5, 10, 16 and 22 all refer to the landscape appearing more "natural," with 2 and 25 particularly noting the effect of trees being left where they fall instead of being cleared. This relates to some respondents embracing the "rough and ready" (29) and "rugged" (5) appearance of the landscape as a positive attribute. Respondents 20 and 27 express this positive view on the non-uniform landscape by stating "the wilder the better!" (20) and that "nature doesn't need to be neat to be beautiful" (27). The term "beautiful" is used 13 times throughout the responses to describe Knepp's landscape, along with similar terms such as "lovely" (6, 15, 17) and "stunning" (29).

On the other hand, Respondent 19 sees the landscape as an "overgrown wilderness" and a "mess." These sentiments are shared by Respondent 18 who describes the landscape as "Scruffy, somewhat out of control in some areas and not very attractive." These indicate preference for the more uniform rural idyll of agricultural countryside, which was directly stated by Respondents 9, 12 and 24. "I miss the golden corn fields, lush pastures." (9) and "It is lovely at times to walk in a lovely green field in a parkland setting, even seeing hedgerows maintained" (24) reflect that their notion of beauty is firmly tied to the more traditional agricultural setting. Respondent 12 concurs, perceiving Knepp as a "dereliction of the duty of landowners to manage the countryside."

An example of the contrast in desired appearance of the landscape is that Respondent 22 states "I hope it catches on all over the country or world", whereas Respondent 24 thinks "It would be awful if the entire country looked like Knepp." The contrast continues with the discussion of 'natural' as Respondent 3 views Knepp's landscape as "how I think a natural landscape should look like.", whereas Respondent 28 states "It's not natural. There is no natural." Lastly, an interesting remark on the scenery was made by Respondent 21, who states that the landscape is "how I imagine the whole of Sussex would look like if humans were lifted off the face of the earth."

Functionality

All respondents shared opinions on the ways in which Knepp Estate is used, particularly in responding to the question on the role of conservation in farming. Most respondents state that conservation should be, could be or is part of farming, with only three (5, 6 and 13) viewing it as separate. Focusing on Knepp, Respondent 13 describes it as "an environmental project and business," and Respondent 5 states that "The meat sold is in effect a by-product of the conservation." On the other hand, there was a difference between those who view conservation as possibly being or becoming part of farming and those who view it as already an integral part of farming. For example, Respondents 1, 2, and 3 all note that conservation "should be part of farming" with Respondent 1 stating that "farmers have a duty to look after the land". However, Respondent 3 acknowledges that "the country needs feeding so it's not practicable for every farm to rewild", which was also supported by Respondents 17 and 19. According to others, conservation "is and always has been part of farming" (28). At the other hand, there is a sense that "farming is frowned upon by conservationist[s]!" (15). Respondent 9 concurs, and regarding Knepp, thinks that "This scheme is not appropriate for this area, the fields were once very productive for farming."

Several respondents comment on how they perceive the change in land use from agriculture to rewilding to have influenced the local economy. On the one hand, some consider the effect to be negligible or neutral, with comments such as "I don't think it brings in much to the local economy" (20) and "Businesses based on the estate were evicted to make way for other businesses that are no more nor less valid, so it has hurt the local economy in some ways but benefitted it in others." (28). However, others highlight the contribution to the local economy as one of the Wildland project's benefits, as "It provides jobs for local people" (25) and "It brings visitors and helps local pubs and businesses" (18). As for respondents who identified as part of a local business, they affirm that it is "Bringing more business to the area" (13) and that it is "great for local business" (23). In addition, many respondents describe how they personally use the place, mainly referring to their use of the footpaths at Knepp as it is "Great for going on long walks" (16), but also psychologically as an educational space for learning about wildlife through experience (3, 13, 16, 23, 25, 27 and 29).

Attachment

There are two aspects to the respondents' sense of belonging: firstly, whether they feel the Wildland project belongs to their community, expressed through whether they perceive fellow locals to agree or disagree with their own either positive or negative views; and secondly, whether the respondents themselves feel a sense of belonging to the Wildland project. Regarding the former, the majority of respondents with both positive and negative views of the project consider other locals to agree with them. As such, there was a stark contrast between "everyone I know hates it" (9) and "I have never had a conversation with anyone around here who has had a bad thing to say about the concept of Knepp" (21). However, a few respondents noted "A mixture [of views], but vehement either way" (4). Four respondents expressed a sense of pride in the project. For example, Respondent 18 states "There is a certain pride that the Rewilding project is on our doorstep."

Regarding the latter, there was again a mixture of views. On the one hand, Respondent 11 feels that Knepp Wildland has "Pull[ed] people together and given them a strong centre." On the other hand, Respondent 19 states that "The estate is no longer the focal point of the area", and that "the community that existed has diluted." A couple of respondents clearly feel a strong sense of belonging to the Wildland project, as they used the term "we" to talk about the project's reintroductions of species (2 and 16). In addition, one respondent volunteers at the project (22), another works there in the summers (16) and another wished to be involved in establishing an educational programme (11). Another way in which it became evident to what extent the locals felt they belonged to the project is through their view on access to the land. On the one hand, Respondent 20 says that, "anyone can visit." and a majority of respondents state that they use the public footpaths almost daily. On the other hand, Respondent 9 feels that "The footpaths and bridleways are overgrown, making it difficult to access". Some think that the project does not belong to the local community, due to its attraction of "trendy hippies from London, talking loudly in smug accents, in inappropriate clothes" (28). Since access other than via the footpaths is expensive, Respondent 12 feels that "The impression left is it's all a bit cliquey and elitist!"

Biodiversity

The respondents' comments on the biodiversity of Knepp Wildland demonstrated how the project's environmental focus has generated new meanings pertaining to nature for some whilst challenging these meanings for others. In doing so, it also shed light on which aspects of nature have significant meaning for some respondents. For example, a few indicated a strong affiliation with wildflowers, as one respondent feels "Sad about the loss of bluebells" (23) and another notes that "the cattle walk all over the wildflowers, particularly the wild daffodils" (19). This indicated a sense of 'loss' that was expressed in opposing ways. On one hand, a respondent who had a generally negative view on Knepp Wildland mentioned the "Loss of countryside views" (12). On the other hand, a respondent with a more positive inclination to Knepp said that the project has "made me more aware of what we have lost." (20). However, many respondents stated that the Wildland project had heightened their awareness and appreciation of nature and the environment through being a source of information on wildlife (1, 3, 13, 16, 23, 25, 27 and 29). For example, Respondent 23 shares that "it has given me great pleasure and understanding of a world that I knew little of."

Almost all respondents mention the wildlife of Knepp and describe it as "magical" (8), "really special" (5) and "beautiful" (25). Many specific animals are mentioned, such as "storks" (11, 17, 18), "Cuckoos and Nightingales" (14, 24), indicating how particular species have caught people's attention. However, the project has also broadened some respondents' interest. For example, Respondent 18 claims that "It develops a passion for understanding the wider diversity of bugs, birds, deer and other mammals." In contrast, Respondent 9 feels that Knepp Estate "Previously always had lovely wildlife not increased by rewilding." For those who feel that Knepp Wildland has positively contributed to their meanings of nature, a few provided examples of how this has inspired action or changed their behaviour. A farmer tells of how they "have created small wild areas. Mini Knepps we call them!" (4); and a resident has created "mini natural rural wild parts in my garden now" (10). Lastly, Respondent 16 indicates her change in behaviour by stating that "there was a time where I would stomp on a spider, now I put them outside!"

Risk

Whilst five respondents reported having no concerns about the Knepp Wildland project, for the rest of the respondents there was a wide array of worries. Several (1, 7, 16, 22, 24) mentioned the worry of the site becoming "too busy, [which will] disturb the wildlife and change the natural feel, beauty and peacefulness of the area"(1). Similarly, there was concern about visitors who are not "respectful" (7), and that "There needs to be stronger education of visitors to respect the privacy of residents" (18). In addition, Respondent 20 has the impression that "There are a lot who abuse it", for example Respondent 21's experience where they "had an encounter recently with a rude man on one of my walks." For some respondents, there are certain elements of the project that worry them, such as "foxes" (14), "pigs" (18), and "piling brash up [as it] is just right for wildfires." (15). For Respondent 9, all the free roaming animals "can be dangerous and is simply annoying," which is likely related to their "scary experience with the ponies."

Regarding the local community, Respondent 4 mentions "Community acceptance" and Respondent 26 views "Alienation of locals" as concerns about the project. This worry is compounded by plans that may distance the project from the community. For example, Respondent 13 shares that "Their [Knepp's] plan to open a cafe worries me a little... we hope not too much business is taken away from us..." A couple of respondents (9 and 11) are concerned about the longevity of the project, with Respondent 24 feeling that "there is pressure on the Estate to keep introducing new attractions to maintain interest." Lastly, there is concern about whether the concept of rewilding has "Viability at smaller scale" (4) and "The implication that everyone can do this (rewilding) when not everyone can" (17).

Emotions-in-Place

Several respondents shared their feelings of experiencing Knepp. This category highlights emotional comments that contribute to understanding the respondents' meaningful experiences of being in Knepp Wildland. About 20 of the respondents shared positive emotions, such as feeling “Invigorated, relaxed, optimistic and happy” (3) and “Excited and peaceful” (4). Respondent 29 considers Knepp a “great place to clear my head and just think”; and Respondent 22 says that “It’s a magical place that soothes my soul”. According to Respondent 16, “you feel a million miles away from ‘the real world’”. Similarly, Respondent 21 feels it “Provides wonderful escapism. Knepp feels like a sanctuary”. Several respondents mention feeling “very lucky to live here” (17, also 20, 25 and 27), such as Respondent 3 who says that “living adjacent to Knepp, it’s a dream come true!” and Respondent 21 who states that “It’s my favourite thing about living down here.” Two respondents refer to it as “my happy place” (20 and 22). On the other hand, three respondents share negative emotions, feeling “Sad” (9 and 19), and “Unhappiness combined with curiosity…” (12).

Temporality

Comments that were made about changes in emotions or views over the course of the project were highlighted under this category, along with comments on temporal elements of the landscape, in order to further explore the change-over-time aspect of the research question. Respondent 28 expresses a significant change in views over time, saying that “I used to love, and visit Knepp regularly. For a time, I thought the place looked astonishingly untidy and wasteful. I like it again in appearance now”. Additionally, Respondent 18 states that “Initially we were concerned”, yet now they feel “Very excited about it”. In a similar way, Respondent 13 shares that “To begin with, I’ll be honest, it looked such a mess! But now some 20 years on (!) it has developed into a fantastic project.” The change of landscape is also noted by Respondent 23, who feels that “The Southern Block was initially slightly scary, but we have grown to love it more than we could ever have imagined.” However, for some respondents, the “feeling has always been the same” (7, also 3 and 21); and for others, “The feeling has got better with time.” (8 and 23).

Regarding the temporal nature of the landscape, Respondent 8 describes it as “Ever changing.” Respondent 2 agrees, stating that “everyday is different.” This transient characteristic is reiterated in both a timeless way and a historic way, as Respondent 21 feels that “it never gets old!” Respondent 20 confides that “It always makes me think that what we are seeing there is how the whole country would have been a few hundred years ago.” The change in landscape over the year is noted by Respondent 24, who says the landscape “varies on the season, even footpaths are muddy and virtually impassable in the winter, and lethal in places during the summer as the clay turns to concrete.” The change during the day is mentioned by Respondent 21, along with their temporal preference, as they “go at all times of day to walk my dogs, mostly in the early mornings when it is quiet so I can see deer.”

Discussion

The broad spectrum and facets of place meanings conveyed by respondents demonstrates the variety and complexity of people-place relationships. This section aims to relate the questionnaire results back to previous research on the topics of place meaning, landscape change and rewilding whilst discussing the key implications of those results.

Rewilding as a 'Natural' Landscape

The explicit use of 'natural' to aesthetically describe Knepp's landscape indicates that for many respondents, the place change from agriculture to rewilding is a shift towards a more 'natural' setting. Part of being more 'natural' was demonstrated through place meaning's seasonal and daily temporal variation for a couple of respondents (21 and 24), which aligns with Bleam's results (2018). Furthermore, the respondents who perceived the landscape as more 'natural' held predominantly positive place meanings, shown clearly through their positive emotions-in-place. For example, Respondent 1 views the landscape as "more natural now" and feels "peaceful" in Knepp's landscape. This affirms Kil et al.,'s (2014) claim that the value of a landscape is based on both tangible elements, here the visibly 'natural' setting, and intangible elements, namely emotions-in-place. Therefore, those who perceived the landscape as being more 'natural' considered this a key positive attribute to the landscape change, which contributes to forming a positive place meaning. This supports Lorimer's (2015) claim that rewilding reconnects humans to nature, which is further affirmed by almost all respondents acknowledging, and mostly praising, Knepp's wildlife.

There is overlap between the categories of beauty and biodiversity, as appreciating Knepp's wildlife held a strong influence on meanings pertaining to nature and aesthetics of the rewilding landscape as a whole. Whilst the grazing animals and rare birds were often mentioned as representatives of this wildlife, the reintroduction programmes for specific species captured many respondents attention and kept them interested in the project's wildlife. This is in agreement with Westling et al.,'s (2014) findings, since the results indicated that the opportunity to view wildlife increased respondents' attachment to Knepp Wildland. Due to its recurrence in the results, Knepp's attraction of wildlife can be seen as a key indicator of the difference between the previous and traditional agricultural landscape and the evolving rewilding landscape. Therefore, the perception of an "abundance of wildlife" (1) affects the local community's place meaning by generating a greater interest in other species' existence through the experience-in-place of seeing a greater variety of species. This effect on place meaning can be seen to have grown over time, as the opportunities of viewing more species has increased as the project matures.

It must be noted that Respondent 9 disagrees with the wildlife claim, as they perceive the agricultural landscape to have "always had lovely wildlife [which is] not increased by rewilding." Nevertheless, Respondent 9 implicitly recognises that wildlife is a key argument for the acceptance of rewilding, by not only mentioning it but also discrediting it based on their experience. This provides further indication of the pivotal role that wildlife plays in how rewilding affects people's place meaning. Respondent 9 shares that they had a negative experience with the ponies, and that they miss the previous agricultural landscape. In line with Manzo's (2005) findings, this affirms that both positive and negative experiences contribute to developing place meanings. Due to negative experiences, Respondent 9 has developed a negative place meaning with the current rewilding landscape and reminisces on the positive experiences associated with the previous agricultural landscape.

In relation to the term 'wildlife', several respondents commented on the 'wild' aspect of the rewilding landscape, with contrasting interpretations of whether this is a desirable or undesirable attribute. These align with different points in time of the discourse on the concept of 'wilderness' explored by Cronon (1995). The negative view of it being an "overgrown wilderness." and a "mess" (19) speaks to the biblical pre-19th century view that wilderness should be civilised by visible human management (Cronon, 1995). In contrast, the positive attribution of "the wilder the better!" (20) speaks to the post-19th century discourse shift whereby witnessing wilderness became an awe-inspiring religious experience. This discursive interpretation of wilderness perceives it as being "without the contaminating taint of civilisation" (Cronon, 1995, p. 69) and aligns with Respondent 21's impression of Knepp as an example of a human-less landscape. These contrasting impressions are matched by other respondents' respectively "Sad" (19) and "happy" (20) emotions-in-place, indicating that their perception of 'wilderness' influences how the rewilding place change affects their place meaning.

Cronon's (1995) criticism is directed at the glorification of American national parks, as they present nature as an experience 'outside' our daily human lives. However, there is a distinction here with the process of an evolving rewilding landscape. According to some responses, the rewilding landscape of Knepp can possibly achieve both the awe of a natural 'wilderness' and the practicality of appreciating nature 'inside' our human sphere. This can be seen through the examples of projects inspired by Knepp Wildland, namely Respondent 4's "Mini Knepps!" and Respondent 10's "mini natural rural wild parts in [their] garden". Therefore, the ability of the rewilding landscape to affect a few respondents' actions clearly indicates that this evolving form of 'wildland' affects place meaning by encouraging nature to be 'let in' to our daily human lives. This is supported by the several respondents who reported Knepp to be a source of information on nature and thus a space that is educational instead of solely a holiday destination. However, Knepp is still susceptible to Cronon's (1995) critique on the elitism of wilderness spaces, given Respondents 12 and 28's experience of Knepp as an exclusive and expensive space.

Changing Cultured Naturalness

The lack of consensus on any aspect of place meaning by respondents presents place change as principally affecting the cohesion of communal place meaning by challenging the cultured naturalness formed around a particular way of life. Although this may sound dismal and cause worry that changing land use will only ever lead to conflict, the results also indicate that with time, cohesion can start to reform as cultured naturalness adapts to the new landscape. For example, most respondents feel that other locals agree with their point of view, be it negative or positive, indicating a sense of cohesion more than conflict. Furthermore, positive experiences through place access and information on the place change can be seen as factors that encourage individuals to renegotiate their place meanings. More than half of the respondents expressed positive place meanings for the rewilding landscape of Knepp Estate, a stark contrast from the negativity at the start of the project, as presented by Tree (2018). This indicates not only a mellowing of opposition to the Knep Wildland project but also posits the possibility of the local community embedding rewilding into their changing local identity and evolving way of life as a consequence of their positive place meaning.

Hull et al.,'s (2001) concept of cultured naturalness is particularly interesting when applied to the case of Knepp, as it assumes that those outside the local community value the 'wild' aspects of the natural setting more than the locals' cultural symbols embedded in the scenery. Controversially, the owners voluntarily opted out of the cultural symbols embedded in the agricultural landscape of the rural England 'idyll', in favour of an 'outsider' valuation of a 'wild' landscape. For this reason, it is perhaps unsurprising that the initial reaction by the local community to the project was rather hostile (Tree, 2018). However, these results demonstrate a growing sense of belonging to the Knepp Wildland project among the local community, given the sense of "pride that the Rewilding project is on our doorstep" (18) felt by some residents. This appears to be reciprocated by several respondents either being involved in the project or considering the project to have a sense of communal ownership through the use of 'we.' As such, the attachment aspect of place meaning has strengthened as the project has matured, due to Knepp becoming perceived as "'that famous place'" (16) and as more locals feel "that we now have an understanding of what is going on there" (13).

A particularly interesting effect on place meaning caused by the rewilding landscape change is a dual sense of loss. The sense of ecological loss for many respondents related to Marshall et al.,'s (2019) research on ecological grief, as indicated by comments such as "now I notice the large open fields with no hedges and trees and the thin soil" (1). However, in Marshall et al., (2019)'s study the participants were mourning the familiar reef that is now dying. At Knepp, the sense of ecological loss on the wider scale is soothed by the immediacy and experience of Knepp's successful conservation of species that some locals may have barely noticed were disappearing. On the other hand, there is a sense of 'agricultural loss', that Reboah's (2019) research identified as the loss of agricultural practices and the loss of 'good farmer' status. These results also found that a few respondents in the local community expressed 'agricultural loss' both on an aesthetic level, such as the "Loss of countryside views," (12) and also in terms of identity since "Good farmers can combine producing food and caring for the wildlife." (19). As such, Knepp's rewilding project makes one respondent "Sad that agricultural land that used to grow food and support dairy has been lost" (19).

The fact that significantly more respondents expressed a sense of ecological loss than agricultural loss indicates that they may hold new interpretations of what accepted visible care and stewardship of the land looks like. Lewis's (2008) findings showing the preference of communities for visible care in the landscape are challenged in the case of Knepp. Whilst a few respondents felt that "The land is no longer cared for" (19), the majority expressed support for the rewilding notion of allowing natural processes to care for and repair the environment. Contrary to the visible influence of human management as a sign of stewardship, many respondents seem to have loosened their notion of control, even though Westling et al., (2014) found such control to be a key influence on perceptions to place change. It appears that Knepp has finely balanced visible control of the grazing animals, perhaps through culling and tagging (Tree, 2018), to create a looser sense of care while many in the local community have over time come to appreciate how "The free roaming animals shape the land" (2) Therefore, in this case, place change affects the local community's place meaning by challenging, and for some changing, their perception of what stewardship of the land looks like.

A key indicator of cultured naturalness is access to the land (Hull et al., 2001). Knepp Wildland has a network of public footpaths throughout the estate, which the majority of respondents reported using regularly. For those with negative or ambivalent place meanings to the rewilding landscape, access was a notable point of concern that drew similarities to the notion of human-wildlife conflicts. The notion that human-wildlife conflicts occur when locals feel wildlife is prioritised (Madden, 2004) is relevant for some respondents through negative experiences with the animals (9, 18 and 26) and ragwort (9, 17, 18, 26 and Tree (2018)). However, as Dickman (2010) claims, it is possible that this notion of conflict is influenced more by a human-human conflict. In this case, Drenthen's (2018) clarification on the dichotomy of those preferring the more cultural landscape of agriculture and those in favour of the 'natural' rewilding landscape provides a likely explanation. Yet it is also notable that this dichotomy appears to be fading given the appeasement of the owners through "perimeter mowing" (18) and the questionnaire results indicating an equally prominent concern for too many humans instead of unease about the animals. Therefore, perceptions on access and thereby, ease of collecting experiences-in-place, can be seen as influential on whether place change affects place meanings negatively or positively.

The predominantly positive emotions-in-place provide an emotive insight into the positive place meanings that more than half of the respondents have with the rewilding landscape. Reflections on the project's two decades show that for some, it is only with time that these place meanings have become positive, while for many, this time has strengthened or maintained their positive place meanings. As such, it seems evident that the local community is gradually changing their cultured naturalness to include the values of rewilding alongside their agricultural way of life. This indicates that with time and access, place change has affected many members of the local community's place meaning in two ways: firstly, through reflection on themselves, such as how they personally value other species and the environment and whether they had previously noticed the environmental damage of certain aspects of industrial farming, such as pesticides; and secondly, through adapting their relationship to Knepp Estate as a place, such as through different experiences-in-place and emotions-in-place, as well as opportunities for involvement in the Wildland project. These effects align seamlessly with Greider and Garkovich's (1994) proposition that place change triggers renegotiation of place meanings relating to the self and the landscape.

It must be emphasised that this research asserts that Knepp Wildland is only at the start of changing cultured naturalness, as the research was not exhaustive of the whole community's views and conflicts in opinion are still very evident. In addition, cultured naturalness requires time, so it depends on the Wildland project's ability to secure its own longevity as to whether the place change of rewilding becomes fully embedded as part of the local way of life. Nonetheless, it suggests an interesting addition to the concept of cultured naturalness by including the ability for evolution through local ways of life changing and, as such, the place change becoming a new chapter in the story of local life. In doing so, it also affirms that place change does affect place meaning in the case of Knepp's transition from an agricultural to a rewilding landscape. Such processes of place change transforming place meaning may proliferate in many communities once the Agriculture Act 2020 is fully implemented and encourages a more diverse mosaic of landscapes.

CONCLUSION

In light of anticipated place change due to the Agriculture Act 2020 and the growing role of participatory planning to give local communities a voice in such change, the problem of public acceptance of place change is increasingly ubiquitous. This research contributes to better equipping those involved in such change, by providing insight on the effects of Knepp Estate's place change on the local community's place meaning. In addition, the research contributes to filling the gap in knowledge about whether place change does affect place meaning, by affirming this proposition in the case of transitioning from an agricultural landscape to a rewilding one. The key finding affirms Greider and Garkovich 's (1994) claim that place change affects place meaning by triggering renegotiation of place meaning for people themselves, in this case, respondents' increasing awareness of, among other things: the environment; the negatives of industrial farming; and appreciation of wildlife. As well as the renegotiation of their relationship with the landscape, residents' experience-in-place has had to adapt from a uniform landscape to an "Ever changing" (8) landscape shared with both familiar and unfamiliar species. In other words, this research suggests that place change affects place meaning by causing a change in cultured naturalness over time.

Facets of the change in cultured naturalness that came to light in the questionnaire results were: increasing sense of ecological loss; changing view of visible stewardship; and increasing appreciation of wildlife. Influences on how place change affected place meaning were perceptions of access, wilderness and belonging. Furthermore, the results indicate that over time the place meaning derived from Knepp Wildland is predominantly positive, particularly evident from the emotions-in-place expressed. However, the results also indicate that public consultation should be ongoing, given that several respondents mentioned worries and felt that "on the local level things some things could be improved" (18). There is still much to be researched in this field. For example, the impact of the positive media portrayal on place meanings could hold clues to how place meaning evolves. Another suggestion is to further explore the role of access to information and to experiences-in-place in shaping place meaning. In addition, other forms of place change could profit from similar research, in order to better understand the effects on people-place relationships and contribute to informing future land management.

REFERENCES

Allendorf, T., Swe, K., Oo, T., Htut, Y., Aung, M., Allendorf, K., Hayek, L., Leimgrubek, P. & Wemmer, C. (2006). Community attitudes toward three protected areas in Upper Myanmar (Burma). *Environmental Conservation*. 33(4): 344-352. doi:10.1017/S0376892906003389.

Barkham, P. (2017). 'It is strange to see the British struggling with the beaver': why is rewilding so controversial? *The Observer*. Retrieved from https://www.theguardian.com/environment/2017/jul/01/rewilding-conservation-ecology-national-trust.

Barkham, P. (2018, April 27). Dutch rewilding experiment sparks backlash as thousands of animals starve. *The Guardian*. Retrieved from https://www.theguardian.com/environment/2018/apr/27/dutch-rewilding-experiment-backfires-as-thousands-of-animals-starve.

Bleam, R. M. (2018). Unbounded place meanings and embodied place identities for conservation volunteers in Scottsdale, Arizona. *Journal of Environmental Psychology*, *56*, 76-83. doi: 10.1016/j.jenvp.2018.03.002.

Braun, V., & Clarke, V. (2006). Using thematic analysis in psychology. *Qualitative research in psychology*, *3*(2), 77-101. doi: 10.1191/1478088706qp063oa.

Burrell, C. & Greenaway, T. (2011) Knepp Castle Estate Wildland Project. In *Landscape Archaeology and Ecology - Animal, Man and Treescapes*. (Eds. Rotherham, I.I & Handley, C.) *9*, 146-159.

Chapin III, F. S., & Knapp, C. N. (2015). Sense of place: A process for identifying and negotiating potentially contested visions of sustainability. *Environmental Science & Policy*, *53*, 38-46. doi: 10.1016/j.envsci.2015.04.012.

Cheng, A. S., Kruger, L. E., & Daniels, S. E. (2003). "Place" as an integrating concept in natural resource politics: Propositions for a social science research Agenda. *Society & Natural Resources*, *16*(2), 87-104. doi: 10.1080/08941920309199.

Cronon, W. (1995). The trouble with wilderness: Or getting back to the wrong nature. W. Cronon (Ed.), *Uncommon ground: Rethinking the human place in nature*, W. W. Norton & Co (pp. 69-90).

Davenport, M. A., & Anderson, D. H. (2005). Getting from sense of place to place-based management: An interpretive investigation of place meanings and perceptions of landscape change. *Society and natural resources*, *18*(7), 625-641. doi: 10.1080/08941920590959613.

Davenport, M. A., Baker, M. L., Leahy, J. E., & Anderson, D. H. (2010). Exploring multiple place meanings at an Illinois State Park. *Journal of Park and Recreation Administration*, *28*(1).

de Bellaigue, C. (2020, February 25). The end of farming? *The Guardian*. Retrieved from https://www.theguardian.com/environment/2020/feb/25/the-end-of-farming-rewilding-intensive-agriculture-food-safety.

Devine-Wright, P., & Howes, Y. (2010). Disruption to place attachment and the protection of restorative environments: A wind energy case study. *Journal of environmental psychology*, *30*(3), 271-280. doi: 10.1016/j.jenvp.2010.01.008.

Dickman, A. J. (2010). Complexities of conflict: the importance of considering social factors for effectively resolving human–wildlife conflict. *Animal conservation*, *13*(5), 458-466. doi: 10.1111/j.1469-1795.2010.00368.x.

Dockerty, T., Lovett, A., Appleton, K., Bone, A., & Sünnenberg, G. (2006). Developing scenarios and visualisations to illustrate potential policy and climatic influences on future agricultural landscapes. *Agriculture, Ecosystems & Environment*, *114*(1), 103-120. doi: 10.1016/j.agee.2005.11.008.

Downing, E. & Coe, S. (2018). Brexit: Future UK agriculture policy. *Commons Briefing papers*, CBP 8218.

Drenthen, M. (2018). Rewilding in layered landscapes as a challenge to place identity. *Environmental Values*, 27(4), 405-425. doi: 10.3197/096327118X15251686827732.

Greenaway, T. (2013). *Knepp Wildland Project Annual biodiversity report and monitoring update 2013*. Retrieved from https://static1.squarespace.com/static/595ca91bebbd1a1d0aaab285/t/5a37858853450a1651053ead/1513588106899/Final+Knepp+Monitoring+Update+2013.pdf.

Greider, T., & Garkovich, L. (1994). Landscapes: The social construction of nature and the environment. *Rural sociology*, 59(1), 1-24. doi: 10.1111/j.1549-0831.1994.tb00519.x.

GOV.UK (2020a). *Government unveils path to sustainable farming from 2021.* Retrieved from https://www.gov.uk/government/news/government-unveils-path-to-sustainable-farming-from-2021.

GOV.UK (2020b). *Landmark Agriculture Bill becomes law.* Retrieved from https://www.gov.uk/government/news/landmark-agriculture-bill-becomes-law

Guha, R. (2000). The Southern Challenge. In R. Guha (Ed.), *Environmentalism: A global history* (pp. 98-115). Longman.

Holladay, P. J., & Ormsby, A. A. (2011). A comparative study of local perceptions of ecotourism and conservation at Five Blues Lake National Park, Belize. *Journal of Ecotourism*, *10*(2), 118-134. doi: 10.1080/14724049.2010.529910.

Hull, R. B., Robertson, D. P., & Kendra, A. (2001). Public understandings of nature: A case study of local knowledge about" natural" forest conditions. *Society & Natural Resources*, *14*(4), 325-340. doi: 10.1080/08941920118871.

Jacobs, M. H., & Buijs, A. E. (2011). Understanding stakeholders' attitudes toward water management interventions: Role of place meanings. *Water Resources Research*, *47*(1). doi: 10.1029/2009WR008366.

Jorgensen, B. S., & Stedman, R. C. (2001). Sense of place as an attitude: Lakeshore owners' attitudes toward their properties. *Journal of environmental psychology*, *21*(3), 233-248. doi: 10.1006/jevp.2001.0226.

Jørgensen, D. (2015). Rethinking rewilding. *Geoforum*, *65*, 482-488. doi: 10.1016/j.geoforum. 2014.11.016.

Kil, N., Holland, S. M., & Stein, T. V. (2014). Place meanings and participatory planning intentions. *Society & Natural Resources*, *27*(5), 475-491. doi: 10.1080/08941920. 2013.842274.

Lewis, J. L. (2008). Perceptions of landscape change in a rural British Columbia community. *Landscape and urban planning*, *85*(1), 49-59. doi: 10.1016/j. landurbplan.2007.09.011.

Lorimer, J. (2015). *Wildlife in the Anthropocene: conservation after nature.* Minneapolis. University of Minnesota Press.

Madden, F. (2004). Creating coexistence between humans and wildlife: global perspectives on local efforts to address human–wildlife conflict. *Human dimensions of wildlife*, *9*(4), 247-257. doi: 10.1080/10871200490505675.

Manzo, L. C. (2005). For better or worse: Exploring multiple dimensions of place meaning. *Journal of environmental psychology*, *25*(1), 67-86. doi: 10.1016/ j.jenvp.2005.01.002.

Manzo, L. C., & Perkins, D. D. (2006). Finding common ground: The importance of place attachment to community participation and planning. *Journal of planning literature*, *20*(4), 335-350.doi: 10.1177/0885412205286160.

Marren, P. (2007, May 9). Recall of the wild. *The Guardian*. Retrieved from https://www.theguardian.com/society/2007/may/09/ruralaffairs.conservation.

Marshall, N., Adger, W. N., Benham, C., Brown, K., Curnock, M. I., Gurney, G. G., ... & Thiault, L. (2019). Reef Grief: investigating the relationship between place meanings and place change on the Great Barrier Reef, Australia. *Sustainability Science,* 14(3), 579-587. doi: 10.1007/s11625-019-00666-z.

Meanwell, A. (2015). As a shepherd, I know we have not 'sheepwrecked' Britain's landscape. *The Guardian*. Retrieved from https://www.theguardian.com/ commentisfree/2015/jul/21/farmers-sheep-lake-district-preserve-environmentalists.

Monbiot, G. (2015, July 16). Let's make Britain wild again and find ourselves in nature. *The Guardian.* Retrieved from https://www.theguardian.com/commentisfree/2015/jul/16/britain-wild-nature-rewilding-ecosystems-heal-lives.

Ngonidzashe Mutanga, C., Vengesayi, S., Gandiwa, E., & Muboko, N. (2015). Community perceptions of wildlife conservation and tourism: A case study of communities adjacent to four protected areas in Zimbabwe. *Tropical Conservation Science, 8*(2), 564-582. doi: 10.1177/194008291500800218.

Park, J. J., & Selman, P. (2011). Attitudes toward rural landscape change in England. *Environment and Behavior, 43*(2), 182-206. doi: 10.1177/0013916509355123.

Patterson, M. E., & Williams, D. R. (2005). Maintaining research traditions on place: Diversity of thought and scientific progress. *Journal of environmental psychology, 25*(4), 361-380. doi: 10.1016/j.jenvp.2005.10.001.

Peterson, M. N., Birckhead, J. L., Leong, K., Peterson, M. J., & Peterson, T. R. (2010). Rearticulating the myth of human–wildlife conflict. *Conservation Letters, 3*(2), 74-82. doi: 10.1111/j.1755-263X.2010.00099.x.

Proshansky, H. M., Fabian A. K., and Kaminof R. (1983). Place identity: Physical world and socialization of the self. *J. Environ. Psychol.* 3:57–83. doi: 10.1016/S0272-4944(83)80021-8.

Reboah, C. (2019). Barriers to rewilding on Sussex farmland: socio-psychological implications of rewilding on farmers' Sense of Place. *Swedish University of Agricultural Sciences.* Retrieved from https://stud.epsilon.slu.se/14664/.

Rodríguez-Izquierdo, E., Gavin, M. C., & Macedo-Bravo, M. O. (2010). Barriers and triggers to community participation across different stages of conservation management. *Environmental Conservation, 37*(3), 239-249. doi: 10.1017/S0376892910000500.

Smith, J. W., Davenport, M. A., Anderson, D. H., & Leahy, J. E. (2011). Place meanings and desired management outcomes. *Landscape and urban planning, 101*(4), 359-370. doi: 10.1016/j.landurbplan.2011.03.002.

Schepers, F. & Jepson, P. R. (2016). Rewilding in a European context. *International Journal of Wilderness, 22*(2), 25-30.

Stedman, R. C. (2002). Toward a social psychology of place: Predicting behavior from place-based cognitions, attitude, and identity. *Environment and behavior, 34*(5), 561-581. doi: 10.1177/0013916502034005001.

Thring, O. (2018, April 29). Isabella Tree: the lady of the manor was born to rewild. *The Sunday Times.* Retrieved from https://www.thetimes.co.uk/article/isabella-tree-the-lady-of-the-manor-was-born-to-rewild-p20nlxgrg.

Tree, I. (2018). *Wilding: The Return of Nature to a British Farm.* London. Picador.

Van den Berg, A. E., & Koole, S. L. (2006). New wilderness in the Netherlands: An investigation of visual preferences for nature development landscapes. *Landscape and Urban Planning, 78*(4), 362-372. doi: 10.1016/j.landurb plan.2005.11.006.

Vaske, J. J., & Kobrin, K. C. (2001). Place attachment and environmentally responsible behavior. *The Journal of Environmental Education, 32*(4), 16-21. doi: 10.1080/00958960109598658.

Vorkinn, M., & Riese, H. (2001). Environmental concern in a local context: The significance of place attachment. *Environment and behavior*, *33*(2), 249-263. doi: 10.1177/00139160121972972.

Westling, E. L., Surridge, B. W., Sharp, L., & Lerner, D. N. (2014). Making sense of landscape change: Long-term perceptions among local residents following river restoration. *Journal of hydrology*, 519, 2613-2623. doi: 10.1016/j.jhydrol. 2014.09.029.

Williams, D. R., & Vaske, J. J. (2003). The measurement of place attachment: Validity and generalizability of a psychometric approach. *Forest science*, *49*(6), 830-840. doi: 10.1093/forestscience/49.6.830.

Williams, D. R. (2014). Making sense of 'place': Reflections on pluralism and positionality in place research. *Landscape and Urban Planning*, *131*, 74-82. doi: 10.1016/j.landurbplan.2014.08.002.

APPENDIX

Study Site Map

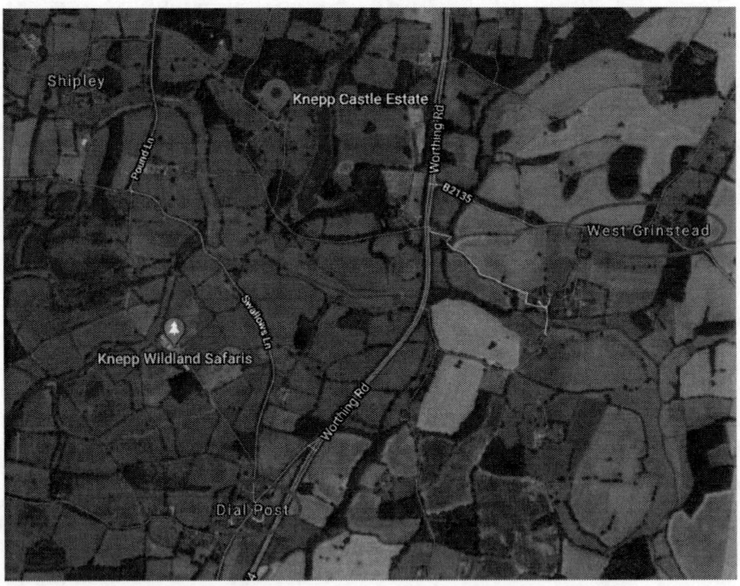

Figure A-1. The bordering villages of Shipley, Dial Post, and West Grinstead are circled in red.

Qualitative Questionnaire Questions

Consent: I understand that the information and opinions provided will be used anonymously for the sole purpose of this research. Quotes may be used but will not be taken out of context.

- Agree

How close do you live to Knepp estate?

- Under 5 miles
- Over 5 miles

How long have you lived near Knepp estate?

- Over 20 years
- 10-20 years
- Under 10 years

Are you a… ?

- Local resident
- Local farmer
- Part of a local business

How often do you visit Knepp estate? e.g., use the footpaths

- (almost) daily
- Weekly
- Monthly
- A few times a year
- Rarely
- Never

How does the landscape make you feel when you visit? Has this feeling changed with time?

Describe your initial reaction to Knepp estate's wilding project.

What do you think of the current landscape at Knepp estate?

From talking to other locals, do their views tend to oppose or agree with your own?

What have been your main concerns with the Knepp project over time?

What do you think are the benefits of Knepp estate's project for the local community?

Has the Knepp project sparked an interest in environmental issues or been a source of information on wildlife for you? If so how?

Has Knepp estate's changing scenery influenced your idea of a 'natural' rural landscape?

Is 'conservation' separate or part of 'farming' from your perspective? Please explain why if possible.

Any additional thoughts on the Knepp wilding project?

If you would like a copy of the final thesis then do submit your email here. (GDPR disclaimer - it will only be used to send you my thesis.)

Poster Advertising Qualitative Questionnaire

Socio-Demographic Characteristics of Respondents

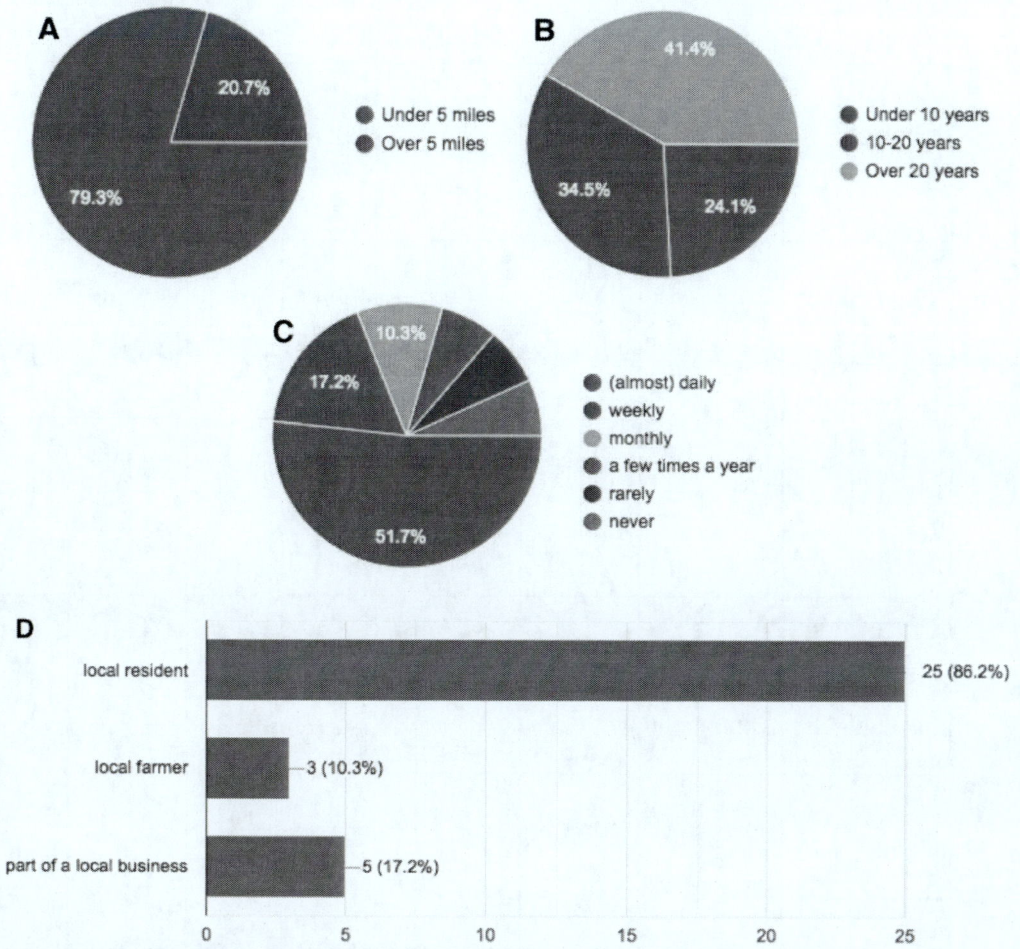

Figure A-2. A: Approximate distance to Knepp estate; B: length of residence in the area; C: regularity of visiting Knepp; D: identification within the community.

Table of Results

Response	Beauty	Functionality	Attachment	Biodiversity	Risk	Feelings-in-Place	Temporal
1	More natural now	more areas in the country should be rewilded	same views as ours	abundance of wildlife; more aware of the environment	more visitors to the estate... change the feel of the place	peaceful	
2	Excellent	[conservation] Should be part of farming	Agree	left natural; encourages more wildlife	No concerns	De stressing	everyday is different
3	Simply beautiful; how I think a natural landscape should look like	must be good for local businesses; country needs feeding; farmers could do more	Many have visited Knepp and share my positive view of the rewilding project	Excited; It's so pleasing to see [animals]; Leave nature to do its thing	None really	Invigorated, relaxed, optimistic and happy; living adjecent to Knepp, it's a dream come true!	feeling not changed, still as excited as when I moved here.
4	A beautiful place	We are becoming less prescriptive; created small wild areas. Mini Knepps we call them!	A mixture, but vehement either way	Rich in wildlife	Community acceptance; long term public financial support. Viability at smaller scale.	Excited and peaceful; Excitement	
5	Rugged natural; scrubby	Totally different as its main aim is not manufacturing a product	most in agreement	Calmed by wildlife; birdsong really special	Managing the public access	Pleased and relaxed	
6	It looks lovely; Beautiful	[Knepp project is] Worthwhile	Yes	wildlife should be paramount	Care off protected wildlife		
7	So open; lakes are amazing; beautiful area	conservation of these areas it will help long term	Definitely agree; Gives everyone something to be proud of	chance to see loads of different wildlife	too busy; not all of them can be respectful of others walking	Grateful	This feeling has always been the same
8	Beautiful... Ever changing	brings visitors to the local pub's and businesses; they go hand in hand	Agree	the wildlife is magical; I think it's how it should be	No concerns	it's such a tranquil place to be..	The feeling has got better with time; [landscape is] Ever changing
9	Rundown; broken fencing	This scheme is not appropriate for this area	difficult to access; everyone I know hates it	Previously always had lovely wildlife not increased by rewilding; not natural	[animals] dangerous and is simply annoying; scary experience with the ponies	Negative	
10	Landscape has a natural rural feel	[Knepp is] as rewilding should be	Agree	privilege to watch; amazing and rewarding	Land.. might clash with the public roaming over it	Excitement!! Inspiring.; amazing and rewarding.	

Response	Beauty	Functionality	Attachment	Biodiversity	Risk	Feelings-in-Place	Temporal
11	Wonderful	Part of farming.	Agree	Interested in the storms [storks]	wondering what else could be added	Wonderful; wish I was part of it	
12	Dereliction of the landscape; Flawed	re-wilding... be more appropriate for wilderness areas	Mixed, but many opposed; cliquey and elitist	appreciate the re-appearance of endangered species	unnatural intrusion; Loss of countryside views	Unhappiness combined with curiosity	
13	[at first] it looked such a mess; [now] beautiful.	Bringing more business to the area	Agree	very much enjoying all the new animals	plan to open a cafe worries me; business is taken away	fantastic	now some 20 years on (!) it ha[s] developed into a fantastic project.
14	A little bit too scrubby.	I encourage it as much as possible on my small farm	A bit of both	I love the wildlife it has encouraged except... foxes	urban foxes will infiltrate and upset the balance.		
15	Predominately lovely	farming is frowned upon by conservationists	Very much agree.	I have always lived on farmland.	piling brash up... just right for wildfires		
16	Landscape now feels more natural left to flourish; looks... healthier	conservation and farming could go hand in hand	beginning to be 'that famous place'	definitely more wildlife; learned to appreciate nature more	the traffic, as the project is getting more known more publicity	very peaceful and calming; feel a million miles away from 'the real world'	
17	Lovely	education opportunities; play off between food production	Mostly agree. There are a few "haters'	we feel excited when we see the storks, or big birds	implication that everyone can do this (rewilding) when not everyone can	Happy, free from worries, healthy, very lucky to live here.	
18	Unsightly; Scruffy, somewhat out of control in some areas and not very attractive.	helps local pubs and businesses; areas of the land management that could be improved	Generally agree; There is a certain pride that the Rewilding project is on our doorstep	develops a passion for understanding the wider diversity of [animals]; stork project is magical.	no longer use the bridleways due to... pigs; ragwort; our own privacy... feel uncomfortable		Initially we were concerned; [current:] Very excited about it
19	A mess; overgrown wilderness;	Good farmers can combine producing food and caring for the wildlife.	Agree; the community that existed has diluted	much scrub has grown up and the cattle walk all over the wildflowers	now just an overgrown wilderness.	Sad	
20	Beautiful. The wilder the better!	hopefully somewhere that won't get built on	anyone can visit; Those who know it seem to love it.	made me more aware of what we have lost.	There are a lot who abuse it	It is my happy place; I marvel at how lucky we are	[it's] how the whole country would have been a few hundred years ago.

Table (Continued)

Response	Beauty	Functionality	Attachment	Biodiversity	Risk	Feelings-in-Place	Temporal
21	Such beauty; Beautiful	rewilding is so much better than allowing another housing development	never had a conversation with anyone around here who has had a bad thing to say	drawn my attention to stalks and butterflies	had an encounter recently with a rude man on one of my walks	It's my favourite thing about living down here; Provides wonderful escapism; feels like a sanctuary	it never gets old! I go at all times of day..., mostly in the early mornings when it is quiet so I can see deer
22	It's so natural and peaceful and beautiful	It would be good if it could be part of farming	I'm now a volunteer	I love walking the estate, savouring all the wildlife	That it will get too popular and become too busy.	a magical place that soothes my soul; It's my happy place	
23	Wonderful	[yes] if farmers re-learn how to work with nature rather than against it.	there are certainly some who will never be happy; presumably people love visiting more	Sad about the loss of bluebells, but overall the gains outweigh the negatives	loss of bluebells	Uplifted; LOVE it	love it more now than before; Southern Block was initially slightly scary, but we have grown to love it
24	Pretty at times, and bare and desolate at others; Not pleasing to the eye	It is a small part of farming	Varies greatly, the entire project can be quite divisive.	increase in wildlife etc. I love hearing the Nightingales of a night.	volume of visitors and walkers it attracts, has turned into a theme park		It varies on the season
25	Beautiful place	provides jobs for local people	Agree	beautiful wildlife which is lovely to see and hear	None	So very very lucky; Amazed and excited	
26	Landscape is not as pristine	loved for nature but should they be growing food too?	Controversial locally	love the wildlife; pure enjoyment of nature.	Alienation of locals; Lack of local employment; aggressive pigs	Very lucky; Excited	
27	Absolutely amazing	don't need to rewild your land to conserve.	Agree	greater understanding of natural processes	Haven't had any [concerns]		
28	[initially] untidy and wasteful. [now] Attractive	not re-wilded... still farmed and managed, but for tourism and nature rather than food	Mostly locals are less favourable to it than I am	It's not natural. There is no natural. We have been meddling for many millennia.	[it's] hurt the local economy in some ways, but benefitted it in others.		I used to love, and visit Knepp regularly... I like it again in appearance now
29	Stunning in a rough and ready kind of way.	One and the same, farming should also be conservation	Agree; a central point of interest	The abundance of life is spilling out to other parts of the community	concerned it would be just a phase.; Uncertainty over the style of farming	Relaxed, great place to clear my head and just think.	[initially] Excited at the prospect of something different

In: Critical Research Techniques in Animal ...
Editors: Michael O'Neal Campbell et al.
ISBN: 978-1-53619-846-1

Chapter 11

ENVIRONMENT EFFECT ON MILKY STORK HABITAT SELECTION IN MATANG MANGROVE FOREST RESERVE, PERAK

Azimah Abd Rahman[1,*] and Aimi Shahirah Mohd Nazri[2]
[1]School of Humanities, University Sains Malaysia, Malaysia
[2]School of Biological Sciences, University Sains Malaysia, Malaysia

ABSTRACT

The global milky stork (*Mycteria cinereal*, Raffles, 1822) population has undergone a rapid decline with a current estimated population of 5,500 individuals remaining worldwide. In Malaysia, milky stork populations have declined more than 90% over the last 20 years. Less than five individuals have been observed recently in Pulau Kelumpang and Pulau Trong, the last known Milky Stork habitats in the Matang Mangrove Forest Reserve. In this study, GIS applications are used to analyse the habitat selection and presence frequency of milky storks in 16 stations in Matang Mangrove Forest Reserve, Perak, Malaysia. The parameters were land use, vegetation and food availability. The findings indicated that the highest density of milky storks was at Pulau Gula and the lowest density was at Pulau Kelumpang, where only one specimen was spotted. Other findings were that these birds are most likely to frequent disturbed areas, especially those with aquaculture activity. Aquaculture activities provided the storks with easier food sources, than available in wild habitats. Captive bred storks from the breeding birds program implemented by the Department of Wildlife and National Parks (DWNP), may have lost their natural foraging ability. Evidence included captive bred stork presence in human presence, with clear expectation of feeding, as reported by the staff of DWNP at the Pulau Gula branch. Environmental factors such as vegetation, food and climate did not have any effects on density of milky storks, differing from published research findings that indicated that released captive birds are able to adapt and

* Corresponding Author's E-mail: azimahrahman@usm.my.

obtain food in natural habitat. These findings make an important contribution to stork conservation and general conservation planning.

Keywords: milky stork, habitat selection, land use, vegetation, food, GIS

INTRODUCTION

The milky stork is a terrestrial, freshwater and marine species, its detailed habitats being listed as "forest, wetlands (inland), marine coastal/supratidal, artificial/terrestrial, artificial/aquatic & marine" (BirdLife International, 2016). This species is a large, magnificent waterbird that attracts tourists, who contribute financially to the Malaysian economy. This attraction of exotic birds for tourists, termed avitourism can be an important factor for foreign exchange earnings for many countries, including Malaysia (Steven et al., 2015; Morrison & Castley, 2015). Avitourism has been defined as "a nature-based tourism subcategory and has been defined as a journey beyond one's usual overnight situation, to see birds in their natural habitat" using observation equipment like binoculars (Nicolaides, 2014; Omar et al., 2019). This provides the economic argument and justification for the conservation of this species. There is also the ecological argument, that the milky stork plays an important ecological role in the coastal ecosystems. The milky stork frequents mangroves along Peninsula Malaysia, Cambodia, island of Sumatra, Bali, Jawa, Sumbawa, Sulawesi, and Baton in Indonesia (Collar, et al., 2001). This species feeds on marine species such as fish, crab, shrimps and mudskippers, hence controlling the populations of these species and maintaining the ecological balance (Delany & Scott, 2006; Ismail & Rahman, 2016). The milky stork is also an important indicator species, that may indicate by its presence the health of local ecosystems, including the ecosystem's stability and balance in several aspects including the stability of mangrove trees, as well as the tropic levels and the ecology of the mangrove forest (Ismail & Rahman, 2012; Ismail & Rahman, 2016).

The milky stork (*Mycteria cinereal*, Raffles, 1822) is listed as an endangered species because of its declining population. It is listed as extant (resident) in Cambodia, Indonesia and Malaysia, extant and introduced (with the seasonality uncertain) in Singapore, and extinct or vagrant in Thailand and Viet Nam (Shepherd & Giyanto, 2009; van Zalinge et al., 2011; BirdLife International, 2016). This decline in the stork population is partly due to the rapid decimation of their coastal habitat. Mangrove habitat losses are partially due to human action and to catastrophic events such as the 2004 tsunami (Sirikulchayanon et al., 2008). BirdLife International (2016) lists the factors for habitat losses as including human intrusions and disturbance (including work and related activities); residential and commercial development, especially housing and urbanization, and the expansion of commercial and industrial areas; terrestrial agriculture (mostly annual and perennial non-timber crops) and marine and freshwater aquaculture, the hunting and trapping of terrestrial animals and logging and wood harvesting from mangrove and other forests. There are also diseases and problematic local species (BirdLife International, 2016).

More direct human actions may contribute to the depletion of the milky stork populations, largely due to the increasing human populations. Stork hunting increases along with human population hunting, as people take eggs, chicks and adult birds for food and trade them in the market. For example, in Indonesia, mangrove forests were destroyed for the development in fish farms and tidal rice cultivation (Brickle, et al., 2015). This contributed to the destruction of the milky stork habitats. In Malaysia, mangrove stands were destroyed for development and agriculture, including for urban development, tourism and aquaculture ((Ismail & Rahman, 2016). These losses are important, as the milky stork uses mangrove trees for nesting (Ismail & Rahman, 2012). As a result, the population of milky storks has declined steadily since the 1980s (Li et al., 2006). Thus, the identification of habitat selection for the milky storks' routine activity is important.

In Malaysia, conservation projects have been attempted, using captive stork breeding and reintroductions, aimed at stork population recovery (Yaacob, 1994; Malaysian Nature Society, 2005). In Malaysia, one of the most important milky stork populations is located in Matang Mangrove Forest Reserve near Kuala Gula, Perak (Ismail & Rahman, 2011). This mangrove forest has provided an opportunity for the Milky Stork Conservation Programme. Pulau Kelumpang in Matang, Perak is an important site associated with milky storks in Malaysia (Ismail & Rahman, 2011). However, monitoring of the populations is problematic, as in some cases the research methods are insufficient for accurate assessments of the stork population variables. The conventional method provided by the Department of Wildlife and National Park (DWNP) is to key in the data manually and sometimes they did not keep it updated. This method is not systematic and errors in the calculation of the present population of milky storks can occur. This may be one contributory factor for the decline in the milky stork populations.

In the current research, Geographic Information Systems (GIS) are used to upgrade the observation methods used by the Department of Wildlife and National Park (DWNP). Using GIS, the observed parameters were those that influenced and enabled the milky stork in its habitat, breeding and feeding choices. GIS was used to add sophistication to these survey methods, by adding the recording of a base map with a geospatial referencing system such as longitude and latitude and the additional of layers of other information (Gislounge, 2012, Justine, 2012) regarding the population, distribution and location of milky storks. Additional information from maps and analysis, valuable insights could be derived from this study. Such GIS oriented studies have been used for the assessment of other stork species in other countries (Gordo et al., 2007; Yamada et al., 2019) and in Malaysia on storks and other species (Abidin et al., 2017; Mohd-Taib et al., 2020) but not in the context of the current study.

The precision, accuracy, analyses and predictive possibilities of GIS technology applications enable more authoritative and usable information from the collected data. Informative data can contribute to the conservation and preservation of the milky stork because stakeholders may learn what type of preference that these birds need for choosing their habitat. For example, with the aid of maps derived from Geographical Information System (GIS) analysis, species hotspots may be derived. Protective strategies, and disturbance management can also be constructed. The latter is particularly important as human action and consequent disturbances have had significant impacts on the habitat of the milky stork. The findings of such research may assist later studies by identifying the exact locations of the milky stork populations in Matang Mangrove Forest Reserve in Perak.

GIS techniques specifically help the understanding of animal movements, habits and distributions (Selkirk et al., 2002). GIS also gives an analytical tool that provides the functionality to perform spatial analysis (Chang et al., 1992). GIS is made up of two types of data, spatial data and attribute data. Spatial data refers to all types of data objects that represent geographical space; this allows the detection of any point on the Earth's surface. Attribute data is the additional information about a spatial feature. In the past GIS was focussed mostly on geographical issues, but recently it has expanded into other areas, such as law enforcement, journalism, environmental relations, agriculture, national development, geology, telecommunications, network management, architecture and engineering (Longley, et al, 2007).

METHODOLOGY

Study Area

Matang Mangrove Forest Reserve in Perak, coordinate 100° 30.00' East 4° 55.00' North with 101,877-hectare was gazetted as a Permanent Forest Reserve in 1906 (Khairul, 2012). Matang Mangrove Forest Reserve is an important site for coastal migratory water birds and some migrant forest birds. It also acts as a natural barrier from tsunamis. The lakes at Pulau Kelumpang and Pulau Terong have been recognized as the most important feeding, roosting and breeding areas for some bird species. Pulau Kelumpang has brackish water that is weakly alkaline (pH 7.65 – 8.5), with high conductivity. Pulau Terong has less brackish water and a pH near neutral (pH 6.73 – 7.6). Both lakes have high nutrient levels (Li, et al., 2006). These two lakes are ecologically linked. During drought on Pulau Kelumpang, dependent bird species move to Pulau Terong to find food. (Li, et al., 2006). The area of Matang Mangrove, Perak is shown as Figure 1 below.

The research was conducted at 16 different stations in Matang Mangrove Forest, Perak. This research was started on 26 November 2015 until 14 January 2016. The 16 different stations that were chosen for this research are listed in Table 1. The research area is a disturbed mangrove forest with local activity such as aquaculture activity and housing development. Most of the stations where the storks were observed have a surrounding aquaculture activity, including Seabass cages and shrimp ponds. At the research site, we found some vegetation along the Matang Mangrove Forest, including *Rhizophora mucronata*, *Sonneratia alba* and *Avicennia* spp. Other bird species were also recorded, such as the Lesser adjutant (*Leptoptilos javanicus*, Horsfield, 1821) and Brahminy kite (*Haliastur indus*, Boddaert, 1783) and some migratory bird species including the Common sandpiper (*Actitis hypoleucos*, Linnaeus, 1758), Common redshank (*Tringa tetanus*, Linnaeus, 1758), Whimbrel (*Numenius phaeopus*, Linnaeus, 1758) and Common greenshank (*Tringa nebularia*, Gunnerus, 1767). They are all foraging their food. Even though the mangrove forest is disturbed with human activity, there are some efforts from agencies to secure the mangrove forest area to their natural cycle. They try to replant the mangrove plant such as Bakau Minyak, Bakau Hitam and others. They also clear the land that they used for aquaculture such as shellfish farming to replant the mangrove vegetation.

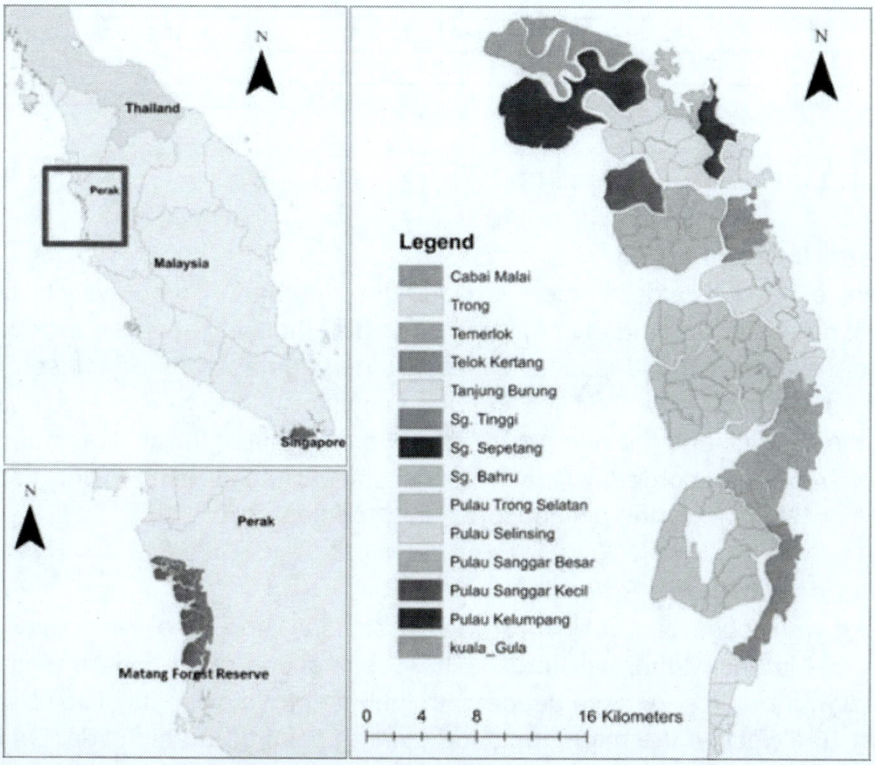

Figure 1. Matang Mangrove forest Reserve.

Table 1a. The twelve stations that we choose with their coordinates and the land use

Longitude	Latitude	Stations	Descriptions
282790.3	546207.32	Cabai Malai 1	Dry shrimp pond
284594.9	545605.79	Cabai Malai 2	Dry shrimp pond
285381.5	545328.16	Cabai Malai 3	Dry shrimp pond
286260.7	545790.87	Cabai Malai 4	Dry shrimp pond
281494.7	545929.69	Pulau Gula 1	Watery shrimp pond
281078.3	544911.72	Pulau Gula 2	Watery shrimp pond
281818.6	544865.45	Pulau Gula 3	Watery shrimp pond
282558.9	544171.37	Pulau Gula 4	Watery shrimp pond

Table 1b. The twelve stations that we choose with their coordinates and the land use

Longitude	Latitude	Stations	Descriptions
281170.8	542089.16	Pulau Kelumpang 1	Seabass cage
279875.2	543384.76	Pulau Kelumpang 2	Seabass cage
277700.5	543060.86	Pulau Kelumpang 3	Seabass cage
281726.7	540099.49	Pulau Kelumpang 4	Seabass cage
282697.8	540423.39	Pulau Kelumpang 5	Seabass cage
282882.8	541626.45	Pulau Kelumpang 6	Seabass cage
283391.8	544356.46	Pulau Kelumpang 7	Seabass cage
282929.1	545559.52	Pulau Kelumpang 8	Seabass cage

Sampling and Data Collection

Parameter Determination

In this research, the parameters are land use, vegetation, food availability and the density of milky stork. For land use, we observed all the local people's activities in that area and the type of these local activities. All the vegetation in the chosen research areas were observed and identified. For food availability, the type of food available at each of the stations and the number of types were identified. Milky stork numbers and density were also recorded, at each of the stations, to determine the impact and analyse how this affects the distribution of this species.

Observation along the River

Storks were observed at each of the 16 stations along the river mouth and the mudflats of Matang Mangrove Forest. The observations were done during the high tide, to allow the boats to move deeper into the mangrove swamps. The observations began at 10 o'clock in the morning, at the starting point of the high tide. On reaching the stations, the coordinates of the locations were taken, and milky storks were counted and observed for their behaviour. The observed stations were active with aquaculture activities such as shrimp pond and seabass cages (four of the former and eight of the latter). The vegetation, other animals and local people activities along the river were also observed. The data that we observed and collected are based on our parameters, which are land use, food availability, and vegetation. All the data were recorded in the database.

Observation along the Road Trip

Road trips were undertaken along the stations located side by side with the mangrove forest. Observations were undertaken along the road at 1 pm, after the observations along the river. The stations were taken along the 4 dry shrimp ponds, which were located near the residential areas, where people were engaged in daily activities. At the survey locations, coordinates and bird counts were again taken.

Observations included the vegetation, local human activities and land use. All the data were recorded in the survey database.

Primary and Secondary Data

Primary data comprised recordings from the sampling research. Secondary data comprised information from the Department of Wildlife and National Park (DWNP). Data was limited to one year, as the project was to last only one year.

Table 2. Primary Data

Data	Source	Year
Population of Milky Stork	Observation	2015-2016
Land Use	Observation	2015-2016
Vegetation	Observation and local people	2015-2016
Food Availability	Observation	2015-2016

Table 3. Secondary Data

Data	Source	Year
Populations of Milky Stork	Department of Wildlife and National Park	2000-2009
Map	Forestry Department of Peninsular Malaysia	2008

Data Analysis

SPSS version 22 was used to analyse the data. A map of the coordinates was created, to map the locations of the population distribution hotspot of the milky stork. The GIS analysis was used to interpret the data in mapping and statistical analysis.

RESULTS AND DISCUSSION

Density of Milky Stork Sightings

Table 4 below shows the density of milky storks at each of 16 different stations with coordinates and the land activity at each of the stations. From Table 4 and Figure 2, a total of 50 individuals of Milky Stork were observed in 16 selected stations in Matang Mangrove. From Figure 3 below; Pulau Gula 1 shows the highest number of individuals founds (7), with Pulau Gula 3 next (6) and both Pulau Kelumpang 3 and Cabai Malai 1 last (5 each). At Pulau Gula 2, 4 milky storks were found. Cabai Malai 4, Pulau Kelumpang 5 and Pulau Kelumpang 7, recorded 3 storks each, while at Cabai Malai 2, Pulau Gula 4 and Pulau Kelumpang 4 recorded 2 milky storks each. Pulau Kelumpang 1, Pulau Kelumpang 2, Pulau Kelumpang 6 and Pulau Kelumpang 8 recorded one stork each.

Table 4. Density of Milky Stork in 16 different stations in Matang Mangrove Forest area

Longitude	Latitude	Stations	Land Activity	Density of Milky Stork
282790.3	546207.32	Cabai Malai 1	Dry shrimp pond	5
284594.9	545605.79	Cabai Malai 2	Dry shrimp pond	2
285381.5	545328.16	Cabai Malai 3	Dry shrimp pond	4
286260.7	545790.87	Cabai Malai 4	Dry shrimp pond	3
281494.7	545929.69	Pulau Gula 1	Watery shrimp pond	7
281078.3	544911.72	Pulau Gula 2	Watery shrimp pond	4
281818.6	544865.45	Pulau Gula 3	Watery shrimp pond	6
282558.9	544171.37	Pulau Gula 4	Watery shrimp pond	2
281170.8	542089.16	Pulau Kelumpang 1	Seabass cage	1
279875.2	543384.76	Pulau Kelumpang 2	Seabass cage	1
277700.5	543060.86	Pulau Kelumpang 3	Seabass cage	5
281726.7	540099.49	Pulau Kelumpang 4	Seabass cage	2
282697.8	540423.39	Pulau Kelumpang 5	Seabass cage	3
282882.8	541626.45	Pulau Kelumpang 6	Seabass cage	1
283391.8	544356.46	Pulau Kelumpang 7	Seabass cage	3
282929.1	545559.52	Pulau Kelumpang 8	Seabass cage	1

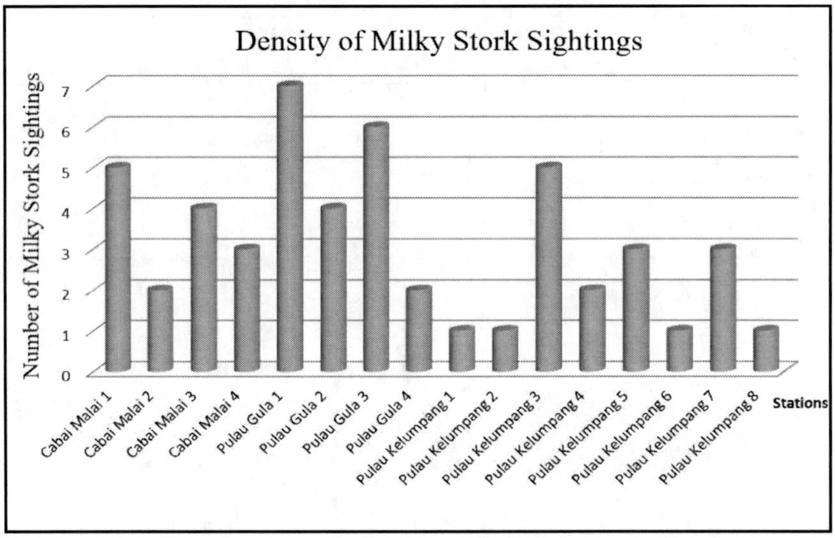

Figure 2. Number of milky storks based on sampling stations.

Land Used

Table 5 displays the activities at each of the selected stations. All were active with aquaculture and were occupied by human activities prior to the development of aquaculture. The total number of stocks at dry shrimp pond was 14, at Watery Shrimp pond 19, and at Sea bass Cage 17. To assess the relationship between the land use and the density of milky storks at each of the stations, a k independent sample test, under nonparametric statistical analysis test using (SPSS Ver.22) was run to investigate if there are significance different in the relationship between the type of land use and the distribution of Milky Stork (Table 6). The land use was divided into three groups, dry shrimp pond, watery shrimp pond and sea bass cage. Each of the groups was denoted as 1, 2, and 3 respectively (Table 7). As shown in Table 8, the significance value is 0.075. We fixed the significance value at $p < 0.05$. With no significant difference between the stations, land use had no impact on the distribution of milky storks.

Table 5. The Density of Milky Storks and Land Use at Each Station

Stations	Land Use	Density
Cabai Malai 1	Dry shrimp pond	5
Cabai Malai 2	Dry shrimp pond	2
Cabai Malai 3	Dry shrimp pond	4
Cabai Malai 4	Dry shrimp pond	3
Pulau Gula 1	Watery shrimp pond	7
Pulau Gula 2	Watery shrimp pond	4
Pulau Gula 3	Watery shrimp pond	6
Pulau Gula 4	Watery shrimp pond	2
Pulau Kelumpang 1	Sea bass cage	1
Pulau Kelumpang 2	Sea bass cage	1
Pulau Kelumpang 3	Sea bass cage	5
Pulau Kelumpang 4	Sea bass cage	2
Pulau Kelumpang 5	Sea bass cage	3
Pulau Kelumpang 6	Sea bass cage	1
Pulau Kelumpang 7	Sea bass cage	3
Pulau Kelumpang 8	Sea bass cage	1

Table 6. Descriptive Statistics

	N	Mean	Std. Deviation	Minimum	Maximum
Density	16	3.13	1.893	1	7
Land Use	16	2.25	0.856	1	3

Table 7. Kruskal-Wallis Test

Land Use	N	Mean Rank
Dry Shrimp Pond	4	10.00
Watery Shrimp Pond	4	12.13
Sea bass Cage	8	5.94

Table 8. Test Statistics

Chi-Square	Df	Asymp. Sig.
5.186	2	0.075

Vegetation

Table 9 below shows the type of vegetation at each of the stations and the density of Milky Stork respectively. The surrounding area within the station vicinities were covered by mangrove vegetation such as *Rhizophora mucronata*, *Avicennia* spp. and *Sonneratia alba*. From Cabai Malai 1 to Pulau Gula 4, the stations are covered by Rhizophora mucronata and Avicennia spp while at Pulau Kelumpang 1 to Pulau Kelumpang 8; the stations are covered with *Avicennia* spp. and *Sonneratia alba*. At the stations that covered with *Rhizophora mucronata* and *Avicennia* spp., the total number of milky storks found was 33 individuals while the total number of milky storks that can be found at the stations that covered with *Avicennia* spp. and *Sonneratia alba* is 17 individuals.

Table 9. Type of vegetation at each of 16 different stations

Stations	Type of Vegetations	Density
Cabai Malai 1	*Rhizophora mucronata,* and *Avicennia* spp.	5
Cabai Malai 2	*Rhizophora mucronata,* and *Avicennia* spp.	2
Cabai Malai 3	*Rhizophora mucronata,* and *Avicennia* spp.	4
Cabai Malai 4	*Rhizophora mucronata,* and *Avicennia* spp.	3
Pulau Gula 1	*Rhizophora mucronata,* and *Avicennia* spp.	7
Pulau Gula 2	*Rhizophora mucronata,* and *Avicennia* spp.	4
Pulau Gula 3	*Rhizophora mucronata,* and *Avicennia* spp.	6
Pulau Gula 4	*Rhizophora mucronata,* and *Avicennia spp*	2
Pulau Kelumpang 1	*Avicennia* spp. and *Sonneratia alba*	1
Pulau Kelumpang 2	*Avicennia* spp. and *Sonneratia alba*	1
Pulau Kelumpang 3	*Avicennia* spp. and *Sonneratia alba*	5
Pulau Kelumpang 4	*Avicennia* spp. and *Sonneratia alba*	2
Pulau Kelumpang 5	*Avicennia* spp. and *Sonneratia alba*	3
Pulau Kelumpang 6	*Avicennia* spp. and *Sonneratia alba*	1
Pulau Kelumpang 7	*Avicennia* spp. and *Sonneratia alba*	3
Pulau Kelumpang 8	*Avicennia* spp. and *Sonneratia alba*	1

Table 10. Test of Homogeneity of Variances

Levene statistic	Significant value
0.312	0.586

To assess the relationship between the type of vegetation and the density of Milky Stork at each of the stations, we run the Levene's Test (Independent sample t-test) statistical analysis test using (SPSS Ver.22). We divide the type of vegetation into two groups which is Group 1 and Group 2. Group 1 is denotes for *Rhizophora mucronata*, and *Avicennia* spp and Group 2 is denotes for *Avicennia* spp. and *Sonneratia alba*. We are suggesting two hypotheses:

Ho: The density of Milky Stork at each station depends on the type of vegetation.

Ha: The density of Milky Stork at each station does not depend on the type of vegetation.

Table 10 shows that the significance value is 0.586. As this is higher than 0.05, it shows that there is a significant difference between the distributions of milky stork and the type of vegetation. So, null hypothesis is rejected. The type of vegetation does not affect the density of milky stork.

Food Availability

Table 11 below shows the food availability at each station and their number of food types.

Table 11. The food availability at sixteen stations and their number type of food

Stations	Source of Food	Number Type of Food	Density of Milky Stork
Cabai Malai 1	small fish	1	5
Cabai Malai 2	small fish	1	2
Cabai Malai 3	small fish	1	4
Cabai Malai 4	small fish	1	3
Pulau Gula 1	shrimp and small fish	2	7
Pulau Gula 2	shrimp and small fish	2	4
Pulau Gula 3	shrimp and small fish	2	6
Pulau Gula 4	shrimp and small fish	2	2
Pulau Kelumpang 1	molluscs, small fish, small crab	3	1
Pulau Kelumpang 2	molluscs, small fish, small crab	3	1
Pulau Kelumpang 3	molluscs, small fish, small crab	3	5
Pulau Kelumpang 4	molluscs, small fish, small crab	3	2
Pulau Kelumpang 5	molluscs, small fish, small crab	3	3
Pulau Kelumpang 6	molluscs, small fish, small crab	3	1
Pulau Kelumpang 7	molluscs, small fish, small crab	3	3
Pulau Kelumpang 8	molluscs, small fish, small crab	3	1

To assess the relationship between the availability of food and the density of Milky Stork at each of the stations, we run the linear regression statistical analysis test using (SPSS Ver.22). The results in Figure 3, show a weak relationship between number of different types of food and the density of milky storks at each of the stations in Kuala Gula. The R^2 value is only 15% and the adjusted R value is -3.87%. These two values are not similar; therefore, the data gathered is not reflecting the actual whole populations of Milky Stork worldwide. We are suggesting two hypotheses:

Ho: Milky storks select their foraging areas based on the number of types of food that are available in the area.

Ha: Milky stork select their foraging areas based on the abundance of available food in the area.

Based on the results presented in Table 13 and Table 14, the significance value is 0.135 which is not significant as the value of significant is $p < 0.05$. This shows that the number of different types of food does not predict the number of Milky Stork. As shown in Table 12, the R^2 value is 0.153. This represents 15%, which shows that the variance in the number of Milky Stork cannot be explained by the number of different types of food. The Beta value for the number of different types of food is – 0.5. This value shows us the slope for our variable is very small and close to zero. The Confidence Interval Value from this test is -2.003 and 0.303, which is a positive interval. Thus, the population studied in Matang Mangrove Forest is not representative of the global milky stork population. With the significant value 0.135, there is enough evidence to reject null hypothesis.

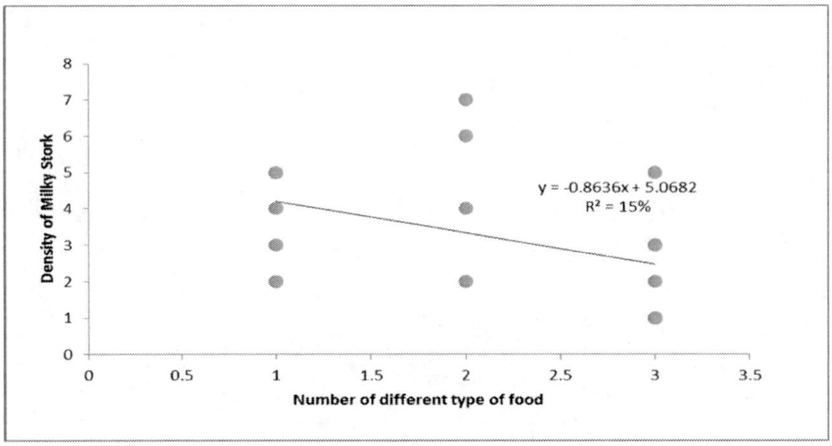

Figure 3. Relationship between food types and stork density in Matang Mangrove Forest.

Table 12. Table of Model Summary

R	R Square	Adjusted R Square	Significant F change
0.391	0.153	0.092	0.135

Table 13. Table of ANOVA

Sum of Squares	Mean Square	Significant value
8.205	8.205	0.135

Table 14. Table of Coefficient

	Unstanderdized Coefficients	Significant Value	95.0% Confidence Interval For B	
	B		Lower Bound	Upper Bound
Constant	5.068	0.002	2.271	7.865
Type of food	-0.864	0.135	-2.003	0.303

To summarize, there is a weak relationship between the number of different types of food and the density of milky storks. The observed storks did not have specific food preferences but responded to the abundance of food available, as shown by their changing of behaviour. The milky storks in our sample were captive bred, hence on reintroduction into the wild, they may have lost their natural foraging ability. They relied more on the comparatively common human-created food. This indicates that more research is required on such scenarios.

Geographical Information System Analysis

The Mapping of Population Distribution of Milky Stork

Figure 4 below shows the map on the distribution of milky stork at 16 stations around Matang Mangrove Forest. Their stopover spot and hot spot are identified at Cabai Malai area, Pulau Kelumpang and Pulau Gula.

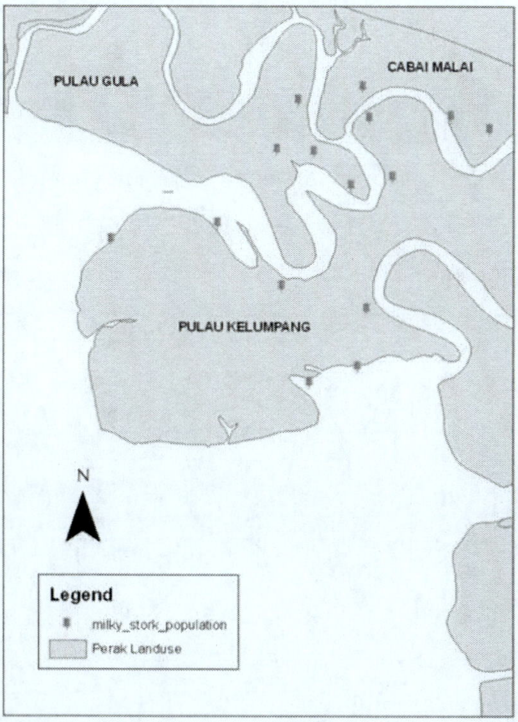

Figure 4. The map of distribution of Milky Stork around Matang Mangrove Forest.

The Mapping of Hotspot Area

Figure 5 shows all the hotspot areas for milky storks around Matang Mangrove Forest. The hotspot area with the highest density, with 7 individuals was at Pulau Gula area. This is shown in Figure 6. A few areas are identified that have the lowest density of the distributions of milky storks, with only one bird counted at each of the stations. These at Pulau Kelumpang (Figure 7).

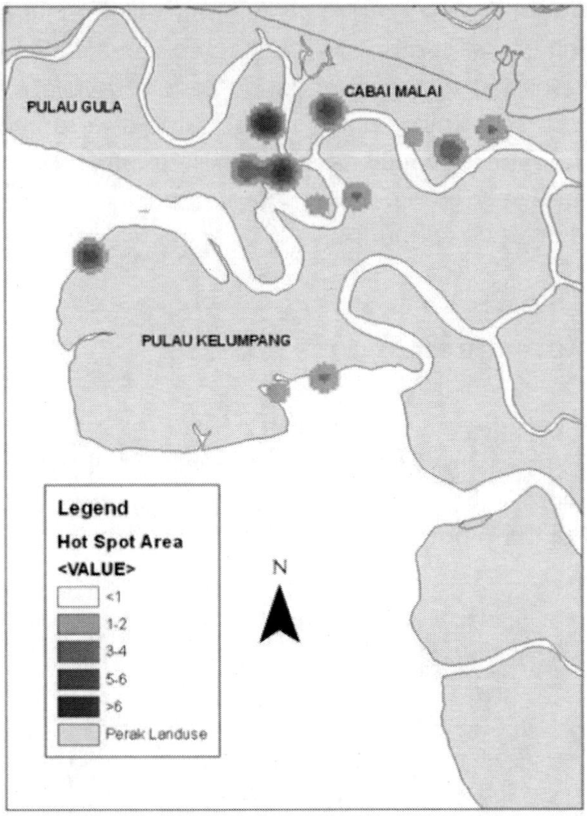

Figure 5. Hotspot Areas for Milky Storks.

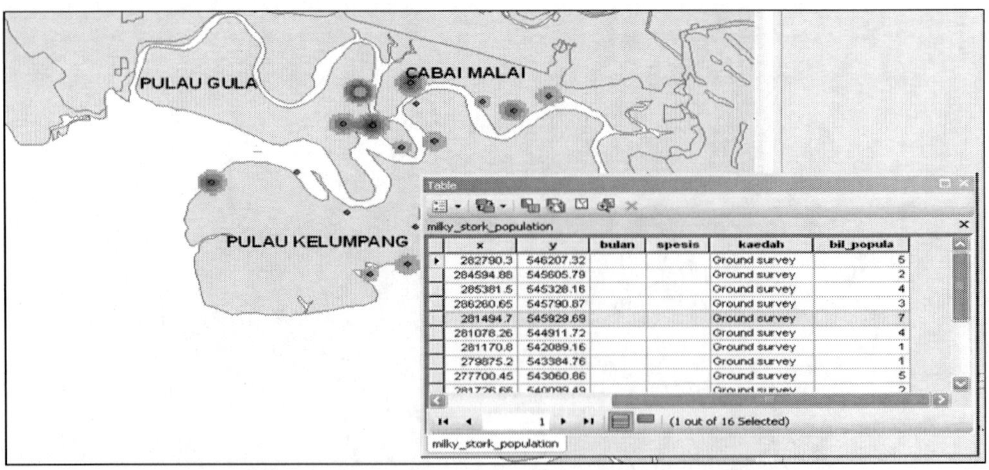

Figure 6. Hot spot area for the highest density of the distribution of Milky Stork.

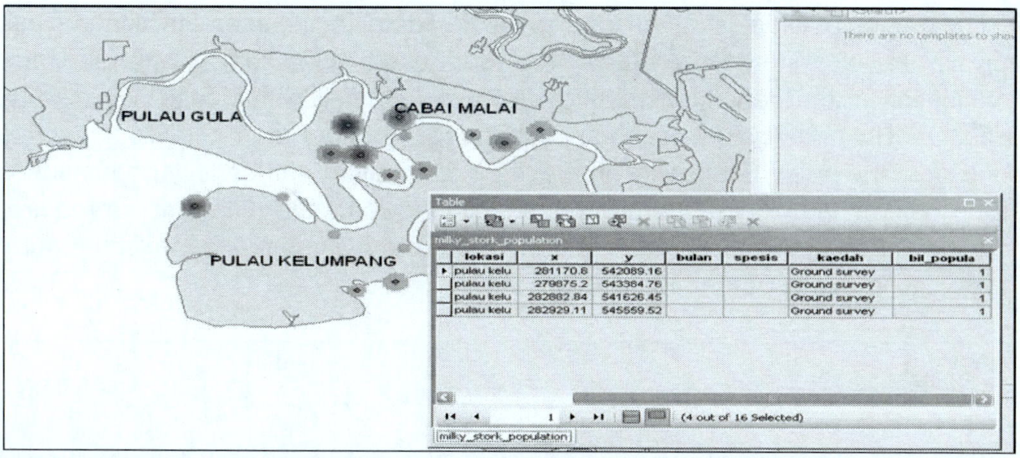

Figure 7. Hot spot area with lowest density of distribution of Milky Stork.

The Mapping of Population Density

Figure 8 below, shows the mapping of the population density of milky storks around Matang Mangrove Forest. Each level of interval of population density, is denoted with different colours to differentiate them. By using GIS applications, we can produce an attribute table from the data.

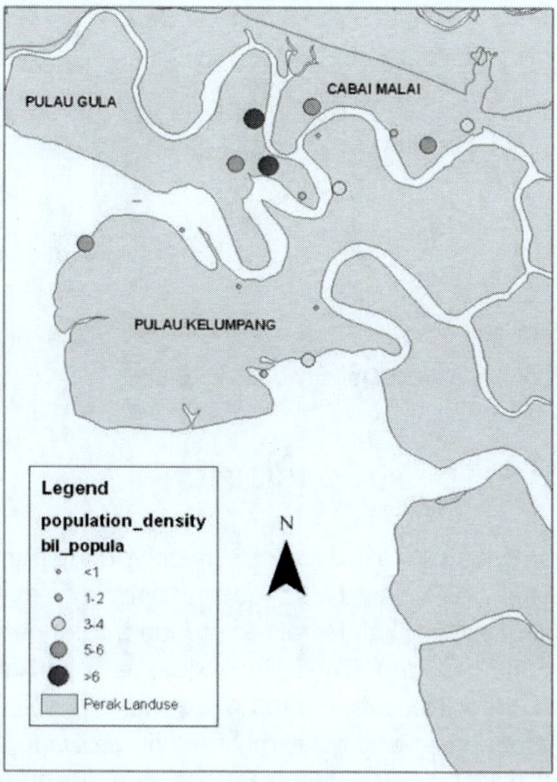

Figure 8. The mapping of population density of Milky Stork.

The attribute table is used to explain the data in the map, facilitating easier analysis. In this analysis, ArcGIS 10.3 was used, with spatial analysis for the identification of the density distributions of milky storks. In this case, we also used Kernel Density analysis under the spatial analysis tool in the ArcGIS 10.3. From this application, the hotspot for milky storks, their distribution and density population also can be determined. As in Figure 9 below, this shows us the data that we require, to show the standard deviation value, mean value, and minimum and maximum number of milky storks found around Matang Mangrove Forest area.

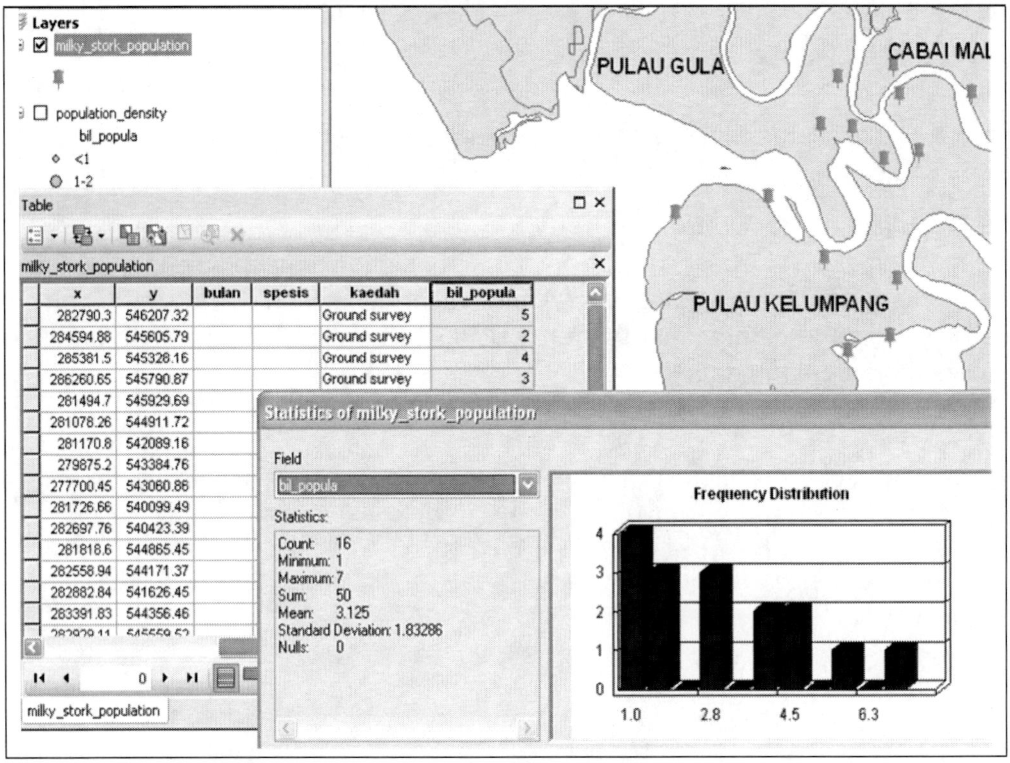

Figure 9. The Mapping of Population Density of Milky Stork.

Conclusion

The milky stork is essentially a coastal species favouring mangroves, mudflats and estuaries (Hancock et al., 1992), and also feeds in rice fields and fishponds (Verheugt 1987). In Matang Mangrove Forest, Perak, we found a total number of 50 specimens of the milky stork around the area. These individuals are part of the captive breeding program, from which the birds are reintroduced into the wild. Sixteen observation stations were chosen for survey and research, the criteria being the stations most likely to contain storks in the Matang Mangrove Forest area. Mangrove forests provide all the ecological and habitat needs for the milky stork, including for breeding, feeding, nesting, and resting.

In the research area, there were three types of land use were identified, these being Dry Shrimp Pond, Watery Shrimp Pond and Sea bass Cage. All of the milky storks were observed in areas where human activities were on-going, rather than in more natural or pristine areas. Consequently, the stork habitat is generally disturbed by aquaculture activities. Although storks are generally highly sensitive towards human presence, in recent times they have built a tolerance towards human presence. For example, a study by Shepherd and Giyanto, 2009, found that the Milky Storks allowed fishermen within relatively close range, but they did not tolerate construction work and associated disturbances that was taking place at the river mouth during visits made in April and May 2007. This may be the results of captive breeding; during captive they become closer and comfortable towards human presence and no longer feeling afraid and protective.

From the population density data analyses of milky storks, we can see that the highest number of milky storks can be found at watery shrimp pond. second highest is at sea bass cage and the lowest is at dry Shrimp Pond. Even though the shrimp pond is dry and there still has the presence of Milky Stork, this is because; the shrimp pond still contains high densities of small fish and eels that can be food sources for milky storks. The Matang Mangrove forest is a reserved forest, but there are high levels of economic development in the area, including aquaculture development and associated activities, such as fishing, boating, shell farming, development of houses and boat restaurant have disturbed the pristine areas and affected the mangrove forests. The Matang Mangrove Forest has been polluted by these activities, the main sources of this pollution being from the residue oils from boats, the logging of mangroves, and the noises and the wastes from human activities. Based on the findings of a study by Rahman et al., (2013), which looked at five surface sediment samples from stork-foraging areas, there is metal contamination in the study area, including Copper (Cu), Zinc (Zn) and Lead (Pb).

In the stations selected for the current research, there three types of mangrove vegetation, namely *Rhizophora mucronata, Avicennia* spp. and *Sonneratia alba*. However, the findings indicated that the storks do not rest or nest in the trees, at least not since the reintroduction started. Consequently, the vegetation type is not a factor for the stork's distribution. This is attributed to the captive breeding, which impedes the ability of storks to breed and nest in a natural state. They have been in captivity from their juvenile life stages, and thus they are unfamiliar with the harsh conditions in the wild. For wild specimens, the mangrove vegetation is important for breeding, nesting and resting. The mangrove species utilised for nesting include *Avicennia marina* and *Rhizopora apiculata*. In Pulau Rambut, Java, most nests are built in *Sterculia foetida* (Mardiastuti, 1992; Imanudin & Mardiastuti 2003). The milky stork generally breeds in mangrove vegetation and the eggs are laid approximately one meter from the ground (Iqbal et al., 2008).

The findings of the current research indicate that the foraging behaviour of the resident storks in Matang Mangrove, Perak is based on selection of easy food sources, rather than on rich or abundance choices. There was a weak statistical relationship between food variety and stork density at the stations. Hence, the variety of food is not the main attraction for milky storks. For example, there were high densities of storks foraging at watery shrimp pond even though only two type of foods were available. Iqbal et al, (2008) found that adult birds preyed on large mudskippers (*Periopthalmus*), also small fish, snakes and frogs (Elliott 1992). They found two species of prawn Indian, white prawn *Penaeus indicus* and banana prawn *P.merguensis* and one clupeid fish species, Dussumier's thryssa *Thryssa dussumieri* in and around nests. Such behaviour results from their captive breeding, where they lost their natural foraging abilities. Ismail et al, (2009) stated that captive birds are fed 2 times a day during the non-breeding season and 3 times a day during the breeding season. The fish are given to the storks in shallow water and in basins by the pools. The basins were used as the feeders mixed the calcium with the fish and other medications were added for deworming and vitamin supplementation. The vitamin would be lost if the fish were placed in the water. These actions encourage the Milky Storks to invade human aquaculture activities, as there are easier possibilities for food.

The GIS analysis produced analytical maps, from which it may be concluded that the hotspot of distribution of milky storks is around the Matang Mangrove Forest. From the map, we also can define the maximum and minimum density of milky storks. The maps are useful especially for future research. This provides a guide for other researchers to find the storks and contribute to help the preservation and conservation of milky storks and their habitat. Based on the map, we also can see the habitat preference of Milky Stork and study more about their habitat. The results of the current research is especially important, as there is not much research done on this species, except in the captive zoo context. With GIS applications, the study of the milky stork can become more international and the data processing will be much more valuable.

BirdLife International (2012) states that the population of milky storks is suspected to be decreasing at a slow to average rate due to hunting, egg collecting, human disturbance at breeding colonies, drainage and agricultural development. In Malaysia Milky Stork populations have decreasing more than 90% over the last 20 years[6], from over 100 individuals in 1984 to 10 in 2005, the last known Milky Stork habitats in the Matang Mangrove Forest, Perak, Malaysia (Ismail et al., 2011). The objective of the current research is to support the urgent need for the conservation of the milky stork in Malaysia to save this species from extinction. However, we are still lucky to have large number of storks in captivity. The Malaysian Zoological Society, Zoo Negara, Wildlife Department Malaysia and Universiti Putera Malaysia are part of the team that trying very hard to conserve and increase the number of milky storks in the wild, particularly in Kuala Gula. This research is also intended to assist the Department of Wildlife and National Parks to keep a track of the number of milky storks and their hotspot for foraging and nesting.

REFERENCES

Abidin, K., Mohd-Taib, F. S., Md-Nor, S. 2017. Distribution and habitat selection of the Asian Openbill (Anastomus oscitans) in Peninsular Malaysia. *Malayan Nature Journal.* 69. 169-181.

BirdLife International 2016. *Mycteria cinerea.* The IUCN Red List of Threatened Species 2016. https://dx.doi.org/10.2305/IUCN.UK.2016-3.RLTS.T22697651 A93627701.en.

Chang, K., Yeo, J. J., Verbyla, D. L. & Li, Z. 1992. Interfacing GIS with wildlife habitat analysis: A case study of Sitka Black-Tailed Deer in southeast Alaska. In *Proceedings of the GIS/LIS'92 Conference*, San Jose, California, 1: 95-104.

Collar, N. J., Andreev, A. V., Chan, S., Crosby, M. J., Subramanya, S., Tobias, J. A., Churchill, S. P., Balslev, H., Forero, E., Luteyn, J. L. & Marsh, H. 2001. *Threatened birds of Asia: The BirdLife International Red Data Book* (No. 598. 2095 T531). Cambridge (RU): Birdlife International.

Delany, S. & Scott, D. 2006. *Waterbird Population Estimates.* Wageningen, NL: Wetlands International.

Gislounge 2012. www.gislounge.com/search-gis-lounge/ (accessed on 30 January 2016).

Gordo, O., Sanz, J. J. & Lobo, J. M. 2007. Spatial patterns of white stork (Ciconia ciconia) migratory phenology in the Iberian Peninsula. *J Ornithol* 148, 293–308. https://doi.org/10.1007/s10336-007-0132-6.

Imanuddin, A. 2003. *Kesuksesan Perkembangbiakan dan Pertumbuhan Anakan Bangau Bluwok (Mycteria cinerea) di Suaka Margasatwa Pulau Rambut Breeding Success and Chick Development of Milky Stork (Mycteria cinerea) in Pulau Rambut Wildlife Sanctuary.*

Iqbal, M. & Hasudungan, F. 2008. Observations of Milky Stork Mycteria cinerea during 2001–2007 in South Sumatra province, Indonesia. *BirdingAsia*, *9*: 97 - 99. Retrieved from https://www.rufford.org/files/BirdingASIA%209%20(2008)%2097 %E2%80%9399.pdf.

Iqbal, M., Ridwan, A., Takari, F. & Mulyono, H. 2008. Rediscovery of a Milky Stork Mycteria cinerea breeding colony in South Sumatra province, Indonesia. *BirdingASIA*, *15*, 62-66. Retrieved from https://www.rufford.org/files/Birding ASIA%2010%20(2008)%2062%E2%80%9366.pdf.

Iqbal, M. 2008. *Survey and Conservation of Milky Stork Mycteria cinerea in Sumatra*, Indonesia. https://www.rufford.org/rsg/projects/muhammad_iqbal.

Iqbal, M. & Ridwan, A. 2008. People's Perspective for Milky Stork: A Case from South Sumatra, Indonesia. *Journal of Wetlands Ecology*, *1*(1), 7-8. https://doi. org/10.3126/jowe.v1i1.1571.

Iqbal, M., Mulyono, H., Riwan, A. & Takari, F. 2008. *An alarming decrease in the Milky Stork Mycteria cinerea population on the east coast of South Sumatra province*, Indonesia. Retrieved from https://www.rufford.org/files/BirdingASIA%2018%20 (2012)%2068%e2%80%9370.pdf.

Ismail, A. & Rahman, F. 2012. An urgent need for Milky Stork study in Malaysia. *Pertanika Journal of Tropical Agricultural Science*, *35*(3): 407-412. http://agris.upm.edu.my:8080/dspace/handle/0/9005.

Ismail, A., Rahman, F., Kin, D. K. S., Ramli, M. N. H. & Ngah, M. 2011. Current Status of the Milky Stork Captive Breeding Program in Zoo Negara and its Importance to the Stork Population in Malaysia. *Tropical Natural History*, 11(1), 75-80. Retrieved from https://li01.tci-thaijo.org/index.php/tnh/article/view/103007.

Ismail, A., Rahman, F., Miyazaki, N. & Naito, Y. 2012. Initial application of bio-logging techniques on captive Milky Stork (*Mycteria cinerea*) in Malaysia. *Tropical Ecology*, *53*(2), 177-181. Retrieved from http://tropecol.com/pdf/open/PDF_ 53_2/Jour-06-F.pdf.

Ismail, A., & Rahman, F. 2016. Current Status of the Milky Stork Re-introduction Programme in Malaysia and Its Challenges. *Tropical life sciences research*, 27(2), 13–24. https://doi.org/10.21315/tlsr2016.27.2.2.

Li, Z. W. D., Yatim, S. H., Howes, J. & Ilias, R. 2006. Status overview and recommendations for the conservation of Milky Stork Mycteria cinerea in Malaysia: Final report of the 2004/2006 Milky Stork field surveys in the Matang Mangrove forest, Perak, Malaysia. *Wetlands International and the Department of Wildlife and National Parks, Peninsular Malaysia, Kuala Lumpur, Malaysia.*

Longley, P., Goodchild, M., Mayuire, D. & Rhind, D., (2001). *Geographic Information Sytems and Science,* New York, NY: John Wiley & Sons.

Malaysian Nature Society 2005. *Report on the Milky Stork captive breeding and re-introduction programme*, Kuala Selangor Nature Park. Kuala Lumpur: Malaysian Nature Society.

Mardiastuti A. 1992. *Habitat and nest site characteristics of waterbirds in Pulau Rambut Nature Reserve*, Jakarta Bay, Indonesia. Unpublished PhD Dissertation. East Lansing, MI: Michigan State University.

Mohd-Taib, F. S., Mohd-Saleh, W., Asyikha, R., Mansor, M. S., Ahmad-Mustapha, M., Mustafa-Bakray, N. A., Mod-Husin, S., Md-Shukor, A., Amat-Darbis, N. D. & Sulaiman, N. 2020. Effects of anthropogenic disturbance on the species assemblages of birds in the back mangrove forests. *Wetlands Ecology and Management*, 28(3): 479–494. https://doi.org/10.1007/s11273-020-09726-z.

Nicolaides, A. 2014. *Stakeholders, purposes and responsibilities: Avitourism in South Africa.* Retrieved from http://hdl.handle.net/10500/20660.

Omar, K., Halim, M., Samsudin, H., Mohamad, Z., Ismail, I. R. & Zulkiffli, S.|N. A. 2019. Self-Engagement in Avitourism: A Case of Taman Negara Bird Count (TNBC 2017) Malaysia. *International Journal of Recent Technology* and Engineering (IJRTE), 8, 1526 – 1530.

Rahman, F., Ismail, A., & Yusof, S. 2013. Metals contamination in the foraging area of Milky Stork: evidence of anthropogenic inputs in the aquatic environment of Kuala Gula, Malaysia. *Toxicological & Environmental Chemistry*, *95*(9): 1499-1505. https://doi.org/10.1080/02772248.2014.892941.

Selkirk, S. W., & Bishop, I. D. 2002. Improving and extending home range and habitat analysis by integration with a geographic information system. *Transactions in GIS*, *6*(2), 151-159. https://doi.org/10.1111/1467-9671.00102.

Shepherd, C. R. & Giyanto. 2009. Observations of Milky Storks Mycteria cinerea in Percut, North Sumatra, Indonesia. *Birding ASIA* 11: 70-72.

Sirikulchayanon, P., Sun, W. & Oyana, T. J. 2008. Assessing the impact of the 2004 tsunami on mangroves using remote sensing and GIS techniques. *International Journal of Remote Sensing*, 29(12): 3553-3576. https://doi.org/10.1080/01431160701646332.

Steven, R., Morrison, C. & Castley, J. G. 2015. Birdwatching and avitourism: a global review of research into its participant markets, distribution and impacts, highlighting future research priorities to inform sustainable avitourism management. *Journal of Sustainable Tourism*, 23(8-9): 1257-1276. https://doi.org/10.1080/09669582.2014.924955.

van Zalinge, R., Visal, S., Phreakdey, S. & Evans, T. 2011. *The status and distribution of large waterbirds in the Tonle Sap Biosphere Reserve*, 2010 update. Wildlife Conservation Society Cambodia Program, Phnom Penh.

Wei, Z. H., Li, Y. K., Xu, P., Qian, F. W., Shan, J. H. & Tu, X. B. 2016. Patterns of change in the population and spatial distribution of oriental white storks (Ciconia boyciana) wintering in Poyang Lake. Zoological Resources, 18; 37(6): 338–346. https://doi.org/10.13918/j.issn.2095-8137.2016.6.338.

Yaacob, M. N. 1994. *Captive-breeding and reintroduction project for the Milky stork Mycteriu cinerea: at Zoo Negara*, Malaysia, 33(1): 39-48. https://doi.org/10.1111/j.1748-1090.1994.tb03553.x.

Yamada, Y., Itagawa, S., Yoshida, T., Fukushima, M., Ishii, J., Nishigaki, M. & Ichinose, T. 2019. *Predicting the distribution of released Oriental White Stork (Ciconia boyciana) in central Japan*, 34(2): 277-285. https://doi.org/10. 1111/1440-1703.1063.

In: Critical Research Techniques in Animal …
Editors: Michael O'Neal Campbell et al.
ISBN: 978-1-53619-846-1

Chapter 12

THE LANDSCAPE ECOLOGY OF URBAN AVIAN SCAVENGERS: AN INTERNATIONAL PERSPECTIVE

Michael O'Neal Campbell*

Simon Fraser University, British Columbia, Canada

ABSTRACT

Urban avian scavengers are increasingly being recognised as important for the ecology of urban environments. For the comparative analysis of these complex dynamics, integrated research methods (IR) may be particularly useful. However, few studies have conducted multi-species, multi-national studies of the role of integrated research methods in urban avian studies. This chapter compares the presence in varied landscapes of international contexts of eight urban scavengers in three case studies, based on published literature: (1) Glasgow UK, Carrion crows *Corvus corone cornix* Linnaeus, 1758, rooks *Corvus frugilegus* Linnaeus, 1758, jackdaws *Corvus monedula* Linnaeus, 1758, magpies *Pica pica* Linnaeus, 1758, Lesser black backed gulls *Larus fuscus graellsii* Linnaeus 1758 and Black-headed gulls *Larus ridibundas* Linnaeus 1776; (2) Kumasi and Accra, Ghana, Pied crow *Corvus albus* Muller 1776, Hooded vulture *Necrosyrtes monarchus* Temminck 1823, and Black kites *Milvus migrans* Boddaer 1783; and (3) Ottawa, Canada, the American crow *Corvus brachyrhynchos* Brehm, 1822, the Turkey vulture *Cathartes aura* Linnaeus 1758, Blue jay *Cyanocitta cristata* Linnaeus 1758, and the Ring-billed gull *Larus delawarensis* G.Ord 1815. It also assesses the changing conceptions of the ecological, biogeographical and socioenvironmental bases for species assessment. The methodology used point counts, landcover measurement and regression analysis of dependent and independent variables taken from the three international contexts. The findings showed similarities between the species, regardless of country in terms of reactions to landscape configuration change (green space area and proximity, water body proximity, urban features), but important inter- and intra- species differences could be discerned through their reactions to human behaviour and local adaptation to

* Corresponding Author's E-mail: ecol55@hotmail.com.

landscape configuration. It is recommended that studies of urban avian scavengers, whether as pests or beneficial scavengers, must emphasize multi-species and multi-city comparisons for effective understanding. This makes an important contribution to urban avian ecology.

INTRODUCTION

The study of avian scavengers and the related theoretical paradigms is a broad topic, as the heterogeneous landscapes and multiple species behaviour of such species require an integrated methodology (Campbell, 2015). Such studies involve the appraisal of human created features, such as green spaces and associated food sources, human behaviour, bird physiology, ecology and the development of a methodology to investigate these dynamics and allowing flexibility for the development of relevant ideas. Several studies have emerged which attempt to analyse these issues have emerged over the last few decades (Tilghman, 1987; Blair, 1996; Belant, 1997; Clergeau et al. 1998; Morneau et al. 1999; Park & Lee, 2000; Marzluff et al. 2001; Cooper, 2002; Francois, 2002; Vuorisalo, 2003; Turner, 2003; Campbell, 2017, 2018). These are situated within the larger issue of socioenvironmental relations (Burger, 1981; Andren, 1992; Hansen & Urban, 1992; Kruger & Lindstrom, 2001; Hostetler & Knowles Yanez, 2003; Lim & Sodhi, 2004; Campbell, 2017).

The application of Integrated Research (IR) methods to intensive, comparative, species assessment is a potentially strong, yet relatively untested area of study. The increased area of human modified landscapes requires that such landscapes are incorporated into regional ecological appraisals as bird habitats on a level already developed for 'natural' landscapes. These issues of avian urban ecologies (including those in urban greenbelts and urban built landscapes) have not been fully examined with international case studies. Common factors for these problems are poor conservation management, narrowly based studies, disciplinary barriers, rapid urbanisation and nationally focused studies (Kutay, 1991; Aikman et al., 1998; Campbell, 2015, 2018).

Wells et al. (1992) argue that for effective ecological assessments, such areas should studied regularly, with multidisciplinary research focused on the landscape ecology, and the cultural and legal issues relevant to the area, using integrated research for the derivation of more quantifiable, detailed and accurate information, for more effective monitoring. The rest of this chapter will look at developments of relevant ecological paradigms, theoretical frameworks and methodologies, the methodology of the field survey for the current research, and the results of the surveys. The findings of the current research are then compared with the established literature and the contribution to urban avian ecologies.

Ecological Paradigms and Applications

Changes in ecological theory in recent decades have contributed to interesting possibilities, from which new tools have emerged that enable assessments of changing ecologies (Campbell, 2018). The primary issues have been the emergence of the new, disequilibrium ecology, which takes a more integrated and flexible approach to ecological and biogeographical change, and conservation biology, as an important link between ecological research and conservation applications. One result of these more integrated approaches to traditional ecological research has been the inclusion of social research methods, for the documentation of relevant landuse changes, creating a more multidisciplinary approach to the cultural biogeography of animal ecology (Campbell, 2017). Urban avian ecologies must be evaluative of variable landcover, from dense urban substrates, through mixed landcover, to dense urban and suburban forests, and the abilities of birds to cope with such variables (Campbell, 2015).

Relevant to this argument, a 'new' ecological paradigm developed towards the end of the 20th century. This emphasises studies of non-linear, non-equilibrium and multi-directional spatial ecological change, either naturally or human induced which may be subject to "multiple interpretations" (Scoones, 1997: 162). These changes may not necessarily be reversible (Solbrig, 1993). The "site histories" of studied contexts are considered important (Zimmerer, 1994: 110), as is the view that chaotic change is intrinsic to the ecological context rather than necessarily due to external interference (Nicolis, 1994). From the perspective of urban avian ecology, this is particularly important because it allows the inclusion of the dynamics of urban landscapes as legitimate subjects within ecosystem studies, rather than external to and wholly destructive of 'natural' ecosystems.

Within such syntheses, the adaption of birds to urban landscapes is crucial, as many hypotheses range from the near extinction of birds (mostly specialist species) to full adaptation (mostly generalist species) (Campbell, 2015). These questions and perspectives have contributed to the creation of more flexible theoretical and analytical tools, complementary to and in some cases supplanting, some of the key ideas of neoclassical ecology (Scoones, 1999). Areas of complex integrated change, where progressive trajectories are difficult to describe and evaluate according to ecological methods may be studied as "non equilibrium or multi-state systems which tend to flow from one state to another", an especially important issue in the study of unpredictable animal responses to human activities within a changing urban ecosystem (Young & Solbrig, 1993: 324).

This non-equilibrium paradigm of ecological change developed slowly mostly through studies of forest change (Solbrig, 1993, Campbell, 1998). Previous to this development, the dominant paradigm was based on the equilibrium hypothesis of classical ecology, emphasizing ecosystem theory, mechanistic succession, vegetation climaxes and human/nature separation, and very importantly separated human created landscapes from 'natural' landscapes, to the general exclusion of human dominated landscapes (Odum, 1971; Pickett et al. 1992; Begon et al. 1996). Entwined with the paradigm of Newtonian physics, this latter paradigm favoured the existence of "deterministic dynamics" of nature which were caused by a "few simple laws" (Solbrig, 1994: 17), generalisation and scientific universalism, with human action generally perceived as independent of such issues (Campbell, 1998).

The wider focus of ecological research has also resulted in a greater awareness of previously neglected features in appraisals of environmental dynamics. Problematic questions emerge: does human action enrich ecosystems or only destroy them, and how may animal adaptation be assessed in this sense? Also, what about those human actions which affect some species more than others. Is such an imbalance 'natural'? In this vein, there has been increased work on the socioenvironmental basis of ecosystem health, which is defined by Karr and Dudley (1981, 21) as "the capability of supporting and maintaining a balanced, integrated, adaptive, community of organisms having species composition, diversity, and functional organization comparable to that of natural habitats of the region." This definition reflects an inclusion of human and spatial perspectives and underscores the social roles in the evaluation and creation of ecosystem dynamics (De Leo & Levin, 1997). The maintenance of environmental health thus relates to environmentally sustainable practices, which are historically quite rare (Ludwig et al. 1993; Campbell, 2017).

Ecologists' acknowledgment of the importance of the natural and human-dominated environments has encouraged new insights within the applied ecological sciences, such as conservation management, urban planning and cultural biogeography. Hence, integrated socio-economic and biological methods have recently become more common (Walters, 1986; Campbell, 2017). These methods acknowledge chaotic variation in complex urban and mixed environments, encouraging the creation and application of new analytical tools, such as modern decision theory, for more dynamic management (Mangel, 1985; Hilborn, 1987; Campbell, 2018). Measurement and analysis of variable impacts of landcover development and resource management methods on different species in the broader ecosystem is a crucial spinoff of these ideas (Andrew & Pepperell, 1992).

Common species may be emphasized in such analyses, but commonalities with rarer species (sometimes previously common), inter- and intra-species, and habitat dynamics entwined with these may be relevant. As argued by Noss (1995: 2) conservation problems with endangered species and precarious ecosystems "arguably could have been prevented if management agencies had taken steps to protect adequate amounts and distributions of habitat before populations declined to where listing was legally required." Community-level conservation strategies focusing on the wider ecological system, especially in landscapes of multiple use are alternative methods. For example, De Leo and Levin (1997) cite the Natural Community Conservation Planning Process of California, which may prevent the "eleventh-hour crises that force choices between losing species and shutting down regional economies" (Mantell, 1992: 12).

A spatial perspective requires attention to the broader landscape and smaller patch scale linkages in variable landscapes within which organisms migrate, interact and evolve in response to human, biological and landscape features. Such connectivity operates within the configuration of such landscapes and is necessarily scaled (Merriam, 1984; Noss, 1991; Keitt et al. 1997; Campbell, 2008, 2009). Keitt et al. (1997) contrast species which may have wide dispersal capabilities, with those with more restricted abilities, acknowledging the former as less affected by changes in landscape configuration (such as fragmentation). A land or other (such as food source) barrier may therefore have selective, scale related effects on different species (Johnson et al. 1992a, b; Keitt et al. 1997). The configuration, variation, patch shape and size of the food source and physical habitats are strong factors for species presence and the extent to which species can connect with foraging and breeding habitats (Gardner et al. 1992; Henein & Merriam 1990; Taylor et al. 1993; Gustafson & Gardner 1996).

Applications to Conservation Biology and Landscape Ecology

Conservation biology and landscape ecology stand as disciplines that have developed and interact with the developments described above. Conservation biology plays an important role in the theoretical and scientific background for avian ecology and enables the creation of measurable parameters for more accurate measurements of environmental dynamics and change. Soule (1986) notes that conservation biology has two main goals: the study of the anthropogenic impacts on biodiversity, and the design of methods for species appraisal and protection. The research areas of conservation biology have expanded during the 1980s to the late 1990s. The earlier focus was on biodiversity, narrowly defined within ecosystem and genetic issues (Soule, 1986). Investigated areas now include processes such as dispersal characteristics, succession and disturbance effects and also predation and competition and other interactions between species (Gibeau, 2000, 22) (see also Beissinger, 1990). There has also been a slow integration of pure and applied research (Campbell, 2017), which has allowed a greater support for the practical issues of wildlife, fish and forestry management (Temple et al. 1988), the planning process (Primack, 1993) and biological conservation (Soule,1986).

An important trend within conservation biology has been the development of ecosystem management (Brunner & Clark 1997), which has complemented the changes in ecological perspectives described above. Social perspectives on conservation (including relatively unexplored areas, such as urban green spaces), as well as biological research, have supported this development (Stanley, 1995). Gibeau (2000: 106) nevertheless argues that the complexity of the socioenvironmental issues that underpin conservation, are reflected in "an arena of multiple, conflicting societal agendas." As noted by Brunner and Clark 1997: 49) in such problematic situations the role of ecosystem management "is not to evaluate action alternatives as if the decision problems were merely technical, but to guide a collective and continuing evolution process that can reduce ambiguities and uncertainties and improve alternatives through the comparative evaluation of practical experience". The objective of such a perspective is to seek and create advancements in the methods of conservation biology by studies of the results of implementation through reflection on the experience that follows decision and action (Brunner & Clark 1997; Gibeau 2000). This requires both the evaluation of parameters for the measurement of related social and ecological dynamics, and assessment of methods of data acquisition and information analysis.

Landscape ecology "specifically includes resource management activities" Risser at al. (1984: 6) and "considers the development and dynamics of spatial heterogeneity, spatial and temporal (our emphasis) interactions and exchanges across heterogeneous landscapes, influences of spatial heterogeneity on biotic and abiotic processes, and management of spatial heterogeneity" (ibid.7). In the literature on landscape ecology, there have been related trends. An examination of the articles in the Journal of Landscape Ecology at the turn of the century (1994-2001) reveals an increased focus on analytical applications, and a greater link between urban and rural dynamics. This was also described by Hess (1994) for the period leading up to the mid-1990s, and this trend has continued until recently (Campbell, 2017). These include applications to landscape management, through the topics of biodiversity conservation, sustainable development and global environmental change and social applications (Naiman, 1996). An important issue in this development is the diversity of the discipline of landscape ecology which provides tools for the interpretation of complex landscape issues, and possibilities for the integration of these results to allow the solution of integrated problems (Weins, 1999; Moss, 2000). There have been increases in both single species and meta-population studies, and studies emphasising management applications, and computer-based methods of analysis.

In terms of the applications to the practical issues of environmental social sciences and planning issues, recent trends in conservation have sought to a greater extent than before to include the study of coevolutionary socioenvironmental relations and to advocate the inclusion of local actors in environmental assessment and management, especially in the case of landscape assessment and the implications for animal ecology. This may unite the issues of development and conservation (Broad, 1994; Agrawal, 2000). Several themes have emerged recently, the main ideas clustering around increased participation by local people, social empowerment and the recognition that these often-neglected people are strong sources of environmental information (Campbell, 2017).

One new approach to ecological management and conservation is adaptive management, a system which developed during the 1970s (Johnson, 1999). This is described as "an innovation that implements policies as experiments...mobilises available information for objectives that are less susceptible to unexpected outcomes" (Agrawal, 1999: 326), and "probes the responses of ecosystems as people's behaviour changes (Lee, 1999: 3). Basically, "adaptive management tries to incorporate the views and knowledge of all interested parties" (Johnson, 1999; 8). Agrawal (2000) gives a comprehensive analysis of adaptive management in transboundary protected areas, taking as an example the Bialowieza National Park and Biosphere reserve in a cross-boundary area of Poland and Belarusia. The contention is that this is the first examination of the role of local residents within conservation strategies and transboundary protected areas. Several researchers have studied the difficulties and prospects for the application of adaptive management within the larger system of ecological management (see Walters 1997a, b; Callicott et al. 1999). The objective is to assess such actions for possible improvement, examining the environmental impacts.

Several issues in adaptive management are important to the assessment of urban and mixed landscapes. Urban green space landuse (human recreational presence, park management of garbage collection, refuse dumping, tree planting and removal, cultivation etc), and the planning of neighbourhoods near dense vegetation and even fresh-water bodies may be assessed as relevant urban animal ecology. Comparative assessment of conventional management and adaptive management hints the former is largely focused on prediction and short-term equilibrium, with linear trajectories of change for flora and fauna. Adaptive management contrasts as a system that integrates knowledge from several sources and is generally experimental with environmental actions. The participation of local actors may be particularly useful for the provision of information on relevant actions (Gunderson, 1999; Agrawal, 2000; Campbell, 2017).

One area where there is a need for more information techniques is the relationship between wildlife management and landscape patch change. This area has developed slowly other the years, with important progress during the period 1980 to 2000 (Campbell, 2017). Risser et al. (1984) argued that at the time of writing there was no existing body of ecological theory, which could be used for analysis of resource management decisions, resulting on reliance on observations. Risser (1995) points out that in the 1980s there was acknowledgement that many ecological management issues had spatial relevance. However, "there were too few connections between the interests and information of the research community and information needs of those who must make natural resource management decisions" (ibid. 130). These issues have not been fully explored, especially in relation to specific examples, despite the major recent developments in this field. Missing in many cases are studies of comparative cases in international locations, which would allow a more integrated approach to species ecology and assessment. (Campbell, 2017, 2018). The next section explores these developments in the narrower topic of urban avian ecology.

URBAN AVIAN ECOLOGY CASE STUDIES

Avian ecology connects with the critical rethinking of the ecological sciences and associated disciplines, as birds are among the most adaptive animals (in terms of landscape change), yet many bird species are also endangered. Birds are more mobile than other vertebrates, hence may be exposed to more landscape variation in multiple contexts. Bird urbanisation, referring to the foraging or breeding presence of birds in urban contexts, has recently developed as an important ecological issue and a planning priority (Belant 1997; Clergeau et al. 1998; Morneau et al. 1999; Park & Lee, 2000; Marzluff et al. 2001; Cooper, 2002; Francois, 2002; Vuorisalo, 2003). This development occurred with increased recognition of the impact of landscape dynamics on avian foraging and ecological needs (Burger, 1981; Andren, 1992; Hansen & Urban, 1992; Kruger & Lindstrom, 2001; Hostetler & Knowles-Yanez, 2003; Turner, 2003; Lim & Sodhi, 2004). Landscape patterns- forest stands, farmland mosaics, built-up areas and/or open water bodies are important factors for avian foraging and nesting behaviour (Burger 1981; Hansen & Urban, 1992; Belant, 1997; Krüger & Lindström, 2001; Francois, 2002; Chace et al. 2003; Hostetler & Knowles-Yanez, 2003; Turner, 2003; Vuorisalo et al. 2003; Lim & Sodhi, 2004). This type of comparative study has been documented for birds of prey (for example, Korpimäki, 1987; Kostrzewa, 1991; Selas, 1997).

Table 1. Bird Species in the Study

Species	**Scientific name**	**Length (cm)**	**Wingspan (cm)**	**Weight (g)**
Crow	*Corvus corone corone,* L., 1758	45-47	93-104	370-650
Rook	*Corvus frugilegus,* L, 1758	44-46	81-99	280-340
Jackdaw	*Corvus monedula,* Linnaeus, 1758	33-34	67-74	180-260
Magpie	*Pica pica,* Linnaeus, 1758	44-46	52-60	180-270
L. black backed gull	*Larus fuscus graellsii*, L. 1758	52- 64	135 -150	620 -1000
Black-headed gull	*Larus ridibundas,* L.1776	34-37	100-110	200-400
Hooded vulture	*Necrosyrtes monachus*, T.1823	60-64	160 - 181	1600 - 2400
Black kite	*Milvus migrans*, Boddaert, 1783	55-60	165-170	700-750
Pied crow	*Corvus albus*, Statius Muller, 1776	45-47	93-104	370-650
Turkey vulture	*Cathartes aura*, Linnaeus 1758	60-64	160 - 181	1600 - 2400
American crow	*Corvus brachyrhynchos*, B. 1822	40 -53	85 - 100	316-620
Ring-billed gull	*Larus delawarensis*, G. Ord 1815	43 - 54	105 -117	300-700
Blue jay	*Cyanocittacristata*, L. 1758	25 - 30	34-43	70-100

Sources: Sterry et al. (2002); Campbell, (2008, 2009, 2015); RSPB (2017). Note L., 1758 refers to Linnaeus, 1758. T. 1823 refers to Temminck, 1823, B. 1822 refers to Brehm, 1822.

In this chapter, case studies are given of avian scavengers in three countries (the United Kingdom of Great Britain (hereafter UK), Ghana and Canada) (Figure 1, Table 1). This study is based on published works by Campbell (1998, 2002, 2007, 2008, 2009a, 2009b, 2013a, 2013b, 2013c, 2013d, 2014, 2015, 2017, 2018) and additional field work that was undertaken around these studies. The study expands the single species, single city format common in the avian urban ecology literature (for example, Turner, 2003) and also contributes to the growing literature on the importance of urban green spaces as general avian (Tilghman, 1987; Blair, 1996; Clergeau at al. 1998; Turner, 2003) vulture (Campbell, 2015, 2017, 2018), corvid (Tatner, 1982; Eden, 1985; Grist, 2000; Skillen, 2000; English Nature, 2003; Vuorisalo et al. 2003) and gull habitats (Rock, 2002, 2003a,b). The study acknowledges the limited number of species, but the selection of these species is adequate to an incisive, illustrative study of the contextualization of urban avian ecology in the study areas, and to the derivation of conclusions applicable other similar areas.

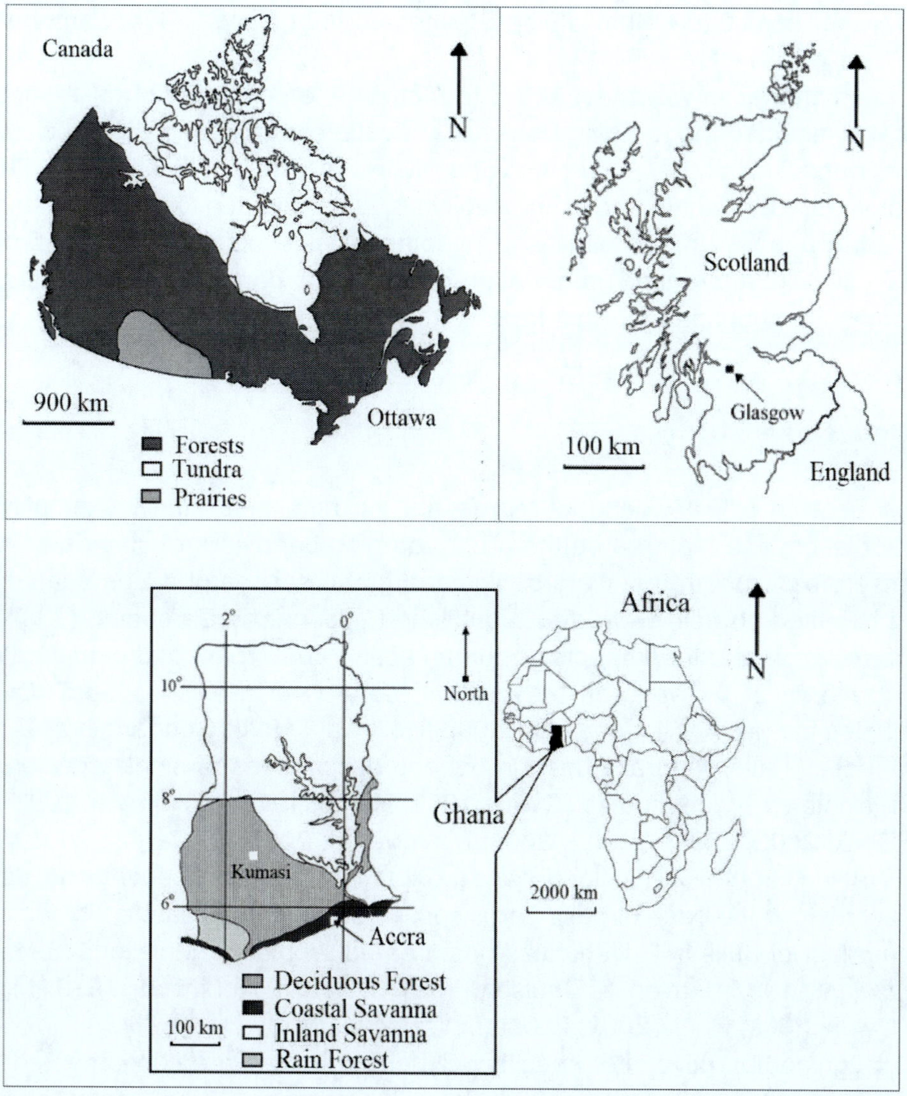

Figure 1. Location of the three case studies. (a) Glasgow, UK; (b) Ottawa, Canada; and (c) Accra and Kumasi, Ghana.

Referring to certain terms used in this chapter, 'landscape configuration' refers to the arrangement and character of cultural and natural features of the landscape- green spaces (treed avenues, lawns, large gardens, golf courses, vegetated graveyards, urban parks, and peri-urban forest patches and plantations), built-up areas (pavements, paved squares, roads, roof ledges) and rivers and small water bodies (artificial lakes, streams and ponds). Urban water bodies (rivers, lakes, ponds) have been increasingly examined as factors for bird foraging presence, including corvids (Donelly & Marzluff, 2004; Campbell, 2008). These are part of the urban landscape configuration, which includes built-up areas (pavements, building ledges, bridges, roofs). Crucially, all these landuse classes are important factors for avian presence in northern and tropical countries (Clergeau et al. 1998; Marzluff et al. 2001; Sterry et al. 2002; Mitchell et al. 2004; Shaw 2004; Stevenson & Fanshawe 2004; Campbell, 2015, 2018).

Three hypotheses were tested: (1) the landscape configuration of the four cities significantly affected the foraging behaviour of vulture, corvid and gull species; (2) the adaptive behaviour of these species varies between different cities even with similar environments; (3) a multi-species, multi-city study contributes to the understanding of urban avian behaviour. To test these hypotheses, the objective of the paper is to provide critically referenced information to derive a comprehensive analysis and assessment contributing to further investigations.

Glasgow Case Study

The Scottish city of Glasgow represents the first case study, exemplifying the impact of landscape features on the urban foraging behaviour of corvids and gulls in the dense urban substrates, waterbodies and green spaces of a European city. This section is based on field work and a published account by Campbell, (2007, 2008). Bird urbanisation and the impacts of human action and urban land configuration (the spatial patterns of the natural and cultural features of the landscape) have been documented in the UK (Gibbons et al. 1993; Holloway, 1996; Goddard et al. 2017), but the emphases have generally been in general descriptions of single species, mostly gulls, in single city case studies (Owen, 1967; Harris, 1970; Lloyd et al. 1991; Raven, 1997; Rock, 2002, 2003a,b; Jha, 2003; Mavor et al. 2003; McCarthy, 2003; Mitchell, 2004; Watson, 2004). Avian landscape adaptation has also become an especially important topic in the case of gulls *Larus* spp. in Great Britain, as the historic increase in the number of gulls in British cities has created problems (Mclellan, 1894; Baxter, 1953; Belant, 1997; Raven & Coulson, 1997; Bolen & Robinson, 1999; McCarthy, 2003; Owen, 2003; Rock, 2002, 2003a,b; Watson, 2004).

Although comparative inter-specific gull behaviour has been documented in some studies (Burger, 1981; Calladine, 1997; Raven & Coulson, 1997; Francois 2002; Rome & Ellis, 2004) and intra-specific competition is also documented (Galvan, 2003), very few studies have focused on general avian studies, examining such bird species in several cases, and their inter-species interactions, especially comparing avian populations in different countries, despite the obvious importance of such studies (Moore, 2002; Sterry et al. 2003; Vuorisalo et al. 2003; Campbell, 2015; Royal Society for the Protection of Birds, 2017). There have also been few studies which document both the competitive foraging habits and landscape adaptation of corvid and gull species in British cities, especially in Scotland (Lloyd et al. 1991; Mavor et al. 2003; Mitchell et al. 2004; Campbell, 2008). For example, the lesser black backed gull *Larus fuscus graellsii* Linnaeus 1758 and the black headed gull *Larus ridibundas* Linnaeus 1776, common in Scotland (Sterry et al. 2002; Mitchell et al. 2004; Shaw 2004), and the commonest gull species in the Glasgow area are seriously under researched in this area (Campbell, 2008).

Glasgow is the largest city and the commercial and industrial capital of Scotland, located on the River Clyde (55.87° N, 4.25° W). Initially a sixth century fishing village, the population reached over one million in the twentieth century. Glasgow's twentieth century growth was marked by the establishment of several large urban parks, in association with dense neighbourhoods, and extensive treed avenues, some linked to rivers, and an extensive surrounding greenbelt of recreational parks and farms. Currently, Glasgow has a population of 621,020 (2017) (Glasgow City Council Area Profile, 2019) and numerous green spaces, which comprise over 20 percent of its area (Campbell, 2008). Smaller water bodies exist in some parks, in addition to the small River Kelvin and the Forth and Clyde Canal, which cut through the city. This area of central lowland Scotland, although a favoured landscape for several corvid species (Thom 1986; Sterry et al. 2002) has been neglected in assessments of avian ecology and especially corvid behaviour (Campbell, 2007, 2008).

The study examines the inter- and intra – species interactions of four corvid species (carrion crow and two gull species (Campbell, 2007, 2008). These are medium sized birds (Table 1). All these species are described as adaptive, intelligent omnivores, primarily inhabitants of mixed woodland and open rural mosaics, with some attraction to water bodies and strong urban colonisation (Royal Society for The Protection of Birds, RSPB, 2017; Hume 2004). However, despite their important role as urban scavengers, there is little documentation of the inter- and intra – species relations, especially in urban areas (Dicerbo, 1991; Royal Society for the Protection of Birds, 2017). There is a need for studies comparing both their habitat preferences and competition, especially their presence in urban areas, which has been described as problematic for human life and planning (Rock, 2003a,b; Campbell, 2008) as informing planning has been described as an important function of landscape ecology (Lovett-Doust & Kuntz, 2002).

Ghana Case Study

For the second case study, the locations were Accra (5.56°N 0.20°W) and Kumasi (6.69°N; 1.63°W) in Ghana (8 00 N, 2 00 W; area 239,460 sq. km; human population 28,102,471 (July 2018 est.) (Central Intelligence Agency, 2019) in Figure 1) (Campbell, 2009b). Accra (population 2,270,000) is within the coastal savanna/thicket on coastal Ghana and Kumasi (population 1,468,609), is within the deciduous forest in south-central Ghana, these being the nation's largest cities (World Population Review, 2019). Avian scavengers in Ghana are rarely studied in detail (Campbell, 2009, 2015). The study focuses on the three dominant avian scavengers, the Hooded vulture, Black kite and Pied crow (Table 1). Although the black kite and hooded vulture may not have exact equivalents in the UK cities, the Pied crow has been compared with Carrion crow in size and diet and to the rook in social habits (Alden et al. 1995; Stevenson & Fanshawe, 2002).

Hooded vultures, Black kites and Pied crows are usually studied in relation to larger vulture species in the Eastern and Southern African plains (Newton, 1979; Ferguson-Lees & Christie, 2001; Wanten & Holz, 2004). Consequently, few studies assess the role of these smaller avian scavengers as dominant urban and rural birds in African contexts outside the range of the larger vultures (Campbell, 2015). These smaller species are generally more adaptive to landcover change and human habitation than the larger African vultures, such as the White-backed vulture (*Gyps africanus*, Salvadori 1865), White-headed vulture (*Trigonoceps occipitalis* Burchell 1824) and Egyptian vulture (*Neophron percnopterus*, Linnaeus 1758) (Campbell, 2015). These are the only avian scavengers common in the southern half of the Ghana, other vultures being largely restricted to the northern half of the country (Bowman, 1996; Campbell, 2015). Due to the relative absence of other avian scavengers in from southern Ghana (including gulls and smaller corvids), the competing scavengers are usually feral dogs (*Canis lupus familiaris*, Linnaeus 1758) (Campbell, 2015).

Considering the abundance of the three smaller scavengers, few detailed, recent studies have investigated the impact land features (abattoirs, rubbish dumps, markets, central business districts and residential areas in urban areas, and rural vegetation, fires, transport networks and farming in rural areas) on their feeding and presence ecology. Most studies briefly discuss mundane topics such as their small size, low competitive status and tolerance for humans (all relative to other avian scavengers, especially the large vultures) (Newton 1979; Ferguson-Lees & Christie, 2001; Houston, 2001; Butchart & Roche, 2003; Wanten & Holz, 2004). There is also mention of traditional cultural perceptions of such birds (Iddi, 2004; Campbell, 2015).

Canada Case Study

For the third case study, the location was Ottawa, Ontario (45.32° N, 75.70° W; population 934,243; Statistics Canada 2016) the political capital and fourth largest city in Canada (Campbell, 2002, 2009b). Ottawa has an important landscape mosaic based on urban developments, green spaces and the largest urban greenbelt in Canada. The greenbelt covers approximately 200 square kilometers of the National Capital Region (NCR) of Ottawa (Wickware & Rubec, 1989), including forests, farms, meadows, rivers, lakes and wetlands (Brunton 1988; NCR 1996). Although Erickson (2004: 199) states that Ottawa is not "well known for progressive contemporary greenway planning efforts" and associated animal ecology, it is also argued to be one of the greenest metropolitan cities in North America, energy into green space maintenance (Campbell, 2002). Such green areas are dominated by mixed deciduous and coniferous associations, the common species being Jack pine (*Pinus banksiana*, Lamb.) Yellow birch (*Betula alleghaniensis* Britt.), Sugar maple (*Acer saccharum* Marsh.), Eastern White cedar (*Thuja occidentalis* L.), White spruce (*Picea glauca* (Moench) Voss), Balsam Fir (*Abies balsamea* (L.) Mill.) and Largetooth Aspen (*Populus grandidentata* Michx.) (Regional Municipality of Ottawa-Carleton RMOC, 1997). The city is also bisected by some water bodies (the Ottawa and Rideau rivers and the Rideau Canal). These waterways form corridors, along which several green spaces are located (Campbell, 2002).

Around the greenbelt, urban satellite towns have developed, (Kanata, Orleans, Leitrum, South Gloucester (Riverside South), Stittsville and South Nepean), linked to Ottawa by urban corridors through the greenbelt (Ottawa 20/20, 2002). A further complication for animal habitats, is this virtual encirclement of the greenbelt by satellite towns (giving the greenbelt the appearance of a large green space within the city, surrounding dense urban substrate). Despite the obvious interest in the possibly changing role and biogeography of urban avian ecology in such a dynamic environment, few studies have been undertaken on this issue. The emphasis has been on the plant ecology of greenbelt (Hough, 1991), and greenbelt change (Brunton, 1988, 1997; City of Ottawa, 1991; Brownell & Larson 1995; RMOC, 1997; Aikman et al. 1998). Other topics have been ice storm impacts (Nielsen, 2001) and mammal human interactions (Brown, 2000), beaver swamp development (Chabigue & Horvath, 2001), deer browsing impacts and the increase in human visitors to the greenbelt and other green spaces (National Capital Commission NCC, 1996, 1999; Brown, 2000).

To fill this gap, the dominant avian scavengers in this area, turkey vultures, American crows, blue jays and ring billed gulls were examined (Table 1). These are omnivores, but the turkey vulture is a full time or obligate scavenger, while the other species are also predators of small animals. Although fulfilling a similar function, possibly because it is not related to the 'Old World' vultures Accipitrid vultures like the hooded vulture, the turkey vulture has rarely been compared with African vultures in terms of its scavenging role (except for its status as one of the few vultures with a sense of smell). The American crow is almost identical in habits to the British Carrion crow and the Pied crow, and it has social flock behaviour like the rook and jackdaw (Maccarone, 1991). The Blue jay is like the British magpie, in that it frequents green spaces and rural contexts (United States Geological Survey, 2004). Similar to the other species in the other case studies, there is little documentation of the inter-species interaction and urban ecology of these species, despite their position as the dominant urban scavengers and avian predators in the cities of southern Ontario (NCC, 1999). This study, like the rest involved a strong field methodology, which is described in the next section.

METHODOLOGY

Although the three case studies used similar techniques, investigating species in similar ecological niches, they are described separately, as there was some variation due to differences in the landscapes under study (Campbell 2008, 2009b, 2009b). All the studies however used the point count method (Hagan et al. 1996; Mattson & Reinhart 1996; Tewksbury et al. 1998; Corkum et al. 1999; Bayne & Hobson 2000; Green & Baker 2003; Lim & Sodhi 2004). The same parameters were also used (green space area, green space proximity in urban areas, and where possible, river proximity in green spaces and urban areas), which are important factors for the ecology and biogeography of urban avian species. The analysis used multiple regression and averages. The presence of the avian species served as the dependent variables which were regressed against the independent variables (green space area, green space proximity, water body proximity). The figures for the individual species were also correlated and regressed against those of other corvid species, to assess the extent to which these species shared foraging patterns. The calculations were tested at the 0.05 two tailed level of significance.

Glasgow Case Study

The selection of the study area was based on its situation on a major river, with large green spaces and large dense human populations, surrounded by complex forest and farm mosaics. The landuse of the green spaces was largely recreational and to a lesser extent agricultural. The corvid and gull survey was based on 93 survey points, which were based on a grid spaced at about 350 metres (with a radius of about 50 metres), were visited for about 20 minutes between 10 am and 4 pm, fortnightly between September 5th 2003 and January 10th 2004, and between 8 am and 7 to 8 pm between January 15th and July 18th 2004. Observations included sightings, vocalizations and nests. Sightings were divided into perching, feeding (including flight to immediate feeding and perching areas, nest feeding and the feeding of large juveniles by adults on the ground). Flyovers were recorded but were not included in the analysis. Human presence was also recorded: the numbers of people present, within 50 metres of each point, the number of bird-feeding actions (both deliberate and accidental) and the number of food sources such as open bins and litter.

Survey points were classified into two groups: those in green spaces and those in urban built-up areas. All the green spaces were similar in general vegetation form (continuous tree cover 20-30 percent, trees with shrub undergrowth 10 – 15 percent, shrubs 10 – 15 percent, grass 40 – 60 percent). Configuration characteristics measured were the size of the of the green spaces, the proximity of green spaces to the urban observation points, the proximity of major rivers (the River Clyde) and the smaller water bodies (the River Kelvin and the Forth and Clyde Canal in Glasgow). The survey in Glasgow covered about 40 square kilometres, including points north and south of the River Clyde, using 34 locations in green spaces and 60 in urban built-up areas. Tree species included the Common juniper (*Juniperus communis*, L.), Scots pine (*Pinus Sylvestris*, L.), ash (*Fraxinus excelsior* L.), oak (*Quercus* L. sp.) and birch (*Betula pendula*, Roth).

Ghana Case Study

Two hundred and twenty observation points were taken in southern Ghana. Each point was visited for at least 15 minutes at least 20 times, during the period September

1985 - September 1993, May 1995- September 1996, and April – June 2000. All forms of bird visualisation were recorded: sightings in flight or perched, and nests. At the feeding areas, the interaction of bird scavengers with each other, and with feral dogs was also monitored. Human population figures were derived from the Ghana Census (2000). The choice of the large towns was deliberate, as these represented the largest concentrations of human population in southern Ghana, each located in a different environment. The points in the urban locations were chosen randomly, but in grid format within the proximity of slaughterhouses or abattoirs, rubbish dumps, markets, business districts, residential areas, green spaces and/or rivers.

Abattoirs varied in size from factory sized buildings killing herds of cattle daily, to smaller butcher's shops, where mostly small livestock (sheep, goats, pigs, chickens) were slaughtered. All types were found in the, while the suburbs. Rubbish dumps in larger cities comprised both large open dumps for landfill (up to eight hectares in area) and smaller neighbourhood truck trailer type dumps. Although, most dumps in smaller settlements were smaller, these tended to have more organic waste, due to the less effective sanitation restrictions. Markets comprised open air social and economic complexes of varying sizes (from ten hectares to less than ten square metres, most producing organic waste, such as fish, meat and rotting vegetables. Business districts and residential areas, usually in the town centre and suburbs, were included, as the green spaces, streets and gutters frequently displayed organic waste (such as dead rats, cats, toads, and market waste).

For the rural locations, eighty random points were also taken in the semi-deciduous forest and 30 in the coastal savanna. These were classified according to: (1) forest; (2) savanna (mixed forest, shrub and grass mosaics); and (3) settlement proximity. The forested areas averaged 50 to 65 percent closed canopy and dense shrubs, and 35 to 50 percent grass. The coastal savanna comprised savanna was 15 to 20 percent treed and over 70 percent grass covered. These were similar proportions to the green areas of Ottawa and Glasgow, although the plant species were different. Although there were large patches of fragmented and contiguous forest in savanna/grassland, and shrub/grassed mosaics within forests, the dominant vegetation within one square kilometre surrounding the point was used for classification. Road kills, bush clearance, forest fires and termite flights were monitored. These are bird attractants (Ferguson-Lees & Christie, 2001; Campbell, 2015).

Forest comprised the common dry deciduous forest of central Ghana. Dominant tree species were *Triplochiton scleroxylon* K. Schum. 1900, *Terminalia ivorensis* A. Chev 1909, *Heriteria utilis* Sprague 1909, *Entandrophragma cylindricum* Sprague 1910, *Samarea saman* 1916, *Ceiba pentandra* (L.) Gaertn. 1791 and *Piptadeniastrum africanum* Hook.f 1955. Savanna comprised the coastal savanna or coastal scrub and grassland vegetation dominated by fragmented stands of deciduous forest trees, mixed with neem *Azadirachta indica* A. Juss 1890 and *Mangifera indica* L. 1753, *Cassia siamea* Lam. 1785, *Ficus thonningii* Blume 1836, *Millettia thonningii* (Schum. & Thonn.) Baker 1948, *Antiaris toxicaria* Lesch.1811, *Fagara zanthoxyloides* Lam. 1758 and grasses (dominated by *Andropogon gayanus* L.1758, *Panicum maximum*

Jacq. 1781, and *Sporobulus* spp. Both vegetation types contained patches of farmland, with some recent growth in areas under recent clearance.

Ottawa Case study

Eighty survey points were used in green spaces and the green belt, and 50 in urban locations, between October 2000 and September 2002. Due to the extreme nature of a Canadian winter, during which most birds migrate, the study period was restricted to the fall (autumn), spring and summer. For the rural locations, the points were within mixed forest, grass and farmland, within conservation areas and public land. The forest comprised upland forest, largely composed of Yellow Birch (*Betula alleghaniensis* Britt.), Sugar maple (*Acer saccharum* Marsh.), Eastern White Cedar (*Thuja occidentalis* L.), White Spruce (*Picea glauca* (Moench) Voss), Balsam Fir (*Abies balsamea* (L.) Mill.) and Largetooth Aspen (*Populus grandidentata* Michx.) including forested wetland (tree covered land subjected to occasional, seasonal or semi-permanent flooding, such as silver maple, cedar and ash swamps, black spruce and larch bogs). Other features in green spaces were meadows and barrens (young shrubbery, sapling growth, open grasslands, bare rock flats, relict prairies and raspberry thickets) fallow cultivated and public parks (cornfields, sod farms, sheep and cattle pasture and weed species). The tree/shrub/grass configuration was like the landscape of Glasgow, despite the larger percentage of conifers in the Ottawa area and the green spaces were predominantly public recreational spaces and farmland. The urban landscape comprised built-up areas, gardens and roads.

RESULTS

Glasgow Case Study

Magpies were the most tied to green spaces and were comparatively rare in urban areas, like the Blue jay in the Ottawa case. In terms of nesting behaviour, only magpie nests were observed (six, all in the largest green space Kelvingrove Park, near the River Kelvin). Crows, rooks and jackdaws were observed nesting outside the study area, in the countryside. Rooks averaged 0.2 per point and jackdaws 0.1 in both green and urban spaces. Crows averaged 1.5 per point in the green spaces, and 1.2 in urban spaces, and magpies averaged 3.3 in green spaces and 1.3 in urban spaces. Magpies in such areas were significantly less mobile than other species. In the observations, 81 percent of observed perching and feeding magpies stayed in the vicinity for the full 20 minutes (crows 64 percent, rooks 56 percent and jackdaws 73 percent). The significant relations between the avian species and with their environments are summarized in Table 2.

Table 2. Land-cover associations of Glasgow birds (summary)

Species	GSA	GSP	WPG	WPU	C	R	M	J	L	B	C	R	M	J	L	B
Land					Green spaces						Urban areas					
Crow	•	•	o	•	X						X					
Rook	•	o	o	•	•	X					o	X				
Magpie	•	•	o	o	•	•	X				•	•	X			
Jackdaw	•	•	o	•	o	•	•	X			o	•	•	X		
LBB gull	•	o	•	•	•	•	o	o	X		o	•	o	o	X	
BH gull	•	o	•	•	•	•	o	o	•	X	•	•	•	•	•	X

• = Significant. o = Insignificant. GSA = Green space area. GSP = Green space proximity. WPG = Water proximity in Green Space. WPU = Water proximity in urban area. C = crow, R = rook, M = magpie, J = jackdaw, L = Lesser black-backed gull, B = Black-headed gull. The right half of the table shows the presence correlations between pairs of species in green spaces and urban areas.

Green space area had the strongest impact on magpies ($r^2 = 0.517$, $F = 33.154$, $p < 0.001$) and crows ($r^2 = 0.512$, $F = 32.557$, $p < 0.001$), and less on rooks ($r^2 = 0.338$, $F = 26.520$, $p = 0.002$), jackdaws ($r^2 = 0.256$, $F = 10.681$, $p = 0.003$), but all were significant. In the urban areas, green space proximity was less significant than that of green space area for all species except magpies (crows $r^2 = 0.137$, $F = 9.22$, $p = 0.004$; rooks $r^2 = 0.076$, $F = 4.784$, $p = 0.033$; jackdaws $r^2 = 0.124$, $F = 8.202$, $p = 0.002$; and magpies $r^2 = 0.546$, $F = 69.811$, $p < 0.001$). For gulls, green space area was less significant than river proximity as a factor for foraging presence. For lesser black backed gulls, urban green space area was significant ($r^2 = 0.155$, $F = 5.69$, $p = 0.023$). For black headed gulls, the green space area was also important ($r^2 = 0.184$, $F = 6.985$, $p = 0.013$). In urban areas, the proximity of green spaces was relatively unimportant for lesser black backed gulls ($r^2 = 0.003$, $F = 0.196$, $p = 0.660$). For black headed gulls, it similarly insignificant ($r^2 = 0.041$, $F = 2.491$, $p = 0.120$). This was despite the large number of park visitors, and the consequently greater availability of organic food sources (on average 21 people within 50 metres of the observation point) which attracted many corvids. The Glasgow Green, Kelvingrove and Alexandra Parks (0.38+ km^2 each), and the Botanical Gardens (0.17 km^2) attracted more human visitors per unit area (an average of 65 per point) than the smaller parks (nine per point).

The proximity of the water bodies had no significant impact on the corvids in green spaces (crows $r^2 = 0.007$, $F = 0.226$, $p = 0.638$; rooks $r^2 = 0.004$, $F = 0.133$, $p = 0.718$; jackdaws $r^2 = 0.046$, $F = 1.507$, $p = 0.229$; magpies $r^2 = 0.069$, $F = 2.310$, $p = 0.139$). In urban areas, the proximity of large rivers had a stronger impact (crows $r^2 = 0.140$, $F = 9.465$, $p = 0.003$; rooks $r^2 = 0.080$, $F = 5.068$, $p = 0.028$; jackdaws $r^2 = 0.099$, $F = 6.374$, $p = 0.014$; magpies $r^2 = 0.058$, $F = 3.544$, $p = 0.065$). Human deposited organic matter was an attractant for birds in these areas (at least 34 percent of untended refuse was along the rivers). However, the availability of scattered food sources away from rivers (trash cans, dropped food on pavements). Within green spaces natural foods (worms, insects, seeds) were also observed being taken, from grassed lawns, and tree foliage, balancing that near water bodies.

By contrast, river proximity had a stronger impact on both gull species. There was a significant relation between river proximity and black backed gull presence in urban areas ($r^2 = .200$, F = 14.545, p <0.001). This importance of large rivers was reduced in green space locations ($r^2 = 0.178$, F = 6.703, p = 0.015). This was also the case for black headed gulls in urban areas in both cities ($r^2 = 0.237$, F = 18.052, p <0.001) and green spaces (0.255, F = 10.588, p = 0.003). Human factors were observed to be largely responsible, as a larger number of organic waste depositions were observed (63%, n = 71) and deliberately proffered food (71%, n = 53) was available on pavements proximate to the rivers, than in green spaces near the large rivers. Nevertheless, it must be assumed that water played a role, as corvids were outnumbered in such areas, despite the availability of food.

There was a moderate relationship between the foraging activities of the four species of corvids. The main issue was the moderately strong relationship between crows and magpies in green spaces ($r^2 = 0.660$, F = 59.327, p <0.001). Both species frequented tree foliage and lawns. There were twelve incidents between these two species, and crows prevailed in eleven, driving the magpies into the foliage, with one draw, with no other conflicts being observed between the other combinations of corvid species. Crows also had a weaker relationship with the rarer rooks ($r^2 = 0.244$, F = 10.002, p = 0.003) and jackdaws ($r^2 = 0.083$, F = 2.800, p = 0.104). Rooks and jackdaws ($r^2 = 0.365$, F = 17.787, p <0.001) also had stronger relations, compared with rooks and magpies ($r^2 = 0.165$, F = 6.109, p = 0.019) and jackdaws and magpies ($r^2 = 0.129$, F = 4.604, p = 0.040). In urban areas, the relations were generally very variable (crows/rooks $r^2 = 0.018$, F = 1.056, p = 0.308; crows/jackdaws $r^2 = 0.120$, F = 7.942, p = 0.308; crows/magpies $r^2 = 0.304$, F = 25.326, p <0.001; rook/jackdaws $r^2 = 0.328$, F = 28.359, p <0.001; rooks/magpie $r^2 = 0.082$, F = 4.901, p = 0.031; jackdaw/magpie $r^2 = 0.237$, F = 17.100, p <0.001). In all cases, rooks and jackdaws had the strongest relations: both were generally in flocks (n = 5+) while crows and magpies were more often solitary or in pairs. Rocks and magpies had consistently low foraging relations, possibly because magpies were intimidated by the rook numbers and size, and hence retreated into denser vegetation (at least 65 percent of cases where both species were present) and as the smaller birds were consistently found in green spaces. There was more specific local behaviour and the species were less often seen together (<50 percent, n = 94).

In green spaces, there was only a weak relationship between corvids and gulls (lesser black backed gulls: crows $r^2 = 0.179$, F = 6.667, p = 0.015; rooks $r^2 = 0.124$, F = 4.375, p = 0.045; jackdaws $r^2 = 0.058$, F = 1.917, p = 0.176; magpies $r^2 = 1.07$, F = 3.737, 0.062. For Black headed gulls: crows $r^2 = 0.0.274$, F = 11.750, p = 0.002; rooks $r^2 = 0.289$, F = 12.616, p = 0.001; jackdaws $r^2 = 0.068$, F = 2.263, p = 0.143; magpies $r^2 = 1.076$, F = 2.537, 0.121). In urban spaces the relation was similarly weak (Lesser black backed gull: crows $r^2 = 0.025$, F = 1.475, p = 0.229; rooks $r^2 = 0.070$, F = 4.358, p = 0.041; jackdaws $r^2 = 0.0313$, F = 1.873, p = 0.176; magpies $r^2 = 0.052$, F = 3.169, 0.080: Black headed gulls: crows $r^2 = 0.141$, F = 9.543, p = 0.003; rooks $r^2 = 0.275$, F = 21.965, p <0.001; jackdaws $r^2 = 0.425$, F = 42.847, p <0.001; magpies $r^2 = 0.093$, F = 5.979, p = 0.018).

The stronger relations (black headed gulls and rooks/jackdaws) was probably due to the social nature of the latter and the smaller size of the gulls (compared with the larger lesser black backed gulls). Although there was a significant correlation in the foraging behaviour of the two gull species (16 aggressive incidents were noted, black backed gulls prevailing in all) this did not translate into significant conflict. The lesser black backed gull appeared a true landscape generalist, present in all landscape components and appeared to dominate several other species. The black headed gull was more selectively present, tending to move in larger numbers in more localised areas. The relation between the two was quite strong in the green spaces ($r = 0.764$, $r^2 = 0.583$, $F = 43.357$, $p < 0.001$), but weaker in the urban areas ($r^2 = 0.048$, $F = 2.900$, $p = 0.094$). This was due to the virtual absence of black headed gulls from the Glasgow City Centre, West End and East End (except in the immediate proximity of the River Clyde). Lesser black backed gulls comprised 98 percent of the gulls observed in the urban areas of Glasgow more distant from the Clyde (100 m +), including areas proximate to small water bodies.

Ghana Case Study

Pied crows (urban = 4.2 per point, forest = 2.8, savanna =1.4) were slightly commoner in urban areas than forest and savanna, as were hooded vultures (urban = 4, forest = 2.1, savanna = 1.6) and black kites (urban = 0.6, forest = 0.7, savanna = 0.3 per point). The significant relations between bird presence and landcover features are shown in Table 3. Hooded Vulture numbers had the strongest (but weak) relation with the proximity of inner city green spaces (hooded vultures 0.261, $r2 = 0.068$, $F = 7.900$, $p < 0.006$; black kite 0.085, $r2 = 0.007$, $F = 0.794$, $p > 0.375$; pied crow 0.139, $r2 = 0.019$, $F = 2.124$, $p > 0.148$). All three species in urban areas showed positive correlations with the density of proximate vegetation. Bird numbers sighted were higher when that vegetation was forest or dense forest-savanna mosaics, as opposed to open, shrub-grass dominated savanna (hooded vulture $r2 = 0.139$, $F = 17.470$, $p < 0.001$; black kite $r2 = 0.091$, $F = 10.756$, $p < 0.001$; pied crow 0.129, $r2 = 0.017$, $F = 1.820$, $p < 0.155$).

Table 3. Land-cover associations of Accra and Kumasi birds

Species	GSA	GSP	WPG	WPU	RVD	UVD	UVP	V	K	C	V	K	C
H. vulture	•	•	o	o	o	•	o	X			X		
B. kite	o	o	o	o	•	o	•	•	X		•	X	
P. crow	o	o	o	o	•	o	•	•	•	X	•	•	X

• = Significant. o = Insignificant. GSA = Green space area. GSP = Green space proximity. WPG = Water proximity in green space. WPU = Water proximity in urban area. RVD = rural vegetation density. UVD = urban vegetation density. UVP = vegetation proximity. V = vulture, K = kite, C = crow.

Passing automobiles and people generally disturbed vultures and crows within about two to three metres, and kites at about 3.5 to four metres. The figure showed that the size of green spaces (even when dominated by dense vegetation) had much less effect on avian distributions than in Glasgow or Ottawa. Possible factors for this included: (1) the higher amount of refuse on urban areas as opposed to green areas (compared with the Glasgow and Ottawa); (2) green areas had virtually no leisure visitors who left food scraps; and (3) as green spaces were rarely weeded or cleared, any dropped food scraps may have been less visible in Ghanaian locations. River proximity had no effect on bird presence in urban areas (vultures $r^2 = 0.001$, $F = 0.115$, $p < 0.735$; kites $r^2 = 0.0001$, $F = 0.0178$, $p < 0.894$; crows $r^2 = 0.014$, $F = 1.562$, $p < 0.214$)

The strongest explanation for variance in bird numbers in urban areas was food deposit variation: abattoirs, rubbish dumps, markets, business districts and residential areas. At the 25 abattoirs monitored, Hooded Vultures were the most common species in such areas (on average 11.1 birds per point, 5.4 perching), compared with Black Kites (0.9) and Pied Crows (4.7). At the twenty-three rubbish dumps surveyed, Hooded Vultures were again the commonest species (4.1 with 1.5 perching, Black Kites 0.7 and Pied Crows 3.2). At the twenty-one markets surveyed, there were an average of 5.5 vultures (1.3 perching), 0.4 kites and 3.2 crows. Fewer birds were recorded in the twenty-two business districts (1.0 vultures, none perching, 0.3 kites, 1.9 crows) and the nineteen residential areas (0.7 vultures with 0.1 perching, 0.4 kites, 2.4 crows).

In the built-up areas, crows exhibited greater agility in accessing scraps from more precarious areas. These included windowsills, balconies, gardens, roads and sidewalks, and rubbish dumps during dumping. They also appeared to have a wider diet; fruits and vegetation, as well as the meat favoured by other scavengers, this possibly explaining their higher presence in urban areas away from the high refuse areas. The larger amounts of refuse explained their slightly higher density compared with Glasgow. However, other factors might be relevant, as the density was like those of the American crow in Ottawa, a much cleaner city.

Ranking the urban areas according to organic matter available (abattoirs = 5, rubbish dumps = 4, markets = 3, business districts = 2, residential areas = 1) and regressing this ranking with bird numbers produced the highest figures for vulture presence ($r^2 = 0.313$, $F = 49.245$, $p < 0.001$) and vulture perching ($r^2 = 0.231$, $F = 32.444$, $p < 0.001$), with lower correlations with Pied Crows ($r^2 = 0.141$, $F = 17.813$, $p < 0.001$) and Black Kites ($r^2 = 0.063$, $F = 7.254$, $p < 0.008$). Where the three species were sharing feeding sites, some conflicts were observed. Approximately twenty percent of the encounters between Hooded Vultures and Pied Crows resulted in conflicts (vultures won 16 percent, crows 2 percent and there were 2 percent draws) as did six percent of the encounters between vultures and Black Kites (all draws). Of the 13 interactions recorded between feral dogs and birds in urban settings, six resulted in conflicts, with dogs prevailing in all. These resulted in winning scraps, but not leaving behaviour.

However, there were only a few records of vultures nesting in urban areas, all of these being in green spaces (urban green spaces 11 percent, proximate peri-urban or rural forests/woodlots 18 percent, and distant (10 km +) forests/savanna wood stands 71 percent). Vulture presence was moderately correlated with green space area ($r^2 = 0.048$, $F = 5.416$, $p < 0.022$) and the density of the treed growths ($r^2 = 0.048$, $F = 5.378$, $p < 0.022$). Sixty eight percent of the non-nesting vultures seen roosting at dusk or dawn were in proximate forests/wooded savanna, 29 percent in distant forests, and three percent were in urban built-up areas (in single trees and roof tops).

All the roosting Black Kites were in distant forests, Pied Crows nested in proximate forest savanna (72 percent) and distant forest (28 percent). Where Pied Crows roosted close to vultures, conflicts were observed. About 23 percent of these resulted in the expulsion of vultures from roosts. In urban roosting areas and well-lit markets and dumps, some vultures and crows remained active in low flight and feeding after dark (7 pm +). There were vultures in flight at dawn from rural roosting areas to urban sites, usually ten to twenty minutes before or after sunrise, around 6 am at low altitudes of five to 15 metres. About 46 percent of these were in the direction of known abattoirs, rubbish dumps and markets. Similar behaviour was observed in crows (about 30 percent), although this was less directional, and occasionally in kites (about six percent of the sightings).

In the rural areas, there was a moderate significant relation between vegetation density and bird numbers, although the figure for vultures was lower than in urban areas near to forest (vultures $r^2 = 0.011$, $F = 1.219$, $p > 0.272$; kites $r^2 = 0.051$, $F = 5.823$, $p < 0.017$; crows $r^2 = 0.0.048$, $F = 5.448$, $p < 0.021$). Numbers of all species increased in rural areas with the proximity of towns (vultures $r^2 = 0.142$, $F = 18.059$, $p < 0.001$; kites $r^2 = 0.033$, $F = 3.671$, $p > 0.058$; crows $r^2 = 0.070$, $F = 8.148$, $p < 0.005$), increasing perching and feeding behaviour, as towns frequently had waste areas and poorer sanitation in peri-urban areas. This was positively correlated with the size of the proximate towns (vultures $r^2 = 0.166$, $F = 21.686$, $p < 0.001$; kites $r^2 = 0.040$, $F = 4.539$, $p < 0.035$; crows $r^2 = 0.044$, $F = 5.066$, $p < 0.026$). River proximity was largely unimportant, but slightly stronger than in the urban spaces (vultures $r^2 = 0.006$, $F = 0.689$, $p < 0.408$; kites $r^2 = 0.022$, $F = 2.426$, $p < 0.122$; crows $r^2 = 0.028$, $F = 3.193$, $p < 0.077$).

In nearby rural areas, vulture perching was more strongly correlated with the presence of other species (vulture/black kite $r^2 = 0.222$, $F = 31.038$, $p < 0.001$; vulture/pied crow $r^2 = 0.239$, $F = 34.195$, $p < 0.001$; black kite/pied crow $r^2 = 0.321$, $F = 51.471$, $p < 0.001$) than in the urban areas, but both were significant. Also, rural areas, 26 percent of the encounters between Hooded Vultures and Pied Crows resulted in conflict, as did eleven percent of the encounters between vultures and Black Kites. Of the eight interactions recorded between feral dogs and vultures in rural settings five resulted in conflicts, with the vultures prevailing in two and the dogs in the rest, with one resulting in the vultures leaving altogether.

Bushfires, termite flights (mostly *Macrotermes* spp.), bush clearance and occasional road kills all attracted larger numbers of crows than vultures or kites. In bushfire areas (started by people for land clearance) the figures were greatest for black kites (3.1 per fire, pied crows 1.8, hooded vultures 1.7), as were termite flights (2.6, Pied Crows 1.2, vultures 0.2). For bush clearance (with machetes and/or tractors) pied crows had the highest presence (5.3, kites 1.5, vultures 0.4). The eight road kills recorded in forest (four maxwell's duikers *Cephalophus maxwellii* H. Smith, 1827, a black duiker *Cephalophus niger* Gray 1846, two cane rats *Thryonomys swinderianus* Temminck 1827 and a palm civet *Nandina binotata* Gray 1830) were attended by 16 pied crows, eight hooded vultures and two black kites. There were two conflicts between the vultures and crows, one an airborne pursuit of a vulture and another a brief ground fight, where the vulture prevailed. In savanna areas the figures were similar, but pied crows were commoner at fires (7.3, kites 0.7, vultures 0.4), termite flights (5.2, kites 0.2, vultures 0.2) and bush clearance (4.1, kites 0.1, vultures 0.1). The six road kills recorded in the coastal savanna (three crowned duikers *Sylvicapra grimmia* Linnaeus 1758 and three cane rats) were attended by twelve crows, four vultures and one kite. There was only one conflict, which saw a vulture drive a crow from the ground to a treed perch.

Ottawa Case Study

In the Ottawa area, turkey vultures averaged 0.34 per point in green spaces and 0.54 in urban areas, like the blue jay (0.56 and 0.46 respectively) Crows were commoner (1.71 and 2.32 respectively), as were ring billed gulls (1.33 and 1.94 respectively). The blue jays were the least mobile, as 66 percent of those observed perching and feeding remained in the vicinity for at least 20 minutes (turkey vultures 11 minutes, crows 6 minutes, gulls 17 minutes average). The significant relations between bird presence and landcover features are shown in Table 4.

Table 4. Land-cover associations of Ottawa birds

Species	GSA	GSP	WPG	WPU	V	C	G	J	V	C	G	J
T. vulture	•	•	•	•	X				X			
Crow	•	•	•	•	•	X			•	X		
R.B. Gull	•	•	o	•	•	•	X		•	•	X	
Blue jay	•	•	o	•	•	•	•	X	•	•	•	X

• = Significant. o = Insignificant. GSA = Green space area. GSP = Green space proximity. WPG = Water proximity in Green Space. WPU = Water proximity in urban area.

The area of the green spaces significantly explained the variance in turkey vulture ($r^2 = 0.349$, $F = 41.734$, $p < 0.001$), crow ($r^2 = 0.441$, $F = 61.516$, $p < 0.001$), blue jay (r^2 0.317, $F = 36.269$, $p < 0.001$) and gull presence ($r^2 = 0.475$, $F = 70.550$, $p < 0.001$), levels similar to the corvids in Glasgow (higher than the gulls). Like Glasgow, large green spaces were more visited by people and this resulted in larger amounts of organic waste. Green space proximity in urban areas was also quite important, more so than for the Glasgow study, for turkey vultures ($r^2 = 0.129$, $F = 7.106$, $p = 0.010$), crows ($r^2 = 0.4226$, $F = 35.135$, $p < 0.001$), blue jays (r^2 0.360, $F = 27.114$, $p < 0.001$) and gulls (r^2 0.399, $F = 31.932$, $p < 0.001$). In terms of nesting, turkey vultures were observed nesting outside the study area (on cliffs and in a few areas tree branches), with no observed nests on the greenbelt or green space areas. Crows and blue jays nested in dense vegetation in the greenbelt and larger green space areas. Green space area had a much stronger impact on gulls than in

Glasgow, despite similar levels of human visitation (61 per point in larger green spaces).

In the green spaces, river proximity was significant for Turkey vultures ($r^2 = 0.090$, $F = 7.747$, $p = 0.007$) and crows ($r^2 = 0.114$, $F = 10.008$, $p = 0.002$), but not for the Blue jay ($r^2 = 0.004$, $F = 0.005$, $p = 0.099$) and Ring-billed gull ($r^2 = 0.040$, $F = 3.298$, $p = 0.073$). The position of gulls in this sense contrasted with Glasgow. In the urban areas, river proximity was significant for turkeys vultures (r^2 0.161, $F = 9.214$, $p = 0.004$), crows ($r^2 = 0.407$, $F = 32.990$, $p <0.001$), blue jays ($r^2 = 0.0.296$, $F = 20.191$, $p <0.001$) and gulls ($r^2 = 0.410$, $F = 33.436$, $p <0.001$), these being comparatively high (crows, gulls) compared with Glasgow. Human deposited waste was important in these areas (about 47 percent of refuse was along the rivers, and when the river levels fell during the drier summer weeks, the banks were foraged for insects and other in vertebrates, and deposited wastes.

There was a significant relation between the foraging activities of the four species. In the green spaces the significance of the relation between vultures and gulls was the highest ($r^2 = 0.401$, $F = 52.480$, $p <0.001$) followed by vultures and crows ($r^2 = 0.280$, $F = 30.434$, $p <0.001$) and lower between vultures and blue jays ($r^2 = 0.093$, $F = 8.074$, $p = 0.006$). Crows and gulls had a strong relationship ($r^2 = 0.508$, $F = 80.152$, $p <0.001$), followed by the crows and blue jays ($r^2 = 0.209$, $F = 20.598$, $p < 0.001$). Gulls and blue jays had a significant relation ($r^2 = 0.229$, $F = 23.126$, $p < 0.001$). The low correlation between blue jays and other species concerned the former to forage in undergrowth and denser vegetation (88 percent of sightings), compared with other species which tended to feed in open spaces (79 percent of sightings).

In the urban areas, the shared presence of crows and turkey vultures was significant ($r^2 = 0.166$, $F = 9.568$, $p = 0.003$), as was that between turkey vultures and blue jays ($r^2 = 0.078$, $F = 3.910$, $p = 0.050$) and that between turkey vultures and gulls (r^2 0.024, $F = 15.186$, $p <0.001$). Blue jays and gulls were also related ($r^2 = 0.248$, $F = 15.802$, $p <0.001$). This was possibly because crows and vultures frequently shared more similar food sources, such as rotting meat, while gulls favoured scraps and jays preferred insects and seeds. Crows also had a strong relation with gulls ($r^2 = 0.486$, $F = 45.296$, $p <0.001$) and with blue jays ($r^2 = 0.310$,$F = 21.542$, $p <0.001$), for the same reasons as in the green spaces. They accessed similar food sources (open rubbish bins, organic matter on riverbanks and invertebrates in grass) which might escape the notice of vultures.

Discussion and Comparison of Cases

Green Space and Water Body Impacts

In all the case studies, the green spaces were variably important factors for the presence of avian scavengers. However, there were distinctions between the species and contexts. Although green space development has been cited as an important factor for avian urbanisation (Vuorisalo et al. 2003) the urban substrate may also serve as an attractant where food is based on human activities ((Tilghman, 1987; Clergeau, 1998; Campbell, 2015). Although in some cases the number of people providing food (especially in Glasgow and in some cases, Ottawa) was a factor, in several cases there was evidence that natural foods such as insects, seeds and worms, and butchers discards in the case were also consumed. The literature gives virtually identical roles for the Pied, Carrion and American crows, and the rook and jackdaw (Alden et al. 1995; Stevenson & Fanshawe, 2002; Hume, 2004) and this was generally supported in this study. The three gull species showed similar behaviour, while the vultures varied slightly.

The focus on more than one urban area, each of different size may have assisted in the investigation of this position. The scaling of both green space distances, urban structures, city size and context, revealed variations what might otherwise have been less noticeable. It also investigated key points not fully examined in the literature: the impact of green space isolation for species all of which were rated as urban generalists (Vurisalo et al. 2003) and a comparison of the urban substrate with the green areas in bird presence. These were seen not only as substrate and corridors, but as valid foraging areas, which might be even more so in small towns closer to rural, green areas (Campbell, 2015).

Water bodies were significant attractants for some species, but this also varied according to the city and whether the points were in green spaces or urban locations. This is a neglected issue, despite the studies which argue for water bodies as avian attractants (Hall 1999; Donelly & Marzluff, 2004). The relative strength of water bodies, relative to green spaces as attractants has also been ignored. In this study water bodies were less important as avian attractants, but there were pronounced variations. Crows and magpies in Glasgow green spaces favoured areas near water bodies, contrasting with the situation in urban spaces. In Ottawa water bodies appeared unimportant in green spaces but were more important in urban spaces, especially for gulls and jays. All the observed species took all forms of organic waste (meat, fish, bread, cake, seeds). Factors that were examined were both of local specific nature (e.g., the presence of organic waste on riverbanks and the presence of human proffered food) and more general (all areas recorded natural food such as insects for corvids) (Campbell, 2015).

Multi-species studies of avian distribution tend to include species that are arguably generalists and specialists (Batten, 1972; Turner, 2003). Few studies examine the relations between closely related species, which may be regarded as generalists in variable contexts (Campbell, 2015). In this case the species had marked similarities, and some differences. The degree of generalist behaviour in such closely related generalists was thus difficult to discern. The configuration factors influenced different associations between the species in foraging activities. The more social rocks, jackdaws and pied crows and the generally solitary blue jays and magpies, and carrion crows and showed that although the results concurred with the literature describing corvids as adaptable generalists (Thom 1986; Sterry et al. 2002; RSPB 2017), the differences in corvid behaviour between the two cities were sufficient to raise questions for a greater contextual awareness in such studies.

Urban Spaces

Urban studies are strong ecological attractants (Werner, 2011), a point demonstrated in the current research. Gulls and corvids have recently expanded their foraging activities into the large cities and smaller towns in Western Europe and North America (Hume, 2004; Sterry et al. 2004) and in some cases they perform the documented roles of vultures in Africa and the Americas (Alden et al. 1995; Stevenson & Fanshawe, 2002). However, although this increase has been documented in the literature, few studies have been conducted of two or more species in more than one case study, hence there is little work to compare with the current study.

In this study, it was shown that landscape features (e.g., green spaces) had a higher impact on turkey vultures than hooded vultures, which were more attracted to urban areas possibly due to the less sanitary conditions in Ghana compared with Ottawa. There has also been insufficient documentation of the impacts of landscape configuration, and human actions. There were significant differences in behaviour of the gull species in the two case studies. Whilst river proximity was a stronger presence predictor of Black headed gulls than the more generalist Lesser black backed gulls in Glasgow, it was a stronger predictor for ring billed gull presence in Ottawa. There was little evidence that this was due to competition from other species, thus pointing to other behavioural factors. Such comparative studies, on urban gull behaviour are recommended for a comparative assessment of gull's urban ecology, which may also be useful for informed planning applications, where gulls and other birds may be either pests or wildlife.

These results for the gulls documented strong similarities and differences between the foraging habits of the three gull species. The study of their interactions showed less evidence of conflict than has been documented in studies of other gull species (e.g., Burger 1981, on the large Herring gulls (*Larus argentatus* Pontoppidan, 1763) and the smaller laughing gulls (*Larus atricilla* Linnaeus, 1758), which are similar in size to the species covered in this paper). Although the Lesser black backed gull has been described as a more aggressive bird than the similarly sized Herring gull (Witherby et al.1941) the low numbers of conflicts observed may be traced to the low numbers of other gull species in Glasgow.

The results for both Ottawa and Glasgow support the general position of the published literature that urban streets and associated waste are prime attractants of gulls (Owen, 2003; Rock, 2002, 2003a,b; Western, 2004; Campbell, 2008), the corvids and vultures (Alden et al. 1995; Stevenson & Fanshaw, 2002; Sterry et al. 2004; Campbell, 2015). However, the findings of the current study differ from some literature sources. The Blue jay, magpie and Turkey vulture were less attracted to the urban substrate, despite similar feeding habitats to some of the other studied species. Urban spaces are frequently described as important for many wild species (Dicerbo, 1991;

Belant 1997; Campbell, 2009; RSPB 2017), were relatively insignificant for these species. However, in their landscape preferences, which varied between the two cities, the gulls showed both similarities and differences, which contrasts with the generalised descriptions in the literature on gulls (Mitchell et al. 2004). There are few studies which document the specific form of foraging behaviour of specific species, commoner issues being the shared general inter-specific similarities such as street colonisation, roof nesting and noise (Horton et al. 1983; Belant, 1997; Mcarthy, 2003; Owen, 2003; Rock, 2002, 2003a,b).

A 'large' gull (Rock 2003b: 1) the Lesser black backed gull is also frequently contrasted with the similarly sized Herring gull (Verbeek 1979; Noordhuis and Spans 1992; Calladine 1997; Francois 2002) rather than its relations with smaller gull species (Campbell, 2008). Rock 2003b) argues that important factors for gull population increase in urban areas are the high food availability, low numbers of predators and minimal ecological competition. In the current study, no large avian predator equivalent to the Black kite existed in the Glasgow and Ottawa studies, although Common buzzards *Buteo buteo* Linnaeus, 1758 and Red-tailed hawks (*Buteo jamaicensis*, J. F. Gmelin, 1788) respectively were sighted outside the study areas.

Most studies however, fail to distinguish how the different gull species react to these factors, with the single species, general genus or coastal focus prevailing (Hunt, 1972; Monaghan 1979; Mudge & Ferns, 1982; Blokpoel, 1983; Blokpoel and Tessier 1984; Patton, 1988; Pons, 1992; Belant & Dolbeer, 1993; Belant et al. 1995; Dwyer et al. 1996; Campbell, 2007, 2010). An important issue for the smaller gulls is whether competition from larger gulls and other common species such as corvids is restricts foraging. In the current study, the smaller Black headed gulls shared their entire foraging range with larger gulls and corvids and despite the low conflicts between gull species and virtually none with corvids (unlike the study by Burger 1981). Further research must therefore investigate the factors for inter-species variation in foraging, especially where there are differences in size, context, locations and species of gulls and associated species, for a fuller picture to emerge on the urban ecology of avian scavengers.

Although the Ghanaian case study was conducted in urban centres with different sanitation levels to those in Canada and Scotland, the results supported those of the other cases: (1) the primary factor for avian scavenger abundance was urban land configuration; (2) urban land configuration was slightly more important in this respect than rural land configuration; (3) these avian scavengers generally, but not entirely, preferred urban areas for feeding; (4) there was some competition between the main bird scavengers; and (5) despite these dynamics, all the recorded nests were in rural rather than urban areas. While some of these points have been briefly mentioned in the general, non-Ghanaian literature (Ferguson-Lees & Christie, 2001), several neglected details have been documented in this study.

The urban survey documented poorer sanitation in the markets, business areas and residential districts of Ghanaian towns than the UK and Canadian cities under study. Lower vehicular traffic also allowed longer periods for perching in small settlements than in the busier, larger towns, an issue noted in the larger green spaces.

The main factor favouring larger towns was the presence of large abattoirs and butcheries, which attracted the largest flocks of vultures, and to a lesser extent Pied crows and Black kites. Although these have been described as significant attractors, they have not been compared with other sites such as markets (Madge & Burn, 1994; Ferguson-Lees & Christie 2001; Obiri-Danso & Jones, 2002). Consequently, a comparison of such cities (in terms of avian scavengers) with butcheries with cleaner cities in the literature is difficult. Parallels could be coastal North American and European cities where dumping of offal might provide parallel environments with butcheries in African towns, which have been mooted in some studies (Burger 1981; Campbell, 2015).

The findings of the current study also produced evidence of different levels of interaction between the species. For example, there was a moderately aggressive relationship between Pied crows and Hooded vultures, and less conflict between Hooded vultures and Black kites. This is rarely mentioned in the literature, the focus instead being on the dominance of other vulture species over Hooded Vultures at feeding sites (Newton, 1979; Ferguson-Lees & Christie, 2001). Interactions included the mobbing of flying vultures by crows, conflicts over scraps on the ground (despite the superior size of the vultures, only marginal dominance was attained in such situations) and slight dietary differences. The greater agility of the Pied crows, the small, more agile scavengers, appeared to fulfil the role played by the Hooded vultures where larger vultures were common (Campbell, 2015) and the role of the jackdaw or rook in Glasgow. However, in the Ghanaian case there was no equivalent to the small and generally solitary magpies (Glasgow) and blue jays (Ottawa), which appear to be intermediate between the larger avian scavengers and the smaller passerine.

The results also show the problems with assessing the ecology of animals in human dominated landscapes. The landscapes in Ottawa and Glasgow were dominated by maintained lawns and planted trees, while those in Ghana included such areas, but also comprised untended areas. Also, there was evidence that the urban built-up areas, in some cases at least, provided more food than the green spaces or more 'natural' environments. This relates to the point of (Zimmerer 1994, 110), who concluded that the site histories of particular contexts are important, and also that of (Nicolis, 1994), who argued that chaotic change is intrinsic to the ecological context rather than necessarily due to external interference (Nicolis, 1994).

Here, would chaotic change include human modifications in the landscape? Therefore, the adaptability of the avian scavengers was also at issue. This also relevant to the case of ecosystem health, which as noted above, was defined by Karr and Dudley (1981: 21) as "the capability of supporting and maintaining a balanced, integrated, adaptive, community of organisms having species composition, diversity, and functional organization comparable to that of natural habitats of the region." In this case, there must be awareness of how or if, urban developments may contribute to the health of an ecosystems, and the extent to which there may be a difference in such an assessment for different species, a point not fully developed in the literature (Ludwig et al. 1993; De Leo & Levin 1997; Campbell, 2017).

Planning Issues and Conservation Biology

The presence of urban birds has been cited as a planning issue (Belant, 1997; Clergeau et al. 1998; Marzluff et al. 2001; Francois, 2002; Vuorisalo 2003) and also an ecological issue (Burger 1981; Hansen & Urban 1992; Kruger & Lindstrom 2001; Hostetler & Knowles Yanes 2003; Turner 2003; Lim & Sodhi 2004). This issue also relates to conservation biology. As this field is concerned as the study of the anthropogenic impacts on biodiversity (Soule, 1986), and the design of methods for species appraisal and protection, the role of urban complexes within rural landscapes is relevant. Conservation generally holds that human impacts may have negative impacts of animal ecology, examples where there may also be positive impacts (as was the case in some points of this study) must be examined. Tourism in green spaces is a powerful attractant, rather than a negative factor for avian ecology. In this sense, it is important to adaptive management. The landuse and human presence, and valuation of land may influence its attractive value for urban birds. Adaptive management, which has been described by several authors as a means of integrating the social decision-making process with the ecological situation on the ground may be flexible enough to deal with such matters (Walters 1997a,b; Callicott et al. 1999; Gunderson, 1999; Agrawal, 2000; Campbell, 2017).

However, the common approach with the detail of single species studies is rarely comparative of species and the planning context. This paper has documented how factors for species presence in one city may be of less importance in another city, raising the unpredictability of avian adaptation. In issues of biological conservation or pest control, this highlights the importance of the consideration of flexible adaptation in avian urbanisation. It also documents the flexible dynamics between human action and landscape configuration (human actions affected the presence of corvids in some areas, but in other similar areas, natural food was more important). Such adaptive dynamics of bird species have not been sufficiently examined in the context of urban planning.

CONCLUSION

This chapter has investigated the effects of the landscape configuration of cities in three countries on the behaviour of corvids, gulls and vultures. The hypotheses were that the landscape configuration had significant impact on the foraging behaviour of the avian scavengers in these cities, and that due to the varying adaptive behaviour between species and locations, a multiple city case study would supply important information. Although the results were supportive of these hypotheses, additional issues have been raised which may have important significance for urban avian ecology. An important limitation would be the socio-cultural, economic and ecological differences between the case studies, especially between Ghana on one side and Canada and Scotland on the other. However, by revealing variations in avian behaviour in distinct, yet structurally similar contexts, the paper contributes to the important argument that multi-city comparative studies are necessary for a fuller understanding of avian scavenger behaviour.

The expanding numbers and urban adaptability of avian scavengers make them exciting subjects for multi-species and multi-city studies, which are necessary for both the development of urban ecology within landscape ecology and for the constructive engagement of these sciences with urban planning. Some existing studies have also emphasised the importance of historical assessments of bird urbanisation and the value of single species studies in double cases. Therefore, more detailed studies will require attention to: (1) the relative strengths of urban configuration and human actions as avian attractors, if possible from a temporal perspective; (2) the comparison of these from intra- and inter-city perspectives combined with inter-species foci; (3) and the implications for urban planning showing awareness of the competing issues of green space development for avian habitats and the social problems raised by increased bird urbanisation, and differences between selected case studies.

There must also be detailed studies comparing urban and rural areas for scavengers, which has also been relatively neglected in the literature. This chapter has highlighted the merit of examining bird behaviour in both urban, semi-rural and rural contexts. The findings illustrated the point that these were not separate behavioural contexts, but there were strong inter-linkages with varying adaptive behaviour from the Hooded Vultures, Black Kites and Pied Crows. The scaling of such comparative analyses to wider contexts will exponentially increase the complexities. This would be necessary, due to the need for understand of the increased rates of urbanisation, human population increase and habitat modification, which may impact on bird scavengers and other raptors in the southern continents.

Concerning the larger theoretical issues of ecology, such studies test the ideas which integrate the study of human influenced environments with the so - called 'natural' environments. Although the relationship between landscape, urban planning and ecology has emerged as an important aspect of applied wildlife studies (e.g., in the journals of *Landscape and Urban Planning* and *Landscape Ecology*), there is a need for further study of how the human modifications of the environment may be integrated into studies of animal ecology, using comparative methodologies. The current study provided a basic approach, by including multiple cases, multiple species and similar parameters. Further steps would increase the case studies and number of species, and the development of more sophisticated research methods and environmental parameters.

References

Agrawal, A. 2000. Adaptive management in transboundary protected areas: the Bialowieza National Park and Biosphere Reserve as a case study. *Environmental Conservation*, 27 (4): 326-333. https://doi.org/10.1017/S0376892900000370.

Aikman, J. Neil, G. Cannons, A. & Milton, J. 1998. *So, Where Are the Stones and Where Are the Swamps Anyways? A Resource Assessment of Stony Swamp, Nepean, Ontario*. Unpublished Paper. Ottawa: Carleton University.

Alden, P.C., Estes, R.D., Schlitter, D. & McBride, B. 1995. *London: Colin's African Wildlife.* Harper Collins Publishers.

Andren, H. 1992. Corvid density and nest predation in relation to forest fragmentation – landscape perspective. *Ecology* 73: 794 – 804. https://doi.org/10.2307/1940158.

Andrew, N. L. & Pepperell. J. G. 1992. The bycatch of shrimp trawl fisheries. *Oceanography and Marine Biology. An Annual Review* 30: 527-565.

Archibald, W.R., Ellis, R. & Hamilton. A.N. 1987. Responses of grizzly bears to logging truck traffic in the Kimsquit River Valley, British Columbia. *International Conference on Bear Research and Management* 7:251-257.

Batten, L.A. 1972. Breeding bird species diversity in relation to increasing urbanisation. *Bird Study*, 19: 157 – 166. https://doi.org/10.1080/00063657209476337.

Baxter, E.V. & Rintoul, L.J. 1953. *The Birds of Scotland.* Vol 2. Edinburgh: Oliver and Boyd.

Bayne, E. & Hobson, K. 2000. Relative use of contiguous and fragmented boreal forest by red squirrels (*Tamiasciurus hudsonicus*). *Canadian Journal of Zoology*, 78: 359-365. https://doi.org/10.1139/z99-219.

Begon, M., Harper, J.L. & Townsend, C.R. 1996. *Ecology: Individuals, Populations and Communities.* (3rd Ed.), Oxford, UK: Blackwell Science Ltd.

Beissinger, S. R. 1990. On the limits and direction of conservation biology. *Bioscience* 40: 456-457.

Belant, J.L. 1997. Gull in urban environment: landscape – level management to reduce conflict. *Landscape and Urban Planning* 38: 245 – 258. https://doi.org/10.1016/S0169-2046(97)00037-6.

Belant, J.L., Seamans, T.W., Gabrey, S.W. & Dolbeer, R.A. 1995. Abundance of gulls and other birds at landfills in Northern Ohio. *American Midland Naturalist*, 134: 30 – 40. https://digitalcommons.unl.edu/icwdm_usdanwrc/136.

Belant, J.L. & Dolbeer, R.A. 1993. Population status of nesting laughing gulls in the United States 1977 – 1991. *American Birds* 47: 220 – 224. Retrieved from https://sora.unm.edu/sites/default/ files/journals/nab/v047n02/p00220-p00224.pdf

Bengston, D.N., Fan, D.P & Celarier, D.N. 1999. A new approach to monitoring the social environment for natural resource management and policy: The case of us national forest benefits and values. *Journal of Environmental Management* 56 (3): 181-193. https://naldc.nal.usda.gov/download/35816/PDF.

Bensen, P.C.1999a. An evaluation of some techniques and problems associated with marking of Accipitrid vultures. In Boschoff, A.F., Anderson, M.D. & Borello, W. (Eds.) *Proc. Vultures in the 21st Century – Vulture Research and Conservation in Southern Africa.* 148-150. Kimberley: Northern Cape Nature Conservation Service.

Benson, P.C. 1996b. Status of vultures in the Northern Province South Africa. In Boschoff, A.F., Anderson, M.D. & Borello, W. (Eds.) *Proc. Vultures in the 21st*

Century – Vulture Research and Conservation in Southern Africa. Kimberley: Northern Cape Nature Conservation Service.

Bezanson, T. 2001. *Ottawa Chapter Develops Habitat Demonstration Project*. Ruffed Grouse Society of Canada and Ministry of Natural Resources, Ottawa.

Biggs, D. 2001. Observations of eagle congregations at a red billed quelea colony in the Kruger National Park. *Bird Numbers*, 10 (1): 7.

Blair, R.B. 1996. Landuse and avian species diversity along an urban gradient. *Ecological Applications* 6: 506 – 519. Retrieved from https://www.esf.edu/cue/documents/Blair_LandUse-AvianSpecDiv-UrbGradient_1996_000.pdf.

Blokpoel, H. 1983. *Gull Problems in Ontario*. Ottawa: Canadian Wildlife Service Information Leaflet.

Blokpoel, H. & Tessier, G.D. 1986. The ring billed gull in Ontario: a review of a new problem species. *Canadian Wildlife Service Occasional Paper* 57: 34.

Bolen, E.G. & Robinson, W.L. 1999. *Wildlife Ecology and Management*. Upper Saddle River: Prentice Hall.

Borcard, D., Legendre, P. & Drapeau, P. 1992. Partialling out the spatial component of ecological variation. *Ecology*, 73: 1045-1055. https://doi.org/10.2307/1940179.

Bowman, M. 1996. *Trip Report: Ghana, November 18 - December 6, 1996*.

Broad, R. 1994. The poor and the environment: Friends or foes? *World Development*, 22 (6): 881-893. Retrieved from https://www.american.edu/sis/faculty/ upload/the-poor-and-the-environment_broad.pdf.

Brunton, D.F. 1988. *Nature and Natural Areas in Canada's Capital: Introductory Guide for the Ottawa-Hull Area*. Ottawa: The Ottawa Citizen and the Ottawa Field-Naturalists Club.

Brunton, D.F. 1997. *Planning our Region: Summary Natural Area Reports for Natural Areas West of the Rideau River (500 series)*. Regional Municipality of Ottawa-Carleton, Ottawa.

Brown, C. 2000. *Deer Collisions*. Ottawa: City of Ottawa.

Brownell, V.R. & Larson, M.H.B. 1995. *Planning our Region: An Evaluation Framework for Natural Areas in the Regional Municipality of Ottawa-Carleton Vol.1*. Ottawa: Region Municipality of Ottawa-Carleton.

Brunner, R.D. & Clark T.W. 1997. A practice-based approach to ecosystem management. *Conservation Biology*, 11: 48-58. https://www.jstor.org/stable/2387275.

Burger, J. 1981. Feeding competition between laughing gulls and herring gulls at a sanitary landfill. *Condor* 83: 328-335. Retrieved from https://sora.unm.edu/sites/default/files/journals/condor/v083n04/p0328-p0335.pdf

Calladine, J. 1997. A comparison of herring gull *Larus Argentatus* and lesser black backed gull *Larus fuscus* nest sites: Their characteristics and relationships with breeding success. *Bird Study* 44 (3): 318 – 326. https://doi.org/10.1080/00063659709461067.

Callicott, J.B., Crowder, L.B. & Mumford, K. 1999. Current normative concepts in conservation. *Conservation Biology*, 13:22-35. Retrieved from http://www.pelagicos.net/MARS6400/ readings/Callicott_et_al_1999.pdf.

Campbell, M. 1998. *Interactions Between Biogeography and Rural Livelihoods in the Coastal Savanna of Ghana.* Unpublished PhD Thesis, (Imperial College, University of London.

Campbell, M. 2002. The Assessment of Biogeographical Change in the Western Greenbelt Of Ottawa. M.A. Thesis. Carleton University, Ottawa, Canada

Campbell, M. 2007. An animal geography of avian ecology in Glasgow. *Applied Geography* 27(2):78-88. https://doi.org/10.1016/j.apgeog. 2006.12.001

Campbell, M. 2008. The impact of vegetation, river, and urban features on waterbird ecology in Glasgow, Scotland. *Journal of Coastal Research*, 24(4): 239-245. http://www.doi: 10.2112/07-0835.1

Campbell, M. 2009a. The impact of habitat characteristics on bird presence and the implications for wildlife management in the environs of Ottawa, Canada. Urban Forestry and Urban Greening 8: 87–95. https://doi.org/10.1016/j.ufug.2008. 12.003.

Campbell, M. 2009b. Factors for the presence of avian scavengers in Accra and Kumasi, Ghana. *Area*, 41(3): 341-349. https://doi. org/10.1111/j.1475-4762.2008.00870.x.

Campbell, M. 2010. An animal geography of avian foraging competition on the Sussex coast of England. *Journal of Coastal Research*, 26(1): 44 – 52. https://doi.org/10.2112/08-1087.1

Campbell, M. 2013a. An animal geography of the dominant urban avian scavenger in contrasting case studies. In Ruiz, L. and Iglesias, F. (eds.) *Birds: Evolution and Behavior, Breeding Strategies, Migration and Spread of Disease*, 67 - 82. Nova Science Publishers, New York.

Campbell, M. 2013d. Biogeographical change and environmental policy in the Western Greenbelt of Ottawa, Canada. In Creighton, E. and Danovich, P. (eds.) *Environmental Policy: Management, Legal Issues and Health Aspects*, 147 - 164. New York, NY: Nova Science Publishers.

Campbell, M. 2013c. The Impact of landscape configuration and competitors on the Hooded Vulture Necrosyrtes monarchus Temminck 1823 in Southern Ghana. In Ruiz, L. & Iglesias, F. (Eds.) *Birds: Evolution and Behavior, Breeding Strategies, Migration and Spread of Disease*, 157 - 166. New York, NY: Nova Science Publishers.

Campbell, M. 2013b. Issues of importance in waterbird ecology, conservation and environmental policy. In Creighton, E. & Danovich, P. (Eds.) *Environmental Policy: Management, Legal Issues and Health Aspects*, 129 - 146. New York, NY: Nova Science Publishers.

Campbell, M. 2014. Advances in zoogeography: micro-spaces of avian foraging and human presence. In Jenkins, O.P. (ed.) *Advances in Zoology Research Vol. 6*, 51 – 64. New York, NY: Nova Science Publishers.

Campbell, M. 2015. *Vultures: Their Evolution, Ecology and Conservation*. New York, NY: Taylor and Francis, CRC Press.

Campbell, M. 2017. *Biological Conservation for the 21st Century: A Conservation Biology of Large Wildlife*. New York, NY: Nova Science Publishers.

Campbell, M. 2018. *Geomatics and Conservation Biology*. New York, NY: Nova Science Publishers.

Carr, L. & Koh, S. 2001. *Forest Stand Evaluations in the City of Ottawa; Greenbelt, South March and Marlborough Forests*. Ottawa: National Capital Commission.

Central Intelligence Agency 2019. The World Factbook; Africa, Ghana. Retrieved from https://www.cia.gov/library/publications/the-world-factbook/geos/gh.html.

Chace, J.F., Walsh, J.J., Cruz, A., Prather, J.W. & Swanson, H.M. 2003. Spatial and temporal activity patterns of the brood parasitic brown headed cowbird at an urban/wildland interface. *Landscape and Urban Planning*, 64: 179 – 190. https://doi.org/10.1016/S0169-2046(02)002 20-7.

Chabigue, T. & Horvath, D. 2001. *Beaver Map of the Greenbelt*. Ottawa: NCC.

City of Ottawa 1991. *Vegetation Cover Classification.* Ottawa: City of Ottawa.

City of Ottawa 2003. *Report on the Ottawa Road Safety*. Ottawa: City of Ottawa.

Clergeau, P., Savard, J-.P.L., Mennechez, G., Sauvage, A. & Falardeau, G. 1998. Bird abundance and diversity along an urban –rural gradient: A comparative study between two cities on different continents. *Condo*r, 100: 413 – 425. Retrieved from https://sora.unm.edu/sites/default/files/journals/condor/v100n03/p0413-p0425.pdf.

Cooper, D.S. 2002. Geographic associations of breeding bird distribution in an urban open space. *Biological Conservation* 104, 205 – 210. https://doi.org/10.1016/S0006-3207(01)00166-5/

Corkum, C.V., Fisher, J.T. & Boutin, S. 1999. Investigating influences of landscape structure on small mammal abundance in Alberta's boreal mixed-wood forest. In Veeman, T.S., Smith, D.W., Purdy, B.G., Salkie, F.J. & Larkin, G.A. (Eds.) *Proceedings of the 1999 Conference of The Sustainable Forest Management Network*. Edmonton.

Davis, J.W.F. & Dunn, E.K. 1976. Intra-specific predation and colonial breeding in lesser black backed gulls *Larus Fuscus*. *Ibis*, 118: 65-77. https://doi.org/10.1111/j.1474-919X.1976. tb02011.x.

De Angelis, D.L. & Waterhouse, J.C. 1981. Equilibrium and Nonequilibrium Concepts in Ecological Models. *Ecological Monographs*, 57: 1-21. https://doi.org/10.2307/1942636.

De Leo, G. A. & Levin. S. 1997. The multifaceted aspects of ecosystem integrity. *Conservation Ecology* [Online]1(1): 3. Retrieved from http://www.consecol.org/Vol1/Iss1/Art3

Demographia 2003. *Glasgow: Population and Density 1891-2001*. Glasgow.

Dicerbo, E. 1991. *The Birds of Scotland*. Moffat: Lohar Publishing.

Dickson, K.B. & Benneh, G. 1995. *A New Geography of Ghana.* London: Longman.

Diduck, A. 1999. Critical education in resource and environmental management: Learning and empowerment for a sustainable future. *Journal of Environmental Management*, 57 (2): 85-97. https://doi.org/10.1006/jema.1999.0289.

Donelly, R. & Marzluff, J.M. 2004. Importance of context to urban bird conservation. *Conservation Biology*, 18 (3): 733 – 745. https://doi.org/10.1111/j.1523-1739.2004.00032.x.

Dwyer, C.P., Belant, J.L. & Dolbeer, R.A. 1996. Distribution and abundance of roof nesting gulls in the Great Lakes Regions of The United States. *Ohio Journal of Science*, 96: 9 – 12. https://digitalcommons.unl.edu/cgi/viewcontent.cgi?article=1124&context=icwdm_usdanwr.

Eden, S.F.C. 1985. The comparative breeding biology of *magpies Pica pica* in urban and rural habitat (*Aves: Covidae*) *Journal of Zoology*, 205: 325 – 334. https://doi.org/10.1111/j.1469-7998.1985.tb05620.x.

English Nature, 2003. *English Nature*. London.

Erickson, D.L. 2004. The relationship of historic city form and contemporary greenway implementation: a comparison of Milwaukee, Wisconsin (USA) and Ottawa, Ontario (Canada). *Landscape and Urban Planning*, 68(2–3):199–221. https://www.doi.org/10.1016/ S0169-2046(03)00160-9.

Ferguson-Lees, J. & Christie, D.A. 2001. *Raptors of the World*. London, UK: Christopher Helm.

Francois, R. 2002. Numbers and behaviour of roof – nesting herring gulls *Larus argentatus* and lesser black backed gulls *Larus fuscus* in Belgium. *Natuur. Oriolus*, 68 (3): 123 – 126.

Galvan I. 2003. Intra-specific kleptoparasitism in Lesser black backed gulls wintering inland in Spain *BioOne*, 26(3): 325 – 330. https://doi.org/10.1675/1524-4695(2003)026[0325:IKILBG] 2.0.CO;2/.

Gardner, R.H., Turner, M.G. O'Neill, R.V. & Lavorel. S. 1992. Simulation of the scale-dependent effects of landscape boundaries on species persistence and dispersal. In Holland, M.M., Risser, P.G. & Naiman, R. J. (Eds.), *The Role of Landscape Boundaries in The Management and Restoration of Changing Environments,* 76-89. New York, NY: Chapman and Hall.

Gibbons, D.W., Reid, J.B. & Chapman, R.A. 1993. *The New Atlas of Breeding Birds in Britain and Ireland 1988 – 1991*. London: T & A.D. Poyser.

Gibeau, M.L. 2000. *A Conservation Biology Approach to Management of Grizzly Bears in Banff National Park, Alberta*. Unpublished PhD Thesis, University of Calgary, Calgary.

Glasgow City Council Area Profile. 2019. National Archives of Scotland. *Mid-2018 Population Estimates by Council Area in Scotland.* Retrieved from https://www.nrscotland.gov.uk/files/statistics/council-area-data-sheets/glasgow-city-council-profile.html

Goddard, M.A., Ikin, K. & Lerman, S.B. 2017. Ecological and social factors determining the diversity of birds in residential yards and gardens. In Murgui, E. & Hedblom, M. (Eds.), *Ecology and Conservation of Birds in Urban Environments*, 371-397. Cham, Switzerland: Springer International Publishing AG. https://doi.org/10.1007/978-3-319.

Green, D.M. & Baker, M.G. 2003. Urbanisation impacts on habitat and bird communities in a Sonoran desert ecosystem. *Landscape and Urban Planning*, 63: 225 – 239. https://doi.org/10. 1016/S0169-2046(02)00195-0.

Grist, N.R. 2000. Urban magpies and crows. *The Glasgow Naturalist*, 23 (5): 60.

Gunderson, L. 1999. Resilience, flexibility and adaptive management antidotes for spurious certitude? *Conservation Ecology* 3 (1): 7 [Online] Retrieved from http://www.consecol.org/vol3/iss1/art7.

Gustafson, E. J. & Gardner, R.H. 1996. The effect of landscape heterogeneity on the probability of patch colonization. *Ecology*, 77 (1): 94-107. https://doi.org/10.2307/2265659.

Hagan, J.M., Vander Haegen, W.M. & Mckinley. P.S. 1996. The early development of forest fragmentation effects on birds. *Conservation Biology*, 10: 188-202. https://doi.org/10.1046/j.1523-1739.1996.10010188.x.

Hall, T.C. 1994. *Magpies.* University of Nebraska, Lincoln.

Hansen, A.J. & Urban, D.L. 1992. Avian response to landscape pattern: The role of species' life histories. *Landscape Ecology*, 7 (3): 163 – 180. https://doi.org/10.1007/BF00133308.

Harris, M.P. 1970. Rates and causes of increases of some british gulls' populations. *Bird Study*, 17: 325 – 335. https://doi.org/10.1080/00063657009476275.

Henein, K., & Merriam. G. 1990. The elements of connectivity where corridor quality is variable. *Landscape Ecology*, 4: 157-170. https://doi.org/10.1007/BF00132858.

Hess, G. 1994. Pattern and error in landscape ecology. *Landscape Ecology*, 9 (10) 3-5. https://doi.org/10.1007/BF00135074.

Hilborn, R. 1987. Living with uncertainty in resource-based systems. *North American Journal of Fisheries Management* 7: 1- 5. https://doi.org/10.1577/1548-8659(1987)7<1:LWUIRM>2.0. CO;2.

Holloway, S.C. 1996. *The Historical Atlas of Breeding Birds in Britain and Ireland 1875 – 1900.* London: T & A.D. Poyser.

Horton, N., Brough, T. & Rochard, J.B.A. 1983. The Importance of Refuse Tips to Gulls Wintering in an Inland Area of South-East England. Journal of Applied Ecology, 20(3): 751-765. https://www.jstor.org/stable/2403124/

Hostetler, M. & Knowles-Yanez, K. 2003. Land use, scale and bird distributions in the Phoenix Metropolitan Area. *Landscape and Urban Planning* 62: 55 - 68. https://doi.org/10.1016/ S0169-2046(02)00096-8.

Hough, M. 1984. *City Form and Process.* London, UK: Routledge.

Hough, J. 1991. Social impact assessment: its role in protected area planning and management. In West, P. & Brechin, S. (Eds.), *Resident Peoples and National Parks: Social Dilemmas and Strategies in International Conservation*, 274-283. Tucson, AZ: University of Arizona Press.

Houston, D. 2001. *Vultures and Condors.* Granton-On-Spy: Colin Baxter.

Hume, R. 2004. *Complete Birds of Britain and Europe.* London: Dorling Kindersley Ltd.

Hunt, G.L. Jr. 1972. Influence of food distribution and human distribution and human disturbance on reproductive success of herring gulls. *Ecology*, 53: 1951 – 1061. https://doi.org/ 10.2307/1935417.

Iddi, K.N. 2004. *Sustainable Use of Birds in Ghana and Mitigation of Constraints to Conservation of Threatened Species.* London, UK: Whitley Laing Foundation.

Innes, J.L. & Koch, B. 1998. Forest biodiversity and its assessment by remote sensing. *Global Biogeography Letters* 7: 397-419. https://doi.org/10.1046/j.1466-822X.1998.00314.x.

Jha, A. 2003. Is it time to start culling seagulls? *The Guardian* August 14.

Johnson, A.R., Milne, B.T. & Wiens. J.A. 1992a. Diffusion in fractal landscapes: simulations and experimental studies of Tenebrionid beetle movements. *Ecology*, 73 (6): 1968-1983. https://doi.org/10.2307/1941448.

Johnson, A.R., Wiens, J.A., Milne, B.T. & Crist, T.O. 1992b. Animal movements and population dynamics in heterogeneous landscapes. *Landscape Ecology*, 7 (1): 63-75. https://doi.org/10.1007/BF02573958.

Johnson, B.L. 1999. The role of adaptive management as an operational approach for resource management agencies. *Conservation Ecology* 3(2): 8. Retrieved from http://www.consecol.org/vol3/ iss2/art8.

Karr, J. R. & Dudley, D.R. 1981. Ecological perspective on water quality goals. *Environmental Management*, 5: 55-68. https://doi.org/10.1007/BF01866609.

Kasworm, W.F., & Manley, T.L. 1990. Road and trail influences on grizzly bears and black bears in northwest Montana. *International Conference on Bear Research and Management*, 8: 79-84. https://doi.org/10.2307/3872905.

Keitt, T.H., Urban, D.L. & Milne, B.T. 1997. Detecting critical scales in fragmented landscapes. *Conservation Ecology* 1(1): 4. Retrieved from http://www.consecol.org/vol1/iss1/art4/.

Krüger, O. & Lindström, J. 2001. Habitat heterogeneity affect population in Goshawk *Accipiter Gentilis. Journal of Animal Ecology*, 70: 173 – 181. https://doi.org/10.1111/j.1365-2656.2001. 00481.x.

Kutay, K. 1991. Cahuita National Park, Costa Rica: A case study in living cultures and national park management. In: West, P. & Brechin, S. (Eds.), *Resident Peoples and National Parks: Social Dilemmas and Strategies in International Conservation*, 114-129. Tuscon: University of Arizona Press.

Lee, K. N. 1999. Appraising adaptive management. *Conservation Ecology* 3 (2): 3. Retrieved from http://www.consecol.org/vol3/ iss2/art3.

Lim, H.C. & Sodhi, N.S. 2004. Responses of avian guilds to urbanisation in a tropical city. *Landscape and Urban Planning*, 66 (4): 199-215. https://doi.org/10.1016/S0169-2046(03) 00111-7.

Lloyd, C., Tasker, M.L. & Partridge, K. 1991. *The Status of Seabirds in Britain and Ireland*. London, UK: T & Ad Poyser.

Lovett-Doust, J. & Kuntz, K. 2002. Land ownership and other landscape - level effects on biodiversity in Southern Ontario's Niagara Escarpment Biosphere Reserve, Canada. *Landscape Ecology* 16(8): 743 – 755. https://doi.org/10.1023/A:1014416900653.

Ludwig, D., Hilborn, R. & Walters, C. 1993. Uncertainty, resource exploitation, and conservation: Lesson from history. *Science*, 260: 17, 36. https://doi.org/10.1126/science.260.5104.17.

Maccarone, A. D. 1991. Seasonal use of foraging habitats by American crows in central New Jersey. *Bird Behaviour*, 9(1-2): 34-40.

Madge, S. & Burn, H. 1994. *Crows and Jays*. London: Christopher Helm.

Mangel, M. 1985. *Decision and Control in Uncertain Resource Systems*. New York, NY: Academic Press.

Mantell, M.A.1992. *The key is habitat, not lone species*. Los Angeles Times, 26 April 1992.

Marzluff, J.M., Bowman, R. & Donnelly, R. (Eds.) 2001. *Avian Ecology and Conservation in An Urbanising World*. Boston, MA: Kluwer Academic Publishers.

Mattson, D.J. & Reinhart, D.P. 1996. Indicators of red squirrel (*Tamiasciurus hudsonicus*) abundance in the Whitebark Pine Zone. *Great Basin Nature*, 56: 272-275. https://scholarsarchive.byu.edu/cgi/viewcontent.cgi?article=2923&context=gbn.

Maver, I. 1998. Glasgow's public parks and the community, 1850 – 1914: A case study in Scottish civic interventionism. *Urban History*, 25 (3): 323 - 347. https://doi.org/10.1017/S096392680 0012943.

Mavor, R.A., Parsons, M., Heubeck, M., Pickerell, G. & Schmidtt, S. 2003. *Seabird Numbers and Breeding Success in Britain and Ireland 2002*. Peterborough, ON: Joint Nature Conservation Committee.

Mclellan, D. 1894. *Glasgow Public Parks*. Glasgow: John Smith & Son.

Merriam, G. 1984. Connectivity: A fundamental ecological characteristic of landscape pattern. In Brandt, J. & Agger, P. (Eds) *Proceedings of The First International Seminar on Methodology in Landscape Ecological Research and Planning. Theme I: Landscape Ecological Concepts*, 1: 5-6. Roskilde: Roskilde University Centre Book Company.

Mitchell, I.P., Newton, S.F., Ratcliffe, N. & Dunn, T.E. 2004. *Seabird Populations of Britain and Ireland*. London, UK: T and A.D. Poyser.

Monaghan, P.C. 1979. Aspects of the breeding biology of herring gulls *Larus Argentatus* in urban colonies. *Ibis*,121: 475 – 481. https://doi.org/10.1111/j.1474-919X.1979.tb06687.x.

Moore, P.G. 2002. Ravens (*Corvus Corax Corax* L.) in British landscape: A thousand years of ecological biogeography in place names. *Journal of Biogeography*, 29: 1039- 1054. https://doi.org/10.1046/j.1365-2699.2002.00750.x.

Morneau, F., Décarie, R., Pelletier, R., Lambert, D., Desgranges, J.L. & Savard, J.P. 1999. Changes in breeding bird richness and abundance in Montreal parks over a period of 15 years. *Landscape and Urban Planning*, 44: 111-121. https://doi.org/10.1016/S0169-2046(99)00002-X.

Moss, M.R. 2000. Interdisciplinarity, landscape ecology and transformation of the agricultural landscape. *Landscape Ecology* 15: 303-311. https://doi.org/10.1023/A:1008144402408.

Mudge, G.P. & Ferns, P.N. 1982. The feeding ecology of five species of gulls (Aves: Larini) in the inner Bristol channel. *Journal of Zoology*, 197: 497 – 510.

Naiman, R.J. 1996. Water, society and landscape ecology. *Landscape Ecology*, 11(4): 193-196. https://doi.org/10.1007/BF02071809.

National Capital Commission, 1996. *Greenbelt Master Plan*. Ottawa: NCC.

National Capital Commission, 1999. *Project of Research on Gatineau and the Greenbelt: Study on Traffic in Stony Swamp-Summer Season*. Ottawa: NCC.

Nicolis, G. 1994. Dynamical systems, biological complexity and global change. In Solbrig, O.T., van Emden, H.M. & van Oordt, P.G.W.J. (Eds.) *Biodiversity and Global Change*, 21-32. Wallingford, UK; CAB International.

Nielsen, C. 2001 Ice Storm 1998-Forest Research Conference. *The Forestry Chronicle*. 77(1): 16 -18.

Newton, I. 1979. *Population Ecology of Raptors*. Berkhamsted: T & A.D. Poyser.

Nicolis, G. 1994. Dynamical systems, biological complexity and global change. In Solbrig, O.T., Van Emden, H.M. & Oordt, P.G.W.J. (Eds.), *Biodiversity and Global Change*, 21-31, Wallingford: Cab International.

Noordhuis, R. & A.L. Spaans 1992. Interspecific competition for food between Herring *Larus argentatus* and Lesser Black-backed Gulls *L. fuscus* in the Dutch Wadden Sea area. *Ardea* 80:115-132.

Noss, R. 1995. Maintaining Ecological Integrity in Representative Reserve Net Studies. *Discussion Paper.* World Wildlife Fund Canada/World Wildlife Fund United States.

Noss, R. F. 1991. Landscape connectivity: Different functions and different scales. In W. E. Hudson, (Eds) *Landscape Linkages and* Biodiversity, 27-39. Washington, D.C.: Island Press.

Obiri-Danso, K. & Jones, K. 2002. Campylobacters and faecal indicators in vultures at Kumasi abattoir, Ghana. Abstract for Poster at Society for Applied Microbiology Summer Conference on Pathogens. *The Environment and Changing Ecosystems*, Nottingham,

Odum, E.P. 1971. *Fundamentals of Ecology* 3rd Ed. Philadelphia: Saunders.

Ottawa 20/20 2002. *Background – Development in Ottawa: A History*. Ottawa: City of Ottawa.

Owen, B.A. 1967. Lesser black backed gulls nesting on warehouse roofs inland. *British Birds* 60: 4 –16.

Owen, J. 2003. Aggressive seagulls menacing urban Britain. *National Geographic News*. http://www.nationalgeographic.com.

Park, C.R. & Lee, W.S. 2000. Relationship between species composition and area in breeding birds of urban woods in Seoul, Korea. *Landscape and Urban Planning*, 51: 29 –36. https://doi.org/10.1016/S0169-2046(00)00094-3.

Patton, S.R. 1988. Abundance of gulls at Tampa Bay, Florida. *Florida Field Naturalist*, 12: 49 – 57.

Pickett, S.T.A., Parker, V.T. & Fiedler, P.L. 1992 The new paradigm in ecology: implications for conservation biology above the species level. In Fiedler, P.L. &d Jain, K., (Eds.) *Conservation Biology: The Theory and Practice of Nature Conservation, Preservation and Management*, 66-87. New York: Chapman and Hall.

Pons, J. 1992. Effects of changes in the availability of human refuse on breeding parameters in a herring gull *Larus argentatus* population in Brittany, France. *Ardea*, 80: 143 – 150.

Primack, R. B. 1993. *Essentials of Conservation Biology.* Sunderland: Sinauer Associates.

Raven, S.J. & Coulson, J.C. 1997. The distribution and abundance of *Larus* gulls nesting on buildings in Britain and Ireland. *Bird Study*, 44: 13-34. https://doi.org/10.1080/000636 59709 461035.

Regional Municipality of Ottawa-Carleton 1997. *Candidate Natural Area Evaluation. Regional Municipality of Ottawa-Carleton.* Ottawa.

Reinhart, D. P. & Mattson. D. J. 1990. Bear use of cutthroat trout spawning streams in Yellowstone National Park. *International Conference on Bear Research and Management* 8: 343-350.

Risser, P.G., Karr, J.R., Forman, R.T.T. 1984. *Landscape Ecology: Directions and Approaches.* Illinois Natural History Survey Special Publ. 2. Urbana: University of Illinois.

Rock, P. 2002. *Roof Nesting Gulls in Bath. Follow Up Survey Conducted in April 2002 And Summarising Surveys From 1995. Report to The Department of Environmental Services.* Bath: Northeast Somerset Council.

Rock, P. 2003a. Birds of a feather flock together. *Environmental Health Journal* (5): 132-135.

Rock, P. 2003b. *The Status and Numbers of Roof-Nesting Gulls in Worcester.* Retrieved from http://www.wbrc.org.uk/WorcRecd/Issue%2015/Gulls.htm

Rome, M.S. & Ellis, J.C. 2004. Foraging ecology and interactions between herring gulls and great black backed gulls in new England. *Waterbirds*, 27 (2): 200 – 210. https://www.jstor.org/ stable/1522435.

Royal Society for The Protection of Birds 2017. *The State of The UK's Birds*. Retrieved from https://www.rspb.org.uk/globalassets/downloads/documents/conservation-science/220-0653-17-18-sukb-2017-web-20-3-18.pdf

Savard, J.P.L., Clergeau, P. & Mennechez, G. 2000. Biodiversity concepts and urban ecosystems. *Landscape and Urban Planning*, 48: 131 – 142. Retrieved from http://citeseerx.ist.psu.edu/viewdoc/download?doi=10.1.1.463.4565&rep=rep1&type=pdf.

Scoones, I. 1997. The dynamics of soil fertility change: Historical perspectives on environmental transformation from Zimbabwe. *The Geographical Journal*, 63(2):161-169. https://www.jstor.org/stable/3060179.

Scoones, I. 1999. New ecology and social sciences: What prospects for a fruitful engagement. *Annual Review of Anthropology*, 28: 479-507. https://doi.org/10.1146/annurev.anthro. 28.1.479.

Selas, V. 1997. Nest–site selection by four sympatric forest raptors in southern Norway. *Journal of Raptor Research*, 31: 16 – 25. https://sora.unm.edu/sites/default/files/journals/jrr/v031n01/ p00016-p00025.pdf.

Shaw, P. 2004. Information: Natural Heritage Trends. Edinburgh: Scottish Natural Heritage.

Skillen, B.S. 2000. Magpie behaviour in Kelvingrove Park. *The Glasgow Naturalist*, 23 (5): 61.

Solbrig, O.T. 1993. Ecological constraints to savanna land use. In Young, M.D. & Solbrig (Eds.) *The World's Savannas: Economic Driving Forces, Ecological Constraints and Policy Options for Sustainable Land Use*, 21-47. Paris: The Pantheon Publishing Group.

Solbrig, O.T. 1994 Biodiversity: An introduction. In Solbrig, O.T. & Van Emden, H.M. & Van Oordt, P.G.W. (Eds.), *Biodiversity and Global Change* 13-17. Wallingford: Cab International.

Soule, M. E. 1986. Conservation biology and the real world. In Soule M. E. (Ed.) *Conservation Biology: The Science of Scarcity and Diversity*, 1-12. Sunderland, MA: Sinauer Associates.

Stanley, T.R., Jr. 1995. Ecosystem management and the arrogance of humanism. *Conservation Biology*, 9: 255-262. https://doi.org/10.1046/j.1523-1739.1995.9020255.x.

Statistics Canada 2016. *Census Profile, 2016*. Census. Retrieved from https://www12.statcan.gc.ca/census-recensement/2016/dp-pd/prof/details/page.cfm?Lang=E&Geo1=CSD&Geo2=PR&Code2=01&Data=Count&SearchType=Begins&SearchPR= 01&TABID=1&B1=All&Code1=3506008&SearchText=ottawa

Sterry, P., Cleave, A., Goodfellow, P. & Clements, A. 2002. *Field Guide to the Birds of Britain and Europe.* Windsor: Automobile Association Developments Ltd.

Stevenson, T. & Fanshawe, J. 2002. Birds of East Africa. London: Christopher Helm.

Tatner, P.C. 1982. The breeding biology of magpies *Pica Pica* in an urban environment. *Journal of Zoology* 197: 559 – 581.

Taylor, P.D., Fahrig, L. Henein, K. & Merriam. G. 1993. Connectivity is a vital element of landscape structure. *Oikos* 68 (3): 571-573. http://www.jstor.org/stable/3544927?origin= crossref.

Temple, S.A., Brussard, P.F., Bolen, E.G., Salwasser, H., Soule, M.E. & Teer, J. G. 1988. What's so new about conservation biology*? Transactions of The North American Wildlife and Natural Resources Conference* 53: 609-612.

Tewksbury, J., Hejl, S.J. & Martin, T.E. 1998. Breeding productivity does not decline with increasing fragmentation in a western landscape. *Ecology* 79: 2890-2903. https://doi.org/10.1890/0012-9658(1998)079[2890:BPDNDW]2.0.CO;2.

Thom, V.M. 1986. *Birds in Scotland.* Calton: T&Ad Poyser.

Tilghman, N. 1987. Characteristics of urban woodlands affecting breeding bird diversity and abundance. *Landscape and Urban Planning*, 14: 481 – 495. https://doi.org/10.1016/0169-2046(87)90061-2.

Turner, W.R. 2003. City biological monitoring as a tool for ecology and conservation in urban landscapes: The case of the Tucson bird count. *Landscape and Urban Planning*, 65: 149 – 166.

United States Geological Survey 2004. Blue jay *Cyanocitta cristata.* Retrieved from http://www.mbr-pwrc.usgs.gov/id/framlst/ i4770id.html.

Vuorisalo, T., Andersson, H. Hugg, T., Lahtinen, R., Laaksonen, H. & Lehikoinen, E. 2003. Urban development from an avian perspective: Causes of hooded crow (*Corvus Corone Cornix*) urbanization in the two Finnish cities. *Landscape and*

Urban Planning, 62: 69 – 87. Retrieved https://www.academia.edu/13832733/Urban_development_from_an_avian_perspective_causes_of_hooded_crow_Corvus_corone_cornix_urbanisation_in_two_Finnish_cities.

Walters, C. 1986. *Adaptive Management of Renewable Resources*. New York, NY: Macmillan.

Walters, C. 1997a. Challenges in adaptive management of riparian and coastal ecosystems. *Conservation Ecology* 1 (2): 1 retrieved from http://www.consecol.org/vol1/iss2/art1.

Walters, C. J. 1997b. Adaptive policy design: thinking at larger spatial scales. In Bissonette, J.A. (Ed.) *Wildlife and Landscape Ecology*, 386-395. New, NY: Springer-Verlag.

Wanten, J. & Holz, R. 2004. *Species Accounts; Ghana: A Report from Birdtours.Co.Uk*. Retrieved from http://www.birdtours.co.uk/tripreports/ghana/ghana1/list.htm.

Watson, J. 2004. *Gulls Strike Terror in the Streets*. Retrieved from https://www.scotsman.com/ news/environment/gulls-strike-terror-in-the-streets-1-1386070

Weins, J.A. 1999. Landscape ecology: The science and action. *Landscape Ecology*, 15 (4): 303-311.

Wells, M., Brandon, K., & Hannah, L. 1992. *People and Parks, Linking Protected Area Management with Local Communities*. Washington DC.: The World Bank.

Werner, P. 2011. The ecology of urban areas and their functions for species diversity. *Landscape and Ecological Engineering*, 7(2): 231-240. http://www.doi:10.1007/s11355-011-0153-4.

Wickware, G.M. & Rubec, C.D.A. 1989. Ecoregions of Ontario. *Ecological Land Classification Series, 26*. Ottawa: Environment Canada, Sustainable Development Branch.

Witherby, H.F., Jourdain, F.C.R., Ticehurst, N.F. & Tucker, B.W. 1941. *The Hand-book of British Birds*, 5. London: H.F. & G. Witherby Ltd.

World Population Review 2019. *Ghana Population 2019*. Retrieved from http://worldpopulationreview.com/countries/ghana-population/

Young, M.D. & Solbrig, O.T. 1993. Providing an environmentally sustainable, economically profitable and socially equitable future for the world's savannas. In Young, M.D. & Solbrig O.T. (Eds.), *The World's Savannas: Economic Driving Forces, Ecological Constraints and Policy Options for Sustainable Land Use*, 321-344. Paris: Pantheon Publishing Group.

Zimmerer, K.S. 1994. Human geography and the 'new ecology': The prospect of integration. *Annals of The Association of American Geographers*, 84(1):108-125. https://doi.org/10.1111/j.1467-8306.1994.tb01731.x.

EDITORS' CONTACT INFORMATION

Michael O'Neal Campbell

Lecturer

Simon Fraser University

British Columbia, Canada

ecol55@hotmail.com

Kaushalendra Kumar Jha

Professor

Tropical Forest Ecosystem and Environment Management Unit,

Technical Forestry Faculty Area,

Indian Institute of Forest Management, Bhopal, India

Jhakk1959@gmail.com

INDEX

B

C

F

G

H

M

N

O

P

V

W

Z